STATISTICS
IN THE
PHARMACEUTICAL
INDUSTRY

STATISTICS: Textbooks and Monographs

A Series Edited by

D. B. Owen, Founding Editor, 1972–1991

W. R. Schucany, Coordinating Editor
Department of Statistics
Southern Methodist University
Dallas, Texas

R. G. Cornell, Associate Editor
for Biostatistics
University of Michigan

W. J. Kennedy, Associate Editor
for Statistical Computing
Iowa State University

A. M. Kshirsagar, Associate Editor
for Multivariate Analysis and
Experimental Design
University of Michigan

E. G. Schilling, Associate Editor
for Statistical Quality Control
Rochester Institute of Technology

Additional Volumes in Preparation

STATISTICS IN THE PHARMACEUTICAL INDUSTRY

Second Edition, Revised and Expanded

edited by

C. RALPH BUNCHER
Department of Biostatistics
University of Cincinnati Medical Center
Cincinnati, Ohio

JIA-YEONG TSAY
Clinical Development Department
Organon Inc.
AKZO Pharma Group
West Orange, New Jersey

MARCEL

DEKKER

MARCEL DEKKER, INC. NEW YORK · BASEL

Library of Congress Cataloging-in-Publication Data

Statistics in the pharmaceutical industry / edited by C. Ralph
  Buncher, Jia-Yeong Tsay, — 2nd ed.
     p. cm. — (Statistics, textbooks and monographs ; v. 140)
  ISBN: 0-8247-9073-1 (acid-free paper)
  1. Pharmacy—Statistical methods. 2. Pharmaceutical industry—
Statistical methods. 3. Drug—Testing. I. Buncher, C. Ralph
(Charles Ralph). II. Tsay, Jia-Yeong. III. Series.
RS57.S8 1994
615'.1901'072—dc20   93-6362
                                                               CIP

The publisher offers discounts on this book when ordered in bulk
quantities. For more information, write to Special Sales/Professional
Marketing at the address below.

This book is printed on acid-free paper.

Marcel Dekker, Inc.
270 Madison Avenue, New York, New York, 10016

Current Printing (last digit)
10 9 8 7 6 5 4 3 2

PRINTED IN THE UNITED STATES OF AMERICA

To
Maxine
and
Tung-Ying

Preface to the Second Edition

This volume started out as a simple revision which was to become the second edition of our book *Statistics in the Pharmaceutical Industry*. Then the appreciation of the enormous changes in the industry in recent years emerged.

In a decade of rapid change in this field, have come, in large numbers, the new initials of PCs (personal computers), CANDAs (computer-assisted new drug applications), CROs (contract research organizations), AIDS (acquired immunodeficiency syndrome), as well as biotechnology, biomonitoring, pharmacoepidemiology, molecular biology, meta-analysis, designer drugs, international planning, work stations, and so forth. Diseases for which there was no pharmaceutical therapy are now routinely treated, and the expectation is that new genetic and molecular knowledge will make possible treatment for virtually all diseases.

The result is that this book not only was completely revised by updating every chapter from the first edition, but also was expanded by adding many chapters to cover new areas that were still in their infancy during the development of the first edition. We believe that reading this book will enlighten readers—be they students in statistics, faculty members, statisticians, or non-statisticians—about the pharmaceutical industry and the many roles played by the biostatistician within that industry. Even those readers with background in the industry may find interesting ideas in the words of others whose experience may differ from their own.

We wish to thank the authors of these chapters who used their practical knowledge to make this revision possible. They have many years of experience in the pharmaceutical industry, whether working for government or industry or both. Their efforts will enable the reader to speed the process of building experience and understanding.

We thank the many people at Marcel Dekker, Inc., who have worked on this book and helped bring it to fruition. We also thank Brenda Riggins in the Division of Biostatistics at the University of Cincinnati for her special efforts which facilitated this volume.

C. Ralph Buncher
Jia-Yeong Tsay

Preface to the First Edition

This book is the culmination of many years of difficult work by numerous people. The idea of a book devoted to the lessons learned by those working in the pharmaceutical field had its genesis in 1973. After an initial flare of interest, the idea lay dormant for three or four years. Then the current group was assembled to bring the thought to fruition. Even this final stage has taken a number of years for all of the usual reasons associated with busy persons trying to put together a volume while also having the rest of their duties undiminished. We are proud that this book fulfills most of our dreams.

AUDIENCE FOR THIS BOOK

For whom is this book written? The editors and authors have discussed this question extensively, especially since this is the first book that we know of devoted to applied statistics in the pharmaceutical industry. We agreed to keep in mind four audiences in particular as the chapters were written.

The principal audience for this book is the graduate student in statistics or biostatistics. We believe that these students can use this volume to find out much more than is currently available about opportunities in the pharmaceutical industry, which is one of the major employers of biostatisticians. It is hoped that some academic institutions will choose to use this volume as a textbook in a seminar type of course, perhaps held jointly with students in pharmacology, pharmacy, or research methods. Secondly, the book has been sprinkled with topics that the authors believe would be good thesis subjects for those who are looking for work which will be of use and interest to others. Finally, we believe that this book provides much general information on applied statistics which will help enlarge the perspective of the student training for a career in the field of statistics, whether that field is on the applied side or in the theoretical world. Many new employees in the pharmaceutical industry also fit into this category.

A second audience for this book consists of the statistics faculty members who have had little or no acquaintance with the pharmaceutical industry. We believe that these statisticians, who have a wider perspective and a greater depth of experience than the students, will be able to sort out which problems in the industry are similar to problems that they are already working on and which problems are relatively unique to the industry. We all agree that there are many valuable contributions that academic statisticians could make toward solving problems facing the pharmaceutical industry. We hope that this book will help provide a greater bridge between industry and academia.

A third audience for this book consists of other persons interested in the pharmaceutical industry who are not statisticians. Many of the chapters in this book can be read and understood by those with no statistical training or only minimal training in statistics. We are thinking of those working in clinical fields, pharmacology, chemistry, quality control, company management, legal departments, and data managers. Each of these persons can find a wealth of experience summarized and presented. These lessons are explained in the hope that other professionals will not have to relearn problems and pitfalls that exist for almost every drug and almost every company. Clearly, members of regulatory agencies and persons wishing to know more about the inner workings of the pharmaceutical industry also fit into this category.

Finally, we believe that the pharmaceutical industry currently documents its statistical and other evidence better than any other segment of our society. As more and more comparable problems—for example, concerns with the value and safety of nontherapeutic chemicals and other environmental agents—become a greater problem for industry and regulatory agencies, we foresee that parts of industry and government other than the pharmaceutical industry and the Food and Drug Administration will be doing the same extensive work and documentation that has been pioneered in the pharmaceutical industry. We think that those other industrial/governmental interfaces will profit by learning the lessons of the pharmaceutical industry and applying them in their own spheres as appropriate.

READING THIS BOOK

Chapter 1 describes the development of a chemical into a new drug for those unfamiliar with the process or with the vocabulary of the industry. The other chapters are in chronological order with respect to drug development.

This book has been written with the idea that many if not most readers will read only a few chapters. We recommend that those unfamiliar with the industry read the introductory chapter and then a few of current interest. It is hoped that after learning some of the material, the reader will find additional chapters to be of interest and will pick up the volume again to gain additional insight and perspective.

This book has been designed to show many of the problems and solutions of the statistical portion of the biopharmaceutical industry. The principal emphasis throughout is on the problems that occur and the solutions that are used "in practice" rather than "in theory." The chapters in this book are written with the personal experience of the authors in the pharmaceutical industry in mind. Thus the book manifests a diversity in subject matter as well as style and attitude toward statistical problems.

An attempt has been made to explain the problems discussed in each chapter before describing some of the solutions. As in most areas of human endeavor, it is easy to miss some of the subtleties of the problem if one has never been exposed to it. Thus the book does require some sophistication in understanding the problems of finding out the truth about a drug, forming impressions of the truth when only part of the data are at hand, convincing others that your understanding of the situation is the correct one, and documenting the material so that an outside reviewer (even one who may be a doubter) can be convinced that a pharmaceutical product is sufficiently safe and efficacious or, in other words, sufficiently well understood to be suitable for marketing.

We thank the authors of these chapters for their cooperation and help during this project. Special thanks go to Charlie Dunnett for key suggestions in this process. Two other persons not mentioned elsewhere were essential to the successful completion of this project. They are Hannah Aron and Elaine Sirkoski who are Staff Assistants in the Division of Biostatistics and Epidemiology at the University of Cincinnati. They did much of the organizing and communicating and all of the typing of this final copy. We thank Hannah and Elaine for a superior job well done.

C. Ralph Buncher
Jia-Yeong Tsay

Contents

Contributors

Jeffrey B. Aldrich Manager, Clinical Data Center, The Upjohn Company, Kalamazoo, Michigan

Mirza W. Ali, Ph.D.[*] Division of Biometrics, Center for Drug Evaluation and Research, United States Food and Drug Administration, Rockville, Maryland

Joseph R. Assenzo, Ph.D. Executive Director, U.S. Pharmaceutical Regulatory Affairs, The Upjohn Company, Kalamazoo, Michigan

C. Ralph Buncher, Sc.D. Professor and Director of Biostatistics and Epidemiology, Department of Biostatistics, University of Cincinnati Medical Center, Cincinnati, Ohio

José F. Calimlim, M.D. School of Medicine and Dentistry, University of Rochester, Rochester, New York

William S. Cash, Ph.D. Director, Clinical Data Operations, Central Research Division, Pfizer Pharmaceuticals, Pfizer Inc., New York, New York

---

[*] *Current affiliation*: Pfizer Inc., Groton, Connecticut

Vita A. Cassese, M.B.A. Vice President, Marketing Research and Pharmaceutical Systems, Pfizer Pharmaceuticals, Pfizer Inc., New York, New York

T. Timothy Chen, Ph.D. Mathematical Statistician, Biometric Research Branch, Cancer Therapy Evaluation Program, Division of Cancer Treatment, National Cancer Institute, National Institutes of Health, Bethesda, Maryland

Owen L. Davies, Ph.D. Professor Emeritus, Department of Statistics, University College of Wales, Aberystwyth, Wales

Robert L. Davis, Ph.D.[*] Director, Clinical Biostatistics, Merck Research Laboratories, Inc., West Point, Pennsylvania

Satya D. Dubey, Ph.D. Chief, Statistical Evaluation and Research Branch, Center for Drug Evaluation and Research, United States Food and Drug Administration, Rockville, Maryland

Charles W. Dunnett, D.Sc. Professor Emeritus, Department of Mathematics and Statistics, and Department of Clinical Epidemiology and Biostatistics, McMaster University, Hamilton, Ontario, Canada

Kenneth H. Falter, Ph.D. Mathematical Statistician, Division of Immunization, Centers for Disease Control and Prevention, Atlanta, Georgia

Alan C. Fisher, Dr.P.H. Director, Department of Biostatistics, Immunobiology Research Institute, Annandale, New Jersey

Roger E. Flora, Ph.D. Vice President and Director, Biostatistics and Information Services, Pharmaceutical Research Associates, Inc., Charlottesville, Virginia

Charles H. Goldsmith, Ph.D. Professor of Biostatistics, Department of Clinical Epidemiology and Biostatistics, McMaster University, Hamilton, Ontario, Canada

Jack W. Green, Ph.D. Senior Vice President, Biostatistics and Data Management, G. H. Besselaar Associates, Princeton, New Jersey

Arthur E. Hearron, B.B.A., M.P.H. Biostatistics Senior Scientist, Upjohn Laboratories, The Upjohn Company, Kalamazoo, Michigan

---

[*]*Current affiliation:* Director, Biostatistics, Astra/Merck Group of Merck & Company, Wayne, Pennsylvania

Harry E. Hudson, Ph.D. Head of Drug Stability Studies, Pharmaceutical Department, Imperial Chemical Industries Ltd., Macclesfield, Cheshire, England

Irving K. Hwang, Ph.D. Senior Director, Clinical Biostatistics and Research Data Systems, Merck Research Laboratories, Inc., West Point, Pennsylvania

Roswitha E. Kelly, M.S. Mathematical Statistician, Division of Biometrics, Center for Drug Evaluation and Research, United States Food and Drug Administration, Rockville, Maryland

Karl K. Lin, Ph.D. Supervisory Mathematical Statistician, Division of Biometrics, Center for Drug Evaluation and Research, United States Food and Drug Administration, Rockville, Maryland

Tsae-Yun D. Lin, Ph.D. Mathematical Statistician, Division of Biometrics, Center for Drug Evaluation and Research, United States Food and Drug Administration, Rockville, Maryland

John R. Murphy, Ph.D. Research Scientist, Statistical and Mathematical Sciences, Eli Lilly and Company, Indianapolis, Indiana

Robert T. O'Neill, Ph.D. Director, Division of Biometrics, Center for Drug Evaluation and Research, United States Food and Drug Administration, Rockville, Maryland

Gordon W. Pledger, Ph.D. Senior Research Fellow in Biostatistics, Department of Medical Biostatistics, R. W. Johnson Pharmaceutical Research Institute, Raritan, New Jersey

Charles E. Redman, Ph.D.[*] Director, Medical Information Systems, Lilly Research Laboratories, Indianapolis, Indiana

Charles B. Sampson, Ph.D. Director, Department of Decision Sciences, Eli Lilly and Company, Indianapolis, Indiana

Roy L. Sanford, Ph.D. Vice President, Quality Management, Department of Corporate Quality, Baxter Healthcare Corporation, Deerfield, Illinois

---

[*]*Current affiliation*: HCA Medical Research, Nashville, Tennessee

John R. Schultz, Ph.D. Executive Director, Department of Biostatistics and Clinical Data Management, The Upjohn Company, Kalamazoo, Michigan

Jia-Yeong Tsay, Ph.D[*] Department of Biostatisitics, University of Cincinnati Medical Center, Cincinnati, Ohio

Sylvan Wallenstein, Ph.D. Associate Professor, Department of Biomathematical Sciences, The Mount Sinai Medical Center, New York, New York

Daniel L. Weiner, Ph.D. Vice President and Director, Institute for Research Data Management, Syntex Development Research, Palo Alto, California

Michael Weintraub, M.D. School of Medicine and Dentistry, University of Rochester, Rochester, New York

Bob Wilkinson Principal Associate, Robert Wilkinson Associates, Blauvelt, New York

Kathy Karpenter Wille, Ph.D.[†] Merrell Dow Research Institute, Cincinnati, Ohio

Karen E. Woodin, Ph.D. Manager, Field Monitoring and Surveillance, The Upjohn Company, Kalamazoo, Michigan

Chiao Yeh, Ph.D. Special Projects Manager, Department of Clinical and Medical Affairs, Zeneca Pharmaceuticals Group, Wilmington, Delaware

Liang Yuh, Ph.D. Associate Director, Department of Biometrics, Parke-Davis Pharmaceutical Research Division, Warner-Lambert Company, Ann Arbor, Michigan

[*]*Current affiliation:* Associate Director of Biometrics, Clinical Development Department, Organon Inc., AKZO Pharma Group, West Orange, New Jersey
[†]*Current affiliation*: Biostatistician, The Procter & Gamble Company, Cincinnati, Ohio

STATISTICS
IN THE
PHARMACEUTICAL
INDUSTRY

1

Introduction to the Evolution of Pharmaceutical Products

C. Ralph Buncher and Jia-Yeong Tsay[*]

*University of Cincinnati Medical Center,
Cincinnati, Ohio*

I. TRENDS IN THE PHARMACEUTICAL INDUSTRY

Excitement is the descriptive word for the pharmaceutical industry today. Statisticians are influential participants among those riding that wave of excitement. Knowledge is the seed from which new drug preparations are grown. Molecular studies in biology are blossoming forth at such exceptional speed that new knowledge is created daily. The enterprising and astute are capable of using this new knowledge as a springboard towards new pharmaceutical products. The new products mean better health for the public and possible prosperity for the company which develops the product.

Risks are very large in this industry. The often quoted number is that $231 million (in 1987 dollars) are needed to develop a new pharmaceutical entity (DiMasi et al., 1991). In round numbers one-quarter of a billion dollars is spent by the typical sponsor (pharmaceutical company) before the first dollar of income starts. Many more chemical entities will enter testing than will ever become medications. Therefore, a key early concern in most development programs is to decide whether a particular chemical is likely to be a winner or if it is one of many in which we shall invest some time, resources, and effort but with no final achievement. The greatest achievable efficiency will result if we

[*] *Current affiliation*: Organon Inc., AKZO Pharma Group, West Orange, New Jersey

terminate the research effort as soon as you know the medication will not be a success. Success means that it must have demonstrable effectiveness, an acceptable level of side effects, and all manufacturing and other problems are solved.

Thus the key decision points are: (a) whether we (the sponsor) are convinced that we should test this new chemical in humans based on results from animal studies, (b) whether we are convinced that the drug works and is safe in humans, (c) whether we now have sufficient information to convince the Food and Drug Administration (FDA) and the medical community that this medication as we now manufacture it is safe and effective, and (d) whether the FDA will approve the drug for marketing.

II. MOLECULAR BIOLOGY

In the last decade, the ability to understand the role of individual chemicals which can cause or can prevent disease has increased faster than at any time in human history. Much of this knowledge has been generated through studies of the genetic material, the proteins created from genes, and the role played by each. Collectively, this work may be called molecular biology. Those who work in molecular biology are converting these theoretical concepts into realities and statisticians are measuring the progress.

Most statisticians working in the industry get to learn and be proficient in some of the biologic and medical terminology even if they have never learned it during their prior education. One can read statements such as "until recently, it was common practice for a pharmaceutical company to market a chiral drug as the racemate." (Nugent et al., 1993). In this case one needs to know that if there are four different chemical groups attached to a single carbon atom, then two mirror-image forms called enantiomers or optical isomers are possible; these are chiral compounds. Generally, nature creates just one of them but human manufacturing usually creates both in equal numbers which is a racemic mixture. Typically only one of them is of therapeutic value but both can cause side effects. If the manufacturing process is able to produce only the form that is therapeutic, then the incidence of side effects will be reduced.

This explosion in molecular biology is one of the five great changes in the last decade in the pharmaceutical industry that we have identified. Another by-product of molecular biology is the proliferation of small companies consisting of only a small number of researchers who are expert in a niche area of medicine and chemistry. These biotechnology companies usually have great expertise in their particular area of research. Through molecular biologic techniques, these researchers can produce a new candidate drug. With the help of other companies, they can develop this candidate into a new drug or they can sell the candidate drug to another, usually larger, company which will undertake the approval process. Alternatively, the company may be taken over by a larger company because of the value of the candidate drug.

A second major change in the last decade is the use of computing equipment. Statisticians used to think first of a mainframe computer, with all the associated characteristics of sharing a big machine with dozens to hundreds of others. In addition, even when things were going well the machine needed one or more caretakers who did nothing but take care of the machine and its associated equipment. Today, statisticians always have a computer on their own desks either in an independent configuration or networked with others. Most of these are personal computers but many are now workstations with power, speed, and capacity that make the statistician of today a super-statistician compared to the predecessor of two decades ago. The new machines, and the wondrous software also now available, give the statistician instant ability to explore data sets, to display the data graphically, to calculate descriptive results, and to do other complex inferential analyses.

The third change in this decade is the level of sophistication in statistical analysis. This is much higher in the last decade because of the ability to do more extensive analysis and because of the desire to see those alternatives. Thus one can easily present alternative analyses with difficult patients both in and out to see if there are any qualitative differences, one can differentially weight the observations, one can omit an investigator from a combined analysis, and one can do formal or informal meta-analysis of the results.

Moreover, the ability to store and retrieve alternative analyses makes the system effective. Any statistician who had to store piles of computer output in the old days pauses every now and then to marvel at the ease of storing much more data and information inside a computer and to be able to retrieve it at will. These abilities have led in turn to the desire to transmit all of this information intact from the sponsor's statisticians to those of the FDA. Statisticians usually want to pursue their own alternatives in analyzing a data set rather than just checking the analysis of others. This sequence of events has led to the concept of a Computer-Assisted New Drug Application or CANDA (see also Chapter 19).

A fourth trend is the ability to monitor drug levels and the effects of medications in various body fluids and tissues. More and more, physicians are able to take into account the characteristics of individual patients and thus to tailor doses to just the right level for each. The sources of variation in optimal dose can be genetic differences (e.g., different varieties of the same enzyme) or different physical characteristics (e.g., differences in body weight). While statisticians find the variance to be informative, physicians find that variation is mostly a problem in optimal therapy. By tailoring doses of medication, the physicians improve their ability to treat but at times make the statistician's job a little more challenging. This ability to treat individually the genetic differences of people, such as the special cases of those who rapidly metabolize a drug and those who slowly metabolize a drug, promises to enhance the average therapeutic value and reduce side effects. These abilities to make informative extra measurements and the resulting explosion of data to be analyzed also make the statistician's job more interesting.

Finally, the fifth trend is seen to be the changes in the structure of the industry. The pharmaceutical industry no longer is focused within individual countries but rather it has become an international marketplace with efficacy and safety research focused on the world market. Much of Europe is now one unit rather than many separate countries. Thus the old information problem was how do we best get safety and efficacy information in this one country. Now the question is more like the following: in which country should we do each piece of research to optimize our opportunities to do research, our costs, the information we obtain, and the information necessary to convince the authorities in many lands to let us market the medication. Optimizing the research has also resulted in restructuring within the companies. Thus companies which used to do all of the work themselves will now frequently (if not always) use outside contractors to carry out some or all of the research and statistical functions that used to be internal. A whole new subindustry has grown up of Contract Research Organizations (CROs) which include, among other capabilities, all of the statistical functions necessary in clinical trial research (see also Chapter 16).

III.   ISSUES IN DRUG DEVELOPMENT

The remainder of this first chapter attempts to accomplish three tasks. The first is to outline briefly the steps involved in the pharmaceutical industry from creation of a new chemical compound in the laboratory until, for some tiny fraction of those compounds, a new drug is available on the market. This brief description is primarily to allow those unfamiliar with the process to have a better idea of the many interrelated steps involved in this long and frequently unsuccessful effort. Second, the outline will emphasize the role of the statistician in each of these phases, since that is a purpose of this volume. Finally, the chapters in this book will be introduced at the point that the description of the development of a new drug relates to that chapter.

There are three main issues in drug development: safety, efficacy, and manufacturing. Safety must first be proven in cell based and whole animal research before a drug is permitted to be used in humans. Then the safety must again be proven in humans to justify long-term clinical rather than experimental use of a drug. After the drug has been approved for marketing, investigators will search for rare side effects of a drug in those patients who have used the drug. Efficacy must be proven in clinical testing of a drug for the medical purpose intended in typical groups of patients. Prior to this time, a chemical has been selected because it has been found to be active in some subhuman biologic screen or because of considerations in molecular biology or chemical structural analysis. After success in screening, this chemical must be sufficiently tested in animals so that one can infer that the drug is likely to be clinically useful in humans. Finally, a drug must be manufactured. What was once a newly active chemical created by a chemist in a laboratory must be produced in a pilot plant operation and then later manufactured in large batches with careful quality control so that each individual dose of the medication will exhibit the high standards of safety and efficacy expected.

Obviously, these developments are not made independently of each other. A drug which does not dissolve as intended may show restricted efficacy, for example, relief of pain for only 2 hours rather than the intended 8 hours. Reformulation of the medication might serve to improve the efficacy. Drug side effects may disappear if the medication is given at mealtime or at bedtime and as a result enhance efficacy. A drug which has been found to be highly efficacious and easy to manufacture may, through lifetime toxicity testing, turn out to cause malignant tumors in rats and thus abruptly end a research program.

Currently one thinks of a typical duration of time from creation of the chemical in the laboratory until a drug is marketed of the order of 7 to 12 years. Safety, efficacy, and manufacturing are each studied for a majority of that period; however, proving safety requires the most time. On the time scale, the lifetime of a drug may be divided into preclinical time, the period from creation of the chemical to its first use in humans; clinical studies, during which time the drug is being tested in humans; and finally postapproval during which time the drug is being sold commercially.

One illustration of this process is shown in Figure 1.1. The figure outlines the successive steps in the form of a pie chart and also shows the events which overlap in time. The pie shape also demonstrates the enlarging program as one proceeds towards a New Drug Application.

In the preclinical stage, one must learn about the characteristics of the drug to such an extent that it makes good sense to the sponsor and to the FDA to try this drug in human beings. In order to reach this stage, the sponsor must be reasonably sure that the drug will be safe, as shown in short-term animal toxicity testing in at least two species. Also, the sponsor will want to know that there is a reasonable indication that the drug will have the desired positive effect as predicted by tests in animal species. Finally, the sponsor will have to be able to manufacture test lots of the proposed medication so questions of dosage form, amount, and procedure for the preparation must be resolved. Typically these experimental quantities of the drug will be made in a pilot plant operation or in special laboratories that make sufficient quantities of the drug for experimental purposes. As a by-product of this research, the sponsor will have studied the metabolism of the drug in animals to know whether it accumulates in the

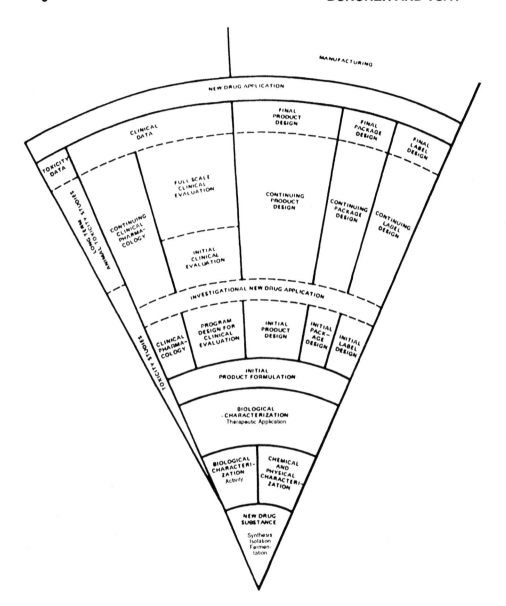

Figure 1.1 Evolution of a pharmaceutical product. (Courtesy of Eli Lilly and Company.)

tissues or whether it is excreted rapidly. Likewise, questions about the active form of the drug—whether it is the parent compound or some metabolite of it—will have been answered, at least tentatively. Doses which have been proven effective in animals will be extrapolated to the likely therapeutic human dose and then to a fraction of that dose to provide a margin of safety for initial testing.

All of this material is carefully written up by the sponsor as an Investigational New Drug (IND) application and submitted to the FDA to ask for an exemption so that the chemical may be tested in humans. Regulations allow the FDA 30 days in which to deny the IND or to ask additional questions which were not adequately presented in the submission.

The clinical period is divided into Phase I, Phase II, and Phase III research. Phase I studies are the earliest studies in humans, involving perhaps 20 to 40 subjects in total. Usually these persons are healthy volunteers. Questions to be answered concern the short-term toxicity of the drug in clinical pharmacology studies which provide data concerning absorption, distribution, metabolism, and excretion (usually called ADME) of the drug and which establish the safe dosage range for the drug as well as likely side effects at higher doses; occasionally some inferences regarding effectiveness may be made. These studies are characterized statistically by few patients but multiple measurements per patient.

Phase II studies involve perhaps 100 to 200 patients with the disease of interest who are studied in carefully supervised controlled clinical trials. These studies show the drug's fundamental effectiveness in restricted circumstances. One usually obtains dose-response curves in humans for effectiveness and for side effects. Common adverse effects can be detected during Phase II. Tests for the serum level or other level of the medication may be incorporated into the research especially if there is sizeable genetic variation in metabolism of the medication.

Phase III trials involve proving efficacy and safety in typical patients. During this phase, various levels of the severity of the disease are studied and patients using various concomitant medications provide information on a more clinical and less experimental usage. The total number of persons studied in this phase rarely exceeds 3000 patients and frequently is much smaller.

During this phase, efficacy is proven conclusively and safety (with the exception of rare adverse effects) is also demonstrated. A sponsor must rapidly notify the FDA of any serious adverse effects (e.g., death or an effect that is permanently disabling) which implies both close monitoring of the data as well as statistical tests of various results from clinical observations of safety.

All of the data on the three clinical Phases with respect to human research is submitted as part of the New Drug Application (NDA). This submission includes original case reports on each of the patients, which forms the great bulk of the NDA submission. More efficient methods for submission of this material in the electronic age are being actively explored including Computer-Assisted NDAs (CANDAs).

The NDA will contain results of the animal pharmacologic and toxicologic studies as well as the human pharmacology studies and the adequate and well-controlled clinical studies demonstrating the drug's efficacy and side effects. Data from long-term animal toxicity testing—for example, lifetime studies in rats lasting about 2 years—are included in this submission. All of the manufacturing information must also be contained in the submission indicating all of the ingredients that go into the drug, and whether the ingredients are active or inactive, included for the purposes of taste, color, physical characteristics of the tablet, or packaging (as in the case of a capsule).

The total submission can easily be equivalent to 500 volumes, each one up to 2 inches thick. The period of preparing an NDA by the sponsor, reviewing the NDA by the FDA, and then reaching a resolution about points for which there is insufficient information often involves several years. By law, the FDA is to accept or reject a submission in 60 days for completeness; however, the usual review time of NDAs is to be one year by 1997. A separate fast track has been added for those drugs which are especially innovative or of use in life threatening diseases for which few or no alternatives are available.

An important part of the submission and of the final NDA approval is the precise labeling to be used with the drug. In the labeling, the many thousands of pages of research are compressed into fewer than 2 dozen paragraphs which summarize the research with the drug.

After the drug has been approved by the FDA, the sponsor is permitted to manufacture and sell it. During this postmarketing period, called Phase IV, a number of other questions are usually answered. These questions concern relative efficacy of the new drug compared to others for the same or similar purpose. Also likely to be answered is the question of the effects of prolonged use of the medication and whether any rare side effects can be discovered. In some instances, approval is made contingent on doing particular Phase IV studies.

IV. PRECLINICAL TESTING

Frequently, the effect of drugs can be investigated by using an animal or portion of an animal as a test system with the characteristic that increasing doses will produce increasing effects. A particular concern is whether the drug will cause cancer in its use. Drs. Lin and Ali report in the next chapter on how the FDA reviews animal toxicity data on the oncogenic potential of a new medication. The more general measurement of the effects of medication on an animal is termed bioassay. Bioassays are particularly good ways of telling how potent a new drug is relative to a standard drug or treatment. Dr. Yeh discusses the dose response relationship in bioassays in Chapter 3. Bioassay procedures are particularly important in the preclinical phase of drug development, but also have great importance in further animal and human testing during the clinical phases of research and in quality control.

A vital area in pharmaceutical research is the area of animal pathology and toxicology. Procedures have been formalized in response to rules and regulations about good laboratory practices. One part of the good laboratory practices refers to the recording and analysis of toxicity data. In addition, there is activity on optimal experimental designs to be used in the practical world of toxicity testing. In that world, animals die for causes unrelated to the experiment, particular samples are sometimes lost through technical error, and practical matters of cost limit the size of experiments. Thus, what is needed are experimental designs that are both powerful (in the statistical sense of being able to observe a difference if it is truly there) and robust (in the statistical sense that if assumptions, such as a particular variable being normally distributed, are not met, the analysis is still valid, and in the laboratory sense that the loss of a few test animals or samples should not invalidate the experiment).

V. TOXICITY TESTING

There are numerous methods for testing toxicity of potential drugs. The first major factor is whether the test is to be of acute exposure or chronic exposure. If of acute exposure, then one can administer a single dose to an animal and find out whether there are any apparent toxic effects. Actually, several different doses are administered. Alternatively, a small number of doses may be given and tested for toxicity. In chronic toxicity tests, the drug is given on a continuous basis, perhaps over the lifetime of the test animal. Numerous unsolved problems are involved with this procedure. If one is simply trying to determine the effects on a test animal, the above procedure is reasonable as it stands, though limited by the problems of sampling error, size of experiment, and so forth. If, however, one is interested in using a test animal as a surrogate for humans, then it is implicit that the test animal handle the drug biochemically and pharmacologically in a manner similar (if not identical) to the human. Thus, a test animal that metabolizes a drug in a different manner than does the human is not likely to be a valid surrogate.

There are many statistical and practical problems in these tests. Short-term acute experiments can be done during the preclinical testing phase. Lifetime experiments, on the other hand, require at least 2 years of observation in rats, a frequently used test animal, and then perhaps another year for finishing the experiment, preparing the numerous slides, reading and evaluating the slides, and producing a statistical analysis of the resulting data. Thus, a chronic rat study can be thought of as requiring of the order of 3 years in duration.

Practicality suggests that such studies should be done only after one is reasonably sure that the drug is going to be used in humans. Statistical problems in these tests involve mortality and sampling. It would be reasonable to schedule a certain number of animals to begin a study and then to sacrifice a fixed, randomly selected proportion at each of several checkpoints in the study. Unfortunately, some animals may die from natural causes or there may be laboratory problems assumed to be unrelated to the drug or exposure. Thus, any statistical design must be robust with respect to these anticipated untoward effects.

Questions about optimal number of animals are also of great importance. Since animal experiments for toxicity are extremely expensive, they should be done in the most efficient manner possible. The statistician can save pharmaceutical companies a great deal of resources (employee time, animal space, and money) with an optimally designed experiment; of course, concern for sensitivity always dictates as large an experiment as possible.

More rapid and inexpensive tests for carcinogenicity have been developed that are based on tests for mutagenicity in bacterial systems and other cell and tissue systems. These tests depend on mutagenicity being a predictor for carcinogenicity, which has been amply demonstrated for groups of chemicals but is not necessarily true in any individual case. More years of experience with these tests are required before their role in the pharmaceutical industry will be well understood.

VI. CLINICAL TESTING

Finally, after drugs have been tested extensively in experimental animals and the FDA has issued an Investigational New Drug (IND) exemption, the drug can be tested for the first time in Phase I human trials. Choosing the proper doses to use in humans is an interesting statistical problem. One can assume that on a fixed number of milligrams of drug per kilogram of body weight, the effects of the drug are constant. For anticancer drugs this is frequently amended to use the unit of milligrams of drug per square meter of body-surface area of the animal. In these or other projections (usually extrapolations, since the experimental animals are much smaller than the humans about to be tested) there is ample room for more statistical work to predict what dose in humans will have the same effect as a dose shown to have been effective in an experimental animal. Differences in metabolic processes, in disease processes, in species-specific modifying factors, in genetic characteristics, in diet and nutrition, and other factors are such that some of the experimental data in animals may be totally inappropriate to use in such a projection. Obviously, portions of this problem are beyond the role of the statistician; however, the statistical problem involves making estimates of an effective dose in humans that is not unduly affected by meaningless data points from a different animal species.

Finally, the eventful day arrives and the drug is used for the first time in humans in a Phase I trial. Initial doses are chosen to be especially safe and usually include double blind, placebo controls. The experimental program in humans reproduces that which was done in animals. First, acute single-dose studies, then short-term studies of more than one dose, and finally studies of several different doses on a longer-term basis are done. The goal of the initial studies is to find out about the toxicity of the drug in humans. What side effects, if any, are found in persons taking what is thought to be a large dose of the drug? What are the characteristics of these side effects? In order to be as certain as possible about any adverse effects, each volunteer or patient is given an extensive physical examination before taking the drug and then again after taking the drug, and for longer-term experiments at various intervals while taking the drug. These examinations include liver function tests, kidney tests, blood chemistry, urine chemistry, eye testing, and various other studies designed to tell more about the systems which might be adversely affected by the drug. An extensive battery of laboratory tests is usually included in these initial clinical pharmacology studies. Most of these tests will not be necessary in later studies when more is known about the clinical pharmacology of the drug.

The statistical characteristic of these initial studies is that they include a large number of observations on a small number of persons. Thus, the ability to be precise in characterizing the effect of the drug on the handful of persons who have taken it is quite good, based on repeated measures/correlated observations in the study persons. On the other hand, the small number of such subjects in these early trials means that inferences about the next persons to take the drug are subject to large prediction errors.

Dr. Dubey discusses the role of the FDA statisticians in the efficacy and safety evaluation of new drugs in Chapter 4. That chapter includes the operating rules for these drug trials based on published federal rules, regulations, and guidelines. Moreover, the results of the studies must be of a quality to satisfy the FDA statisticians; therefore, it is necessary to understand the criteria used in making these evaluations.

In Chapter 5, Drs. Buncher and Tsay sketch a number of the important points that should be considered when developing clinical trial designs. This area of statistical work involves many statisticians in the pharmaceutical industry. Accordingly, a number of other chapters are devoted to various aspects of clinical trials.

Patients who are studied in a clinical trial are supposed to be representative of those persons who will later take the drug. Studies have made apparent the great selectivity involved concerning the patients who actually take part in modern regulated pharmaceutical research. Drs. Weintraub and Calimlim write in Chapter 6 about the selection of both inpatients and outpatients participating in clinical trials.

Each different class of drugs involves special problems with respect to doing clinical trials. For example, antibiotics generally involve short-term trials, while drugs for the cardiovascular system involve tests over months and years. Trials with geriatric patients differ from those with persons in the middle of life. A major area of research is pharmaceutical preparations to be used to control or cure cancer. Dr. Chen discusses these specialized trials in Chapter 7. Another field of interest is medications for persons who suffer epileptic seizures. Interestingly, this is an example of how the government interacts with industry; Drs. Tsay and Pledger explain this program in Chapter 8. Another special area is discussed in Chapter 9 in which Dr. Green discusses the design and analysis of clinical trials of analgesic drugs.

A fairly common trial design is one in which the patient receives two or more different treatments, usually in a pattern that balances the order of treatment and allows each patient to be compared with himself or herself. In Chapter 10, Drs. Fisher and Wallenstein discuss these crossover designs. Dr. Buncher points out in Chapter 11 some of the characteristics of clinical trials in AIDS patients that are especially challenging to statisticians. Finally, Drs. Weiner and Yuh discuss the important area of the design and analysis of bioavailability studies in Chapter 12.

VII.   PLACEBO EFFECTS AND OTHER TOPICS

In order to know the effects of a drug, one must separate the pharmacological effects from the medical aura or placebo effects. The accepted way to do this is to compare the active drug with a pharmacologically inert substance. The substance is designated a placebo from the Latin for "I shall please." The placebo is well known to be a good analgesic; it cures or reduces headaches, backaches, postoperative pain, etc. A vast catalog of effects could be cited, but one interesting case might help prove the point: about one-tenth of all women who were anovulatory ovulated following administration of a placebo under study conditions (Johnson et al, 1966).

Side effects from placebo therapy are even more extensive than the list of conditions that are aided by placebo. Headaches, nausea, vomiting, dizziness, and so forth have all been caused by the administration of placebos. One study reported the following conclusion: "Virtually no toxic effects were reported from 'known' control pills containing lactose, but the exactly similar 'unknown' control pills, which were thought by the subjects to contain iron, produced as many side effects as the pills which did, in fact, contain it" (Kerr and Davidson, 1958). One should also remember that sometimes placebo proves to be the significantly better treatment (Echt et al., 1991). The placebo effect is discussed in Chapter 13 by Dr. Sanford.

Studies used to be completed first and then were subject to statistical analysis; however, many researchers prefer to know how the study results are proceeding even as the study is ongoing for ethical, financial, and scientific reasons. A specialized area in statistics concerning these interim analyses has developed. Drs. Davis and Hwang explain the methodology involved and Dr. O'Neill provides the regulatory view of these methods in Chapter 14. There are also many studies of medications that start after the NDA has been granted; Dr. Wille discusses these postmarketing studies in Chapter 15. A group of companies known collectively as Contract Research Organizations have sprouted to provide statistical and clinical trial management services for the sponsors of medications. In Chapter 16, Dr. Flora discusses the role of these organizations in drug development.

VIII. DOCUMENTATION

Many potential problems arise because a statistician in the pharmaceutical industry produces data that are to be evaluated by other statisticians, in particular those at the FDA. Anyone who has ever tried to review a major work of another statistical analyst realizes that there are important points to be resolved. The first major point concerns the ability to follow a complex analysis, since most statistical work is only reported in a skeletal outline. One needs to be able to follow exactly which patients were included in the analysis. Were all data points used or were some outliers rejected, presumably for a standard set of reasons? Are the statistical methods used standard methodology or are they subject to particular artifacts? Since for the pharmaceutical statistician the company side—proving that the drug is better than the comparison—is advantageous, the pharmaceutical statistician must frequently be certain that the conclusions drawn would be reported even if one was antagonistic to the drug. Thus besides the usual issues of doing the proper statistical analysis, there is also the problem of convincing the reader, especially a skeptical reader, of the correctness of the statistical conclusions.

Dr. Schultz and colleagues address the questions raised in documenting the results of a study in Chapter 17. Academic institutions teach their students how to analyze a study but rarely provide much instruction in documenting results to put the records in an orderly fashion and how to include sufficient details to convince another of the correctness of the analysis. This role of documentation is also important in other industries that find themselves trying to convince potentially skeptical audiences.

Dr. Falter examines a related problem in Chapter 18 on data quality assurance. Pharmaceutical statisticians are frequently responsible for putting together thousands of different numbers into documents which will be used in an analysis and will be reviewed for accuracy. As an example, the investigating physician may have recorded data incorrectly. Avoiding errors and finding the errors committed by others is a distinct challenge. Methods of quality control in handling vast quantities of data are essential in the pharmaceutical industry. To replace the submission of only hard copies of the data in the computer age, the problems of submitting Computer-Assisted New Drug Applications (CANDAs) are being solved. Dr. Cash and Mr. Cassese explain how to manage these CANDA submissions in Chapter 19.

IX. MANUFACTURING

Everyone is familiar with the concept of thousands of tiny capsules or tablets being carefully produced by a pharmaceutical manufacturer. With a little reflection we realize that this sort of production requires a tremendous amount of development before it becomes a reality. The chemical that has been tested in animals and found to be active must be given to humans. If the chemical is to be given in tablet form, the tablet must dissolve, typically in the stomach of the person taking the medication. The tablet must not break up into chunks in some people and dissolve neatly in other people. Therefore, other ingredients must be added to the tablet to give it proper disintegration and dissolution characteristics, to hold the tablet together before it is taken, to be less affected by temperature and humidity, and to yield various other favorable properties. This is a part of the field of drug formulation. There are many methods of drug delivery and these are still improving (Langer, 1990).

   Another part of the formulation process involves the human reaction rather than physical reactions. For example, what does the tablet taste like? Perhaps a sweetener must be added to avoid a bitter taste. Perhaps something must be added to prevent the tablet from feeling chalky. Other ingredients will be added to change the color of the tablet. In the case of capsules, a gelatin will be used with the addition of food coloring to give the capsule a particular identifying color or set of colors.

   More than one dosage size will exist for many pharmaceuticals, even in the early stages of testing. Thus a second problem is to create different doses of the same chemical (for example, a 10-mg and a 25-mg tablet) for early studies, especially for testing dose response in humans. If one is going to test the dose response in a blind trial, then there must also exist a placebo for the drug. The placebo must look and feel exactly the same as the drug and should also taste the same. Thus one is faced with the formulation of not just one product, but often several sizes as well as comparable placebos.

   After the initial formulation work is completed, the drug is tested in humans in the original Phase I and Phase II studies. During these early studies some problem with the formulation may be discovered. Meanwhile, pilot plant preparation of the drug is being worked on. At later stages full-scale manufacture of the drug will be planned and accomplished. Required changes in the drug at any of these stages will require a restart of many of the formulation steps. Checking procedures and other steps preparatory to a formal quality control program must also be worked out.

The statistician works with other employees in the quality control field to be certain that the drug is manufactured to the best standards possible. This field is active partly because of Food and Drug Administration regulations (FDA, 1980). Drs. Murphy and Sampson discuss many of these points in Chapter 20 on quality control of the manufacturing process.

Another practically important and statistically interesting question concerns the stability of a drug. For how long after the drug is manufactured can it be considered clinically adequate? These time periods are typically measured in years. Thus, we have the choice of waiting for years for the answers or testing in such a way that we can predict (extrapolate) what will happen. In Chapter 21, Drs. Lin, Lin, and Kelly discuss stability testing in room temperature studies while Drs. Davies and Hudson, consider this problem using higher temperatures.

X.   OTHER STATISTICAL ISSUES

A classical statistical problem is that of multiple comparisons which must be considered when there are more than two treatment groups in the experiment or more than two measurements on each subject simultaneously. If a drug may have 1 of 4 activities and we wish to claim only 1 of them, then we must take account of the fact when setting a "0.05 level" that there are 4 random chances that the drug will be shown to be effective rather than just 1. In a similar manner, if there are 10 chemicals competing to become a drug, the possibility that at least 1 of them will be better than placebo by chance is certainly enhanced by the fact that it is 1 of 10. Again, the probability levels must be properly adjusted. Drs. Dunnett and Goldsmith discuss this problem and some of the solutions in Chapter 22 on when and how to do multiple comparisons. In Chapter 23, Dr. Dubey adds some views on adjusting P-values for correlated symptoms in pharmaceutical testing.

One step in the evolution of a drug is the screening of newly created chemical compounds; biological screens are tests designed to separate those chemicals which are apparently inert with respect to the tested biological systems from those chemicals which have desired effects. In designing such a screen, one must keep in mind that of all chemicals created, far fewer than 1% will ever pass through the various stages of showing sufficient efficacy and safety in cell based and test animal systems to ever be given to any human.

This realization implies that any worthwhile system must rapidly and economically eliminate inert compounds. Then a greater proportion of the test effort can be spent on those few compounds that have been shown to be of interest. Obviously, a drug tested on only three mice or two dogs, even if it could later be shown to be of great therapeutic value, might through random variation not show activity in the small number of test animals. This act of balancing the risks of missing a worthwhile drug versus excessive testing of useless chemicals is one of the most important roles of a statistician in the pharmaceutical industry. The opportunity for saving resources for the research facility are truly prodigious. In many of the other roles of a statistician, the statistician asks for additional expenditure of resources; this role is one in which the statistician can demonstrate how statistical thinking creates savings.

The statistical problem in screening chemicals for potential new drugs is to determine how many animals in each screen should be tested with each chemical compound. Should the screen be a simple one-stage screen or should there be several levels of success that must be passed before the chemical is ready for additional testing? Should an effect be statistically significant before the chemical is tested in later screens or is an indication good enough? These and numerous other statistical questions must be resolved before any routine screen is established and used.

Drs. Redman and Dunnett cover some of these points in Chapter 24 on screening compounds for clinically active drugs. The concept of a screen is that of a mesh that will hold back most of the useless chemicals and permit through those that are more likely to be useful. The actual screen consists of a chemical measurement, physiological reaction, or behavioral reaction in some cell, organ, or whole-animal system or computer model that has been shown to be a mimic of some desired action in humans. Obviously, screens are specific to the desired outcome. Frequently a number of different screens are being run simultaneously at any given organization. Usually, a particular company specializes in some smaller number of fields rather than covering all potential medical aspects.

Finally, some survey data on the situation for statisticians working in United States pharmaceutical companies are presented by Dr. Buncher and Mr. Wilkinson in Chapter 25.

REFERENCES

DiMasi, J. A., Hansen, R. W., Grabowski, H. G., and Lasagna, L. (1991), Cost of innovation in the pharmaceutical industry. *J. Health Econ.*, **10**:107-142.

Echt, D. S., Liebson, P. R., Mitchell, L. B., et al. (1991), Mortality and morbidity in patients receiving encainide, flecainide, or placebo. The Cardiac Arrhythmia Suppression Trial. *N. Engl. J. Med.*, **324**:781-788.

Food and Drug Administration Regulations (1980). *Current Good Manufacturing Practices in Manufacturing, Processing, Packing, or Holding of Drugs.* Code of Federal Regulations.

Johnson, E. E., Jr., Cohen, M. R., Goldfarb, A. F., Rakoff, A. E., Kistner, R. W., Plotz, E. J., and Vorys, N. (1966), The efficacy of clomiphene citrate for induction of ovulation. *Internat. J. Fertility,* **11**:265-270.

Kerr, D. N. S. and Davidson, S. (1958). Gastrointestinal intolerance to oral iron preparation. *Lancet,* **2**:489-492.

Langer, R. (1990), New methods of drug delivery, *Science,* **249**:1527-1533.

Nugent, W. A., RajanBabu, T. V., and Burk, M. J. (1993). Beyond nature's chiral pool: Enantioselective catalysis in industry, *Science,* **259**:479-483.

2

Statistical Review and Evaluation of Animal Tumorigenicity Studies

Karl K. Lin and Mirza W. Ali *

*United States Food and Drug Administration,
Rockville, Maryland*

1. INTRODUCTION

The risk assessment of a new drug exposure in humans usually begins with an assessment of the risk of the drug in animals. It is required by law that the sponsor of a new drug conduct nonclinical studies in animals to assess the pharmacological actions, the toxicological effects, and the pharmacokinetic properties of the drug in relation to its proposed therapeutic indications or clinical uses. Studies in animals designed for assessment of toxicological effects of the drug include acute, subacute, subchronic, and chronic toxicity studies, tumorigenicity studies, reproduction studies, and pharmacokinetic studies.

The Statistical Application and Research Branch (SARB), Division of Biometrics, Center for Drug Evaluation and Research (CDER), Food and Drug Administration (FDA) is responsible for statistical reviews of results of long-term (or chronic) animal tumorigenicity experiments submitted by drug sponsors to FDA as parts of their investigational new drug (IND) or new drug application (NDA) submissions. Long-term animal tumorigenicity studies usually are conducted on both sexes of mice and rats for the majority of the normal lifespans of those animals. The primary purpose of these studies is to determine the oncogenic potential of the new drug. There are different ways to use the results of long-term animal tumorigenicity studies in the determination of oncogenic potential of chemical compounds. The first way is to use the results merely for screening of unsafe chemical compounds. The second way is to do risk assess-

---

* *Current affiliation*: Pfizer Inc., Groton, Connecticut

ments of chemicals in humans, which involves extrapolations of results from animals to humans, and from high to low doses. The third way is to verify scientific hypotheses about the mechanisms of carcinogenesis.

Statisticians in CDER are not involved in extrapolating animal tumorigenicity study findings beyond the ranges of doses studied or to species other than those studied, nor are they involved in investigation of mechanisms of carcinogenesis. Accordingly, statisticians in the CDER develop a quantitative assessment of the risk of a drug for each species and sex of rodent, and the reviewing pharmacologists and medical officers apply their knowledge of mammalian similarities and interspecies differences to extrapolate qualitatively from rodent to humans (Fairweather, 1988).

The statistical interest in detecting oncogenic potential of a new drug is to test if there are statistically significant positive dose-response relationships in tumor incidence rates induced by the new drug. The phrase "positive dose-response relationship" in this chapter refers to the increasing linear component of the effect of treatment, but not necessarily to a strictly increasing tumor rate as dose increases. However, the review and evaluation of the results of long-term animal tumorigenicity experiments studying the oncogenic potential of a new drug is a complex process. The final interpretation of the study results involves issues which require statistical as well as nonstatistical biomedical judgements. The statistical issues in carrying out an animal tumorigenicity experiment include the validity of design of the experiment, the appropriateness of methods of statistical analysis of experimental data, adjustment for the effect of multiple tests, and the use of comparable historical data in the final interpretation of the results.

There is a vast amount of statistical literature on these issues. The techniques or methods of analysis, and decision rules adopted by FDA statisticians in their reviews are based on current literature, consultations with outside experts, our research, and best scientific judgements, although we recognize that some of the issues are still without consensus of opinion among experts in evaluation of animal tumorigenicity studies.

In addition to reviewing the reports submitted by the sponsors, statisticians in the CDER also need tumor data on computer readable media from drug sponsors to perform additional statistical analyses which they believe are appropriate and necessary to evaluate the analyses and conclusions contained in the reports. The Division of Biometrics has issued guidelines for formats and specifications for submission of animal tumorigenicity study data. To expedite the statistical review, sponsors are urged to submit the tumor data on computer readable media using the Division of Biometrics formats and specifications or the STUDIES formats (National Technical Information Service, 1990), which became available through a joint effort of several regulatory agencies within the government, along with their original, initial submissions of the hardcopy NDA or IND.

The purpose of this chapter is to provide some guidance in the design of animal tumorigenicity experiments, method of statistical analysis of tumor data, interpretation of study results, presentation of data and results in reports, and submission of tumor data to FDA statistical reviewers that drug sponsors can follow in their preparations for the nonclinical parts of IND and NDA submissions. A discussion on the validity of the design of experiment is given in Section II. This is followed by an extensive discussion on methods of statistical analysis in Section III. In Section IV, a discussion on how the results should be interpreted is given. Discussions on data presentation and submission are given in Section V. Finally, some concluding remarks are given in Section VI.

II. VALIDITY OF THE DESIGN

In the evaluation of the validity of experimental designs, statistical reviewers check if randomization methods are used in allocating animals to treatment groups to avoid possible biases caused by animal selections, and if a sufficient number of animals is used in an experiment to insure reasonable power in statistical tests used. It has been recommended that, in a standard four-treatment-group experiment, each dose group and concurrent control group should contain at least 50 animals of each sex. If interim sacrifices are planned, the initial number should be increased by the number of animals scheduled for the interim sacrifices.

In general, based on the results of sponsors' single dose, short-term subchronic toxicity studies, FDA statisticians, reviewing pharmacologists, medical officers, and CDER Carcinogenicity Assessment Committee (CAC) members will evaluate the appropriateness of the doses used in animal tumorigenicity experiments. However, in negative studies (i.e., studies in which no significant positive dose-response relationships or drug related increases in tumor incidence rates were detected) the statistical reviewers working with other FDA scientists will perform an additional evaluation on the validity of the designs of experiment to see if there are sufficient animals living long enough to get an adequate exposure to the chemical and to be at risk of forming late-developing tumors. Also of concern is whether the doses used are high enough and close enough to the maximum tolerated dose (MTD) to present a reasonable tumor challenge to the tested animals.

The adequacy of the number of animals surviving, the length of exposure, and the appropriate dose strength depend on species and strains of animals employed, routes of administration, and other factors (see the discussion in Haseman, 1984). A general rule is that a 50% survival rate in any group between weeks 80–90 of a two-year study will be considered as a sufficient number and an adequate exposure. However, the percentage can be lower or higher if the number of animals used in each treatment/sex group is larger or smaller than 50 so that there will be between 20 to 30 animals still alive during these weeks. In consultations with reviewing pharmacologists and medical officers, FDA statistical reviewers often followed the criteria proposed in Chu, Ceuto, and Ward (1981) in their evaluation to see if the high dose used is close to the MTD and presents a reasonable tumor challenge to the animals. Based on results of 200 National Cancer Institute carcinogen bioassays, these investigators considered a high dose to be close to the MTD if: a) there was a detectable weight loss of up to 10% in the dosed group relative to the controls, and/or b) the animals exhibit clinical signs or severe histopathologic toxic effects that was attributed to the chemical in the dosed animals, and/or c) there was a slightly increased mortality in the dosed animals compared to the controls.

The appropriateness of the high dose is always addressed in the CDER/CAC meetings during the final determinations of the oncogenic potential of new drugs under review at the FDA. It is an important, controversial, and complicated issue in the evaluation of validity of designs of animal experiments. Information about body weight gain, mortality, and clinical signs and histopathologic toxic effects still are used to resolve the issue. Other information such as pharmacokinetic and metabolic data are also often needed in evaluation of dose selection.

III. METHODS OF STATISTICAL ANALYSIS

A. Test of Intercurrent Mortality Data

Intercurrent mortality refers to all deaths not related to the development of the particular type or class of tumors that are being studied for evidence of carcinogenicity. Like human beings, older rodents have a many times higher probability of developing or dying of tumors than those of younger ages. Therefore, it is essential to identify and adjust the possible differences in intercurrent mortality (or longevity) among treatment groups to eliminate or reduce biases caused by the differences. It is pointed out that "the effects of differences in longevity on numbers of tumor-bearing animals can be very substantial, and so, whether or not they appear to be, they should routinely be corrected for when presenting experimental results" (Peto et al., 1980). The following examples clearly demonstrate the above important point.

Example 1. (Peto et al., 1980). Consider an experiment consisting of one control group and one treated group of 100 mice each. A very toxic but not tumorigenic new drug was administered to the animals in the diet for two years. Assume that the spontaneous incidental tumor rates for both groups are 30% at 15 months and 80% at 18 months of age and that the mortality rates at 15

Table 2.1 Data for Example 1.

	Control			Treated		
	T	D	%	T	D	%
15 Months	6	20	30	18	60	30
18 Months	64	80	80	32	40	80
Totals	70	100	70	50	100	50

T = Incidental tumors found at necropsy.
D = Deaths

months for the control and the treated groups are 20% and 60%, respectively, due to the toxicity of the drug. The results of the experiment are summarized in Table 2.1. If one looks only at the overall tumor incidence rates of the control and the treated groups (70% and 50%, respectively) without considering the significantly higher early deaths in the treated group caused by the toxicity of the drug, one will conclude erroneously that there is a significant ($p = 0.002$, one-tailed) negative dose-response relationship in this tumor type (i.e., the new drug prevents tumor occurrences). The one-tailed p-value is 0.5 when the survival-adjusted prevalence method is used (Peto et al., 1980).

Example 2 (Gart et al., 1986). Assume that the design used in this experiment is the same as the one used in the experiment of Example 1. However, we assume that the treated group has a much higher early mortality than the control (20% versus 90%) before 15 months, and that the drug in this example induces an incidental tumor which does not cause the animal's death, either directly or indirectly. Also assume that the incidental tumor prevalence rates for the control and treated groups are 5% and 20%, respectively, before 15 months of age, and 30% and 70%, respectively, after 15 months of age. The results of this experiment are summarized in Table 2.2. Note that the age-specific tumor incidence rates are significantly higher in the treated group than those in the control group. The survival-adjusted prevalence method yielded a one-tailed p-value of 0.003; this shows a clear tumorigenic effect of the new drug. However, the overall tumor incidence rates are 25% for the two groups. Without considering the significantly higher early mortality in the treated group, one would conclude that the positive dose-response relationship is not significant.

Before analyzing the tumor data, the intercurrent mortality data are routinely tested first by FDA statisticians to see if the survival distributions of the treatment groups are significantly different or if there exist significant dose-response relationships. Cox's Test (Cox, 1972; Thomas, Breslow, and Gart, 1977; Gart et al., 1986), the generalized Wilcoxon or Kruskal-Wallis test (Bres-

Table 2.2 Data for Example 2.

	Control			Treated		
	T	D	%	T	D	%
Before 15 months	1	20	5	18	90	20
After 15 months	24	80	30	7	10	70
Totals	25	100	25	25	100	25

T = Incidental Tumors Found at Necropsy.
D = Deaths

low, 1970; Gehan, 1965; Thomas, Breslow, and Gart, 1977), and the Tarone trend tests (Cox, 1959; Peto et al., 1980; Tarone, 1975) are routinely used to test the heterogeneity in survival distributions and significant dose-response relationship (trend) in mortality.

There is an issue on the use of the results from tests of intercurrent mortality data in the determination whether a survival-adjusted method should be used in the analyses of tumor data. If we treat the test for heterogeneity in survival distributions or dose-response relationship in mortality as a preliminary test of significance (Bancroft, 1964), then a level of significance larger than 0.05 should be used. A very large level of significance used in the preliminary test means that survival-adjusted methods should always be used in the subsequent analyses of the tumor data.

B. Contexts of Observation of Tumor Types

The choice of a survival-adjusted method to analyze tumor data depends on the role which a tumor plays in causing the animal's death. Tumors can be classified as "fatal", "mortality-independent (or observable)", and "incidental" according to the contexts of observation described in Peto et al. (1980). Tumors which kill the animal either directly or indirectly are said to have been observed in a fatal context. Tumors which are not directly or indirectly responsible for the animal's death, but are merely observed at the autopsy of the animal after it has died of some unrelated causes are said to have been observed in an incidental context. Tumors, such as skin tumors, whose times of criterion attainment (i.e., detection of the tumor at a standard point of their development) other than the times or causes of death are the primary interest of analyses are said to have been observed in a mortality-independent (or observable) context. To apply a survival-adjusted method correctly, it is essential that the context of observation of a tumor be determined as accurately as possible.

Different statistical techniques have been proposed for analyzing data of tumors observed in different contexts of observation. For example, the death rate method, the onset-rate method, and the prevalence method are recommended for analyzing data of tumors observed in fatal, mortality-independent, and incidental contexts of observation, respectively (Peto et al. 1980). Peto et al. also demonstrate the possible biases resulting from misclassifications of incidental tumors as fatal tumors, or fatal tumors as incidental tumors.

C. Statistical Analyses of Incidental Tumors

The prevalence method described by Peto et al. (1980) is routinely used by FDA statisticians in testing for a positive dose-response relationship in prevalence rates of incidental tumors. Briefly, this method focuses on the age-specific tumor prevalence rates to correct for intercurrent mortality differences among treatment groups in the test for positive dose-response relationships in incidental tumors. The experimental period is partitioned into a set of intervals plus interim sacrifices (if any) and terminal sacrifices. The incidental tumors are then stratified by those intervals of survival times. The selection of the partition of the experiment period does not matter very much as long as the intervals "are not so short that the prevalence of incidental tumors in the autopsies they contain is not stable, nor yet so large that the real prevalence in the first half of one interval could differ markedly from the real prevalence in the second half" (Peto et al., 1980).

In each time interval and for each group, the observed number of animals with a particular tumor type found in necropsies is compared with the number of animals which died in the time interval and expected to have the tumor type found in the necropsies under the null hypothesis that there is no dose-response relationship. Finally, the differences between the observed and the expected numbers of animals found with the tumor type after their deaths are combined across all time intervals to yield an overall test statistic using the method described in Mantel and Haenszel (1959).

The following derivation of the Peto prevalence test statistic uses the notations in Table 2.3. Let the experimental period be partitioned into the following M intervals $I_1, I_2, ..., I_M$. As mentioned before, interim sacrifices (if any) and terminal sacrifices should be treated as separate intervals. The following partitions (in weeks) are used most often by FDA statisticians in two-year studies: a) 0–50, 51–80, 81–104, interim sacrifice (if any), and terminal sacrifice, b) 0–52, 53–78, 79–92, 93–104, interim sacrifice (if any), and terminal sacrifice (proposed by National Toxicology Program), or c) Partition determined by the "ad hoc runs" procedure described in Peto et al. (1980).

This method uses a normal approximation in the test for a positive dose-response relationship in tumor prevalence rates. The accuracy of the normal

Table 2.3 Notations Used in the Derivation of Peto Prevelance Test Statisitic

Group	Dose	Interval								
		I_1		I_2		...	I_k		...	I_M
		R_1		R_2		...	R_k		...	R_M
0	D_0	O_{01}	P_{01}	O_{02}	P_{02}	...	O_{0k}	P_{0k}	...	O_{0M} P_{0M}
1	D_1	O_{11}	P_{11}	O_{12}	P_{12}	...	O_{1k}	P_{1k}	...	O_{1M} P_{1M}
.	.	.	.	.	.	...	.	.	...	. .
.	.	.	.	.	.	...	.	.	...	. .
.	.	.	.	.	.	...	.	.	...	. .
i	D_i	O_{i1}	P_{i1}	O_{i2}	P_{i2}	...	O_{ik}	P_{ik}	...	O_{iM} P_{iM}
.	.	.	.	.	.	...	.	.	...	. .
.	.	.	.	.	.	...	.	.	...	. .
.	.	.	.	.	.	...	.	.	...	. .
r	D_r	O_{r1}	P_{r1}	O_{r2}	P_{r2}	...	O_{rk}	P_{rk}	...	O_{rM} P_{rM}
Sum		$O_{.1}$	$P_{.1}$	$O_{.2}$	$P_{.2}$	...	$O_{.k}$	$P_{.k}$	...	$O_{.M}$ $P_{.M}$

R_k: Number of animals that have not died of the tumor type of interest but come to autopsy in the time interval k.
P_{ik}: Proportion of R_k in group i.
O_{ik}: Observed number of autopsied animals in group i and interval k found to have the incidental tumor type.
$O_k = \sum_i O_{ik}$

approximation depends on the numbers of tumor occurrences in each group in each interval, the number of intervals used in the partition, and the mortality patterns. However, it is known that under the regularity conditions, the approximation will not be stable and reliable when the numbers of tumor occurrences across treatment groups are small. In this situation, an exact permutation trend test based on an extension of the hypergeometric distribution (discussed in Section III-F) is used to test the positive dose-response relationship in tumor prevalence rates.

Although Peto et al. (1980) proposed general guidelines for partitioning the experimental period into intervals in the prevalence method, there is no unique way to do the partition. Test results could be different when different sets of intervals are used. Dinse and Haseman (1986) applied 10 different sets of intervals to the same tumor data set and got 10 different p-values ranging from 0.001 to 0.261. Because of the lack of a unique way to partition the experimental period, some regression-type methods have been proposed as alternatives for analyzing incidental tumor data from animal carcinogenicity experiments. The logistic regression method (Dinse, 1985; Dinse and Haseman, 1986; Dinse and Lagakos, 1983; Lin and Reschke, 1987) and the Cochran-Armitage trend test methods (Armitage, 1955, 1971) are two of those proposed alternatives. The main advantage of the regression type methods is that they adjust for the differences in intercurrent mortality by including the survival time as a continuous regression variable. This makes it unnecessary to partition the experimental period into intervals. Another advantage of these methods is that other variables having effects on the prevalence rates, such as body weight and cage location, can also be incorporated into the model as covariates.

The logistic regression model is defined as

$$E(Y_i) = \frac{e^{a + bD_i}}{1 + e^{a + bD_i}}$$

without adjustment for intercurrent mortality differences, and as

$$E(Y_i) = \frac{e^{a + bD_i + F(t_i)}}{1 + e^{a + bD_i + F(t_i)}}$$

with adjustment for intercurrent mortality differences, where $E(Y_i)$, D_i, and t_i are the expected value of Y_i, the dose level, and survival time, respectively, of animal i and

$$F(t) = c_1 t_1 + c_2 t_2 + \ldots + c_p t_p$$

The following statistic

$$Z = \frac{\hat{b}}{\hat{V}(\hat{b})^{1/2}}$$

which is approximately distributed as a standard normal, is used to test the positive dose-response relationship in a specific incidental tumor. The term $\hat{V}(\hat{b})$ in the above equation is the variance of the estimated regression coefficient $\hat{b}$.

However, there is another issue in using the logistic regression method. The functional form of $F(t_i)$ has to be specified in the logistic regression model to indicate the effect of survival time on tumor prevalence rate. Like partitioning the experimental period into intervals in the Peto prevalence method, there is no unique way of determining the functional form and different functional forms of $F(t_i)$ can yield different results.

Table 2.4 Analysis of Variance Table

Source of variation	D.F.	Sum of squares
Treatment	r	$S_1 + S_2$
Linear	1	S_1
Departure from linearity	$r - 1$	S_2
Error	$T - r - 1$	S_3
Total	$T - 1$	$S_1 + S_2 + S_3$

T is the total number of animals used in the study. There are $r + 1$ treatment groups including the control.

Armitage (1955) applies the one-way analysis of variance model to the dependent variable Y, individual animal tumor status (i.e., $Y = 1$ if an animal developed the tumor of interest; $Y = 0$ otherwise) using the dose variable as the grouping variable to obtain the sum of square components of various sources of variation as shown in Table 2.4.

However, because the dependent variable Y assumes only values 0 and 1, the test procedure for the linear contrast in regular analysis of variance has to be modified. Armitage suggested the use of the following alternative statistic

$$\chi_0^2 = \frac{S_1}{(S_1 + S_2 + S_3)/T}$$

which is distributed approximately as with one degree of freedom under the null hypothesis of no positive dose-response relationship. The above analysis of variance approach to the trend test is equivalent to the test of significant positive slope of the regression equation of Y on X. Here the score variable X can take the values $X_1 = -r/2$, $X_2 = -(r - 2)/2$, . . ., $X_r = r/2$, or any set of $r + 1$ equally spaced numbers for the case of r+1 groups. If the fitted regression equation is expressed as

$$Y_i = \hat{a} + \hat{b}X_i$$

then the test statistic

$$Z = \frac{\hat{b}}{\hat{V}(\hat{b})^{1/2}}$$

which is distributed approximately as a standard normal is used to test the positive dose-response relationship.

The Cochran-Armitage regression methods are survival-unadjusted. The results from the unadjusted methods are reasonably unbiased if the intercurrent mortalities among the treatment groups are not significantly different. For experiments experiencing significant differences in intercurrent mortality, the Cochran-Armitage trend test procedures can be modified to adjust the effect of the survival differences. Two different modifications can be made. The first is to use the survival time as a covariate and perform the analysis of covariance; the second is to include the linear term or quadratic term or both in the regression analysis as other independent variables in addition to the score variable X.

The computations in the modified Cochran-Armitage regression method are much simpler than those in the logistic regression method. However, it does not satisfy the condition of constant variance in regression analysis. The modified Cochran-Armitage regression method also has the a shortcoming similar to the logistic regression method; that is, there is no unique way to determine the functional relationship between survival time and tumor incidence.

Lin (1988) conducted an empirical study using tumor data from three experiments to compare the Peto prevalence method with the logistic regression and the modified Cochran-Armitage regression type test procedures, with the following results. The p-values from the logistic regression methods assuming the effect of survival time on tumor incidence rate were linear, and both linear and quadratic forms were similar in the three studies used. However, this is not true in the case of the modified Cochran-Armitage regression method. The p-values from the model including only a linear term of survival time were in general appreciably larger than those from the model including both the linear and the quadratic terms.

The p-values from the Cochran-Armitage method using linear survival time as a covariate or as an independent regression variable were close to those from the logistic regression method also adjusted by the linear term of survival time, although they were somewhat larger. There was no clear pattern in p-values from these two test procedures when both the linear and the quadratic terms of survival time are included.

The p-values from the Peto prevalence method were in general smaller than those from the Cochran-Armitage regression method adjusted by the linear term of survival time. There was no clear pattern in p-values between the two methods when the quadratic term of survival time was added to the Cochran-Armitage method. There was no clear pattern in p-values when the Peto prevalence method was compared with the logistic regression method.

The p-values from the unadjusted logistic and the unadjusted Cochran-Armitage test procedures were virtually identical, and were not very different from the p-values from the Peto prevalence method in the study in which there is no significant difference in mortality.

Finally, in terms of decision making, the Peto prevalence, the adjusted logistic regression, and the adjusted Cochran-Armitage regression methods reached consistent conclusions (either all methods reject or accept, at a given level of significance, the null hypothesis of no positive dose-response relationship in the tumors tested in the three studies).

Before the issues related to the functional form of the effect of survival time on tumor incidence rate and the power and the conservativeness of the logistic regression and the modified Cochran-Armitage regression test procedures are fully studied, FDA statisticians will continue to recommend the Peto prevalence method in analyzing incidental tumor data from animal experiments.

D.   Statistical Analyses of Fatal Tumors

In their reviews and analyses of animal carcinogenicity study data, FDA statisticians routinely use the death-rate method described in Peto et al. (1980) to test the positive dose-response relationship in tumors observed in a fatal context.

The notations of Section III-C with some modifications will be used in this section to derive the test statistic of the death-rate method. Now let $t_1 <$ $t_2 < ... < t_M$ be the time points when one or more animals died. Use these time points to replace the intervals used in the prevalence method. The notations in Table 2.3 are redefined as follows:

R_k: The number of animals of all groups just before t_k.
P_{ik}:   The proportion of R_k in group i (the same as in the prevalence method).
O_{ik}:   Observed number of animals in group i just before t_k found to have the fatal tumor.
$O_{.k}$: $\sum_i O_{ik} O_{ik}$

As in the prevalence method, the test statistic T for the positive dose-response relationship in the fatal tumor is defined as:

$$T = \sum_i D_i(O_i - E_i)$$

with estimated variance

$$\hat{V}(T) = \sum_i \sum_j D_i D_j V_{ij}$$

where D_i, O_i, E_i, and V_{ij} are defined similarly as in Section III-C. Under the null hypothesis of equal death rates among the treatment groups, the statistic

$$Z = \frac{T}{\hat{V}(T)^{\frac{1}{2}}}$$

is distributed approximately as a standard normal.

E. Statistical Analyses of Tumors Observed in Both Incidental and Fatal Contexts

When a tumor was observed in a fatal context for a set of animals and was observed in an incidental context for the remaining animals in an experiment, data should be analyzed separately by the prevalence and the death-rate methods. Results from the different methods can then be combined to yield an overall result. The combined overall result can be obtained by simply adding together either the separate observed and expected frequencies and variances, or the separate T statistics and their variances.

F. Exact Analysis

As mentioned in the previous sections, the prevalence and death-rate methods use a normal approximation in the test for positive dose-response relationship (trend) in tumor rates. The adequacy of the normal approximation may depend on factors such as the number of tumor bearing animals, scores assigned to the treatment groups, number of intervals used in partitioning the study period, etc. It is particularly true that when the number of tumor bearing animals is "small", the normal approximation is unreliable and tends mostly to underestimate the exact p-values (Ali, 1990). Under this situation, the use of an exact permutation trend test is suggested (Gart et al., 1986) to test for dose-response relationship in tumor rates. The exact trend test is a generalization of the Fisher exact test to sequences of $2 \times (r + 1)$ tables.

1. The Exact Method

The exact method is derived by conditioning on the row and column marginal totals of each of the $2 \times (r + 1)$ tables formed from the partitioned data set of Table 2.1. Consider the kth interval I_k (in Table 2.1) and write it as in Table 2.5. Now let the column totals $C_{0k}, C_{1k}, ..., C_{rk}$, and the row totals $O_{.k}$ and A_k be fixed. Define

$$P_{jk} = \frac{C_{jk}}{R_k}$$

Table 2.5 The Data in the kth Time Interval I_k Is Written as a $2 \times (r+1)$ Table

Group	n	1	...	i	...	r	
Dose	D_n	D_1	...	D_i	...	D_r	Total
Number with tumor	O_{0k}	O_{1k}	...	O_{ik}	...	O_{rk}	$O_{.k}$
Number without tumor	A_{0k}	A_{1k}	...	A_{ik}	...	A_{rk}	$A_{.k}$
Total	C_{0k}	C_{1k}	...	C_{ik}	...	C_{rk}	R_k

Then the quantities $E_{ik} = O_{.k}P_{ik}$, $V_{ijk} = P_{ik}(\delta_{ik} - P_{jk})$, and $V(t)$ (defined in Section 3.3) are all known constants.

Now let z be the observed value of Z. Then (under conditioning on the column and row marginal totals in each table) the observed significance level or

$$\text{p-value} = P(Z \geq z) = P\left[\frac{\Sigma D_i(O_i - E_i)}{\sqrt{V(T)}} \geq z\right] = P[\Sigma D_i O_i \geq y]$$

$$= P(\Sigma_i D_i \, \Sigma_k O_{ik} \geq y) = P(\Sigma_k \Sigma_i D_i O_{ik} \geq y)$$

$$P(\Sigma Y_k \geq y) = P(Y \geq y)$$

where $Y = \Sigma Y_k = \Sigma_i D_i O_{ik}$ and $Y = \Sigma y_k$, the observed value of Y.

We compute this p-value $[P(Y > = y)]$ from the exact permutational distribution of Y. Given the observed row and column marginal totals in a $2 \times (r+1)$ table, generate all possible tables having the same marginal totals. Let S_k $(k = 1, 2, ..., K)$ be the set of all such tables generated from the kth observed table. Form a set of K tables taking one from each S_k. Assuming independence between the K tables, the above expression for the p-value can now be written as

$$\text{p-value} = \Sigma[P(Y_1 = Yy_1) \ldots P(Y_K = y_K)]$$

where $y_k = \Sigma_i D_i o_{ik}$ $(k = 1, 2, ..., K)$, the sum is over all sets of K tables such that $y_1 + y_2 + ... + y_k \geq y$, the observed value of Y, and $P(Y_k = y_k)$ is the conditional probability given the marginal totals in the kth table,

Table 2.6 Tumor Count Table

Time interval (weeks)		0	1	2	Total
		Dose Levels			
0–50	O	0	0	0	0
	C	1	3	3	7
51–80	O	0	0	0	0
	C	4	5	7	16
81–104	O	0	0	2	2
	C	10	12	15	37
Terminal sacrifice	O	0	1	0	1
	C	35	30	25	90

O = observed tumor count, C = number of animals necropsied

$$P(Y_k = y_k) = \frac{\binom{C_{0k}}{O_{0k}} \binom{C_{1k}}{O_{1k}} \cdots \binom{C_{rk}}{O_{rk}}}{\binom{R_k}{O_{.k}}}$$

Example. Consider an experiment with 3 treatment groups (control, low, and high) with dose levels $D_0 = 0$, $D_1 = 1$, and $D_2 = 2$, respectively. Suppose the study period is partitioned into the intervals 0–50, 51–80, 81–104 weeks, and the terminal sacrifice week. Consider a tumor type (classified as incidental) with the data shown in Table 2.6.

Since all the observed tumor counts (i.e., O's) in the first two time intervals are zeros, the data for these intervals will not contribute anything to the test statistic and we may neglect these intervals. The observed tables formed from the last two intervals are as follows:

		Observed Table 1					Observed Table 2			
Dose	0	1	2	Total		Dose	0	1	2	Total
O	0	0	2	$2 = o_k$		O	0	1	0	$1 = o_{.2}$
A	10	12	13	$35 = a_{.1}$		A	35	29	25	$89 = a_{.2}$
C	10	12	15	$37 = R_1$		C	35	30	25	$90 = R_2$

Table 2.7 All Possible Configurations of $o_{.1}$ and the Corresponding Hypergeometric Probabilities

Configurations			y_1	$P(Y_1 = y_1)$
0	0	2	4	.15766
0	2	0	2	.09910
2	0	0	0	.06757
0	1	1	3	.27027
1	0	1	2	.22523
1	1	0	1	.18018

We will now generate all possible tables from observed Table 2.3. Since the marginal totals are fixed, we may generate these tables by distributing the total tumor frequency $o_{.1}$ ($= 2$) among the four dose groups. Thus each table will correspond to a configuration of this distribution of $o_{.1}$. The configurations, the values of y_1, and the $P(Y_1 = y_1)$ are shown in Table 2.7.

To illustrate the computation of y_1 and $P(Y_1 = y_1)$ consider the last row. Here $y_1 = (D_0 \times 1) + (D_1 \times 1) + (D_2 \times 0) = (0 \times 1) + (1 \times 1) + (2 \times 0) = 1$, and

$$P(Y_1 = 1) = \frac{\binom{10}{1}\binom{12}{1}\binom{15}{0}}{\binom{37}{2}} = \frac{10 \times 12 \times 2}{37 \times 36} = .18018$$

The configurations and probabilities obtained from Observed Table 2 are given in Table 2.8.

Note that the first configuration (0, 0, 2) in Table 2.3 corresponds to the Observed Table 1 with a value of $y_1 = (0 \times 0) + (1 \times 0) + (2 \times 2) = 4$ and a probability of .15766, and the second configuration (0, 1, 0) in Table 2.4 corresponds

Table 2.8 All Possible Configurations of $o_{.2}$ and the Corresponding Hypergeometric Probabilities

Configuration			y_1	$P(Y_1 = y_1)$
0	0	1	2	.27778
0	1	0	1	.33333
1	0	0	0	.38889

to the Observed Table 2 with a value of $y_2 = (0 \times 0) + (1 \times 1) + (0 \times 0) = 1$ and a probability of .33333. Thus the observed value of $y = y_1 + y_2 = 4 + 1 = 5$. Now the exact

$$\text{p–value (right–tailed)} = P(Y = Y_1 + Y_2 > = 5)$$

$$= P(Y_1 = 4, Y_2 = 1) + P(Y_1 = 4, Y_2 = 2) + P(Y_1 = 3, Y_2 = 2)$$

$$= (.15766 \times .33333) + (.15766 \times .27778) + (.27027 \times .27778)$$

$$= .17142$$

For the purpose of comparison it may be noted that the normal approximated p-value for the data set in the above example is .0927.

2. Combined Analysis of Tumor Types Observed in Both Fatal and Incidental Contexts

When a tumor type is observed in a fatal context in some animals and in an incidental context in other animals, the appropriate method is to combine each analysis by a pooled Z statistic (as described in Section 3.5). A parallel exact method may be adopted when the total number of fatal and incidental tumors (of the type of interest) is small. Tumor count tables will be formed corresponding to each context of observation and the exact p-value will be computed based on the combined collection of tables from both contexts. For example, suppose that in the time interval 51–80 weeks the tumor type was observed in an incidental context while in the time interval 81–104 weeks it was observed both in fatal and incidental contexts. Then we will have three observed tumor count tables for this tumor type—two tables for incidental context in 51–80 and 81–104 weeks, and one for fatal context in 81–104 weeks. The exact p-value will now be computed using these three observed tables according to the method described in the previous subsection.

3. Comparison of Exact and Approximate Methods

As mentioned before, the use of exact p-values has been suggested when the tumor frequency is small. However, the magnitude of this "smallness" is not known. Mantel (1980) suggested the use of the exact procedure whenever the total tumor frequency is 5 or less. However, a simulation study by Ali (1990) showed that, in a four group experiment with 50 animals in each, the normal approximated p-value may severely underestimate the exact p-value even when the total number of tumor bearing animals is as large as 10. In Ali's simulation, survival data for the four groups were generated under the proportional hazard assumption with a baseline Weibull model for the control group, and the tumor-bearing animals were distributed in one or more of the four survival time intervals: 0–50, 51–80, 81–104, and over 104 weeks (i.e., the terminal sacrifice week).

FDA reviewers routinely apply the exact trend test whenever the total number of animals bearing tumor type of interest across treatment groups is 12 or less.

An inherent feature of the exact method (as described above) is that p-values are computed from the (conditional) null distribution which is discrete. Depending on the extent of this discreetness, the exact method will result in a conservative test in the sense that its actual significance level will, usually, be smaller than the nominal level. When performing multiple tests of significance in an experiment designed to test an 'overall experimentwise' hypothesis, the extent of this conservativeness may play an important role in determining the experimentwise Type I error rate (also referred to as the false-positive rate). Under such circumstances, it is useful to gain knowledge of the actual significance levels of the individual tests.

The scenario just described fits an animal tumorigenicity study. In a typical animal study, four parallel experiments (two species each with two sexes) are run. In each experiment, a combination of 20 or more organ/tissue types with several lesion types are tested for positive linear trend in tumor rates across the treatment groups. Thus the number of tests performed per experiment could be as high as 60 (or even higher). Since, for many tumor types, the incidence is a relatively rare event, it is usually the case that each of a large class of (relatively rare) tumor types will be observed in only a few animals. Hence the number of exact trend tests performed will also be large. Thus the experimentwise false-positive rates will depend heavily on the actual significance levels of the individual exact tests. In addition to the issue of false-positive rates, the question of false-negative rates also arises in a parallel context.

Some knowledge about the Type I and Type II error rates of an exact trend test compared to approximate tests can be found in the results of a simulation study by Ali (1990). In this study, the actual significance levels and power of the exact trend test was compared with three approximate tests for the special case of small number of tumor-bearing animals. Data for the simulation were generated under various Weibull models for survival time, and time to tumor, and tests were computed using four different score sets for the treatment groups. For details on the results of this study the reader is referred to the paper cited above. Here we will state only the main results comparing the exact test and its normal approximation version.

The actual attained significance levels (as estimated by 10,000 simulated experiments) were compared to the nominal 5% and 1% levels. Five Weibull models each with three score sets resulted in 15 cases to consider. The average number of tumor bearing animals among these fifteen cases ranged from 2.5 to 7.9. The attained significance levels of the exact test ranged from .82% to 1.7%

when the nominal level was 5%, and from .08% to .32% when it was 1%. Hence it is clear that the exact test was always very conservative in rejecting the null hypothesis of "no trend" when the tumor prevalence rates across treatment groups were equal. On the other hand, the significance levels attained by the normal approximated test ranged from 3.01% to 8.36% corresponding to a nominal level of 5%, and from .31% to 2.09% when the nominal level was 1%. It is seen that the normal approximation was very unstable in the sense that the significance levels fluctuated above and below the nominal level.

Ali (1990) also performed power comparisons between the two tests. The power was computed under various Weibull alternatives for tumor prevalence functions. The average number of tumor bearing animals ranged from 4.4 to 9.5. The power of the exact test corresponding to a 5% nominal level ranged from 2.14% to 15.09%, and between .35% and 4.46% when the nominal level was 1%. Thus, in case of very low total tumor rates, it is almost impossible for the exact test to detect increasing tumor prevalence across the treatment groups. In case of the normal approximation test, the power ranged from 6.2% to 33% corresponding to 5% nominal level, and between 1% and 12.8% when the nominal level was 1%. Hence, although the normal approximation improved the power, it was not high enough to make a real difference.

[A computer program to perform exact trend tests in sequences of $2 \times (r + 1)$ tables is available from the authors on written request.]

IV. INTERPRETATION OF STUDY RESULTS

Interpreting results of tumorigenicity experiments in an overall evaluation of the tumorigenic potential of a new drug is a complex process. Because of inherent limitations—such as small number of animals used, low tumor incidence rates, and biological variation—a tumorigenic drug may not be detected (i.e., a false negative error is committed). Also because of a large number of statistical tests performed on the data (usually 2 species, 2 sexes, 20–30 tissues examined, and 4 dose levels), there is a great potential that statistically significant positive dose-response relationships in some tumor types are purely due to chance of random variation alone (i.e. a false positive error is committed). Controlling these two types of error is the central element in the interpretation of study results and involves statistical and nonstatistical biological judgements. Therefore, it is important that an overall evaluation of the tumorigenic potential of a drug should be made based on knowledge of statistical significance of positive dose-response relationships, and information of biological relevance.

The controls of the two types of error are also directly related to tests of statistical significance used. Should one test for heterogeneity or positive dose-response relationship (trend) with respect to dose? Peto et al. (1980) make the following recommendation:

If two or more dose levels are studied, statistical tests for positive trend with respect to the actual dose-levels tested will usually be more sensitive than the standard alternative statistical methods would be to any real carcinogenic effects that may exist. . . . In other words, when there is a fairly consistent positive trend in the experimental results, the p-value yielded by a test for heterogeneity will tend to be less impressive than the p-value yielded by a test for trend (pp. 338, 339).

In general, FDA statistical reviewers follow this recommendation and test for a positive dose-response relationship in tumor incidence rates in their reviews.

Based on biological information, the overall false positive error in animal tumorigenicity studies caused by the effect of multiple tests of statistical significance can be controlled by reducing the number of variables evaluated. This can be achieved by combining certain tumor types. McConnell et al. (1986) proposed the following guidelines for combining tumors: a) tumors of the same histomorphogenic type with substantial evidence of progression from benign to malignant stage; b) tumors, such as hyperplasia and benign tumors, in which criteria for differentiating them become unclear; c) tumors in other organs/tissues but of the same histomorphogenic type; and d) tumors of different morphologic classification but with comparable histomorphogenesis.

There are statistical methods proposed for controlling the overall false positive error rates. The first group of those methods uses a Bonferroni type of adjustment for the effect of multiple comparisons (Gart, Chu, and Tarone, 1979; Mantel, 1980; Tarone, 1990). This group of methods takes into consideration the fact that all tests performed on data pertaining to different tumors at the same or different sites are not independent, and/or that significant results are not possible in some of the tests. The above modifications to the Bonferroni adjustment reduce the number of multiple tests performed and thus increase the power of the tests.

Tarone (1990) proposed a modification of the Bonferroni method for discrete data. Since the statistical tests (trend or pairwise comparison tests) are based on discrete null distributions of the test statistics, Tarone's modified Bonferroni method is particularly suitable for correcting the effect of multiple tests in tumor data analysis. Tarone's modification method is conditional on the marginal totals of 2 by 2 or 2 by c tables. The method is described here (using Tarone's notation).

Suppose that there are I sites (i.e., tissue/tumor combinations) for which a significance test can be performed. Let α_i be the minimum achievable significance level at site i, i = 1, 2, ..., I. The minimum achievable significance level is the minimum of the observed p-values under all possible permutations of the animals of the given sex in the given experiment. For each integer k, let m(k) =

number of the I sites for which $\alpha_i < \alpha$, where α is the nominal significance level. Let K be the smallest value of k such that $m(k)/k \leq 1$, and let R_k denote the set of indices satisfying $K\alpha_i < \alpha$. A statistical test at site i will be considered to yield a significant result only if i is contained in R_k and $P_i < \alpha/K$, where P_i is the observed p-value for site i. Note that K is the modified Bonferroni correction factor. It can be readily seen that the overall false positive error rate (i.e., the probability of rejecting the null hypothesis, say, of no trend at any site) is bounded by α.

Tarone has suggested a further refinement of the modification method by considering the fact that, in most cases, the total probability in the rejection regions (as defined above) of the m(k) tests will be less than α. Under this situation, it may be possible to expand one or more of the m(k) rejections, or even outside the set R_k by adding points until the overall false positive error rate does not exceed α.

In the same spirit, Fears, Tarone, and Chu (1977) showed that, in animal carcinogenicity studies, the issue of multiple tests is a problem only for the tumor types with high incidence rates. Since the majority of the tumor types in animal studies of human drugs have very low incidence rates and the final determination of the oncogenic potential of a new drug is based on results of statistical tests as well as relevant biologic and pathological information, it is argued (Fears, Tarone, and Chu, 1977; Haseman, 1977, 1983, 1984) that the false error rates in animal carcinogenicity studies are not as large as some people previously thought (Salsburg, 1977). Haseman (1983) showed that if a comparison of tumor rates in high dose versus control groups is carried out at the 0.01 level for all commonly occurring tumors and at the 0.05 level for all rare tumors, then the overall false positive error rate associated with this approach in NCI/NTP carcinogenicity studies appears to be no more than 7 to 8%.

Farrar and Crump (1988, 1990) proposed an alternative method to adjust for the effect of multiple tests, and the effect of dependencies that may exist between tumors on the overall false positive error. In the proposed method, simple functions of p-values from conventional tests applied to each individual tumor (approximation or exact permutation, pairwise comparisons or trend tests) are evaluated for statistical significance using a Monte Carlo procedure that treats individual animals as units of variation. The functions of p-values of individual pairwise and trend tests can be the minimum p-value or the product of a fixed number, K, of the smallest p-values. For material that causes tumors at only a single site, the minimum p-value may be a meaningful summary statistic, and the test based on this statistic may also be more powerful. However, for less specific carcinogens, the product of the K smallest p-values, which combines information from K sites, may be more appropriate.

As mentioned above, the statistical significance of a chosen function of the p-values used as the test statistic is then evaluated using a Monte-Carlo randomization (permutation) procedure. Animals are randomly assigned to treatment groups with the number of animals assigned to each treatment group being preserved. The test statistic is recomputed for each reassignment. The proportion of the statistics which are at least as extreme as the observed minimum p-value (or the product of the K smallest p-values) computed from the original data is used as the estimated overall false positive error.

A method related to the Farrar-Crump method but independently developed by Heyse and Rom (1988) deals exclusively with the use of the minimum of the p-values from all exact permutation trend tests and random permutations in the adjustment for the effect of multiple statistical tests. In this method, the overall false positive error is estimated by the following formula (using the authors' notation)

$$P^*_{[1]} = 1 - \prod_{i=1}^{\cdot \, r} (1 - P^*_{(i)})$$

$$= P^*_{[1]} - P^*_{(i)}P^*_{(j)} + \ldots + (-1)^{n+1} P^*_{(i)}P^*_{(j)} \ldots P^*_{(r)}$$

where

$P^*_{[1]}$ = estimated overall false positive error.

$P^*_{(i)}$ = $\Pr(S_{(i)} \geq S^*_{(i)})$ = the largest p-value which is attainable (with given number of tumors at site i) and is smaller than or equal to P_1.

$S_{(i)}$ = A random variable assuming score of measuring trend from the exact permutation trend test.

$S^*_{(i)}$ = The observed value of $S_{(i)}$ which satisfies the definition of $P^*_{(i)}$above.

P_1 = Minimum of the p-values from the exact permutation trend tests on individual sites and tumors.

n = The number of $P^*_{(i)}$components in each term of $P^*_{[i]}$.

$$P^*_{(i)}P^*_{(j)} \ldots P^*_{(r)} = \Pr(S_{(i)} \geq S^*_{(i)} \text{ and } \ldots \text{ and } S_{(t)} \geq S^*_{(r)})$$

The above probabilities of joint events are calculated from multivariate randomization distributions of trend measure scores, $S_{(i)}$s.

r = the number of site/lesion combinations tested.

The above formula considers possible dependencies between sites and tumors. However, the authors showed empirically that "the independence assumption may prove to be a biologically reasonable approximation for the data of this sort".

Westfall and Young (1989) proposed another method for controlling the experimentwise false error rate. In this method, all p-values are adjusted for the multiplicity of testing using vector-based bootstrap resampling method. In the test for positive dose-response relationship in tumor incidence rates using the survival-unadjusted Cochran-Armitage linear trend test (Armitage 1955, 1971), the p-values can be adjusted for the effect of multiple tests by the above method as follows:

1. Assume the observed data are $x_{11}, ..., x_{1n_1}, ..., x_{g1}, ..., x_{gn_g}$, where each x_{ij} is a $k_x 1$ vector, and g is the number of treatment groups.
2. Compute the k unadjusted p-values, pv_k, for all lesion/site combinations using the Cochran-Armitage trend test.
3. Generate a prespecified number (with desired accuracy), say, 10,000, of replicate samples of the observed data, $x_{11}, ..., x_{1n_1}, ..., x_{g1}, ..., x_{1n_g}$, with bootstrap resampling. Let $x_{11}^*, ..., x_{1n_1}^*, ..., x_{g1}^*, ..., x_{1n_g}^*$ denote a replicate sample.
4. Calculate the new set of p-values, pv_k^*, by applying the Cochran-Armitage trend test to each of the replicate samples, find the smallest, min pv*, of the k p-values, pv_k^*, calculated from the replicate sample.
5. Calculate the adjusted p-values, apv_k, for each k using the proportions of samples for which min pv_k^* is equal to or less than pv_k.

The authors conducted a simulation study comparing the bootstrap method with the permutation method by Farrar and Crump (1988, 1990), and Heyse and Rom (1988). They reported the following simulation results: a) the bootstrap method approximates nominal significance levels more closely than the permutation method, and b) the bootstrap method has more power than the permutation method.

In the tests for the positive dose-response relationship in tumor incidence rates, FDA statistical reviewers currently use data of the concurrent control groups and comparable historical control data to classify common and rare tumors, and to adopt and modify the following decision rule in their evaluation: A positive dose-response relationship is considered not to occur by chance of variation alone if the p-value is less than 0.01 for a common tumor, and 0.05 for a rare tumor.

There is a concern that application of the NTP decision rule (which was derived based on pairwise comparisons to trend tests) could result in excessive overall false positive errors; however, since data of all treatment groups are used in the trend test, it is more sensitive. The magnitude of the overall false positive errors when the NTP rule is used in the trend test is still unknown and needs further study. However, Haseman (personal communication) gave a rough estimate of about 15% overall false positive error (doubling the 7–8% reported in his papers) in the strains and species of animals used in NTP studies. If one is interested in maintaining the 7–8% overall false positive error level and also applying the NTP rule to trend tests, then Haseman suggested that his proposed levels of significance be cut by half (i.e., 0.025 for rare tumors and 0.005 for common tumors). Again, this rule applies only to F344 rats and B6C3F1 mice, and may have to be modified when they are applied to experiments using different strains of animals and having different study durations.

There is another factor which Haseman failed to consider in his rough estimate of the overall false positive error when the decision rule he proposed is applied to trend tests. When trend tests are used, the total number of tests is only one third of the number of total tests when pairwise comparison tests are used. Therefore, the effects of higher sensitivity, which could increase the overall false positive error, and of smaller number of tests, which could decrease the overall false positive error, could balance out. This may make the overall false positive errors resulting from both the tests for positive trend and the tests for heterogeneity with respect to dose compatible.

The false negative error issue in animal carcinogenicity study, although equally important as the false positive error issue, has not received as much attention as has the false positive error issue. This may be in part due to the following two reasons.

1. This issue is less familiar to people. Statistically, the theory of the false negative error issue is more complicated than that of the false positive error issue. The false negative error is a function of alternative hypotheses one is interested in testing. The statistical distributions used in the evaluation of false negative errors are complicated and involve noncentrality parameters.
2. Because of the high cost involved in developing a new drug, the drug sponsor will pay more attention to false positive errors than to false negative errors.

As mentioned at the beginning of this section, the large false negative error that occurs in animal carcinogenicity study is caused by the inherent limitations of small numbers of animals used and by the low incidence rates in the majority of tumors examined. Due to the above limitations, the power of statistical tests for positive dose-response relationship is going to be small. That is, the false negative errors are expected to be large. A study by Ali (1990) shows that under the conditions he simulated (which assumed tumor incidence rates following Weibull models), the powers of the exact permutation trend test, the Peto prevalence test for trend, and some modified forms of Peto prevalence test are no more than 0.25. That is, the false negative errors are greater than 0.75. If the above simulation results reflect the general magnitudes of the power of statistical trend tests, then the false negative error issue should cause concern

Table 2.9 Tumor Rates (%) Needed to be Induced in the Treated Group in Order to Achieve Levels of Power of 0.50 and 0.90.

Spontaneous tumor rate in control	$\alpha = 0.05$		$\alpha = 0.01$	
	Power = 0.5	Power = 0.9	Power = 0.5	Power = 0.9
0.1%	9.5%	15.8%	13.5%	20.5%
1.0	11.0	18.4	15.1	23.4
3.0	14.0	22.9	18.9	29.0
5.0	17.0	27.0	22.5	33.3
10.0	24.2	35.7	30.2	41.9
20.0	36.8	49.0	43.2	56.0
30.0	48.1	61.1	54.8	67.0

to investigators and be weighted at least equally with the false positive error issue in the overall evaluation of results of an animal carcinogenicity study.

Table 2.9 contains some of Haseman's calculations (1984) of tumor rates needed to be induced in the treated group in order to achieve certain levels of power in the Fisher's exact test at 0.05 and 0.01 levels of significance under various assumed spontaneous rates in the control group (assuming 50 animals in the treated group and in the control).

Statistically, there are at least three ways to increase the power of tests to ensure that the overall false negative errors are not excessive. The most obvious way is to increase group sizes. However, the increase in power probably won't be significant unless the group size is drastically increased, say, from 50–250 animals per group. This approach to increasing power may not be financially or logistically feasible.

The second way to ensure adequate power in statistical tests of positive dose-response relationship in tumor rates is to administer to treated animals dose levels which are high enough to induce tumors. As mentioned in Section II, the determination of a dose close to MTD for treating animals in the high dose group is an important, controversial, and complicated issue. Information about clinical signs, histopathological toxic effects, body weight gain, mortality, as well as pharmacokinetic and metabolic data are needed for the evaluation of MTD.

Haseman used results of some NTP studies to emphasize the importance of using dose levels which provide an adequate tumor challenge to the treated animals. He found that half of the carcinogens tested in those studies would be judged as noncarcinogens if half of the MTD's were used as the highest dose. Under the current four-group design in which a medium group was added as a cushion for cases where the high dose used may be over MTD, it is feasible to take a greater risk of using the highest possible dose level to ensure adequate power in statistical tests.

The third way to increase power in statistical tests is to assume a larger overall false positive error. One may have to be willing to assume an overall false positive error in the 15–20% range in order to balance out the low power of statistical tests.

If one wants to control one of the two types of error to a small magnitude, then he or she has to pay the price for committing a large magnitude of the other type of error. In the general case, a statistical test is performed at a prespecified level of false positive error, usually at 0.05, and a decision rule is derived to maximize the power (or to minimize the false negative error) of the test under the alternative hypothesis tested. However, because of the intertwining and conflicting relationship between the magnitudes of the false positive error and false negative error that one is willing to assume, the choice has to be determined by the cost-risk (or cost-risk-benefit) factor in new drug evaluation. For drug products, such as cancer and AIDS drugs that are intended for treating terminally ill patients, one may take a greater risk (false negative error) by taking a smaller overall false positive error. This will be especially true when there is no alternative drug available in the market. On the other hand, for drug products for treating common illnesses that can be treated with other available approved alternative drugs and which will be used by a larger population, one can be more cautious about the overall false negative error. To ensure that the false negative error is not excessive, one may have to assume a larger overall false positive error. It is true that limited resources should not be wasted by rejecting an effective drug, but for the protection of the health of general public, it is equally important that drugs with carcinogenic potential should not be misinterpreted as safe and allowed to enter the market.

Although concurrent control groups are the most relevant controls in testing drug related increases in tumors in a study, there are situations in which historical control data from previous comparable studies can be useful in the overall evaluation of the results of the study. One of the situations is to use the comparable historical control information to define rare tumors (which have less effect on overall false positive error) and therefore can be tested at higher levels of significance. Another situation is to check if a marginally significant finding is really drug related or purely due to chance of variation. A third situation is to use historical control data to check if a study was conducted properly.

In the first situation, a tumor is defined as rare if it was so classified by reviewing pharmacologists and pathologists, or if the background spontaneous incidence rate is less than 1%. In the second situation, the incidence rates of the treated groups are compared with the incidence rates of the historical control data. The significant finding will not be considered as biologically meaningful if the incidence rates of the treated groups are within the ranges of historical control incidence rates. In the third situation, a question about the quality of the study will be raised if incidence rates of tumors of the study are not consistent with those in the comparable historical control data. "However, before historical control data can be used in a formal testing framework, a number of issues must first be considered" (Haseman, 1984).

These issues include the nomenclature conventions and diagnostic criteria used by pathologists, conducting laboratories, study durations, strains and species of animals used, and time (calendar year) when a study was conducted. It is important that the historical control data can be useful only if it is comparable with the concurrent control data. The comparability includes identical nomenclature conventions and diagnostic criteria, same species/strain/sex, same source of supplier, same testing laboratory, comparable survival and age at termination, comparable time frame of studies (within 5 years), and comparable food consumption and body weight gain.

FDA statisticians routinely perform tests for positive dose-response relationship (trend) in incidence rates in individual or pooled site/tumor combinations using the proposed NTP decision rule (i.e., testing common tumors at 0.01 and rare tumors at 0.05 levels of significance). Comparable historical control data, when available and reliable, are used to assist in classifying common and rare tumors, and in deciding if significant findings are biologically relevant. As mentioned at the beginning of the chapter, the adoption of the Bonferroni type of adjustment for the effect of multiple tests by FDA statisticians is based on review of current literature, consultations of outside experts, our own research, and our best scientific judgement.

To make sure that the false negative error committed is not excessive, statistical reviewers collaborate with the reviewing pharmacologists, pathologists, and medical officers to evaluate the adequacy of the gross and histological examination of both control and treated groups, the adequacy of dose selection, and the durations of experiments in relation to the normal life span of the tested animals.

V. DATA PRESENTATION AND SUBMISSION

To facilitate statistical reviews, sponsors should follow the guidance described in the FDA *Guideline for the Format and Content of the Nonclinical/Pharmacology/Toxicology Section of An Application.* They should present their data in the reports in such a way that the reviewers are able to verify the sponsors' calculations, to validate their statistical methods as being appropriate to the way the data were generated, to trace back the sponsors' conclusions through their summaries and analyses to the raw data, and to reanalyze the data, if necessary, in order to explore alternatives or to gain greater insight into the relationships between various events of the studies (Fairweather, 1988).

In addition to reviewing the reports submitted by sponsors, statisticians at FDA also perform additional statistical analyses which they believe are appropriate and necessary to evaluate the analyses and conclusions contained in the reports. Therefore sponsors should make the raw data easily accessible in an appropriate format to the statistical reviewers. Statistical reviews are delayed when data are not accessible or not submitted in appropriate formats. For filing, mailing, and other management purposes, sponsors can consider putting the tumor data onto computer diskettes using the Division of Biometrics Formats and Specifications for Submission of Animal Tumorigenicity Study Data. A copy of the revised formats and specifications is given in the Appendix to this chapter. To expedite the statistical reviews, sponsors are advised that the tumor data on computer readable forms be submitted with their original initial submissions of the hardcopy NDA or IND.

Through cooperation and discussions between the FDA, the U.S. Environmental Protection Agency (EPA), the National Center for Toxicological Research (NCTR), and the Consumer Product Safety Commission (CPSC), a set of formats called STUDIES for the electronic transfer of individual animal toxicological data generated from long-term rodent tumorigenicity studies have been issued (Federal Register, December 19, 1990, Vol. 55, page 52096, and October 24, 1991, Vol. 56, pp. 55129–55130). The acronym STUDIES stands for Submitters Toxicological Uniform Data Information Exchange Standard. Copies of the document "STUDIES/CHRONIC Data Formats for Chronic/ Carcinogenicity Rodent Bioassays" can be ordered (order number PB90-213885) from the National Technical Information Service (NTIS), U.S. Department of Commerce, 5285 Port Royal Rd., Springfield, VA 22161. These standard computer formats are to be used by industry to submit their data from long-term rodent tumorigenicity studies to the FDA, EPA, and other agencies for statistical analysis and quantitative risk assessment.

As an integrated part of the development in Computer-Assisted NDA Review (CANDAR), a joint effort between the FDA and the Pharmaceutical Manufacturers Association (PMA) to use computers in reviewing new drug applications, the Center and a PMA working group are in the process of developing a program called Computer-Assisted Pharmacology Evaluation Review (CAPER). The program will allow the exchange of animal tumorigenicity study data between pharmaceutical companies and FDA reviewers using standard formats and specifications. However, this more advanced method of direct data submission through computer network transmission is still in the preliminary experimental stage. Before the method becomes operational, the current method of submitting tumor data to the FDA for statistical review will continue.

VI. CONCLUDING REMARKS

In designing an experiment, randomization methods should be used in allocating animals to treatment groups to avoid possible biases caused in animal selection. A sufficient number of animals should be used in the experiment to ensure reasonable power in the statistical tests used. In negative studies in which results of the analysis show no significant positive dose-response relationships in tumor incidence rates, a further evaluation on the validity of the designs of experiment should be performed to see if there are sufficient numbers of animals that lived long enough to get adequate exposure to the chemical and to be at risk of forming late-developing tumors, and if the doses used are high enough and close to the MTD to present a reasonable tumor challenge to the tested animals.

In the review and evaluation of methods of statistical analysis in an animal tumorigenicity study submission, the statistical reviewers in FDA examine the appropriateness of the statistical methods used by the sponsor and perform additional independent analyses to evaluate and verify the sponsor's conclusions. Appropriate statistical analyses of animal tumorigenicity study data should include the following areas:

The intercurrent mortality data should be evaluated first to see if the survival distributions of the treatment groups are significantly different and if the dose-response relationship in mortality is significant. Since the effects of differences in intercurrent mortality on number of tumor bearing animals can be substantial, survival-adjusted methods should be used in tests for positive dose-response relationships in tumor incidence rates.

The determination of survival-adjusted methods to be used in tests for positive dose-response relationships in tumor incidence rates should be based on the contexts of observation of the tumors whose data are to be analyzed. The death-rate method and the prevalence method should be used to analyze data of tumors observed in fatal and incidental contexts of observation, respectively.

When the number of tumor occurrences across treatment groups is small, the test results of the death-rate method and the prevalence method which use the normal approximation are not stable and reliable. In this circumstance, exact permutation methods should be used to replace the above methods in tests for positive dose-response relationships in tumor incidence rates.

Controlling the overall false positive error and the overall false negative error to acceptable levels is the central element in the interpretation of study results. The control of the two types of error involves both statistical and nonstatistical issues which require statistical as well as biological judgements. Therefore, it is important that an overall evaluation of the tumorigenic potential of a drug should be made based on knowledge of statistical

significance of positive dose-response relationship and information of biological relevance.

To facilitate the FDA's statistical review, sponsors should present their data in
the reports in such a way that the reviewers should be able to verify the
their calculations, to validate their statistical methods, to trace back the
their conclusions through their summaries and analyses to the raw data.
The sponsors should make the raw data easily accessible in an appropriate
format to the statistical reviewers. Statistical reviews are delayed when
data are not accessible. Sponsors are advised that the tumor data on
computer diskettes be submitted with their original initial submissions of
the hardcopy NDA or IND.

ACKNOWLEDGMENT

The authors thank Dr. William R. Fairweather, Chief, Statistical Applications
and Research Branch, Division of Biometrics, CDER, for his encouragement in
writing the chapter; and Dr. Judith L. Weissinger, Assistant Director (Pharmacology), Office of Drug Evaluation II, CDER, for her thoughtful comments and
suggestions.

The views expressed in this chapter are those of the authors and are not
necessarily those of the Food and Drug Administration. This work has been
produced by the authors, Karl K. Lin and Mirza W. Ali, in the capacity of
federal government employees, as part of their official duty, and is in the public
domain, and is not subject to copyright.

APPENDIX: DIVISION OF BIOMETRICS FORMATS AND SPECIFICATIONS FOR SUBMISSION OF ANIMAL TUMORIGENICITY STUDY DATA, REVISED APRIL 19, 1989 (WITH SOME CHANGES AND MODIFICATIONS SUGGESTED BY DRUG SPONSORS AND ANIMAL RESEARCH LABORATORIES)

A. Introduction

In addition to reviewing animal tumorigenicity study reports submitted by
sponsors, statisticians at the FDA also perform independent statistical analyses
which they believe are appropriate and necessary to evaluate the analyses and
conclusions contained in the reports. To expedite the statistical reviews, the
sponsors are urged to submit the tumor data in computer readable media using
the following formats and specifications along with their original, initial submissions of the hardcopy NDA or IND. The specifications and formats provided
in this document are recommendations based on the computer facilities currently

in use within FDA. References to commercial products herein should not be interpreted as an official FDA endorsement.

The Division of Biometrics has received valuable responses from drug sponsors and animal research laboratories since the April 19, 1989 revised guideline was issued. Some suggested changes and modifications are incorporated into the current version included in this appendix.

B. Media Of Data Submission

I. Submission of Data on Floppy or Hard Diskettes

Previously, the Division of Biometrics requested sponsors to submit animal tumorigenicity data on computer readable magnetic tapes. However, we feel that this medium has been surpassed for ease of handling by floppy or hard diskettes, and other storage media. For filing, mailing, and other management purposes, sponsors are urged to submit data on floppy or hard diskettes using the formats in the Data Formats section. The use of this medium of data submission will result in a faster review turn-around time. The diskettes used should be IBM-PC usable, 5.25 inch, and two-sided, or IBM-PS/2 usable, 3.5 inch, two-sided. Either low- or high-density diskettes are acceptable.

Additional information that should be supplied by the company to expedite processing of floppy diskettes includes:

1. The brand and model of computer (e.g., IBM PS/2) on which data were generated.
2. The type of program, routine, or language used to created the dataset.
3. For a SAS dataset, the results of PROC CONTENTS.
4. A code sheet describing the records and fields in the dataset.
5. A printout of the data, or if a large database is sent, at least 50 records.
6. The name and telephone number of the person responsible for creating the computer files.

II. Submission of Data on Computer Readable Magnetic Tapes:
Requirements for Computer-Readable Data Tapes and Cartridges

The technical requirements for tapes and cartridges compatible with FDA's IBM 3081 model K with operating system XA OS/VS2 facility are as follows:

9-Track Tapes
BPI - 800, 1600, or 6250
Character Set - IBM-EBCDIC or ASCII
Label - IBM-standard label or unlabeled tapes.
Notes: 1.   The subparameter Bypass Label Processing (BLP) is not allowed.
2.   Non-IBM system generated tapes should be IBM compatible with no label (i.e., no headers, just data and tape mark).

Indicate which of the above options are used along with the following information:

Volume Number (VOL=SER)
Record Length (LRECL)
Record Format (RECFM)
Blocksize (BLKSIZE)
Data-set name (DSN)

If the tape contains more than one dataset, provide the file number and the above information for each dataset.

Additional information that should be supplied by the company to expedite processing of tapes includes:

1.   The type of computer on which data were generated.
2.   Type of program, routine or language used to created the dataset (i.e., is it similar to IBM utilities IEBGENER or IEBCOPY?).
3.   For a SAS dataset, the results of PROC CONTENTS.
4.   A code sheet describing the records and fields in the dataset.
5.   A printout of the data, or if a large database is sent, at least 50 records.
6.   The name and telephone number of person at the installation responsible for creating the computer files.
7.   Mark the tape(s) on the outside with the firm's name and address, reference number (NDA, IND, PLA, etc), name of the person to whom tape(s) should be returned.

C.   Data Submission Formats

These formats represent a change from earlier ones for submission of animal tumorigenicity study data. The data submission formats were designed primarily for the use of SAS list input (free format data reading). To be able to use SAS list input, the following requirements must be met: a) input values have to be separated from each other by at least one blank (by at least two blanks for character input values with one or more single embedded blanks), and b) periods, rather than blanks, should be used to represent missing values. The periods should always be separated by two blanks. There is no prespecified number of bytes (or columns) in each field (or variable) in the following formats. Sponsors can use as many bytes as needed in their databases.

1.   Format of Animal Dataset

Record 1 to Record 9 contain header information of the dataset (identifications and descriptions of the experiment from which the dataset was generated).

Record number	Variable name	Description
1	SPONNAME	Name of sponsor
2	LABNAME	Name of laboratory that conducted the study.
3	STTDATE	Date of start of study: YY/MM/DD
	ENDDATE	Date of end of study: YY/MM/DD
4	INDNUM	IND number
	NDANUM	NDA number
	PROTONUM	Protocol number
	STUDYNUM	Study number
5	GENNAME	Generic name of the drug
	TRADNAME	Trade name of the drug
6	SPECIES	Species name
	STRAIN	Strain name
7	ROUTEADM	Route of administration
8	DOSEUNIT	Dose unit (mg/kg/day, etc)
9	DLEVELS	Dose levels for control, low, medium, and high groups

Record 10 to last record of this dataset contain tumor data of individual animals.

Record number	Variable name	Description	Codes
10	ANIMALNU	Animal number	
	SEX	Sex:	M = Male
			F = Female
	DOSEGP	Dose group:	0 = Control group one
			1 = Low dose group
			2 = Medium dose group
			3 = High dose group
			4 = Control group two (if any)
	DEATHWK	Week of death or sacrifice	
	DTHSACST	Death or sacrifice status:	1 = Natural death (death in study)
			2 = Terminal sacrifice
			3 = Intermittent sacrifice
			4 = Moribund sacrifice
			5 = Accidental kill animal (e.g., gavage death)
	EVEREXAM	Ever examined code	1 = At least one tissue was examined
			2 = No tissues were examined
	RECORDS	Number of records for in the animal (See the note below)	
	TUMORTYP	Tumor type code	

Record number	Variable name	Description	Codes
10(cont)	ORGCODE	Organ/tissue code	
	DETECTWK	Week of detection	
	MALIGNST	Malignancy status	1 = Malignant
			2 = Benign
			3 = Undetermined
	DEATHCAU	Cause of death	1 = Tumor caused death
			2 = Tumor did not cause death
			3 = Undetermined
	AUTOLYSI	Autolysis code	1 = Organ/tissue was usable
			2 = Organ/tissue was not usable
			(e.g. autolyzed)

(Additional fields as appropriate can be added on the same line.)

End of Record # i0.

Record # 11: (Same format as Record # 10). Either next animal data or data of second tumor or data of an unusable or a missing organ/tissue of the same animal. In the latter case, repeat the information contained in the first 7 variables of Record 10 for the same animal and add the data of the second tumor or the data of the unusable or missing organ/tissue to the last six variables.

Note: The number of records for each animal will be the sum of the number of tumors and the number of organs/tissues that were unusable or missing. There will be one record for each animal with zero or one tumor but without unusable or missing organs/tissues. There will be no data for the last six variables of the record of an animal with no tumors on a unusable organ/tissue. For an animal with unusable or missing organs/tissues, a record should be generated for each such organ/tissue identified by the ORGCODE and AUTOLYSI variables.

II.   Format of Tumor Type Code Data Set

Variable Name	Description
TUMORCOD	Tumor type code
TUMORNAM	Tumor type name

(One record for each tumor type).

III. Format of Organ/Tissue Code Data Set

Variable Name	Description
ORGCOD	Organ/tissue code
ORGNAME	Organ/tissue name

(One record for each organ/tissue).

D. Contact

For questions or comments regarding these specifications and formats, call (301) 443-4710 or write to

William R. Fairweather, Ph.D. or Karl K. Lin, Ph.D.
Statistical Application and Research Branch, (HFD-715)
Division of Biometrics, US Food & Drug Administration
5600 Fishers Lane
Rockville, MD 20857

REFERENCES

Ali, M. W. (1990), A comparison of power between exact and approximate tests of trend of tumor prevalence when tumor rates are low, *Proc. 1990 Biopharmaceut. Sect., Amer. Stat. Assn.*, American Statistical Association, Alexandria, VA, pp. 66-71.

Armitage, P. (1955), Tests for linear trends in proportions and frequencies, *Biometrics*, **11**:375-386.

Armitage, P. (1971), *Statistical Methods in Medical Research*, John Wiley & Sons, New York

Bancroft, T. A. (1964), Analysis and inference for incompletely specified models involving the use of preliminary test(s) of significance, *Biometrics*, **20**:427-442.

Breslow, N. (1970), A generalized Kruskal-Wallis test for comparing K samples subject to unequal patterns of censorship, *Biometrics*, **57**:579-594.

Chu, K. C., Cueto, C., and Ward, J. M. (1981), Factors in the evaluation of 200 National Cancer Institute carcinogen bioassays, *J. Toxicol. Environ. Health*, **8**:251-280.

Cox, D. R. (1959), The analysis of exponentially distributed life-times with two types of failures, *J. Roy. Stat. Soc. Ser. B*, **21**:412-421.

Cox, D. R. (1972), Regression models and life tables (with discussion), *J. Roy. Stat. Soc. Ser. B*, **34**:187-220.

Dinse, G. E. (1985), Testing for trend in tumor prevalence rates: I. Nonlethal tumors, *Biometrics*, **41**:751-770.

Dinse, G. E. and Haseman, J. K. (1986), Logistic regression analysis of incidental-tumor data from animal carcinogenicity experiments, *Fundament. and Appl. Toxicol.*, **6**:44-52.

Dinse, G. E. and Lagakos, S. W. (1983), Regression analysis of tumor prevalence data, *J. Roy. Stat. Soc. Ser. C*, **32**:236-248.

Fairweather, W. R. (1988), Statistical considerations in tumorigenicity study review (Abstract). Presented at the Drug Information Association Meeting, Toronto, Canada, July 12, 1988.

Farrar, D. B. and Crump, K. S. (1988), Exact statistical tests for any carcinogenic effect in animal bioassays, *Fundament. and Appl. Toxicol.*, **11**:652-663.

Farrar, D. B. and Crump, K. S. (1990), Exact statistical tests for any carcinogenic effect in animal bioassays, II. Age-adjusted tests, *Fundament. and Appl. Toxicol.*, **15**:710-721.

Fears, T. R., Tarone, R. E., and Chu, K. C. (1977), False-positive and false-negative rates for carcinogenicity screens, *Cancer Res.*, **37**:1941-1945.

Food and Drug Administration (1987), *Guideline for the Format and Content of the Nonclinical/Pharmacology/Toxicology Section of An Application*. U.S. Department of Health and Human Services, Rockville, MD.

Gart, J. J., Chu, K. C., and Tarone, R. E. (1979), Statistical issues in interpretation of chronic bioassays for carcinogenicity, *J. Natl. Cancer Inst.*, **62**:957-974.

Gart, J. J., Krewski, D., Lee, P. N., Tarone, R. E., and Wahrendorf, J. (1986), *Statistical Methods in Cancer Research, Volume III - The Design and Analysis of Long-Term Animal Experiments*, International Agency for Research on Cancer, World Health Organization. Lyon, France.

Gehan, E. A. (1965), A generalized Wilcoxon test for comparing k samples subject to unequal patterns of censorship, *Biometrika*, **52**:203-223.

Goldberg, K. M. (1985), An algorithm for computing an exact trend test for multiple 2 x k contingency tables. Presented at Symposium on Long-Term Animal Carcinogenicity Studies. Washington, DC.

Haseman, J. K. (1977), Response to 'Use of statistics when examining lifetime studies in rodents to detect carcinogenicity', *J. Toxicol. Environment. Health*, **3**:633-636.

Haseman, J. K. (1983), A reexamination of false-positive rates for carcinogenesis studies, *Fundament. and Appl. Toxicol.*, **3**:334-339.

Haseman, J. K. (1984), Statistical issues in the design, analysis and interpretation of animal carcinogenicity studies, *Environmental Health Perspective*, **58**:385-392.

Heyse, J. F. and Rom, D. (1988), Adjusting for multiplicity of statistical tests in the analysis of carcinogenicity studies, *Biometrical J.*, **30**:883-896.

Lin, K. K. (1988), Peto prevalence method versus regression methods in analyzing incidental tumor data from animal carcinogenicity experiments: An empirical study, *Proc. 1988 Biopharmaceut. Sect., Amer. Stat. Assn.*, American Statistical Association, Alexandria, VA, pp. 95-100.

Lin, K. K. and Reschke, M. F. (1987), The use of the logistic model in space motion sickness prediction, *Aviation, Space, and Environmental Med.*, **Aug.**:A9-A15.

Mantel, N. (1980), Assessing laboratory evidence for neoplastic activity, *Biometrics*, **36**:381-399.

Mantel, N. and Haenszel, W. (1959), Statistical aspects of the analysis of data from retrospective studies of disease, *J. Natl. Cancer Res.*, **22**:719-748.

McConnell, E. E., Solleveld, H. A., Swenberg, J. A., and Boorman, G. A. (1986), Guidelines for combining neoplasms for evaluation of rodent carcinogenesis studies, *J. Natl. Cancer Inst.*, **76**:283-289.

National Technical Information Service (1990), *Studies/Chronic, Data Formats for Chronic/Oncogenicity Rodent Bioassays*, PB90-213885, U.S. Department of Commerce, Springfield, VA.

Office of the Federal Register, (1985), Chemical carcinogens; A review of the science and its associated principles in *Part II, Office of Science and Technology Policy*, Federal Register, 47-58. (Note: The paper was also published with the same title and authorized by U.S. Interagency Staff Group on Carcinogens in *Environmental Health Perspectives*, **67**:201-282).

Peto, R. Pike, M. C., Day, N. E., Gray, R. G., Lee, P. N., Parish, S., Peto, J., RIchrds, S. and Wahrendorf, J. (1980), Guidelines for simple, sensitive significance tests for carcinogenic effects in long-term animal experiments. In *Long-Term and Short-Term Screening Assays for Carcinogens: An Critical Appraisal*, International Agency for Research on Cancer, Lyon, France. IARC Monographs Supplement 2, pp. 311-426.

Salsburg, D. S. (1977), Use of statistics when examining lifetime studies in rodents to detect carcinogenicity, *J. Toxicol. and Environmental Health*, **3**:611-628.

Tarone, R. E. (1975), Tests for trend in life table analysis, *Biometrika*, **62**:679-682.

Tarone, R. E. (1990), A modified bonferroni method for discrete data, *Biometrics*, **46:515-522.**

Thomas, D. G., Breslow, N., and Gart, J. J. (1977), Trend and homogeneity analyses of proportions and life table data, *Computer and Biomed. Res.*, **10**:373-381.

Westfall, P. H. and Young, S. S. (1989), P value adjustments for multiple tests in multivariate binomial models, *J. Amer. Stat. Assoc.*, **84**:780-786.

SUGGESTED ADDITIONAL READING

Ali, M. W. (1990), Exact versus asymptotic tests of trend of tumor prevalence in tumorigenicity experiments: a comparison of p-values for small frequency of tumors, *Drug Information J.*, **24**:727-737.

Ames, B. N., Gold, L. S. (1990), Too many rodent carcinogens: mitogenesis increases mutagenesis, *Science*, **240**:970-971

Bickis, M., and Krewski, D. (1989), Statistical issues in the analysis of the long-term carcinogenicity bioassay in small rodents: an empirical evaluation of statistical decision rules, *Fundamental and Applied Toxicology*, **12**:202-221.

Brown, C. C., and Fears, T. R. (1981), Exact significance levels for multiple binomial testing with application to carcinogenicity screens, *Biometrics*, **37**:763-774.

Charles River Company (undated), Spontaneous neoplastic lesions in the Crl:CD BR rat, Wilmington, MA., (Information complied by P. L. Lang and included in the class notes of Comparative Pathology, T. P. O'Neill, FDA/CDER Staff College).

Charles River Company (1987), Spontaneous neoplastic lesions in the Crl:CD-1 (ICR) BR mouse, Wilmington, MA., (Information compiled by P. L. Lang and included in the class notes of Comparative Pathology, T. P. O'Neill, FDA/CDER Staff College).

Charles River Company (1989), Spontaneous neoplastic lesions in the B6C3Fl/CrIBR mouse, Wilmington, MA., (Information compiled by P. L. Lang and included in the class notes of Comparative Pathology, T. P. O'Neill, FDA/CDER Staff College).

Charles River Company (1990), Spontaneous neoplastic lesions in the CDF (F-344)/ CrIBR rat, Wilmington, MA., (Information compiled by P. L. Lang and included in the class notes of Comparative Pathology, T. P. O'Neill, FDA/CDER Staff College).

Chen, J. J., and Gaylor, D. W. (1987), The upper percentiles of the distribution of the logrank statistics for small numbers of tumors, unpublished report, National Center for Toxicological Research, FDA.

Cohen, S. M., and Ellwein, L. B. (1990), Cell proliferation in carcinogenesis, *Science,* **249**:1007-1011.

Fears, T. R., and Tarone, R. E. (1977), Response to use of statistics when examining lifetime studies in rodents to detect carcinogenicity, *J. Toxicol. Environ. Health,* **3**:627-632.

Gart, J. J., and Tarone, R. E. (1987), On the efficiency of age-adjusted tests in animal carcinogenicity experiments, *Biometrics,* **43**:235-244.

Gaylor, D. W., and Hoel, D. G. (1981), Statistical analysis of carcinogenesis data from chronic animal studies, in *Carcinogens in Industry and the Environment,* (Sontag, J. M., Ed.), Marcel Dekker, New York.

Hardisty, J. F., and Eustis, S. L. (1990), Toxicological pathology: a critical stage in study interpretation, in *Progress in Predictive Toxicology,* (Clayson, D. B., Munro, I. C., Shubik, P., and Swenberg, J. A., Eds.), Elsevier Science Publisher B.V. (Biomedical Division), New York.

Haseman, J. K. (1985), Issues in carcinogenicity testing: dose selection, *Fundamental and Applied Toxicology,* **5**:66-78.

Haseman, J. K. (1990), Use of statistical decision rules for evaluating laboratory animal carcinagenicity studies, *Fundamental and Applied Toxicology,* **14**:637-648.

Haseman, J. K., Huff, J. and Boorman, G. A. (1984), Use of historical control data in carcinogenicity studies in rodents, *Toxicologic Pathology,* **12**:126-135.

Haseman, J. K., Huff, J., Rao, G. N., and Eustis, S. L. (1989), Sources of variability in rodent carcinogenicity studies, *Fundamental and Applied Toxicology,* **12**, 793-804.

Haseman, J. K., Winbush, J. S., and O'Donnell, M. W. (1986), Use of dual control groups to estimate false positive rates in laboratory animal carcinogenicity studies, *Fundamental and Applied Toxicology,* **7**:573-584.

Hoel, D. G., and Walburg, H. E. (1972), Statistical analysis of survival experiments, *J. Nat. Cancer Inst.,* **49**:361-372.

Holm, S. (1979), A simple sequentially rejective multiple test procedure, *Scand. J. Statistics,* **6**:65-70.

Kodell, R. L., Haskin, M. G., Shaw, G. W. and Gaylor, D. W. (1983), CHRONIC: a SAS procedure for statistical analysis of carcinogenesis studies, *J. Statistical Computation Simul.,* **16**:287-310.

Kuritz, S. J., Landis, J. R. and Koch, G. G. (1988), A general overview of mantel-haenszel methods: applications and recent developments, *Ann. Rev. Public Health,* **9**:123-160.

Lin, K. K. (1989), Methods of statistical analysis of animal tumorigenicity studies, *American Statistical Association 1989 Proceedings of the Biopharmaceutical Section,* Alexandria, VA, pp 142-147.

Salsburg, D. S. (1983), The lifetime feeding study in mice and rats—an examination of its validity as a bioassay for human carcinogens, *Fundamental and Applied Toxicology,* **3**:63-67.

Tarone, R. E. (1982), The use of historical control information in testing for a trend in proportions, *Biometrics*, **38**:215-220.

Westfall, P. H. (1985), Simultaneous small-sample multivariate Bernoulli confidence intervals, *Biometrics*, **41**:1001-1013.

Westfall, P. H. and Young, S. S. (1993), *Resampling-Based Multiple Testing. Examples and Methods for p-value Adjustment*, John Wiley and Sons, Inc., New York.

3

Uses of Bioassay in Drug Development: Dose-Response Relationships

Chiao Yeh

Zeneca Pharmaceuticals Group,
Wilmington, Delaware

I. INTRODUCTION

The object of a biological assay or bioassay in drug development is to measure the potency of new compounds relative to some standard drugs in terms of the magnitudes of their effects on responses from living subjects. In the preclinical phase of drug development bioassay plays an important role in evaluating the pharmacological and toxic effects for those compounds. In addition, it can be used in the preclinical phase for screening drug combinations, clinical research, and quality control.

There are three elements involved in an assay, namely: 1) a substance or a stimulus (e.g., a drug); 2) a subject (an animal, a bacterial culture, a patient, or a piece of animal culture); and 3) a response.

All the methods to be discussed (except those in Section VI) assume that the standard preparation (S) and test preparation (T) act like the same substance but at different concentrations. Such assays are called analytic dilution assays. When this condition is not fulfilled, the assay is called a comparative assay. One difficulty with the comparative assay is that the estimate of potency may not be a constant.

Based on the ways of response measurements and the types of dose-response relationships, there are several kinds of bioassays. Three main types of assays are in common use for estimation of potencies: 1) direct assay, 2) indirect assays based on quantitative responses, and 3) indirect assays based on quantal (all-or-nothing) responses.

In the direct assay, the animal (e.g., guinea pig) is given a drug (e.g., digitalis) slowly so that its threshold dose for some well defined reaction (e.g., arrest of heartbeat) can be determined. Frequently, this procedure cannot be done and an animal is simply administered a specified dose of the preparation. However, it is still possible to assess potency by employing the methodology of indirect assay. In this assay several dose levels of the test and standard preparations are given to separate groups of animals.

In some assay systems the responses will be quantitatively measured, with average responses linearly related to the logarithms of the doses of either preparation. This type of assay is called a parallel line assay. When the average response is linearly related to the dose for either preparation, it is called a slope ratio assay. This situation arises particularly in microbiological assays where the response is a turbidimetric measure of the growth of microorganisms.

In the indirect quantal assay a prespecified dose is given to the animal and then one observes whether or not a well defined effect is apparent in the animal. It is similar to the quantitative assay in the form of statistical analysis. Both make use of dose-response regression relationships, though the calculations are more tedious for quantal response due to the need for differential weighting of the observations. In logical structure, however, quantal assays are more closely related to direct assay than quantitative assay.

Statistics will be used: 1) to obtain the "best" estimate (usually by least squares) of the potency of the test preparation, 2) to construct fiducial limits or confidence limits for its true value, and 3) to test as many assumptions involved in the assay as possible.

The general principles of bioassay were established during the 1930s and 1940s. A good survey of the principal methods is found in Chapter 3 (by D. J. Finney) of Burn et al. (1952) and a comprehensive account in Finney (1978). For the *United States Pharmacopeia* (USP) bioassays, the interested reader may find a concise account of biometrical procedures in the chapter on design and analysis of biological assays in USP (19th revision).

In this chapter, only parallel line assay (Section II) and quantal assay (Section V) will be discussed in detail. Section III briefly discusses the case when the dose-response relationship is parabolic. Section IV is concerned with slope ratio assays while Section VI discusses the comparative assays and nonparallel dose-response lines. Section VII focuses on the application of radioimmunoassay to determine drug concentrations.

II. PARALLEL LINE ASSAYS

The technique of parallel line assay has been widely used for drug screening programs in the pharmaceutical industry. Potency estimation is a way of telling how potent a new compound is relative to a standard drug. In addition, this assay technique has been employed successfully in the clinical evaluation of analgesic effectiveness (see Wallenstein and Houde, 1975, and Chapter 9 of this volume). In this section, statistical estimation procedures and statistical viewpoints of planning will be discussed. Additionally, an exact data analysis technique will be illustrated using a numerical example from a randomized block experiment with missing observations.

A. Statistical Model

Suppose that the dose of the standard preparation administered to an animal is denoted by z_s and the log dose by $x_S = \log z_s$. The response measurement for standard doses will be denoted by y_s. The response measurement will be taken to be a random variable with linear regression on x_S:

$$E(y_S \mid x_S) = \alpha + \beta x_S \tag{3.1}$$

We will consider only the case of homogeneous variance for y_S at each x_S. Further, assume that all y_S are independently distributed.

A test preparation (T) acts like a dilution of the standard preparation (S), that is, one unit of the test preparation is as effective as ρ units of the standard preparation. The constant ρ is called the potency of the test preparation relative to the standard. if z_T units of T will have f% inhibition, $z_S = \rho z_T$ units of S will have the same f% inhibition. Denote the response for a dose of T by y_T. Then the potency relationship implies

$$E(y_T \mid z_T) = (Ey_S \mid \rho z_T) \tag{3.2}$$

or

$$E(y_T \mid x_T) = \alpha + \beta \log \rho + \beta x_T \tag{3.3}$$

This is the algebraic statement of the condition of similarity, a prerequisite of all dilution assays. Therefore, the regression lines for S and T have different intercepts (α, $\alpha + \beta \log \rho$), but the same slope and hence the name parallel line assay. The variance of y_T will be equal to the variance of y_S and all y_T values are mutually independent and independent of all y_S values. The model, as given thus far, is sufficient for obtaining unbiased—or at least consistent—estimations with standard errors and also for planning experiments. To obtain more precise probability statements, for both confidence limits and tests of significance, some sort of distribution theory is necessary. The procedure to be given can be justified by taking the distribution of y_S and y_T values to be normally distributed, or by appealing to limit theorems from randomization theory.

B. Estimation

Rewrite Equations 3.1 and 3.2 to facilitate the statistical reasoning:

$$E(y_S \mid x_S) = \alpha_S + \beta(x_S - \bar{x}_S) \tag{3.4}$$

$$E(y_T \mid x_T) = \alpha_T + \beta(x_T - \bar{x}_T) \tag{3.5}$$

where the intercepts are

$$\alpha_S = \alpha + \beta\bar{x}_S \tag{3.6}$$

$$\alpha_T = \alpha + \beta\bar{x}_T + \beta \log \rho \tag{3.7}$$

Thus, the log potency ($\log \rho$) is related to the intercepts and slope of the parallel lines by the expression

$$\log \rho = (\bar{x}_S - \bar{x}_T) - \frac{\alpha_S - \alpha_T}{\beta} \tag{3.8}$$

The best linear unbiased estimates of α_S, α_T, and β can be obtained by least-squares analysis. These can then be used to obtain a consistent, though not unbiased, estimate of ρ by substituting in Equation 3.8. The least-squares estimate b of β is a weighted average of the least-squares estimates of slope from the standard drug data and the test drug data. Therefore, the estimate M of the log potency is

$$M = \log \hat{\rho} = (\bar{x}_S - \bar{x}_T) - \frac{\bar{y}_S - \bar{y}_T}{b} \tag{3.9}$$

The geometrical meaning of the right-hand side is illustrated in Figure 3.1.

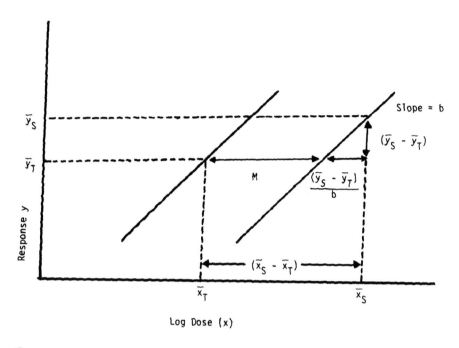

Figure 3.1. Geometrical meaning of Equation 3.9.

Thus, the log potency is the horizontal distance between the parallel lines. If $M = x_S - x_T > 0$, then $\hat{\rho} > 1$ and T is more potent than S. If $M = x_S - x_T < 0$, then $\hat{\rho} < 1$ and T is less potent than S.

Suppose that there are N_S and N_T observations on S and T and k_S and k_T are number of dose levels for S and T, respectively. Applying Fieller's theorem (see Finney, 1978) with $V_{12} = 0$ to obtain limits for $(\alpha_T - \alpha_s)/\beta$, the lower and upper $(1 - \alpha)$ confidence limits (A,B) may be written:

$$A,B = \frac{1}{1-g}\left[\frac{\bar{y}_T - \bar{y}_S}{b} \pm \frac{ts}{b}\sqrt{(1-g)\left(\frac{1}{N_S} + \frac{1}{N_T}\right) + \left(\frac{\bar{y}_T - \bar{y}_S}{b}\right)^2\frac{1}{\sum SSX}}\right]$$

(3.10)

where t is $(1 - \alpha/2)^{th}$ percentile of the Student t distribution with $N_S + N_T - k_S - k_T$ degrees of freedom, $g = t^2s^2 / (b^2 \sum SSX)$ and $\sum SSX$ is the pooled within-preparation sum of squares of x. Subtracting $\bar{x}_T - \bar{x}_S$ from A and B to obtain limits for log ρ and substituting $M + (\bar{x}_T - \bar{x}_S)$ for $(\bar{y}_T - \bar{y}_S)/b$ the limits for log ρ, denoted by log ρ_L and log ρ_U, are

$$\log \rho_L, \log \rho_U = (\bar{x}_S - \bar{x}_T) +$$

$$\frac{1}{1-g} \left\{ M + (\bar{x}_T - \bar{x}_S) \pm \frac{ts}{b} \sqrt{(1-g)\left(\frac{1}{N_S} + \frac{1}{N_T}\right) + \frac{[M + (\bar{x}_T - \bar{x}_S)]^2}{\sum SSX}} \right\} \tag{3.11}$$

If $g = 0$, as will be the case in most applications, and if the doses are chosen so that the mean responses of the two preparations are approximately equal $[\bar{y}_S \approx \bar{y}_T$ or $\hat{\alpha}_S = \hat{\alpha}_T$, or $M = (\bar{x}_S - \bar{x}_T)]$, the confidence limits for log ρ can be simplified as

$$M \pm \frac{ts}{b} \sqrt{\frac{1}{N_T} + \frac{1}{N_S}} \tag{3.12}$$

C. Analysis of Variance

There are several reasons to analyze the data by the technique of analysis of variance (ANOVA). First, the estimate of σ^2 (i.e., s^2) is easily obtained in this way and can be used in the formulae for confidence limits. Secondly, tests of validity of some of the assumptions in the model are obtained; and thirdly, the results yield a simple summary to be used in the design of further assays of the same type.

Table 3.1 shows the ANOVA for parallel line assay. The total sum of squares is divided into components of "between doses" and "error" (within doses). Furthermore, the sum of squares between doses can be separated into: 1) linear regression, 2) deviation from parallelism, 3) difference between S and

Table 3.1 Analysis of Variance for Parallel Line Assay

Source of variation	Degrees of freedom
Between doses	$k_S + k_T - 1$
Regression	1
Parallelism	1
Preparations	1
Deviations from linearity	$k_S + k_T - 4$
Within doses	$N_S + N_T - k_S - k_T$
Total	$N_S + N_T - 1$

T, and 4) if there are more than four doses, the remainder can be lumped together as deviations from linearity.

A hypothesis is formulated for each of the four components as follows:

1. Linear regression: The null hypothesis that the slope of the response-log dose line is zero is tested. Obviously, the assay is invalid unless this hypothesis is rejected. In any properly conducted assay, the mean square of linear regression ought to be relatively large. If there were no significant regressions, the dose-response relationship is of no use for the estimation of potency.

2. *Deviation from parallelism*: The null hypothesis that the slopes for S and T are equal is tested. If this hypothesis is rejected, the assay must be considered invalid.

3. *Difference between standard and test preparations*: This is not usually of great interest in itself, though it improves precision if y_S and y_T are not too different (see Eq. 3.12).

4. *Deviations from linearity*: A relatively large mean square to the error term would indicate that the model is invalid.

The calculation procedure for sum of squares is very similar to that for the data on a simple regression, as found in any elementary statistics textbook.

D. Symmetrical Parallel Line Assays

Symmetry in the number and spacing of doses and in the allocation of animals to doses has advantages both for ease of computation and precision. We define the following quantities:

n = number of responses at each of the dose levels
N = 2nk = total number of responses
k_S = k_T = k = number of dose levels for S and T
D = ratio between each dose and the one below it (the same for all doses, and for S and T, so that the doses are equally spaced by log D on the logarithmic scale).

The great advantage of symmetrical assays is to test for nonvalidity by using orthogonal contrasts and to simplify calculation of the relative potency and its confidence limits.

In the six-point assay, for instance, the sum of squares of each single component can be calculated easily by using the coefficients of orthogonal contrasts listed in Table 3.2. The contrasts may conveniently be referred to as L_p (preparations), L_1 (regression), L_1' (parallelism), L_2 (quadratic curvature), L_2' (difference of quadratics). The subdivision of a sum of squares using orthog-

Table 3.2 Coefficients of Orthogonal Contrasts for Six-Point Parallel Line Assay

| | Dose | | | | | | |
	S_1	S_2	S_3	T_1	T_2	T_3	Divisor
L_p	−1	−1	−1	1	1	1	6n
L_1	−1	0	−1	−1	0	1	4n
L_1'	1	0	−1	−1	0	1	4n
L_2	1	−2	1	1	−2	1	12n
L_2'	−1	2	−1	1	−2	1	12n

onal contrasts is a general process described, for example, by Brownlee (1965, p. 517) and Cochran and Cox (1957, p. 77).

In symmetrical assays, to simplify calculations it is better to use logarithms to a base different than 10, such as γ in general (e.g., $\gamma = \sqrt{D}$ for a four-point assay, $\gamma = D$ for a six-point assay). Since antilogs are available only for base-10 logarithms, it is necessary to convert to base 10 before looking up antilogs. To serve this purpose, keep in mind that

$$\log_{10} R = (\log_{\gamma} R)(\log_{10} \gamma) = M \log_{10} \gamma \qquad (3.13)$$

The potency estimation can be simplified by using the following formulae

$$\bar{y}_S - \bar{y}_T = -\frac{L_p}{nk} \qquad (3.14)$$

$$b = \frac{3L_1}{2N} \qquad (3.15)$$

$$R = \hat{\rho} = \frac{Z_{S1}}{Z_{T1}} \text{antilog}_{10} \left(\frac{4L_p}{3L_1} \log_{10} D \right) \qquad (3.16)$$

where Z_{S1} and Z_{T1} are the smallest doses of S and T, respectively. Confidence limits for potency are

$$\frac{Z_{S1}}{Z_{T1}} \text{ antilog}_{10} \left\{ \left[\frac{4L_p}{3L_1(1-g)} \pm \frac{4st}{3L_1(1-g)} \sqrt{N(1-g) + \frac{2N}{3}\left(\frac{L_p}{L_1}\right)^2} \right] \log_{10} D \right\}$$

$$(3.17)$$

where

$$g = \frac{2Ns^2t^2}{3L_1^2}$$

$$(3.18)$$

Using these simplified formulae, a numerical example of symmetric parallel line assay is given in Chapter 9.

E. A Numerical Example from a Randomized Block Experiment with Missing Observations

In some assays, the potential responses can be assembled with homogeneous sets (e.g., by day, by litter, etc.) in advance of the experiment. The differences between sets are segregated later so that they do not adversely affect either the computed potency or its confidence limits. In order to eliminate the effects of interset differences on the precision of the assay, a randomized block design is usually adopted. Furthermore, even in the most carefully conducted experiment, an accident may occur and cause the loss of observations and destroy the symmetry of the design. It is important to know how to analyze data taking into account this type of real situation.

The following example is adapted from Colquhoun (1971). The data in Table 3.3 are measures of the tension developed by the isolated guinea pig ileum in response to pure histamine (S) and to a solution of a test preparation (T) containing various impurities and an unknown amount of histamine. Five replicates of each of the six doses were give, all to the same tissue, so there is a danger that one response may affect the next response. Therefore, the order of the doses was arranged randomly for each tissue. Three observations were missing.

The exact analysis of the assay, taking account of the three missing observations, is complicated by the impossibility of using Table 3.2 or any other simple alternative in the construction of orthogonal components of an analysis of variance. If the method of fitting constants for the analysis of nonorthogonal data was employed, one must solve linear equations for four parameters representing differences among the five tissues and five parameters representing differences among six doses (see Yates, 1933; Snedecor, 1956, Section 12.17).

Since the main interest is focused on five special contrasts among the six doses and differences between blocks are of no direct concern, an easier method

Table 3.3 Response of the Isolated Guinea Pig Ileum

(Tissue) block	Standard histamine dose (ng/ml)			Test dose (ng/ml)			Total
	S_1 (4)	S_2 (8)	S_3 (16)	T_1 (8)	T_2 (16)	T_3 (32)	
1	20.5	27.0	38.0	18.5	30.0	35.0	169.0
2	18.5	31.5	44.0	15.0	24.0	34.5	167.5
3	20.0	26.0	35.5	13.0	—	38.0	132.5
4	18.0	23.5	41.5	13.5	26.0	—	122.5
5	20.0	—	38.5	12.0	25.0	32.0	127.5
Totals:	97.0	108.0	197.5	72.0	105.0	139.5	719.0

is to set up equations for the estimation of these five contrasts alone. This may be done formally by multiple regression. Introduce five independent variates x_1, x_2, x_3, x_4, x_5, defined to take the following values at the six dose levels:

	S_1	S_2	S_3	T_1	T_2	T_3
$x_1 =$	-1	-1	-1	1	1	1
$x_2 =$	-1	0	1	-1	0	1
$x_3 =$	1	0	-1	-1	0	1
$x_4 =$	1	-2	1	1	-2	1
$x_5 =$	-1	2	-1	1	-2	1

Thus, the five variates have the same values as the orthogonal coefficients in Table 3.2. The regression coefficients b_1, b_2, b_3, b_4, and b_5 are obtained as follows:

$$
\begin{bmatrix} b_1 \\ b_2 \\ b_3 \\ b_4 \\ b_5 \end{bmatrix} =
\begin{bmatrix} \sum x_1^2 & \sum x_1 x_2 & \sum x_1 x_3 & \sum x_1 x_4 & \sum x_1 x_5 \\ & \sum x_2^2 & \sum x_2 x_3 & \sum x_2 x_4 & \sum x_2 x_5 \\ & & \sum x_3^2 & \sum x_3 x_4 & \sum x_3 x_5 \\ \text{Symmetric} & & & \sum x_4^2 & \sum x_4 x_5 \\ & & & & \sum x_5^2 \end{bmatrix}^{-1}
\begin{bmatrix} \sum x_1 y \\ \sum x_2 y \\ \sum x_3 y \\ \sum x_4 y \\ \sum x_5 y \end{bmatrix}
$$

(3.19)

where Σx_i^2, $\Sigma x_i x_j$, and $\Sigma x_i y$ are the sums over the blocks of the corrected sums of squares and products within blocks. For example,

$$\sum x_1 y = (-20.5 - 27.0 - 38.0 + 18.5 + 30.0 + 35.0 + \ldots$$

$$-20.0 - 38.5 + 12.0 + 25.0 + 32.0) - \frac{(-132.5 - 122.5 + 127.5)}{5}$$

$$= -60.5$$

The reader may check in detail the following other items:

$$\sum x_1^2 = 26.4$$

$$\sum x_1 x_2 = \sum x_1 x_3 = \sum x_1 x_4 = \sum x_2 x_3 = \sum x_2 x_4$$

$$= \sum x_2 x_5 = \sum x_3 x_4 = \sum x_3 x_5 = \sum x_4 x_5 = -1.2$$

$$\sum x_1 x_5 = 3.6$$

$$\sum x_2^2 = \sum x_3^2 = 18.8$$

$$\sum x_4^2 = \sum x_5^2 = 49.2$$

$$\sum x_2 y = 192.5; \ \sum x_3 y = -8.5; \ \sum x_4 y = 0.5; \ \sum x_5 y = -54.5$$

By solving Equation 3.19 with numerical values, one obtains

$$\begin{bmatrix} b_1 \\ b_2 \\ b_3 \\ b_4 \\ b_5 \end{bmatrix} = \begin{bmatrix} -1.7220 \\ 10.0986 \\ 0.0486 \\ 0.1979 \\ -0.7294 \end{bmatrix} \text{ with covariance matrix}$$

$$\text{cov}(\underline{b}) = \begin{bmatrix} 3.8517 & 0.2512 & 0.2512 & 0.0997 & -0.2671 \\ & 5.3768 & 0.3768 & 0.1495 & 0.1256 \\ & & 5.3768 & 0.1495 & 0.1256 \\ & & & 2.0436 & 0.0498 \\ & & & & 2.0591 \end{bmatrix}, \ 10^{-2} s^2$$

$$(3.20)$$

Table 3.4   Analysis of Variance for Data in Table 3.3

Source of variation	df	Sum of squares	Mean square
Blocks ignoring doses	4	53.26	
Doses adjusted for block	5	2081.60	
Regression on log dose	1	1971.08	
Preparations, parallelism, and deviation from linearity	4	116.52	
Error:	17	115.94	6.82

The sum of squares of doses adjusted for blocks is equivalent to the sum of squares accounted for by this regression with five variants

$$-60.5b_1 + 192.5b_2 - 8.5b_3 + 0.5b_4 - 54.5b_5 = 2087.60 \tag{3.21}$$

The regression on log dose alone would account for a square obtained by omission of x_1, x_3, x_4, and x_5, namely

$$\frac{(192.5)^2}{18.8} = 1971.08 \tag{3.22}$$

The difference (Eqs. 3.21–3.22), with four degrees of freedom, is a composite test of preparations, parallelism, and deviations from linearity, which cannot be separated completely because of nonorthogonality. Tests of individual components could be done by omitting other components or by adjusting for others, but the portions so obtained would not be independent and additive (see Snedecor, 1956, Chapter 14). The sum of squares of blocks ignoring doses can be obtained by $(169^2 + 167.5^2)/6 + (132.5^2 + 122.5^2 + 127.5^2)/5 - (719)^2/27 = 53.26$. The sum of squares of error is defined in the usual manner. The results of ANOVA are constructed in Table 3.4.

Now b_2 is an estimate of the regression of response on log dose, while b_1 is an estimate of one-half the difference in mean responses for S and T, the quantity usually called $(\bar{y}_T - \bar{y}_S)$. The method of estimation has ensured that these estimates are adjusted for the nonorthogonality. Hence,

$$M = \bar{x}_S - \bar{x}_T - \frac{2 \times 1.72197}{10.09862} = \bar{x}_S - \bar{x}_T - 0.34103$$

To construct the confidence limits for M, the variances and covariance of $2b_1$ and b_2 are needed. One can obtain from Equation 3.20 covariance matrix of $\underline{b}$ as

$$\text{var}(2b_1) = 4 \times 3.8517 \times 10^{-2}\, s^2 = 0.15407\, s^2$$

$$\text{var}(b_2) = 0.05377\, s^2$$

$$\text{cov}(2b_1, b_2) = 2 \times 0.2512 \times 10^{-2}\, s^2 = 0.00502\, s^2$$

Hence,

$$g = \frac{(2.11)^2 \times 0.05377 \times 6.82}{(10.09862)^2} = 0.01601$$

Fieller's theorem is applied to give confidence limits for $(\alpha_T - \alpha_S)/\beta$; these are

$$-0.34103 \pm \frac{2.11 \times 2.6115}{10.0986}$$

$$\times \sqrt{0.15407 + 2 \times 0.34103 \times 0.00502 + (0.34103)^2 \times 0.05377}$$

$$= -0.34103 \pm 0.54564 \times 0.40466$$

$$= -0.56183, -0.12023$$

Using Equation 3.13,

$$R = \text{antilog}\,(\bar{x}_S - \bar{x}_T \times 0.34103 \times \log 2)$$

$$= \text{antilog}\,(-0.30103 - 0.34103 \times 0.30103)$$

$$= \text{antilog}\,(-0.40369)$$

$$= 0.3947$$

and similarly

$$R_L = \text{antilog}\,(-0.30103 - 0.56183 \times 0.30103)$$

$$= 0.3387$$

$$R_U = \text{antilog}\,(-0.30103 - 0.12023 \times 0.30103)$$

$$= 0.4600$$

F. Planning a Completely Randomized Parallel Line Assay

The statistical aspects of planning an assay are concerned with the decision as to: 1) the number of dose levels of S and T, 2) the placement of the doses, and 3) the number of observations to be taken at each dose level.

Simple answers to the first question are as follows:

1. If the assumptions of linearity and parallelism do not need checking and the standard line is well established, then one dose level of T is sufficient and no additional data on S need be obtained.
2. If the assumptions of linearity and parallelism do not need checking and the slope of the standard line is stable but the intercept of the standard line is not known, then a dose level of S and a dose level of T must be used.
3. If the assumptions of linearity and parallelism do not need checking but the parameters of the standard line are not known, then two doses of S and one dose level of T will be needed.
4. If the assumption of linearity can be taken for granted but the assumption of parallelism must be checked, then it is necessary to use two dose levels for each preparation.
5. If the conditions of linearity and parallelism must be checked, then it is necessary to have at least three doses for each preparation.

The above rules are simple and obvious, but difficult to apply in practice. It is customary to use a number of doses and discard those outside the region of approximate linearity.

The choice of dose location is based on the criteria: a) $\bar{x}_S - \bar{x}_T = M$, b) $\bar{y}_S = \bar{y}_T$, and c) minimize g and maximize Σ SSX. Some knowledge of ρ is often obtained from a small pilot experiment.

If the doses can be specified, and enough information on β and σ^2 are available, the number of observations necessary to assure a desired precision can be anticipated. If we would like $\hat{\rho}$ to be within 100E% of the true value of ρ with 95% confidence, then the total number of observations is taken as

$$N \geq \frac{16\hat{\sigma}^2}{\hat{\beta}^2 \log(1+E)} \quad \text{if } N_S = N_T = \frac{N}{2} \tag{3.23}$$

If we want to assure an interval of expected length L for M, then the total number of observations is approximately

$$N = \frac{16t^2 \hat{\sigma}^2}{\hat{\beta} L^2}$$

(3.24)

where t is the $(1 - \alpha/2)^{th}$ percentile of the Student t distribution with the same degrees of freedom as $\hat{\sigma}^2$.

Equations 3.23 and 3.24 can be derived easily by using the variance of log ρ (see Eqs. 3.10–3.12 and Brown, 1964).

III. PARALLEL CURVE ASSAYS

Suppose that the regression equation for the standard preparation is quadratic rather than linear, so that

$$E(y_S \mid x_S) = \alpha + \beta x_S + \gamma x_S^2$$

(3.25)

then for the test preparation

$$E(y_T \mid x_T) = \alpha + \beta(x_T + \log \rho) + \gamma(x_T + \log \rho)^2$$

(3.26)

Elston (1965) noted that in complement fixation tests and gonadotrophin assays, a curved relationship is usual. Such cases have been examined by Bliss (1957) and by Elston (1965). Elston has proposed a simple method for estimating relative potency from a quadratic log dose assay. An equivalent alternative and somewhat simpler method has been described by Cox (1972). Obtaining an estimate of potency by their methods does not require a transformation or repeating the assay. Furthermore, their methods can also be used to estimate relative potency when the underlying log dose response relationship is a general polynomial of degree n.

IV. SLOPE RATIO ASSAYS

In the slope ratio assay, the response to a drug is linearly related to the dose itself, or to some power of the dose. This application is used frequently in microbiological assays (e.g., for vitamins) and is seen occasionally in other fields. Usually, several test preparations are assayed at one time with a single standard preparation and a control (zero dose), so called blanks.

The regression of S is

$$E(y_s \mid Z_s) = \alpha + \beta_s Z_s$$

(3.27)

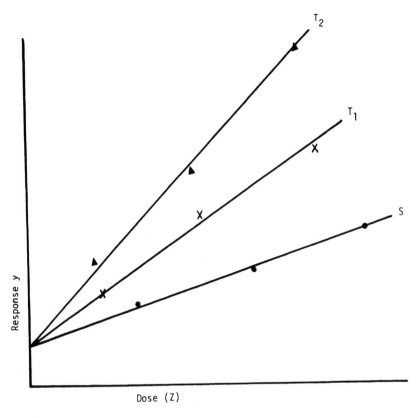

Figure 3.2 Slope ratio assay, regression lines for S, T$_1$, and T$_2$.

Considering a dose, Z_{Ti} of the ith test preparation to act like $Z_s = \rho_i Z_{Ti}$ units of S, then the response of dose Z_{Ti} is

$$E(y_{Ti} \mid Z_{Ti}) = E(y_s \mid \rho_i Z_{Ti})$$
$$= \alpha + \beta_s(\rho_i Z_{Ti}) \quad \text{for } i = 1, 2, \ldots v - 1 \qquad (3.28)$$
$$= \alpha + \beta_{Ti} Z_{Ti}$$

Equations 3.27 and 3.28 represent v straight lines with the same intercept α and with slopes in the ratio 1:ρ_i, hence the name slope ratio assays (Figure 3.2).

In the analysis of observations from a slope ratio assay, the observed response at various doses Z_s and at doses Z_{Ti} of T_i must be fitted by v regression lines, restricted to pass through the same intercept on the vertical axis. The problem can be conveniently regarded as one of multiple regression, such that

$$E(y \mid Z_s, Z_{Ti}) = \alpha + \beta_s Z_s + \beta_{T1} Z_{T1} + \beta_{T2} Z_{T2} + \dots \qquad (3.29)$$

where y is the dependent variable and Z_s and Z_{Ti} are predictor variables. For any observation on S, Z_s is nonzero, and for all the doses of test preparations, $Z_{Ti} = 0$; for observations on T_i, all doses are zero except Z_{Ti} is nonzero; for control observations without either S or T, $Z_S = Z_{Ti}$ (for all i) = 0. If we assume the residual variation is approximately normal and has equal variance, by the standard estimation procedures for multiple linear regression we can obtain the estimated intercept and variance slopes b_{Ti} and their estimated covariance matrix. Thus, the potency (ρ_i) can be estimated by

$$R_i = \frac{b_{Ti}}{b_s} \qquad (3.30)$$

Using Fieller's theorem and the covariance matrix, confidence limits for ρ_i can be obtained.

As in parallel line assays, an analysis of variance is an aid to examination of the validity of the assay. Table 3.5 shows the ANOVA for slope ratio assay. The subdivision of the components for the sum of squares of doses can be separated into: 1) linear regression, 2) intercepts, 3) blank, and 4) deviations from linearity.

Linearity is a requirement for statistical validity; thus in any good assay the mean square of linear regression ought to be relatively large. The sum of

Table 3.5 Analysis of Variance for Slope Ratio Assay

Source of variation	df
Between doses	vk
Regression	v
Intercept	v − 1
Blank	1
Deviations from linearity	v(k − 2)
Within doses	(1 + vk)(n − 1)
Total:	n(1 + vk) − 1

v = number of preparations; k = number of nonzero dose levels of each preparation; n = number of replications.

squares of intercept indicates the equality of the intercepts of v regression lines (disregarding the blank data). Intersection of the v lines is a requirement for fundamental validity, analogous to parallelism in parallel line assay, since it is derived from the condition of similarity. Significance means invalidity of the assay system. The sum of squares of blank expresses the equality of the blank expectation and the common intercept of the v lines. A significant F test for blank might indicate nonlinearity for very low doses; if the remaining validity tests were satisfactory the assay could still be analyzed adequately by omitting the blank data. The mean square deviation from linearity comprises the curvature components for the v preparations separately (excluding zero dose). Further details, with examples, can be found in Clarke (1952), Barraclough (1955), and Finney (1978, Chapters 7 and 8).

In addition to its being used in microbiology, the slope ratio assay has been applied in cardiology to analyze S-T segment depression data during exercise. Patients with documented angina were exercised before and after 1 or more weeks administration of an antianginal drug (e.g., isosorbide dinitrate). Patients were observed first after the control (no drug) period and then after drug. In this application, S = control, T = drug, and doses are the exercise periods (e.g., 30, 60, 90, ..., 300 sec). The relative potency can be interpreted as a relative efficacy over control.

V. QUANTAL ASSAYS

A. Quantal Dose-Response Curves and the Tolerance Distribution

In quantal assay, a prespecified dose is given to the animal and one observes whether or not a well defined effect is apparent in the animal. Examples of biological quantal response are dead/alive, presence/absence of convulsion, and sick/cured. For any one preparation, the response curve relating the percentage of animals responding to the log dose yields a smooth sigmoid curve, which rises rather steeply in its center portion and changes but little at its upper and lower tails (Figure 3.3). The response curve, rising from 0 to 1 on the vertical scale, may be regarded as the cumulative distribution function of a random variable which can be called the tolerance of animal to the preparation under experiment. The tolerances are exactly the critical doses in a direct assay. For example, if a dose X corresponds to a response P, we can interpret this as expressing that a proportion P of animals have a tolerance less than X, and 1 − P have a tolerance greater than X. The response function P is thus the distribution function of tolerance; the corresponding density function is called the tolerance distribution.

To approach such a dose-response curve statistically, one must find a mathematical model that matches the main features of the experimental situation.

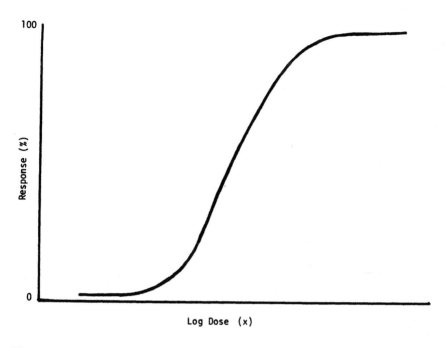

Figure 3.3 Quantal dose-response curve.

Two main models provide this match; the integrated normal curve and the logistic curve. By suitable linear transformation, the sigmoid curve reduces to a straight line so that ordinary regression procedures can be used to estimate the parameters. As the quantal dose-response curve matches the curves, so do their linear transformations.

The best sigmoid curves are listed below, together with their linear transformations. Here P is the probability of a response at dose X, and $Y = \varphi^{-1} = \alpha + \beta X$ is an inverse transformation for each of the models:

1. Integrated Normal Curve

$$P = \frac{1}{\sqrt{2\pi}} \int_{-\infty}^{\alpha + \beta x} \exp\left(\frac{-t^2}{2}\right) dt, \quad -\infty < x < \infty \tag{3.31}$$

$$Y = \varphi^{-1}(P) = \text{Normal equivalent deviate (NED) of P} \tag{3.32}$$

The probit is defined as NED increased by 5 to avoid negative values, that is,

$$\text{Probit} = \text{NED} + 5 \tag{3.33}$$

2. *Logistic Curve*

$$P = [1 + \exp(-\alpha - \beta x)]^{-1}, \quad -\infty < x < \infty \tag{3.34}$$

The logit P is given by

$$Y = \text{logit } P = \log \frac{P}{1 - P} \tag{3.35}$$

where $\log = \log_e$.

3. *Angular Distribution*

$$P = \sin^2\left(\alpha + \beta x + \frac{\pi}{4}\right), \quad -\frac{\pi}{4} \le \alpha + \beta x \le \frac{\pi}{4} \tag{3.36}$$

$$Y = \sin^{-1}\sqrt{P} - \frac{\pi}{4} \tag{3.37}$$

4. *Uniform Distribution*

$$P = \alpha + \beta x + \frac{1}{2}, \quad -\frac{1}{2} \le \alpha + \beta x \le \frac{1}{2} \tag{3.38}$$

$$Y = P - \frac{1}{2} \tag{3.39}$$

The four curves are substantially alike over the range generally used in bioassay problems.

B. Estimation Procedures

Two procedures will be discussed to estimate the intercept α and the slope β.

A particular quantity of interest in practical application is the median effective dose (usually denoted ED_{50}) or the median lethal dose (denoted LD_{50}). This is the dose for which the response (or mortality) is 50%; it is given by

$$\gamma = \log ED_{50} = \frac{(5 - \alpha)}{\beta} \text{ (from probit)} \tag{3.40}$$

$$\gamma = \log \text{ED}_{50} = -\frac{\alpha}{\beta} \quad \text{(from logit)}$$

(3.41)

In an assay of S and T, the linearized quantal dose-response curve for Y can be expressed in Equations 3.1 and 3.3 in parallel line assay:

$$Y_S = \alpha_S + \beta x_S$$

(3.42)

$$Y_T = \alpha_T + \beta x_T$$

(3.43)

where $\alpha_T = \alpha_S + \beta \log \rho$. The problem is one of estimating $\log \rho = (\alpha_T - \alpha_S)/\beta$.

1. Maximum Likelihood Method

If n_i animals were tested at log dose x_i, each reacting independently, the probability that r_i respond is obtained from the binomial distribution. If k doses were tested, the log-likelihood is

$$\log L = \sum_i \log C_{r_i}^{n_i} + \sum_i r_i \log P_i + \sum_i (n_i - r_i) Q_i$$

(3.44)

Following standard iterative maximum likelihood estimation procedure (see Finney, 1971), if y follows the integrated normal curve, then $Y_1 = a_1 + b_1 x$, where a_1, b_1 are initial guess values for α and β denoting $W = Z^2/PQ$, where Z is the ordinate to the normal curve at the point with abscissa Y, and $y = Y_1 + (p - P)/Z$, where p is the observed proportion of response, and a_2, b_2, the first cycle estimates, can be obtained from a regular weighted regression procedure:

$$b_2 = \frac{\sum nW(x - \bar{x})(y - \bar{y})}{\sum nW(x - \bar{x})^2}$$

(3.45)

and

$$a_2 = \bar{y} - b_2 \bar{x}$$

(3.46)

The revised $Y_2 = a_2 + b_2 x$ is calculated as the weighted linear regression of y on x, the new weights being nW. Now going to the new cycle, all arithmetic can be replaced with Y_2 in place of Y_1. Usually, successive iterates converge rapidly even if Y_1 is a poor estimate. The asymptotic variances for a and b are

$$\text{var}(a) = \frac{1}{\sum nW} + \frac{\bar{x}^2}{\sum nW(x - \bar{x})^2}$$

(3.47)

$$\text{var}(b) = \frac{1}{\sum nW(x - \bar{x})^2}$$

(3.48)

If the logistic curve is adopted for P, some degree of simplification can be offered. Writing PQ = W in Equations 3.45–3.48, we obtain maximum likelihood estimates for the logit.

2. Noniterative Minimum χ^2 Method

In the noniterative minimum χ^2 method the aim is to minimize the sum of weighted square differences between the observed and estimated expected values, that is,

$$\sum n_i' W_i' (p_i - \hat{p}_i)^2 = \sum n_i' W_i' (y_i - \hat{\alpha} - \hat{\beta} x_i)^2$$

(3.49)

It has been shown by Taylor (1953) that the procedure is a regular best asymptotically normal procedure (RBAN), and hence is asymptotically equivalent to the maximum likelihood procedure. This method is also equivalent to that of obtaining a least-squares solution for α and β using weights W_i'. The explicit solution is

$$b = \frac{\sum nW'(y - \bar{y})(x - \bar{x})}{\sum nW'(x - \bar{x})^2}$$

(3.50)

$$a = \bar{y} - b\bar{x}$$

where

$$\bar{y} = \frac{\sum nW'y}{\sum nW'}$$

(3.51)

$$\bar{x} = \frac{\sum nW'x}{\sum nW'}$$

The weight of the integrated normal curve is $W_i' = z_i^2 p_i q_i$, where $z_i = \varphi^{-1}(y_i)$, while the logistic is $W_i' = p_i q_i$. If p_i is zero or 1, $y_i = \varphi^{-1}(p_i)$ is undefined. In this case Berkson (1955) recommended the use of 1/2n or $1 - 1/2n$, respectively. This practice does not alter the asymptotic properties of the estimates.

3.   Comparison of Maximum Likelihood and Minimum χ^2 Methods

The maximum likelihood and minimum χ^2 methods are asymptotically equivalent. Little is known theoretically of the relative merits of the estimates in small sample sizes although in practice they have been shown to be almost identical.

In deriving the estimates by the method of minimum χ^2, only the expectation nP is required, whereas to derive maximum likelihood estimates it is necessary to assume that the observed proportion p is distributed binomially. This is an important methodological advantage of minimum χ^2 estimates over maximum likelihood estimates.

The logistic estimates, both those obtained by the method of maximum likelihood and those obtained by minimum χ^2, are sufficient as well as efficient. Those obtained by using maximum likelihood and the integrated normal curve are not sufficient.

C.   Goodness of Fit

To test the goodness of fit of the estimated line to the observations, χ^2 tests can be used. The χ^2 is the weighted sum of squares of the difference between empirical and estimated values,

$$\chi^2 = \sum nW(y - a - bx)^2 \qquad (3.52)$$

with $(k - 2)$ degrees of freedom and where k is the number of doses. A significantly large value indicates that all the weights have been overestimated by a factor of $\chi^2/(k - 2)$. Compensation for this must be made by multiplying all the variances by the "heterogeneity factor" (see Finney, 1971, p. 72). However, a large χ^2 throws doubt on the linearity of the regression.

D.   Estimations of ED_{50} and Relative Potency

Denoting the estimates of α and β by a and b (which can be obtained by either the maximum likelihood procedure or by the minimum χ^2 procedure), then log ED_{50} can be estimated as

$$C = -\frac{a}{b} \qquad (3.53)$$

with

$$\hat{\text{var}}(C) = [\hat{\text{var}}(a) + C^2\hat{\text{var}}(b) + 2C\hat{\text{cov}}(a,b)]/b^2 \qquad (3.54)$$

$$\hat{\text{ED}}_{50} = \text{antilog } C \qquad (3.55)$$

$$\hat{\text{var}}(\hat{\text{ED}}_{50}) = (\hat{\text{ED}}_{50} \times \log 10)^2\, \hat{\text{var}}(c) \qquad (3.56)$$

where $\log = \log e$. Using Fieller's theorem, confidence limits on $\log \text{ED}_{50}$ can be obtained.

To estimate the relative potency $\log \rho$, three estimates are needed (see Eqs. 3.42 and 3.43). The results of the estimates have the same form as the estimates in the parallel line assay except for the meaning of the weight.

If the two regression lines of S and T are parallel, the horizontal distance between S and T is the estimated $\log \rho$. If the lines are not parallel, a reasonable convention is to use the 50% response point as a standard so that respective $\hat{\text{ED}}_{50}$ values are being compared.

Berkson (1953) mentioned that the tests for parallelism (either χ^2 or F) are worthless since their "power" is small in practice. He recommended use of the ratio of the two $\hat{\text{ED}}_{50}$ values whether the lines are parallel or not. Detailed calculation of relative potency may be found in Finney (1971, Chapter 6).

E.   Conversion of Quantitative Data to Quantal Data

Biostatisticians in the pharmaceutical industry frequently encounter people who are reluctant to use measurements as quantitative data. Rather, they are inclined to establish some sort of cutoff value of the measurements such that, for example, measurements lower than this value are defined as "responses" while measurements higher than the value are called "nonresponses." Thus, a dose-response curve is fitted to the percentage data and an ED_{50} is estimated. To establish a reliable cutoff point, it is necessary to study the distribution of measurements in the control animals (no drug). Sometimes it is found that the measurements of control animals follow the log-normal distribution very well. If it is a good assay and the cutoff point is appropriate, then the relative potency calculated from parallel line assay should be equal to relative ED_{50} calculated from quantal responses. The conversion of quantitative data to quantal data should be used cautiously. In general, the estimate from quantitative measurements would be more precise than the estimate from quantal data.

F.   A Numerical Example

A hypothetical example of the results of a quantal bioassay is given in Table 3.6.

Table 3.6 A Hypothetical Example of Quantal Assay

Preparation	Z_i dose (mg/kg)	X_i log dose	No. animals	No. dead	Mortality (%)
Standard	1	0	20	1	5
	2	0.30103	20	5	25
	4	0.60206	20	9	45
	8	0.90309	20	16	80
	16	1.20412	20	19	95
Test	1	0	20	4	20
	2	0.30103	20	8	40
	4	0.60206	20	15	75
	8	0.90309	20	18	90

BMD03S Probit Analysis and Procedure Probit of SAS (Statistical Analysis System) can be employed to estimate LD_{50} using maximum likelihood and the integrated normal curve. Maximum likelihood logit estimates can be obtained from the computer program listed in Waud's (1972) paper. Minimum logit χ^2 estimates are easy to calculate following the regular weighted regression procedure. In practice, all the estimates are essentially equal. Table 3.7 gives the comparison of log LD_{50} estimates from minimum logit χ^2 and probit maximum likelihood (ML).

Employing the minimum logit χ^2 method, the estimates of β_S and β_T are 4.51 and 4.11, respectively, Hence, it is reasonable to conclude that the logit-log dose lines of S and T are parallel. Estimates of Equations 3.42 and 3.43 are

Table 3.7 Comparison of log LD_{50} Estimates

Preparations	Methods	Log LD_{50}	95% Confidence limits	χ^2 Testing goodness of fit	df
Standard	Mimimum logit χ^2	0.60	0.48–0.73	0.55	3
	Probit ML (SAS)	0.60	0.39–0.82	0.41	3
Test	Minimum logit χ^2	0.36	0.21–0.50	0.20	2
	Probit ML (SAS)	0.36	−0.54–0.84	0.25	2

$$\hat{\alpha}_S = -2.60, \quad \hat{\alpha}_T = -1.58, \quad \text{and} \quad \hat{\beta} = 4.34$$

It follows that an estimate for log ρ is log $\hat{\rho} = \hat{\alpha}_T - \hat{\alpha}_S/\hat{\beta} = 1.02/4.34 = 0.235$. Therefore, the potency of the test preparation relative to the standard preparation is 1.72.

G. Uses of Quantal Assay

In drug-screening programs, the techniques of quantal assay are always used. To calculate a therapeutic index, LD_x and ED_y need to be calculated (see Chapter 24). The assay technique has been used to measure the potencies of drug mixtures (Hewlett, 1969). In addition, most authorities agree that the probit analysis is the most useful standard approach for comparisons of burn mortality (Waisbren et al., 1975). In this application, the dose is the percentage of burn.

VI. COMPARATIVE BIOASSAYS AND NONPARALLEL DOSE-RESPONSE LINES

In analytic dilution assays, the relative potency r does not depend on the choice of animals, the nature of the biological response, the dose levels at which the tests are conducted, or any other experimental conditions. Assays in which r is not an absolute constant are often referred to as comparative. A variable r can occur in the comparison of test and standard preparations that are qualitatively different, either because they are chemically different or for some other reasons. This ρ is understood to depend on experimental conditions. In such an assay relative potency is the ratio of the amounts of S and T required to give the same response and not the ratio of the amounts of a single "effective constitute."

Consider linear log dose-response lines for S and T with possible unequal slopes,

$$E(y_s \mid x_S) = \alpha_S + \beta_S x_S \tag{3.57}$$

$$E(y_T \mid x_T) = \alpha_T + \beta_T x_T \tag{3.58}$$

Given any x_T, the value of x_S which gives the same response [i.e., $E(y_S \mid x_s) = E(y_T \mid x_T) = y_0$] is found to be

$$x_S = \frac{\alpha_T - \alpha_S}{\beta_S} + \frac{\beta_T}{\beta_S} x_T \tag{3.59}$$

Remember that the log ρ in previous sections is the difference between log doses giving equal responses; namely,

$$\log \rho = x_S - x_T = \frac{\alpha_T - \alpha_S}{\beta_S} + \left(\frac{\beta_T}{\beta_S} - 1\right)x_T \tag{3.60}$$

This means log ρ is a linear function of x_T, the log dose of T. Thus, the range of relative potency can be found by inserting the values of the interested dose range of the test preparation into Equation 3.60. Similarly, we can express log relative potency as a linear function of response y. Thus, using Equation 3.58 we can express x_T in terms of y:

$$\log \rho = \frac{\alpha_T}{\beta_T} - \frac{\alpha_S}{\beta_S} + \left(\frac{1}{\beta_S} - \frac{1}{\beta_T}\right)y \tag{3.61}$$

This expression is also very useful in practice. After we restrict y over the particular range of interest, the knowledge of relative potency will be obtained.

Cornfield (1964) reviewed the basic concepts of analytic dilution assays. He proposed that there are no invalid assays but only invalid inferences. Thus, nonparallel log dose-response lines are interpreted to mean that relative potency is defined as the ratio of doses leading to equal response. He gave the estimation formulae including confidence limits for log p. His numerical example was taken from a nonparallel log dose-response line assay of two different corticosteroid drugs.

VII.   RADIOIMMUNOASSAY

A.   Introduction

The literature on radioimmunoassay (RIA) statistics and data processing is large and growing rapidly. Yallow and Berson (1970) provided an excellent survey of the technical respects of radioimmunoassays in the book *Statistics in Endocrinology*. Immunoassay, unlike the traditional bioassay, is dependent on specific chemical reactions that obey the law of mass action and that are not subject to errors introduced by the biological variability of test systems. Nevertheless, they are so similar in structure that they need consideration from the viewpoint of bioassay.

Landon and Moffat (1976) gave a comprehensive survey of drug immunoassay. RIA gives reliable measure drug concentrations in the low nanogram/milliliter range or below and has ability to process large numbers of samples in a relatively short time. The technique has been applied in virtually all areas

Table 3.8 Schematic Illustration of the Composition of Tubes in a Radioimmunoassay of Drug A

---

1. Composition of tubes prior to incubation:

	Standards			Test
Drug A concentration (ng/ml)	0.039 Ag	0.078 Ag	20.0 Ag	? Ag
Constant amount of ^{125}I-drug A	Ag*	Ag* ...	Ag*	Ag*
Fixed amount of drug A antibody	Ab	Ab	Ab	Ab

2. Incubation under identical conditions to equilibration:

$$Ag + Ab \; \underset{k_2}{\overset{k_1}{\longleftrightarrow}} \; Ag\text{-}Ab \; + \; Ag$$

$$Ag* + Ab \; \underset{k_2}{\overset{k_1}{\longleftrightarrow}} \; Ag*\text{-}Ab \; + \; Ag*$$

Bound	Free
Fraction	Fraction

3. Separation of the bound and free ^{125}I-drug A.

4. The radioactivity of the bound fraction is measured using a gamma counter.

---

of preclinical and clinical pharmacology, including pharmacokinetics and pharmacodynamics. Thus, this section is focused on the application of RIA to determine drug concentrations.

Table 3.8 presents the experimental procedure of an actual RIA of drug A. In such immunoassay, the antigen Ag being assayed (which in Table 3.8 is drug A) and the appropriate antibody Ab, when brought together in the same system, lead to a reaction in which antigen and antibody bind to form an antigen-antibody complex. The competition for a limiting number of antibody binding sites between a fixed amount of labeled drug A (Ag*) and increasing concentrations of unlabeled drug A (Ag), reaches to the definition of a standard curve of the type shown in Figure 3.4. This reaction is governed by the law of mass action, so that the greater the concentration of unlabeled standard drug (column 1 in Table 3.9) in the reaction mixture, the less labeled drug is bound to antibody after incubation (column 2 in Table 3.9).

Statistics are involved in: 1) choosing a calibration function, the "standard curve" with which unknown potency (drug concentration) can be compared,

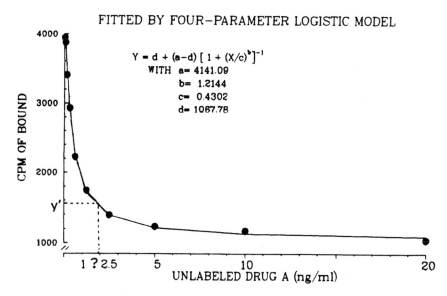

Figure 3.4 Typical RIA standard curve. (Plot of data in Table 3.9 fitted by four-parameter logistic model.)

Table 3.9. Mean Counts of the Standard Preparation in a RIA of Drug A

(1)	(2)	(3)	(4)
Concentration of unlabeled drug A (ng/ml)	Mean counts per minute of bound ^{125}I-drug A (with duplicate counts)	Estimated bound (cpm) of ^{125}I-drug A	% Deviation from observed
0.039	3942.5	3983.1	1.0
0.078	3869.5	3797.9	1.9
0.156	3403.0	3447.0	1.3
0.312	2932.5	2900.5	1.1
0.625	2225.0	2261.8	1.7
1.25	1742.5	1728.4	0.8
2.5	1393.5	1392.2	0.1
5	1232.0	1216.5	1.3
10	1173.5	1133.7	3.4
20	1044.0	1096.5	5.0

2) fitting this curve to the data, and 3) estimating inversely the unknown (test) antigen potency from results of (1) and (2).

B. Dose-Response Relationships in RIA

The widely used symbols in RIA are that B and F denote bound and free counts for the labeled drug at any dose (concentration) and T (= B + F) is the total count. At zero dose, B_0 is used for the bound count and at infinite dose the nonspecific count is N.

There are a large number of calibration formulae that have been proposed. Rodbard (1978) classified 13 methods into three categories: 1) empirical methods, 2) exact expression of mass action law, and 3) approximations to the first-order mass action law.

Empirical methods include linear (e.g., T/B versus dose, B/T or B versus log dose and logit (B/B0) versus log dose), polynomial (e.g., B, F, B/T or B/F versus dose or log dose), spline functions, exponentials, log-log and ratio of two polynomials. They have been very successful in practice. However, they lack a theoretical basis to provide information for the design of subsequent assays.

Exact expression of mass action law methods were developed by Yallow and Berson (1968) and Ekins et al., (1968). McHugh and Meinert (1970) derived a theoretical model for statistical inference in isotope displacement immunoassay. In practice, these methods may break down, because the real world systems may fail to satisfy the required assumptions.

The final category of curve-fitting methods is called approximations to the first-order mass action law. They have some theoretical justifications why they should fit calibration data. The most popular method is the four parameter logistic model, introduced by Heally (1972), as below:

$$y = d + (a - d)\left[1 + \left(\frac{X}{c}\right)^b\right]^{-1} \tag{3.62}$$

where y is bound count (B), X is dose and a, b, c, d are parameters. a and d correspond to B_0 and N, respectively, c is a dose for which the expected count is halfway between the two limits and b is related to the slope of the center of the curve. The curve can be rewritten as

$$\ln\left[\frac{y - d}{a - y}\right] = b \ln c - b \ln X \tag{3.63}$$

The model is applied to RIA, immunoradiometric assay (IRMA or labeled antibody assay), two-site IRMA, enzyme linked immunosorbant assay (ELISA), and enzyme multiplied immunological techniques (EMIT). This model is chosen to be the "standard curve" in this section.

C. Estimation of Four-Parameter Logistic Model

If a large body of data indicates that the variance of a count is proportional to U, that is,

$$\text{var}(u) = VU^J \tag{3.64}$$

where u is an observed count, U is the expected mean of u and the index J between 1.0 and 2.0 takes account of the rate at which variance increases with count, then the problem of heterogeneity of variance cannot be neglected.

Unweighted (assuming equal variance of raw count) and weighted (using reciprocal of var(u)) least squares estimates of a, b, c, and d can be produced from one of the commonly-used nonlinear curve-fitting computer programs. The initial values of a and d (a_0 and d_0) can be easily guessed from data, while the initial values of b and c can be obtained through fitting straight line $\ln [(y - d_0)/(a_0 - y)]$ versus $\ln X$. (see Eq. 3.63).

Using SAS PROC NLIN DUD method with initial values of $a_0 = 4000$, $b_0 = 1.2413$, $c_0 = 0.6487$, and $d_0 = 1000$, to fit data in Table 3.9, unweighted least squares estimates were obtained after 8 iterations:

$\hat{a} = 4141.09$ with SE 76.57 $\hat{b} = 1.2144$ with SE 0.0886
$\hat{c} = 0.4302$ with SE 0.0267 $\hat{d} = 1067.78$ with SE 40.81

The estimated bound count and percent deviation from observed are given in columns 3 and 4, respectively. The fitted curve is also shown in Figure 3.4.

Very similar results were found by using var(u) with J = 1.0, 1.5, and 2.0, to obtain weighted least squares estimates.

D. Potency Estimation

The major purpose of this particular RIA of drug A is to estimate inversely the concentration of drug A in test samples. If the observed bound count in one test sample is y′ cpm, then the concentration of drug A can be estimated as:

$$\hat{X} \text{ (ng/ml)} = \sqrt[\hat{b}]{(\hat{a} - y')/(y' - \hat{d})}\,\hat{c} \tag{3.65}$$

Suppose the test sample in Table 3.8 yields 2832 cpm, then the concentration of drug A is 0.337 ng/ml. The reader is referred to Finney (1976) for in-depth discussions on estimation of four-parameter logistic model and relative potency estimation for additional purposes. If a systemic pattern is shown in the residual plots, this implies that the chosen model was unsatisfactory. Considering the effects of sequential dilution errors, Racine-Poon et al. (1991) have developed a Bayesian procedure to increase the precision of the potency estimation.

REFERENCES

Barraclough, C. G. (1955). Statistical analysis of multiple slope ratio assays. *Biometrics*, **11**:186-200.

Berkson, J. (1953). A statistically precise and relatively simple method of estimating the bioassay with quantal response based on the logistic function. *J. Amer. Stat. Assoc.*, **48**:565-599.

Berkson, J. (1955). Maximum likelihood and minimum χ^2 estimates of the logistic function. *J. Amer. Stat. Assoc.*, **50**:130-162

Bliss, C. I. (1957). Bioassay from a parabola. *Biometrics*, **13**:35-50.

Brown, B. W. (1964). Lecture print on statistical procedures in biological assay. Stanford University.

Brownlee, R. A. (1965). *Statistical Theory and Methodology in Science and Engineering, 2nd ed.*, Wiley, New York.

Burn, J. H., Finney, D. J., and Goodwin, L. G. (1952). *Biological Standardization, 2nd ed.*, Oxford University Press, London.

Clarke, P. M. (1952). Statistical analysis of symmetrical slope-ratio assays of any number of test preparations. *Biometrics*, **8**:370-379.

Cochran, W. G., and Cox, G. M. (1957). *Experimental Designs, 2nd ed.*, John Wiley, New York.

Colquhoun, D. (1971). *Lectures on Biostatistics: An Introduction to Statistics with Applications in Biology and Medicine*, Clarendon Press, Oxford. pp. 319-327.

Cornfield, J. (1964). Comparative bioassays and the role of parallelism. *J. Pharmacol. Ther.*, **144**:143-149.

Cox, C. P. (1972). On estimating relative potency from quadratic log-dose response relationships. *Biometrics*, **28**: 875-881.

Ekins, R. P., Newman, G. B., and O'Riordan, J. L. H.(1968). Theoretical aspects of "saturation" and radioimmunoassay. In *Radioisotopes in Medicine, In Vitro Studies*. (Hayes, R. L., Goswitz, F. A., and Murphy, B. E. P., Eds.), U.S. Atomic Energy Commission, Oak Ridge, TN. pp. 59-100.

Elston, R. C. (1965). A simple method of estimating relative potency from two parabolas. *Biometrics*, **21**:140-149.

Finney, D. J. (1971). *Probit Analysis, 3rd ed.*, University Press, Cambridge.

Finney, D. J. (1976). Radioligand assay. *Biometrics*, **32**:721-740

Finney, D. J. (1978). *Statistical Methods in Biological Assay, 3rd ed.*, Hafner, New York.

Heally, M. J. R. (1972). Statistical analysis of radioimmunoassay data. *Biochem. J.,* 130:207-210

Hewlett, P. S. (1969). Measurement of the potencies of drug mixtures. *Biometrics,* 25:477-487.

Landon, J. and Moffat, A. C. (1976). The radioimmunoassay of drugs, a review, *Analyst,* 101:225-243

McHugh, R. B. and Meinert, C. L. (1970). A theoretical model for statistical inference in isotope displacement immunoassay. In *Statistics in Endocrinology,* (McArthur, J. W. and Colton, T., Eds.) MIT Press, Cambridge, MA. pp. 399-410

Racine-Poon, A., Weighs, C., and Smith, A. F. M. (1991). Estimation of the relative potency with sequential dilution errors in radioimmunoassay. *Biometrics,* 47:1235-1246.

Rodbard, D. (1978). Data processing for radioimmunoassay: an overview. In *Clinical Immunochemistry: Chemical and Cellular Bases and Applications in Disease,* (Natelson, S., Pesce, A. J., and Dietz, A. A., Eds.) U.S. Department of Health, Education, and Welfare, National Institute of Health. pp. 477-494

Snedecor, G. W. (1956). *Statistical Methods, 5th ed.,* Iowa State University Press, Ames, IO.

Taylor, W. F. (1953). Distance functions and regular best asymptotically normal estimates. *Ann. Math. Stat.,* 24:85-92.

Waisbren, B. A., Stern, M., and Collentine, G. E. (1975). Methods of burn treatment: Comparisons by probit analysis. *J. Amer. Med. Assoc.,* 231:255-258.

Wallenstein, S. L., and Houde, R. W. (1975). The clinical evaluation of analgesic effectiveness. In *Methods in Narcotic Research,* (Ehrenpreis, S. and Neidle, A., Eds.). Marcel Dekker, New York, pp. 127-145.

Waud, D. R. (1972). On biological assays involving quantal responses. *J. Pharmacol. Exp. Ther.,* 183:577-607.

Yallow, R. S. and Berson, S. A. (1968). General principles of radioimmunoassay. In *Radioisotopes in Medicine, In Vitro Studies.* (Hayes, R. L., Goswitz, F. A., and Murphy, B. E. P., Eds.) U.S. Atomic Energy Commission, Oak Ridge, TN. pp. 7-41

Yallow, R. S. and Berson, S. A. (1970). Radioimmunoassays. In *Statistics in Endocrinology,* (McArthur, J. W. and Colton, T., Eds.) MIT Press, Cambridge, MA. pp. 327-344

Yates, F. (1933). The principles of orthogonality and confounding in replicated experiments. *J. Agric. Sci.,* 23:108-145.

SUGGESTED FURTHER READING

Armitage, P., Bailey, J. M., Petrie, A., Annable, L., and Stack-Dunne, M. P. (1974). Studies in the combination of bioassay results. *Biometrics,* 30:1-9.

Ashford, J. R. and Smith, C. S. (1966). Models for the noninteractive joint action of a mixture of stimuli in biological assays. *Biometrika,* 53:49-60.

Ashton, W. D. (1972). *The Logit Transformation,* Hafner, New York.

Boen, J. R. and Brown, B. W. (1967). Use of prior information to design a routine parallel line assay. *Biometrics*, **23**:257-267.

Brown, B. W. (1966). Planning a quantal assay of potency. *Biometrics*, **22**:322-329.

Finney, D. J. (1965). The meaning of bioassay. *Biometrics*, **21**:785-798.

Govindarajulu, Z. (1988). *Statistical Techniques in Bioassay*, S. Karger Ag, Basel, Switzerland.

Hubert, J. J. (1984). *Bioassay. 2nd ed.*, Kendall-Hunt Pub. Co. Dubuque, Iowa.

Mantel, N. and Schneiderman, M. (1975). Nonparametric interval estimation of relative potency for dilution assays, including the case of nonmonotone dosage response curves. *Biometrics*, **31**:619-632.

Sen, P. K. (1971). Robust statistical procedures in problems of linear regression with special reference to quantitative bioassays. *Rev. Int. Stat. Inst.*, **39**:21.

4

The FDA and the IND/NDA Statistical Review Process

Satya D. Dubey

United States Food and Drug Administration,
Rockville, Maryland

I. THE FDA: WHY?

Food and drug laws have been a necessity to humankind since the beginning of civilization. Early Hebrew and Egyptian laws governed the handling of meat, Greek and Roman laws prohibited adding water to wine and short measures for grain and cooking oil, and in royal households, the "King's taster" protected the monarch from inferior or poisoned food. As civilization advanced, more complex protection became necessary. Apothecaries and food merchants of the Middle Ages organized as trade guilds to combat adulteration by inspecting spices and drugs. With the industrial revolution came an increasing use of chemicals, some of them harmful, such as poisonous food colors containing lead, arsenic, and mercury, and preservatives such as formaldehyde and borax. Such practices led the British Parliament in 1860 to pass the first nationwide general food law of modern times.

Records of the Massachusetts Bay Colony tell of Nicholas Knopf, who in 1630 was sentenced to pay a fine or be whipped for selling "a water of no worth nor value" as a cure for scurvy. Bread inspection laws were enacted by the American colonists soon after their arrival. Most of the early American food laws were for trade promotion. By providing for quality inspection of their salt

93

beef, port, fish, flour, etc., American merchants sought protection from having their exports rejected or downgraded on arrival overseas.

The first general law against food adulteration in the United States was enacted by Massachusetts in 1784; gradually, other states passed a variety of food and drug statutes. As the country expanded, however, it became clear that a national law was needed. Many states had no laws or lacked enforcement. Products that met the requirements of one state could be illegal in adjoining states and variations in labeling requirements became intolerable. From 1879 to 1906 more than 100 food and drug bills were introduced in the U.S. Congress. The first advocates of Federal legislation were state officials who knew the problems and the weaknesses of existing controls. It was the leadership of one remarkable man, Harvey Washington Wiley, Head Chemist for the Department of Agriculture's Bureau of Chemistry, which finally made food and drug protection a function of the Federal Government.

With support and encouragement from various segments of the drug and food industries, state governments, women's groups, writers, business organizations, and a host of crusading individuals, Congress passed the first national legislation designed to control impure and unsafe foods and drugs: The Pure Food and Drug (Wiley) Act of 1906. The administration of the law was assigned to the Bureau of Chemistry, which was headed by Dr. Wiley. Under Wiley's direction the Bureau continued the development of scientific methods of analysis, worked out the legal procedures and techniques of inspection, and applied them in hundreds of hard-fought court cases. They won scores of judicial interpretations which both strengthened the law and disclosed its weaknesses.

Laws and amendments following the Wiley Act have greatly increased the ability of the Federal Government to protect the U.S. consumer and to safeguard this nation's sources of food and drugs. It became quickly evident that this initial legislation did not have the necessary "teeth" to control many of the problems associated with the distribution and consumption of foods and drugs existing in the U.S. during the early part of this century. For example, the Supreme Court ruled in 1911 that the law allowed for false and unproven therapeutic claims as long as all of the ingredients were properly listed. In the following year (1912) in an attempt to correct this omission, Congress passed The Sherley Amendment prohibiting "false and fraudulent label claims." This amendment did not prove effective, however, for it placed the burden of proof as to what constituted fraud on government prosecutors; a distributor only needed to demonstrate that he "believed" that the product produced the advertised effect in order to escape prosecution for fraud.

In an effort to increase the visibility of the organization and to raise needed revenues, the Bureau of Chemistry became the Food, Drug, and Insecticide Administration in 1927. Four years later, in 1931, the name was changed to the Food and Drug Administration.

Motivated by the public clamor resulting from the infamous Elixir Sulfanilamide disaster, Congress passed the Food, Drug and Cosmetic Act in 1938. The Sulfanilamide incident resulted from the marketing of a medication containing diethylene glycol, a common component of antifreeze. The solution was prescribed as an antibiotic but produced fatal kidney failure that killed 107 people. This "elixir" was marketed without toxicological tests.

The Food, Drug and Cosmetic Act of 1938 greatly increased the power and responsibility of the Food and Drug Administration. Marking a basic change in the attitude of the Government to the regulation of drugs, this legislation required the preapproval of drugs by the FDA. In this new system, drug companies were required to submit evidence of drug safety. With this procedure the FDA had from 60 to 180 days to review an application. Failure to disapprove within the time period would lead to marketing.

The 1938 law required that the drug manufacturers list additional warnings and descriptions for use of marketed drugs. In addition, the Act required factory inspections, gave prosecutors the added weapon of court injunctions as a regulatory weapon, and simplified the prosecution of false claims by eliminating the need to prove fraud.

Amendments to the 1938 Act and regulations issued by FDA have further refined and improved the U.S. drug regulatory system. The Humphrey-Durham Amendment in 1951 gave FDA the authority to define and label prescription drugs and prohibit refills.

Under this law, labeling was available on request but was not routinely shipped with the product. In 1960, the FDA issued regulations requiring that detailed information on indications, dosing, and safety be included with drug packaging and in sales literature.

In 1962, in a response to the Thalidomide tragedy, the Kefauver-Harris Amendment was passed, further increasing the regulatory authority of FDA. This amendment required for the first time that drug sponsors demonstrate the efficacy of a drug by providing substantial evidence from controlled trials. The Kefauver-Harris Amendment also eliminated the passive approval system by making it a requirement that the FDA approve drugs prior to marketing. The 1962 Amendment also required that adverse drug reactions be reported to FDA, tightened IND (Investigational New Drug Application) provisions requiring informed consent, and gave FDA the authority to regulate advertising for prescription drugs

In 1983, the Orphan Drug Act was passed to provide incentives for the pharmaceutical industry to develop drugs for relatively rare diseases. The 1984 Price Competition/Patent Term Restoration Act provides for increased patent protection to compensate for patent life lost during the approval process and simplified the approval of generic drugs. In responding to criticisms concerning the length of the review process, the FDA issued the "NDA Rewrite" in 1985, with new and revised drug regulations that were designed to improve the content and format as well as the processing of NDAs (New Drug Applications).

In 1987, recognizing the health crisis brought on by the AIDS epidemic, the Agency issued the "Interim Regulatory Procedures" (Federal Register, Part VI, 21 CFR Parts 312 and 314, page 41516). These Interim Procedures make it possible for more seriously ill patients to receive promising experimental drugs, while preserving appropriate guarantees for safety and effectiveness. These procedures reflect the recognition that the benefits of a drug need to be evaluated in light of the severity of the disease being treated; physicians and patients are willing to accept greater risks from products that treat life-threatening and severely debilitating illnesses than they would accept from products that treat less serious illnesses.

With this rule, the expanded availability of drugs for "immediately life-threatening conditions," can begin near the end of the second phase of human testing. In this way the drug would become available as soon as the initial safety evidence was on-hand and the proper dose had been determined (Phase I) and after some evidence of efficacy had been obtained (Phase II). Under these procedures, it is also possible that, with early evidence for efficacy and safety, drugs for "serious but not immediately life-threatening illnesses" can be approved for expanded use during Phase III trials. If FDA approval is gained on the basis of limited but sufficient evidence from clinical trials, it will usually be important to conduct postmarketing (Phase IV) clinical studies to extend the knowledge of the drug's safety and efficacy, thus allowing physicians to optimize its use (21 CFR, Part 312.85, page 92).

The body of law currently defining the FDA's authority is comprehensive, providing a variety of controls required by the nature of the market, the product and attendant health risk. For many years, the Agency's job consisted almost entirely of inspections aimed at uncovering adulterated and impure products and exposing fraudulent labeling. However, in addition to this traditional monitoring role, the FDA today serves an important role as the "gatekeeper" for new drug technologies; it applies its substantial scientific resource base to the premarket evaluation and approval of new drugs, and to the postmarketing monitoring of drug labeling, advertising, and quality control.

The demand for the FDA's regulatory role in the marketing of drugs, as expressed in U.S. laws, amendments, and regulations, has evolved based on the needs of the drug industry, the public, the scientific community, the courts, and the Federal Government. As this brief historical description on the "Why" of the FDA's existence illustrates, American drug laws and the FDA owe their existence to a fundamental belief that drug companies cannot be trusted to assure the safety and efficacy of their products. Unfortunately, there are enough examples in the past and present of dangerous and ineffective drugs on the market to perpetuate this mistrust. At the same time it should be remembered that a combination of forces—scientific, regulatory, economic, medical, and legal—work together to assure the safety and efficacy of the American drug supply. None of these elements alone, should be considered sufficient to provide the margin of control that the American public expects and demands.

A.   The FDA Today

The FDA has been described as the "principal consumer protection agency of the Federal Government." In fact, it has been said that U.S. consumers spend approximately 25 cents out of every dollar on products regulated by the FDA. People take for granted the wholesomeness of the food they buy and the safety and efficacy of the drugs they use; the FDA's goal is to see that this confidence is warranted by ensuring industry's compliance with Federal laws regulating products in commerce (U.S. Department of Health and Human Services, 1987, page 7).

In simplest terms, the provisions of the food and drug laws are intended to ensure:

1.   Food is safe and wholesome;
2.   Drugs (both human and veterinary), biological products (e.g., vaccines and blood for transfusion), and medical devices are safe and effective
3.   Cosmetics are unadulterated;
4.   The use of radiological products does not result in unnecessary exposure to radiation; and
5.   All of these products are honestly and informatively labeled.

The FDA is an agency of the Public Health Service. It is administered by a Commissioner who is appointed by the Secretary of the Department of Health and Human Services. To perform its mission the Agency is organized into six centers.

The review and evaluation of safety and efficacy of drugs, a primary focus of this discussion, is the responsibility of the Center for Drug Evaluation and Research (CDER); the CDER is divided into 8 offices, including:

Office of the Director;
Office of Management;
Office of Epidemiology and Biostatistics;
Office of Compliance;
Office of Drug Evaluation I;
Office of Drug Evaluation II;
Office of Drug Standards; and
Office of Pharmaceutical Research Resources.

B. The Division of Biometrics

The Division of Biometrics, which is part of the Office of Epidemiology and Biostatistics, provides comprehensive statistical and computational services to all programs of both the Center for Drug Evaluation and Research (CDER). The Division includes two Branches, the Statistical Evaluation and Research Branch (SERB) and the Statistical Application and Research Branch (SARB), and employs more than 60 statisticians and support personnel. Statisticians in the Division of Biometrics:

Review IND, NDA, and nonclinical submissions developed by regulated industry;
Conduct research and development in statistical, biomathematical, computational and other scientific decision making methodologies;
Develop statistical methodologies for analyzing clinical, preclinical and epidemiological data; and
Evaluate, develop, maintain and apply statistical computer systems (hardware, software, and procedures) for the review and evaluation of clinical trials.

Support of CDER programs is allotted to the Division's two Branches. The SERB is primarily responsible for statistical review and research activities pertinent to clinical trials (i.e., IND and NDA submissions). The SARB provides review and consultative services for several CDER programs including preclinical submissions, epidemiological studies of postmarketed drugs involving adverse reaction data and poison and toxic hazards data, bioavailability and bioequivalence studies, and laboratory compliance programs.

The CDER Research Steering Committee was established in order to enable the FDA to participate in clinical research to validate research conducted by others, to develop new methods of drug testing, and to investigate drug issues not of interest to commercial sponsors. Three local university hospitals (Georgetown, Johns Hopkins, and Maryland) have been awarded contracts with the FDA in order to provide the CDER with clinical research facilities for clinical pharmacology.

II. THE IND AND NDA REVIEW

The primary responsibility for ensuring the safety and efficacy of drug products belongs to the manufacturer. The FDA's role is to monitor manufacturers and thereby provide consumers with assurances that the drug industry is meeting its responsibilities to the public. As one of its primary functions, the FDA evaluates sponsor-tested efficacy and safety claims for new drugs. To accomplish this premarketing review function, CDER scientists (medical officers, chemists, pharmacologists, statisticians, etc.) are responsible for the evaluation of IND and NDA submissions.

The IND must be submitted to the FDA prior to the testing of new drugs on humans. The IND contains the plan for the study and it is intended to give a complete picture of the drug including its structural formula, animal test results, manufacturing information, etc.

The NDA, on the other hand, is the application requesting FDA approval to market a new drug for human use in interstate commerce. The NDA must contain substantial evidence from controlled clinical trials and incorporates data for each of the scientific disciplines included in the review system (chemistry, pharmacology, medicine, biopharmaceutics, statistics, and, for antiinfectives, microbiology). The NDA is a comprehensive document designed to present the rigorous proof required to demonstrate the safety and efficacy of a new drug.

It has been estimated that, on average, it takes approximately 100 months to get a new drug on the market, starting from initial synthesis and proceeding through animal testing, Phase I, Phase II, and Phase III trials, and the NDA review process. The average approval time of an NDA after initial submission is 24 months, but it can range from two months to seven years (An FDA Consumer Special Report, 1988, p. 5).

Clinical trials for drug approval are conducted in three phases. Phase I trials are designed to learn more about the safety of a new drug, although they may also collect some information concerning efficacy. These trials are normally done in healthy volunteers who are paid to receive the drug and submit to tests measuring what happens to the drug in the human body—how it is metabolized and excreted, what effects it has on various organs and tissues, etc. Phase II studies are designed to determine whether the drug is effective in treating the disease or condition for which it is intended and short-term side effects and risks in people whose health is impaired. These studies are generally randomized controlled trials.

Much, however, is still left to be learned after the completion of Phase II trials. Phase I and II trials are not designed to provide sufficient information concerning optimum dose rates and schedules, safety and efficacy. Phase III trials—which are randomized, controlled, usually blinded, and are conducted on larger patient populations under conditions that more closely approximate medical practice—provide the replicated scientific evidence required for the approval of a new drug. Recent experience has shown that of 100 drugs for which INDs are submitted to FDA, approximately 25–30 percent successfully complete Phase III trials and are approved by the FDA (An FDA Consumer Special Report, 1988, p. 14).

III. THE STATISTICAL REVIEW PROCESS

With the completion of Phase III trials, sponsors submit the NDA to the FDA for marketing approval. The NDA Rewrite, which went into effect in 1986, established a thorough review process designed to insure the approval of drugs scientifically demonstrated to be safe and effective, and the disapproval of those drugs which are not. One of the measures implemented in the Rewrite was the requirement that sponsors submit separate NDA volumes to six different scientific review disciplines, including statistics. This procedure facilitates the parallel review of NDAs by the disciplines involved in making regulatory decisions on the safety and efficacy of new drugs.

The specific format and content of the technical sections submitted to the various disciplines are discussed in their respective guidelines. The format and content of the statistical section of an NDA are discussed in a 1988 publication entitled "Guideline for the Format and Content of the Clinical and Statistical Sections of New Drug Applications" (U.S. Department of Health and Human Services, 1988).

This guideline illustrates what medical and statistical reviewers would like to see in the submission and how they would like to see these contents presented. Clinical and statistical sections conforming to this guideline greatly facilitates the review of NDAs by medical officers and Statistical Evaluation and Research Branch statistical scientists.

A. The Institutional Statistical Review Process

It is informative to examine the course of a statistical review after it has been received by the Division. When a new NDA is received by the SERB, a group leader assigns the submission to a statistical reviewer and works closely with the reviewer to evaluate the evidence contained in an NDA and to produce the draft review. The draft review, in turn, receives a peer review by the group leader and branch chief.

When an individual reviewer receives an assigned NDA, he or she often starts from scratch. He or she begins by reading the cover letter and index to determine the drug class, proposed indication, and the types of studies included in the NDA submission. Following this initial look, the reviewer locates and examines background information on the disease the drug is intended to treat, the properties of the relevant class of drugs, the sponsor's drug developmental program, and previous correspondence between the sponsor and FDA.

Following the initial review of background materials, the statistical reviewer, generally, meets with the medical reviewer to obtain essential clinical perspective on the disease, the drug, and the studies. From here, he or she examines the Integrated Summary of Effectiveness Data, the Integrated Summary of Safety Data, and summaries of the controlled studies (detailed descriptions of these documents can be found in the Guideline for the Format and Content of the Clinical and Statistical Sections of New Drug Applications). After reviewing these materials and having discussions with the medical officer, the statistical reviewer tentatively identifies those well-controlled studies considered to contain primary scientific evidence relating to the efficacy and safety of the new drug. Such studies may or may not agree with the studies identified by the sponsor as "pivotal."

At this point the reviewer is ready to begin the detailed evaluation (entitled the "Full and Integrated Clinical and Statistical Report of a Controlled Clinical Study" in the Guideline) of the individual study reports and gives priority to the most important studies. From the protocol contained in each of these reports, the reviewer should be able to determine the primary objective of the study, the type of study design, the randomization scheme used, the degree of blinding, dose levels and frequencies, primary efficacy parameters, important safety parameters, and methods of analyses. The reviewer must determine whether or not the study design can support the proposed indication and whether or not the sample size is adequate (particularly for active control trials).

In addition, the reviewer examines the randomization scheme and codes and the extent of error in treatment assignments. To the extent possible, he or she also examines the quality of the data, the extent of noncompliance, missing values, and dropouts. The reviewer also pays special attention to sources of bias and confounding. If sources of bias or confounding are discovered, the reviewer must attempt to quantify these factors and assess the potential impact on the primary efficacy parameters.

The statistical reviewer identifies all of the data sets used by the sponsor in the analysis of primary and supportive secondary efficacy measures. For example, data sets should be identified which include evaluable patients, efficacy patients, patients randomized to a treatment, and those used for intent-to-treat analyses. The reviewer must know the inclusion/exclusion criteria used in forming each of these data sets. If there are questions concerning inclusion/exclusion criteria, differential dropouts and reasons for dropouts, the reviewer must determine the magnitude of potential biases and assess the impact of these biases on efficacy measures.

The reviewer also determines the types and appropriateness of the analyses performed by the sponsor to statistically demonstrate efficacy and safety. The statistical reviewer conducts a reanalysis or alternative analysis, if necessary, on the same or modified data sets.

In the review of evidence relating to safety, the reviewer identifies the data sets used by the sponsor in presenting the safety profile of the drug (e.g., a data set of all treated patients). Paralleling the review of the efficacy data, the SERB statistician must judge the appropriateness of the safety analyses and, if necessary, perform reanalyses or alternative analyses on the same or modified sets of data. Incidentally, an informal FDA survey of 53 NDAs evaluated by SERB showed that reanalyses were performed in 83% of the cases.

The statistical reviewer does this for each of the important studies. The depth of this review varies from study to study and from one NDA to another, and depending on a variety of factors, such as, the quality of the data, the conduct of the trial, the completeness of the individual study report, and any other important clinical or statistical issues.

After all of the important studies and supportive secondary studies have been reviewed, the statistical reviewer assesses the overall safety and effectiveness of the drug for the proposed indication. A statistical review is then written to summarize the findings and draw conclusions based on the strength of statistical evidence and other relevant factors.

The statistician's review is then given to the Group Leader who reviews the report for scientific adequacy and consistency with CDER and Agency policy. Differences in opinion between the reviewing statistician and the Group Leader are discussed. Should the disagreement remain unresolved, the Group Leader records the basis for the disagreement and a recommendation for action by the Branch Chief.

The Branch Chief reviews the report and the Group Leader's recommendations for soundness of the statistical evaluations, conclusions and recommendations, overall scientific content, and adherence to official policies and administrative procedures, and resolves all essential scientific and other issues. The report and all recommendations are then forwarded to the review team leader of the appropriate medical drug review division.

B.   The Computer-Assisted NDA Review

The Computer-Assisted NDA (CANDA) Review Project is an on-going effort by the Agency to make more efficient and effective use of the computer as a regulatory review tool. The CANDA Review Project has two primary goals:

1.   The reduction of review time without sacrificing the quality of review in order to expedite regulatory decisions on new drugs and
2.   The reduction of the paper burden arising from large NDAs by creative applications of computer technology.

The Agency has defined CANDA Review rather generally as "any submission that involves provision of information to FDA or that permits remote access by FDA in any electronic medium."

CANDA Review Project activities have concentrated primarily on the design of prototype microcomputer CANDA Review systems designed to back-up the paper submission of large NDAs. Sponsors, working with FDA reviewers, have implemented a variety of systems designed to enhance the ability of FDA scientists to electronically access and review the evidence presented in NDAs. Features of these systems include, for example:

On-line access to a sponsor's mainframe computers;

Electronic messaging systems;

Access to a sponsor's program and data files on microcomputer data bases;

Optical disk systems for review of case report forms;

Data base management software for review-oriented subsetting and modification of sponsors' data sets;

Statistics, word processing, and graphics software capable of verifying and reanalyzing sponsor's results and producing reviews.

A great deal has been accomplished to date but it is clear that more energy must be devoted if the objectives of the CANDA Review Project are to be reached (see also Chapter 19). With an eye towards the future, it is useful to examine our expectations for this project. They include:

1. The development of a user-friendly, enhanced version of the CANDA system that effectively integrates both the clinical and statistical aspects of the review process for the across-study evaluation of submissions (in accordance with the NDA Rewrite requirements).

2. The development of computer systems to access NDA clinical data, both within and across studies of different designs, should require more a priori attention to what data is collected, how it will be stored, summarized, retrieved, and used by industry and regulatory scientists. This should improve the structure of clinical data systems and the accessibility of any suitable data subset.

3. NDA Rewrite requirements for tables, data display, efficacy analysis, safety evaluation, and a synthesis of effectiveness and safety findings on a new drug should enhance the role of CANDA within the context of both the clinical and statistical review processes.

4. Statistical software should be developed, specifically designed to validate and verify statistical analyses submitted in an NDA.

5. The NDA Rewrite is increasing the intensity of interaction among the primary, secondary, and tertiary statistical reviewers as part of the peer review process. Also, the integrated clinical-statistical guideline is improving interaction among the medical and statistical reviewers. Experimental innovations in the review process, such as "NDA Day," involve intensive interaction among all FDA reviewers of an NDA. Consequently, the CANDA Review process will prove more efficient in performing the institutional review of an NDA, and thereby reduce the review time without sacrificing the quality of statistical reviews.

6. More experience is needed in dealing with the specific concerns of different drug classes with microcomputers. Before we can propose a standardization of NDA data bases or procedures, intensive and extensive interaction between the FDA reviewers and drug sponsors should continue. This

should result in the development of several strategies and designs appropriate to the drug and disease in the light of clinical and statistical review needs.

III. CONCLUSION

Industry and the FDA must, out of necessity, work closely together in conscientiously maintaining and strengthening the public trust in the safety and efficacy of the nation's drugs. It should be prominently noted that without voluntary efforts by a majority of drug sponsors to meet and exceed the requirements of the law, the FDA would never have enough staff or resources to enforce its requirements. It is the responsibility of the FDA and industry statisticians to work together to ensure that drugs are marketed only after rigorous scientific evidence has demonstrated that they are both safe and efficacious. The American public demands and expects this type of professional dedication.

Researchers in the Division of Biometrics have been active in writing, presenting and publishing papers on the various aspects of regulatory statistical research. The list of publications is a long one and even a brief summary of these papers is beyond the scope of this chapter; they are published in pertinent statistical and scientific journals. Readers are encouraged to review these published papers with a view to extending, enlarging, as well as enhancing their perspectives and gaining greater insights into our evaluation process of the IND/NDA submissions.

ACKNOWLEDGMENT

The views expressed in this chapter are those of the author and not necessarily those of the U.S. Food and Drug Administration. This work was produced by the author in his capacity as a Federal employee and, as such, is not subject to copyright.

REFERENCES

Chi, G. Y. H., M. Huque, T-H. Ng, and S. D. Dubey, (1989), A preliminary assessment of two evolving canda review systems from the perspective of FDA statistical reviewers, *1989 Proc. Biopharmaceut. Sect. Amer. Statist. Assoc.*, American Statistical Association, Washington, DC, pp. 36-45.

Dubey, S. D., (1989), Computer assisted new drug applications (CANDA) and statistical reviews, *1989 Proc. Biopharmaceut. Sect. Amer. Statist. Assoc.*, American Statistical Association, Washington, DC, pp. 26-35.

Farley, D., (1988), Benefit vs. risk: How FDA approves new drugs. In *From Test Tube to Patient: New Drug Development in the United States*, FDA Special Consumer Report, U.S. Department of Health and Human Services Publication No. (FDA) 88-3168, Rockville, MD.

Flieger, K., (1988), Testing in real people. In *From Test Tube to Patient: New Drug Development in the United States*, FDA Special Consumer Report, U.S. Department of Health and Human Services Publication No. (FDA) 88-3168, Rockville, MD.

Janssen, W. F., (1986), *The U.S. food and drug law: How it came, how it works*, U.S. Department of Health and Human Services Publication No. (FDA) 86-1054, Rockville, MD.

U.S. Department of Health and Human Services, (1987), *All About FDA—An Orientation Handbook*, Public Health Services, Food and Drug Administration, Rockville, MD.

U.S. Department of Health and Human Services, (1988), *Guideline for the Format and Content of Clinical and Statistical Sections of New Drug Applications*, Rockville, MD.

U.S. Food and Drug Administration, (1985), *Staff Manual Guide*, CDB 1269.4, Rockville, MD.

U.S. Government, (1989), Code of Federal Regulations, 21, Parts 300 to 499, Revised as of April 1, 1989, Washington, DC.

Young, F. E., (1988a), The reality behind the headlines. In *From Test Tube to Patient: New Drug Development in the United States*, U.S. Department of Health and Human Services, Publication No. (FDA) 88-3168, Rockville, MD.

Young, F. E., (1988b), Experimental drugs for the desperately ill. In *From Test Tube to Patient: New Drug Development in the United States*, U.S. Department of Health and Human Services Publication No. (FDA) 88-3168, Rockville, MD.

An FDA Consumer Special Report, *From Test Tube to Patient: New Drug Development in the United States*, HHS Publication No. (FDA) 88-3168, pp. 5, 14, January, 1988, Rockville, MD.

5

Clinical Trial Designs

C. Ralph Buncher and Jia-Yeong Tsay[*]

*University of Cincinnati Medical Center,
Cincinnati, Ohio*

Designing clinical trials of pharmaceutical products shares many characteristics with other types of scientific study. The unique characteristic of clinical trials is that the experimental units are other human beings—and usually sick ones at that. Modern clinical trials must be designed in such a way that the participants can be well-informed about the conduct and purpose of the trial, and that the welfare of each participant remains more important than carrying out the trial as designed. The basic purpose of the trial is to obtain pharmaceutical information that is applicable to the people who will take the drug in the future. In short, the design must be such that the interests of the individual are balanced with the interest of the group (Buncher, 1972).

A century ago volunteers were given diseases such as yellow fever and malaria in experimental conditions and some of these persons lost their lives. This predictable loss of life of a few of the volunteers resulted in the saving of the lives of thousands of other persons because of the more rapid scientific advances made from those experimental studies. Acceptable research behavior, however, has drastically changed because of greatly improved mortality rates, changing moral concepts in the twentieth century, and also as a reaction to some prior abuses in clinical trials. Suffice it to say that at the current time in the United States and other countries with comparable research philosophies, a primary characteristic of all clinical trials is a concern for the safety of the volunteer participants. The general principle of informed consent—that all

---

[*] *Current affiliation*: Organon Inc., AKZO Pharma Group, West Orange, New Jersey

107

participants will be fully informed of the characteristics of the trial—is now universally accepted although there are continuing efforts for improvement (Simel and Feussner, 1992).

Each class of drugs generates unique characteristics for clinical trials. For example, antibacterial drugs tend to have an action that can be measured in days and safety that can be measured in months, while a cardiovascular drug produces an action measured in months and safety that is measured in years. This chapter will discuss some of the fundamental characteristics of all clinical trials and leave to the later chapters the specific traits within a few of the unique fields.

I. CONTROLS

The first characteristic is that any clinical trial must be controlled. In the scientific sense, this means having some objective means of evaluating the effectiveness of the drug under study; it is preferable that the comparison be provided in the same clinical trial (i.e., by a concurrent control). The most common type of control is a placebo control, a type of study in which at least some of the study participants receive a preparation with the physical appearance of the drug but none of its pharmacological properties. This type of trial is considered ethical in those instances where there is no standard treatment and where placebo is well known to cure a high proportion of persons (e.g., pain relief or mood enhancement). Chapter 9 provides more information on analgesic trials.

Alternatively, a study may use a positive control for the disease under study. An example would be an antibiotic or an antihypertensive drug known to be effective which could be used in instances where it would not be ethical to deny treatment to participants. Other control possibilities which are much less frequently used are historical controls and no control.

II. MASKED EVALUATION

Another characteristic of the clinical trials is that all evaluations must be made without bias. Painful experience has proven that patients and clinical evaluators have subjective opinions and views of life that can affect the outcome in most trials. These potential biases must be thwarted by proper experimental design. The most common design is to make the participants and evaluators in a clinical trial act as if they were blindfolded or masked as to the identity of which medication is which in the trial. These trials are most commonly called *double-blind*, short for double-blindfold, although "double-masked" is a preferable term (Ederer, 1975). This is accomplished by making the medications look exactly alike and be alike in all characteristics that might be tested save only the chemical differences.

Moreover, the labeling must be done in such a way that a person having access to the complete set of labels would not know which medication is which and therefore could not tell whether "active" medication would be given to the next participant. This is usually accomplished by numbering the medications sequentially (1, 2, 3, 4, . . .) and having separate codes, which are available only in an emergency situation, to describe which number is equivalent to each medication. A few decades ago, the drug might be labeled A1, A2, A3, etc. and the placebo B1, B2, B3, etc., but now such systems are not considered adequate.

Each container of medication used in interstate commerce must be labeled with the contents, but the labels used in pharmaceutical trials are generally a variety of sealed labels such that one can only determine the medication by breaking the seal (often a glue sealed by heat). If the label is opened, it can not be resealed again. Obviously, the labels are also made so that holding them up to the light does not permit reading what is inside. Thus, return of the sealed labels is one important piece of evidence that no one has discovered which medication was taken by that patient. Clearly, the degree to which the placebo matches the active medication and other characteristics of the trial also contribute to the proof that no one knew which medication was which.

Most trials are double-blind (or double-masked) in the sense that neither the participant nor the evaluating clinical personnel knows which medication is which. It is also possible for trials to be single-blind. In the more common situation the clinician in early trials knows which is the medication and which is the placebo and is therefore more able to monitor progress of the trial and side-effects; however, the participant in the experiment does not know which medication is which. In some single-blind situations—for example, when the medication can't be blinded, as in the comparison of a surgical treatment and a drug treatment—the patient is aware of which treatment is which but the evaluating clinical personnel are "blindfolded." Again, this is a single-blind trial. Finally, some studies are done on an open-label or unblinded basis, for example, when all concerned are aware that an experimental drug is being used.

III. PHASE I AND PHASE II

Phase I and Phase II clinical trials are usually conducted by highly trained clinicians (called clinical pharmacologists) who make the study of therapeutic medications their specialized area in medicine. In these trials the first information obtained concerns the degree of safety of the drug; subsequently, the effectiveness of the drugs in humans is studied. Usually these trials are sufficient to select the optimal dose or doses of the drug for usual clinical use. Also, the common side-effects are discovered on a subjective basis (e.g., dizziness and nausea), as are common biochemical side-effects (e.g., changes in liver enzymes or sodium levels).

There are interesting design problems associated with these early drug trials. Kaitin et al. (1991) provide some useful insight. Since the audience for these trials is small and specialized, we have not devoted much space to them. Instead we shall discuss general problems of trials and several of the problems encountered in later drug trials.

IV. PHASE III

Phase III clinical trials are the definitive studies that prove the effectiveness and safety of the medication; these trials also provide the detailed comparisons which verify the earlier results. In addition, less common side-effects are discovered during these larger, more extensive trials.

For a number of reasons, Phase III trials are usually conducted by general clinicians or specialists in a particular clinical field rather than by those specifically trained in clinical trials. First, there is the desire to use the drug in a more typical environment rather than in the highly controlled situation of a clinical pharmacologist in a research-oriented setting. Second, one must find clinicians who have larger numbers of patients in order to provide the numbers of drug recipients necessary to verify efficacy and safety. Finally, it must be pointed out that later clinical trials are rarely of as much scientific interest as the earlier trials and thus do not as frequently lead to medical journal articles, the basic currency of a "publish or perish" scientific standard.

The clinicians who do participate are motivated by other factors. These factors include financial benefits for the participants and investigators, access to a drug which would not be available outside the research program, the opportunity and status of participating in a research program, and doing a favor for an old friend. Clearly the research design must be robust in order to counter any biases that could creep into a study because of these reasons for participating in it.

One of the issues that makes drug approval at the FDA slower is the question of safety. After a drug is approved and used by thousands or even millions of people, some side-effects that were not known at the time of approval may be discovered. An alternative is to do studies of side-effects after the drug is approved for marketing. These postmarketing surveillance studies are generally called Phase IV studies although this is not an official FDA definition. In fact, interventional clinical postmarketing studies are often called Phase IV to distinguish them from Phase V which consists of observational postmarketing studies (Chapter 15). The observational studies are not randomized and fall more into the realm of epidemiology than biostatistics.

V. RANDOMIZATION

Randomization serves four important purposes: i) it avoids known and unknown biases on the average in the assignments to treatment groups; ii) it balances known and unknown prognostic factors on the average; iii) it helps convince others that the trial was conducted properly; and iv) it is the basis for the statistical theory that allows us to calculate probabilities.

In general, patients are assigned to treatments based on a randomization scheme. There are many ways of allocating drug names to the labels numbered in order. While many methods are suggested in textbooks and pamphlets, those who frequently randomize use a random permutation scheme. Random permutations are the most general scheme for randomization, and they can be tailored to almost any design requested.

Consider the numbers 1, 2, 3, 4. There are 24 different orderings of these four numbers (e.g., 1, 2, 3, 4; 1, 2, 4, 3; 1, 3, 2, 4; and so on through 4, 3, 2, 1). If four random numbers are generated, then the rank of the magnitude of these numbers can be considered as an ordering. For example, suppose the random numbers were 378, 842, 103, and 927; then these random numbers considering their ranks from smallest to largest would have generated the permutation 2, 3, 1, 4. Since each rank had equal chance of occurring in each position, each of the 24 possible orders would have had equal opportunity of being chosen.

To apply the random permutation technique to choose two persons to receive the active drug and two to receive the placebo, one decides before starting that numbers 1 and 2 will be active drug, for example, and 3 and 4 will be placebo. The random permutation then chooses which persons will receive which drugs. The extension of this scheme to larger sizes should be obvious. The scheme can be done by hand with ease in small trials, programmed into a computer, or found in software packages.

Importantly, if one is trying to allocate 50 persons onto drug and 50 onto placebo, this scheme (allocating the lowest 50 random numbers to drug and the higher 50 random numbers to placebo) guarantees that exactly 50 persons will be randomly assigned to each treatment. Moreover, if there are three drugs or there are unequal numbers of persons per drug, then it is still easy the guarantee exactly the results desired. Thus, if one wanted 30 on positive control, 50 on test drug, and 20 patients on placebo, one need only specify that the smallest 30 random numbers will indicate the positions of the patients to receive the positive control, the next 50 will indicate test drug, and largest 20 will indicate placebo.

Changes over time during the course of a designed trial are always a possibility and, for various reasons, a trial may even stop before all patients have been entered. The statistician can make the design more robust against these contingencies by randomizing the total study into blocks or subunits that are balanced at the end of each block. For example, in the three-drug study cited, 10 blocks of 10 patients each would guarantee the desired balance at the end of every 10th patient. In this way, randomization is carried out within each block, which is one form of "restricted randomization." In like manner, in the 100 patient study with two drugs, one could create blocks of size 10 or 20 (or even 8 or 12 with one smaller block) so that the study would be balanced at the end of each of these intervals. From 6 to 20 patients per block are effective block sizes. Frequently, one also randomizes the block sizes to protect against investigators guessing the final treatments in a block. There could be one or multiple blocks per center in a multicenter trial. Of course, one need not tell any of the participants in the study that blocking exists, and given the complexities of the usual trial they will not notice it.

Statisticians have rejected alternatives to randomization for decades, although there is an active field of research in useful modifications to strict randomization. For example, Royall (1991) examined ethical and scientific dilemmas in randomized clinical trials and urged statisticians to improve the statistical methodology of nonrandomized clinically trials when randomization is not ethically or practically feasible. An excellent source is the December, 1988 issue of Controlled Clinical Trials (volume 9, number 4) which features a number of articles about randomization, mostly written by John M. Lachin.

An area of research that is still developing is unequal randomization. Do the two or more groups have to have the same number of patients? This type of clinical trial is not unusual outside of the pharmaceutical industry. Thus if the group is somewhat convinced that treatment A is better than treatment B, it is sensible to randomize so that more patients end up on the better treatment (treatment A). A simple such scheme is a play the winner rule (Wei, 1988). If treatment A results in a success, then treatment A is also used on the next patient. When treatment A results in a failure, then treatment B is used on the next patient. If this process is continued for the whole trial, the result is that more of the patients will have used the better treatment. Some of the literature in this field is found under the title of biased coin or adaptive randomization (Simon, 1977). Although these designs are of great interest to theoreticians and are very useful in certain instances, they are not used very much in the pharmaceutical industry. The schemes are more difficult to use, introduce problems when trying to convince others of the efficacy of the product, and standard designs are adequate.

VI. STATISTICAL DESIGN

The actual design of a trial that explains which patient receives which drug (called "treatment" in the jargon of design of experiments) can be classified in one of two general categories. The trial may be a parallel design in which each patient receives one and only one treatment (although that treatment may be given at more than one dose) or a crossover design in which the same patient is given two or more different drugs. In this latter case, the patient serves as her own control, since the comparison of treatments is made within the patient.

Some time ago, crossover designs were considered among the best in pharmaceutical research. The theoretical advantage of smaller sample size with a crossover design was emphasized and also the advantage of making comparisons within patients, which is especially appealing to clinicians. However, problems with crossover designs are also important. Suppose there is an adverse effect found 1 week after the end of the study. How do we know which drug, if either, should be held responsible? As an extreme example, if 10 patients died after a crossover study, all we would know is that each took each medication. On the other hand, in a parallel study, we might know that 5 patients were on placebo and 5 on the test drug or that 9 patients were on the test drug and only 1 was on placebo. Clearly the parallel design offers the potential of more information about side-effects. The advantages and disadvantages of crossover designs are discussed more completely in Chapter 10 and also in Chapter 12 with regard to bioavailability studies. The U.S. Food and Drug Administration (1992) recommends the use of two-treatment crossover designs for bioequivalence studies.

All the usual designs available in the statistical design of experiments are also candidates for pharmaceutical trial designs, for example, the completely randomized and the randomized block designs in which patients are assembled so that those in a "block" resemble each other more than they resemble patients in other blocks. For example, in a study of pain one might use as a blocking factor the type of operation, or the gender of the patients, or the degree of severity of the pain, or a combination of all three. The extreme example of this type of design is the matched pair design in which two persons are selected who are very similar and then drug and placebo are randomly allocated one to each member of the pair.

Latin square designs are also frequently useful if their size is small enough. A Latin square is a type of randomized block design in which there are two simultaneous blocking factors. Thus, patients could be blocked on degree of pain in three classifications and in three classifications of type of operation. Frequently the second blocking factor is the time period in which the drugs will be given. For example, in a study of the pain of rheumatoid arthritis, one could design three different levels of impairment and then have three pain relievers (new drug, placebo, positive control) given for 1 week each over a total of 3 weeks of treatment. This is also another example of a type of crossover trial.

Efficient experimental designs defined according to certain statistical optimality criteria are desirable in pharmaceutical research. However, most frequently the optimal criteria, for which there is a theory already developed, impose some assumptions which are hard to meet in the clinical setting. Therefore, these designs are little used in clinical trials. There is an opportunity for a great deal of statistical research using current optimality criteria as well as developing new optimality standards to suggest the most efficient designs that will also accomplish the necessary goals of practical research. Among the latter are robustness against frequently encountered problems of clinical trials, balance with respect to all important factors, and the challenge of convincing others of the correctness of the interpretation of the results of the trial. Simpler designs lead to obvious interpretations; complex designs lead to statistical interactions, extraneous but important variables, other unavoidable realities of practical research, and difficult interpretations.

VII. INCOMPLETE BLOCK DESIGNS

Any of the designs with blocking can be incomplete in the sense that not all of the treatments are used in each of the blocks. The usual goal is to balance these incomplete blocks in such a way that the number of treatments within a block is the same in all of the blocks and each treatment appears with each other treatment an equal number of times.

An interesting example of this variety involves the use of three analgesic drugs which were self-administered by pregnant women during labor pains (Stewart, 1949). There were three analgesic mixtures: 50% nitrous oxide and 50% air, 75% nitrous oxide and 25% oxygen, and 0.5% trichlorethylene and 99.5% air. The usual randomized block experiment would have each patient (the blocking factor) use each analgesic in random order. The constraint was that patients were not expected to be able to use more than two of the analgesic mixtures during their labor pains. The design selected was to have each of 150 women use two mixtures such that each of the three pairs of drugs was compared by 50 randomly selected women. Moreover, the 50 women within each block were subjects in a crossover design so that 25 had one mixture first followed by the other second while the other 25 had the second mixture first followed by the first mixture. Each subject was asked, "Which was more effective in relieving the pain of uterine contraction?" in order to evaluate the pair of analgesics. The results showed that the 50% nitrous oxide mixture was the weakest and the other two were about equally effective.

This design is a balanced incomplete block design since each block is missing one analgesic and the blocks are balanced for all possible combinations. This example is mentioned to suggest that the standard designs mentioned are only the simplest of samples out of a vast design warehouse. The simplest designs have the advantage of being easier to finish successfully in practical situations.

VIII.   STUDY PROTOCOL

Before any clinical trial is carried out by a pharmaceutical company, the details of the study are first agreed to and expressed in a written document called the study protocol. At most pharmaceutical companies, the study protocol is the joint effort of the medical experts, the biostatisticians, others with a concern in the study, and the investigator. A general discussion (which is not confined to pharmaceutical trials) has been given by Ederer (1979).

The protocol includes information on the objectives of the study as well as details of the study. The objectives should be stated as precisely as possible, for example, "This study will investigate whether the new drug causes weight loss in diabetic patients of ages 45 to 65 when compared to an identically appearing placebo." The details include the study design (e.g., 8-week parallel trial, control treatment, blinding, randomization), the criteria for inclusion of patients into the study, how patients will be diagnosed for the disease being treated, criteria for exclusion of patients from the study, the clinical and laboratory procedures which will be carried out with each patient, the description of the drug treatment schedule (doses of drugs, route of treatment), description of laboratory and other tests, the criteria that will be used to measure efficacy, the planned statistical analysis, and other statements about how the study will be carried out.

In the statistical procedures section, the statisticians describe before the study is started how the sample size was chosen and how they will analyze the data. Finally, the protocol contains the method of eliciting information about side-effects and the course of action to be taken if side-effects are found. The set of case report forms that will be used for this study is usually attached to the protocol as an appendix.

In summary, the protocol contains all of the directions that explain how the study will be done. The challenge is to maintain a balance between completeness and brevity. The protocol must be long enough to contain meaningful comments on each of the points just mentioned. On the other hand, if the protocol gets to be too long, then it is unlikely that the investigator and other personnel involved with the study will carefully read the details of the protocol. In that case, the investigator and coworkers are likely to violate some of the requirements specified in the protocol. The statistician then has a difficult time trying to analyze the data and interpret the results.

Most studies are reviewed by an Institutional Review Board which considers the ethical issues in the study. Some larger and longer pharmaceutical studies may also have a Safety and Monitoring Committee that is concerned with ending a medication or complete study before the designed conclusion. More and more frequently, studies are reviewed before completion in an interim analysis. Information on this topic can be found in Chapters 14A and 14B, and in Simon (1991), which includes other comments on clinical trial designs.

REFERENCES

Buncher, C. R. (1972). Principles of experimental design for clinical drug studies. In *Perspectives in Clinical Pharmacy*, (Francke, D. E. and Whitney, H. A. K. Jr., Eds.). Drug Intelligence, Hamilton, IL, pp. 504-525.

Ederer, F. (1975). Patient bias, investigator bias, and the double-masked procedure in clinical trials. *Amer. J. Med.*, **58**:295-299.

Ederer, F. (1979). The statistician's role in developing a protocol for a clinical trial. *Amer. Statistician*, **33**:116-119.

FDA. (1992). *Guidance: Statistical Procedures for Bioequivalence Studies using a Standard Two-Treatment Crossover Design*. Food and Drug Administration, July 1, 1992, Rockville, MD.

Kaitin, K. I., Phelan, N. R., Raiford, D., and Morris, B., (1991) Therapeutic ratings and end-of-Phase II conferences: Initiatives to accelerate the availability of important new drugs. *J. Clin. Pharmacol.*, **31**:17-24.

Royall, R. M. (1991). Ethics and statistics in randomized clinical trials. *Statistic. Sci.*, 6: 52-88.

Simel, D. L. and Feussner, J. R. (1992) Clinical trials of informed consent. *Controlled Clin. Trials*, **13**:321-324.

Simon, R. (1977) Adaptive treatment assignment methods and clinical trials. *Biometrics*, **33**:743-749.

Simon, R. (1991). A decade of progress in statistical methodology for clinical trials. *Stat. Med.*, **10**:1789-1817.

Stewart, E. H. (1949). Self-administered analgesia in labor with special reference to trichlorethylene. *Lancet*, **2**:781-783.

Wei, L. J. (1988). Exact two-sample permutation tests based on the randomized play the winner rule. *Biometrika*, **75**: 603-606.

6

Selecting Patients for a Clinical Trial

Michael Weintraub and José F. Calimlim

*School of Medicine and Dentistry, University of Rochester,
Rochester, New York*

Statisticians are taught that the patients in a clinical trial are a sample from some population of potential patients for the medication(s) involved. One key issue is whether the sample is representative of the population to which inferences are being made. We know that pregnant women—and often women who may become pregnant during a trial—have frequently been excluded from clinical trials. Persons addicted to illegal drugs have also been excluded. An important concern then becomes apparent when a clinician wishes to treat patients in one of these groups and discovers that no or few studies have ever involved persons in these classifications.

One example of the lack of representativeness is the use of geriatric patients in research articles. Geriatric patients are underrepresented among patients in clinical trials. For example, Morley et al. (1990) showed that the percentage of articles on humans that included subjects over 65 years of age grew from perhaps 12% in 1966 to 15% in 1986 while discharges from short-stay hospitals of the same persons increased from 17% to 31% in 1986. In addition, geriatric patients take more medications than do younger patients and yet are used in trials far less than their share of consumption of the final product.

More recently, major studies find the description of patient recruitment to be an important part of the study characterization. For example, the methods, strategies, costs, and effectiveness of recruitment for the Lung Health Study have been explained in detail (Lung Heath Study Research Group, 1993). This chapter provides some actual observations on what happens with patient selection between the time a trial is designed and the study patients are started on medication.

Part A: The Outpatient

Michael Weintraub

I. INTRODUCTION

Patient selection affects many aspects of a clinical trial. It determines whether or not the clinical trial can be carried out and how long it will take to complete. It will affect the outcome of the trial and thus will directly influence the regulatory agency. The selection process will also affect the clinician's ability to generalize the findings of the study. Ultimately, then, selection of patients for a trial can even influence drug utilization.

In the next section Dr. Calimlim discusses the selection of inpatients for short-term hospital studies. This section deals with outpatient populations having chronic conditions. To illustrate the problems encountered in patient selection, I will use examples from actual clinical trials of new and old medications. I will show why patients do not enter studies and discuss the implications of this for the investigator's ability to carry out the trial, the effect on outcome, and application of the data to the general patient population. Finally, I will offer some recommendations for improving the selection process.

In a series of studies of anorexiant medications, different eligibility criteria were used which led to different patient populations. Of course, the ability to generalize the results and, indeed, the results themselves were in a sense determined by the participants' characteristics.

In the first case (Weintraub, 1986) all of the participants were women between 18 and 44 years old who were mildly overweight (115–130% of ideal body weight). The basis for these restrictions was an attempt to mirror the typical purchaser of the over-the-counter medication as determined by marketing surveys. Of course, the narrow window for percent over ideal body weight, health, and past history requirements created great difficulties in recruitment. There were 265 women who were screened but did not enter the study. Since 106 did participate, 371 women actually were considered for the study, but only 28.6% of potential participants did in fact begin medication. Our sample size goal, 100 participants, was arrived at by statistical calculations and the clinical requirements.

The weight limits were based on scientific and medical criteria as well an attempt to achieve uniform treatment groups. Below 115% of ideal body weight we felt people should use other methods for lowering their weight, while above 130% we felt that physician involvement would be beneficial. Thus scientific, medical, and representativeness considerations worked to make recruitment difficult. Attempting to achieve a balance between conformity and representativeness caused problems which were overcome, but only with great effort.

The second weight control study presented other problems in both the conflicting goals of uniformity and representativeness and also those of long term weight control study. To investigate the value of anorexiant medications as an adjunct to other forms of weight control therapy, we studied 121 people in a 34-week, double-blind clinical trial of 60 mg extended-release fenfluramine plus 15 mg phentermine resin versus placebo added to behavior modification, caloric restriction, and exercise. Participants weighed 130% to 180% (154% $\pm$ 1.2%, mean $\pm$ SEM) of ideal body weight (Metropolitan Life Tables, 1983) and were in good health. By week 34, participants receiving active medication lost an average of 14.2 $\pm$ 0.9 kg, or 15.9% $\pm$ 0.9% of initial weight (n = 58), versus a loss of 4.6 $\pm$ 0.8 kg or 4.9% $\pm$ 0.9% of initial weight by subjects taking placebo (n = 54; p < 0.001). Blood pressure decreased and pulse remained unchanged in both groups. Dry mouth was the most common adverse effect in subjects receiving fenfluramine plus phentermine; all adverse effects decreased after 4 weeks and only nine participants left the study in the first 34 weeks. Overall, fenfluramine plus phentermine used in conjunction with behavior modification, caloric restriction, and exercise aided weight loss and continued to be efficacious for 34 weeks (Weintraub et al., 1992).

In this case, the National Heart, Lung & Blood Institute funded long-term weight control studies, medications were added to caloric restriction, behavior modification and exercise. Participants were recruited with the expectation that they would remain in the area for the four years of the study. We hoped to include both men and women. Despite the wider window of 130–180% of ideal body weight and the inclusion of men, strict health requirements, past history of adverse effects on the medications, and the long-term nature of the study resulted in only 22% of those expressing a desire to participate actually beginning the study. Once again, we barely exceeded our goal of 120 participants by recruiting 121 who met all of the criteria. To accomplish our recruitment goal in the one month allotted, we used the media, physician referrals, radio talk shows, and paid advertisements.

We were, perhaps, overcautious in excluding potential participants with hypertension and those with past history of adverse effects. However, since we were testing a combination of two agents whose effects in people with hypertension were unknown and whose adverse effects in combination had only been tested on 20 people, we decided to choose safety considerations over representativeness considerations.

II. DESCRIPTION OF OTHER STUDIES

One study was a double-blind, dose-ranging, and efficacy trial in patients with rheumatoid arthritis (Weintraub et al., 1977). One of several doses of the new drug or a placebo was added to the patient's existing standard therapy. The patients selected for the study all had to satisfy 7 of the 11 criteria of the American Rheumatism Association (ARA) for classical, or definite, rheumatoid arthritis (Ropes et al., 1965). Despite full tolerable doses of standard therapy (gold and/or aspirin and/or steroids), the patients had to have active disease defined as: at least three swollen joints, at least six painful joints, at least 45 min of morning stiffness, and an elevated sedimentation rate. Before entering the study, all of the patients were screened for active peptic ulceration, hepatic or renal disease, and other serious abnormalities. After a 2-week placebo run-in period, which was single (patient) blind, all patients were begun on one of the various doses of active medication or an identical-appearing placebo.

In the University of Rochester Rheumatology Unit, the names of patients and their diagnoses are kept in a coded card catalog. It is possible, then, to retrieve quickly a list of all patients having the diagnosis of rheumatoid arthritis. We knew from experience that the diagnosis recorded on the cards should not be accepted as final, since further investigations often reveal a second disease or, with the passage of time, a revised diagnosis may be made but not recorded in this file. Also, the diagnosis appearing on the card may have been based on the physician's clinical impression rather than on the rigid criteria needed in our clinical trial.

For these reasons we hired a medical student to do a feasibility study before beginning the trial, and convinced the sponsoring pharmaceutical company to pay for it. The student reviewed the records and entered diagnostic information on a standard form, assisted the investigators in contacting the patients and finding out whether they still met the diagnostic criteria, and set up appointments for interviews and examinations. The initial survey of the card file provided 300 patient names and unit numbers. The charts of 150 of these patients were not available for review. Several of the patients had died. Of the 150 patients whose charts were available, 101 were rejected after screening for the following reasons:

40% had not been seen for more than 5 years
10% did not meet the diagnostic criteria
10% had recently been examined in the clinic and were reported to be in remission or without symptoms

10% were judged by the rheumatologist caring for them to be unable to respond
to therapy other than surgical replacement of joints
10% had drug toxicity akin to that expected from the study drug
10% were considered unable to follow the protocol reliably

Only 49 patients (16% of the initial sample) remained to be interviewed
and examined. As mentioned more completely elsewhere (Hassar and Weintraub, 1976), 31 patients were removed from consideration for participation
because:

11 did not have disease activity great enough to warrant their inclusion
7 had concurrent disease not evident from the charts of screening, including:
2 with abnormal liver chemistries (SGOT, SGFT, and LDH)
2 with asymptomatic gastric ulcers demonstrated by a screening upper gastrointestinal (GI) series
1 who revealed after long discussion that she had frequent nausea and vomiting
with or without drug therapy
1 with an asymptomatic aortic aneurysm
1 with renal stones

Four were found unacceptable for reasons pertaining to the protocol, namely:

1 in whom pregnancy was diagnosed by the radiologist (before beginning the
upper GI series)
2 who were taking other nonsteroidal, antiinflammatory drugs not noted in their
records and were doing well on them
1 who had been scheduled for joint replacement

Four declined to participate because of inconvenience or distance to the clinic
Five refused outright to participate in the study. The reasons they gave
were interference with their work, with family responsibilities, or with their
"life style." Only one patient gave toxicity as the reason, and that was prompted
by pressure from her husband, who feared that she might develop an ulcer and
be unable to participate in the enjoyment of his retirement. (This patient later
entered an open-label study of the same drug and did, in fact, develop a duodenal
ulcer.)

III. THE SELECTION PROCESS AND GETTING ENOUGH
PATIENTS TO DO THE STUDY—LASAGNA'S LAW AND
ITS COROLLARIES

"Lasagna's law" (Gorringe, 1970) teaches us that the incidence of the disease
under study will drastically decrease once the study begins. It will not return
to its previous level until the completion of study (if completion occurs before

the investigators retire). There are obvious and valuable public health aspects of this law, but it has an undesirable impact on the conduct of clinical trials. The following discussion examines how Lasagna's law operates to diminish acceptable candidates for participation in a clinical trial.

A. The Many Become the Few

One can never know whether there will be enough patients for a clinical trial simply from investigators' estimates of how many patients will meet the diagnostic criteria. Physicians have selective memory of how many patients of a particular type they see. Their interest in the study may cause them to overestimate the number of suitable patients available; files are frequently out of date or lost; patients move, retire, die, or recover; and ideas change over time as to what constitutes the disease entity under study.

B. The Few Become the Fewer

The diagnostic strictures imposed by the clinical trial decrease the number of available patients even further. In clinical practice the diagnostic criteria need not be as rigid as those laid down by, for example, the ARA, the pharmaceutical industry, or the Food and Drug Administration. In clinical practice, the special tests required to fulfill the stringent criteria are neither done nor necessary. Patients with variations on the theme of the disease are included under the basic rubric because such fine distinctions will not often affect therapy. If in studies, however, only the strictest criteria are used, one ends up with the purest-of-the-pure sample, and this distillate will be very small. That this is a much-refined population that may not provide an adequate or fair test of the study medication is only rarely considered. Data based on such a purified sample may not apply to the population at large but may fulfill certain internal or regulatory needs.

C. Avoiding the Lazarus Trap

Once the diagnosis is assured, investigators must make a judgment on disease activity, the stage of disease, and the severity of the disease. An optimum selection would include patients with enough disease present to show a good response to medication but not so much disease that they are unable to respond or have irreversible changes (i.e., "burned-out" disease). The latter patients would not be suitable for participation in a clinical trial because inclusion of their data could cause a Type II error and a rejection of an active medication. Requiring a drug to show its efficacy in patients in whom no other medication has been of value is what has been called the "Lazarus Phenomenon." Including too many of these Lazaruses will bias the study against the drug.

D. All God's Children Get Sick from Time to Time

The next major problem of the selection process is the presence of concurrent disease that has been apparent from the very outset, before the pre-drug-screening tests. Although some diseases obviously require exclusion, what about past conditions that mimic the expected toxicity of the study drug? In the clinical trial of the nonsteroidal agent, the possibility of gastrointestinal toxicity with the study drug alerted us to the need for obtaining the history of such disease in prospective participants. Who should be rejected: patients who had an ulcer 2 years ago? 5 years ago? 10 years ago? Then, too, sick people frequently have other diseases. Some diseases often coexist with, or result from, therapeutic measures used in the disease under study. These are important determinants of the suitability of a patient for inclusion in a clinical trial. Yet if everyone who has the merest touch of another disease is excluded, the study population will shrink even further. The question of what constitutes serious renal, hepatic, or cardiovascular diseases also must be raised. Many diseases and laboratory test changes may occur as part of the natural course of aging (e.g., cataracts, pulmonary changes, electrocardiographic "abnormalities", hypertension, adult onset diabetes, decreased creatinine clearance, and increased globulins). Too often clinical logic does not function in the elimination criteria for studies of drugs brought to Phase II trials. The standard exclusions are used indiscriminately without any modification based on the disease process, the type of medication, toxicity shown in preclinical testing, or the toxicity demonstrated during Phase I trials.

E. The Few Become the Rock-Bottom Fewest

Both Bloomfield (1969) and Lasagna (Gorringe, 1970) have pointed out the therapeutic effect of looking for study participants. As discussed above, the reasons why potential study participants disappear once a clinical trial begins are quite mundane. The main reason is the rigor of diagnostic criteria. Another is that the patient's disease changes; a patient may have had active enough disease during the initial review but improves before the trial gets underway, or during the run-in period, especially if the disease is cyclic. Then during the screening period, potential participants frequently are found to have laboratory pseudo-abnormalities in the form of meaningless deviations from normal values —cholesterol levels that are too low, minor electrocardiographic variations, or even spurious laboratory vagaries (Joubert et al., 1975). In the study under discussion, we made sure that only patients with "cast-iron stomachs" and no laboratory abnormalities would participate (i.e., whose upper GI series results were negative despite high doses of aspirin/or prednisone, and/or gold therapy).

Many times the abnormalities that show up in laboratory tests are due to illnesses unrelated to the disease under study or to other necessary therapeutic intervention. It is more difficult to decide whether to include this latter group in a trial. For example, in the first study we found a patient with increased liver enzymes which were probably induced by aspirin. She of course, was excluded from participation. Similarly, aspirin therapy, especially if the patient takes aspirin intermittently, can result in the shedding of large numbers of renal epithelial cells into the urine. These may be disturbing to the person examining the urinary sediment. Asymptomatic, serious illnesses may be discovered in patients being screened for a clinical trial. We found two patients with asymptomatic ulcers and one with an aortic aneurysm in the NSAID study. At this point in the selection process, investigators may turn their faces to the heavens and, like Job, cry, "What else, Lord, what else?"

F.   Need You Ask?

There are other burdens. In the first study a young woman who had assured us that she was practicing birth control became pregnant. Fortunately, the radiologist was astute enough to question the patient about her menstrual period before performing the upper GI series, and found she had missed her last period, which should have occurred between the screening interview and the time for the X-ray.

Next, the question arises: how will the patient's other therapy effect the outcome of the trial? A certain amount of standard therapy must be permitted in many current clinical trials for ethical reasons. One could not, in good conscience, deprive patients suffering from serious rheumatoid arthritis of all their usual therapy. Investigators must learn a lesson from the early trials of L-dopa, when patients whose anticholinergic therapy was discontinued regressed to the point of severe Parkinson's disease and required months to return to even baseline status (Yahr et al., 1969). Conversely, drugs that obviously interfere— ones capable of causing adverse effects similar to those expected from the study medication—should be discontinued. Competing agents should also be stopped. Other, nondrug treatment modalities can be handled in a variety of ways. For example, ancillary therapy can be forbidden, standardized, or measured and included in the analysis. Background therapy can be categorized (none, minimal, some, maximal) by a set of rules and participants in each classification that will minimize the differences between groups (Taves, 1974).

G.   Participant Psychology: Capricious and Intelligent Noncompliers

Psychological factors play an important role in the physician's assessment of who should participate in a clinical trial. Physicians make judgements about the patient's ability to adhere to the protocol according to the patient's past demonstration of understanding prescription directions. "Capricious compliers" should not be included in clinical trials, since they vary their medication intake from day to day according to ideas not necessarily founded on pharmacokinetic or pharmacodynamic theory. On the other hand, "intelligent noncompliers"— patients who stop medication for rational reasons—should be included if they can be relied upon to notify the study physician (Weintraub, 1976). Patients who accurately report adverse effects may also be known to physicians and would be valuable participants in a clinical trial. The important psychological attributes of participants in a trial are stability coupled with flexibility.

When considering the ethical aspects of the participant selection process, physicians must also analyze the psychological factors that enable a patient to make a reasoned judgment about participation in a clinical trial. The ideal participants would be patients who will carry through the clinical trial and actively interact with the investigators rather than being passive experimental subjects.

H. "Not Unless I Can Be in the Placebo Group, Doctor"

Finally, there are some patients who meet the diagnostic and disease activity criteria, pass the screening tests, can respond to the study drug, and take the correct amount and type of other treatment but who refuse outright to participate. These are rare, however, in studies of treatments for chronic diseases. Patients with active rheumatoid arthritis despite therapy often will want to participate in a study that offers any hope of relief, no matter how remote. However, in studies such as the trial of a postoperative analgesic discussed in Part B by Dr. Calimlim, more patients decline to participate.

Some of the obstacles to participation mentioned by patients in our study may actually have been veiled but valid refusals. Problems with the clinic schedule, travel arrangements, and "cure" during run-in periods may give the patient ways to decline participation without outright refusal. Healthy volunteers often find alterations in the "life style" the most disrupting aspect of participation in a clinical trial (Hassar et al., 1977), and perhaps this is an important deterrent to some patients as well. Fear of toxicity is another. However, in a test of how well the patients in this study recalled the information given them during the consent procedure, we found that very little of the toxicity data was retained (Hassar and Weintraub, 1976). Perhaps because of anxiety about their disease and desire to participate in the study, patients did not really listen to the discussion of the negative aspects of the trial. Other, less anxious patients may have listened and refused to participate.

I. The Selection Process and Regulatory Requirements

The choice of the target population must be made so that the drug has a fighting chance. I would like to term this "Lazarus versus Grendel." As previously discussed, too many "Lazarus" patients can cause even the best-designed study to reject an active, valuable drug. One must balance the availability of fresh, barely treated, or even untreated patients with the ethically and practically more sound practice of seeking difficult responders to participate in initial studies. This is the "Grendel," or worthy opponent, principle. (A lesser opponent than the terrible monster Grendel would not have truly tested Beowulf's courage, diminishing his heroic credentials.) In the chronic arthritis trial discussed here, we elected to use patients with active disease despite full doses of standard therapy—but not patients with end-stage or nonresponsive disease, even if they met the criteria for pain and disease activity. If the drug works in tough but treatable (Grendel) patient populations, one can say, "Great: We have an active valuable agent." If, however, it fails in the improper (Lazarus) patients, it does not mean that the drug could not be effective in less severely ill patients.

J. Between a Rock and a Hard Place

Another goal of clinical trials, one that is required by the regulatory process, is toxicity monitoring. The selection process exerts an influence on this goal also. Screening out every patient who has ever had, or could have, a particular sort of problem leaves a small, select group providing little indication of possible serious toxicity. If, for example, patients likely to develop gastrointestinal disease are included and each toxicity does occur, is it worse than if gastrointestinal lesions appear in patients who have been carefully screened for any possible predisposing factor or presence of disease? If, on the other hand, such toxicity does not occur, can one then assume that it will not occur in the average patient? If all patients are included in a study without any sort of clinical logic being applied, then we will end up including patients who already have some disease ("toxicity") before treatment with the study drug. The results will then make the drug look falsely toxic. Such patients may participate in late (Phase III) studies, where the goals are different and information on general usefulness is being sought. If included earlier, they must be equally distributed among the treatment groups.

VI. PATIENT SELECTION PROCESS

A. Generalization of Data to Other Patient Populations

In following the rigorous selection process outlined above, who finally enters the study? Does this patient population have any relationship to that seen in actual practice? The answer, of course, is that there are many important differences between study populations and patients in general, and that these differences decrease our ability to apply the study results to any other population.

Patients in a clinical trial are usually fairly homogeneous in terms of diagnostic criteria, other treatments, and duration and severity of disease; in clinical practice the population is much more heterogeneous. Furthermore, diagnostic criteria are much more stringent with study patients than with patients in general. Their disease may be more severe or more treatment resistant. Study participants may tolerate adverse reactions because of perceived benefit for the more severe illness. Patients in studies have often been referred to specialists, whereas in actual practice more patients are treated by primary care physicians. Patients participating in Phase II studies usually live in large cities, frequently those with university medical centers and academic investigators, whereas in actual practice there is a mixture of population densities, and physicians are less likely to be academic investigators. Patients in studies may have less restriction on their time; they may be retired, disabled, unemployed, or work for a benevolent company. In a study of a medication for prevention of complications of diabetes mellitus, 20% of participants worked for state, county, or city governments. Patients in clinical trials tend to adhere closely to therapeutic regimens and are good observers. This is not necessarily the case among the general population of patients.

B.   Ethics and Extrapolation

Ethical considerations may also affect the patient selection process. Racial, social, and economic factors have frequently been offered as an important distinction between patients who participate in a trial and those who do not. In the first study used as an example for this chapter, all participants were white and from the middle class. Obviously, this is not representative of the population as a whole. Participation by patients who stand to benefit themselves or for the societal good from the research should be fostered.

In obtaining consent from our patients we found that much of the material on adverse effects was forgotten (Hassar and Weintraub, 1976). Two-thirds of the patients could not remember at the end of the study ever having been told that they could get an ulcer from the medication, despite having been told five times about the ulcerogenic activity of the drug, having been given the written patient information form to take home on two occasions, having had an upper GI series, and having been questioned every two weeks about gastrointestinal symptoms. One-third of the patients incorrectly noted that they had been told that this drug was safer than any other drug for rheumatoid arthritis. Only one-third of the patients reported apprehension about the side effects of the new drug before the study started. This apprehension soon disappeared, however.

We keep our patient information form short and to the point. They are written in what we hope is an easily understood style, although, considering the socioeconomic status and educational level of the patients, much more complex material should have been easily understood. Actually, when tested, our patients retained material contained in the information form, rejecting from memory only the material on adverse effects.

An "add-on" study, in which test medication is added to the patient's current treatment, is frequently more ethical but presents serious difficulties for a clinical trial. Add-on studies alter the target population, in many cases making it broader and making the study more feasible. However, there will be less room for improvement in each individual patient (part of the Lazarus dilemma mentioned above). The resultant decreased experimental sensitivity and decreased patient responsiveness should be taken into account in the creation of "power curves" needed to determine the number of patients who should be in the study. The studies will be "dirtier"; that is, there is likely to be an increased incidence of adverse drug reactions and less clear-cut response attributable to the new agent. Data from add-on trials are easier to apply to the patient population at large. The ethical nature of the add-on studies, as well as the ability to be practical and to extrapolate to the general population, probably outweighs the drawback of results that are harder to interpret.

C. What, Me Worry?

Why should investigators and monitors in the pharmaceutical industry worry about patient selection processes and the effect on extrapolation? The most important reason relates to the possibility of achieving a true result from the study and a valid estimate of common toxicity. In addition, medical students have increasingly been trained in the critical evaluation of the literature. Physicians will downgrade studies in which the selection process appears to have biased the outcome or in which the selection process was so rigid as to preclude extrapolation to their patient population. Then they will be less likely to use the drug except in selected patients. Regulatory bodies carefully analyze the study population's characteristics. Labeling restrictions (or even approval) may thus be affected by participant selection.

V. IMPROVING THE SELECTION PROCESS

Bloomfield (1969) recommended that investigators should check records and do formal pilot studies, assessing the availability and suitability of the patient population at hand for participation in a clinical trial and if the protocol is workable. Investing a small amount of time and money in such prestudy surveys will save the concerned parties much grief. Sponsors, investigators, and regulators must remain flexible in determining selection criteria. Small changes in the criteria may make vast differences in patient availability without materially influencing the outcome of the study or its extrapolatability. For example, in a study of a new hypnotic agent, slightly increasing the age limits for entry resulted in a large increase in potential participants, facilitating completion of the trial.

A corollary of the "flexibility" recommendation is to tailor the criteria to the institutions. Clinical trial logic must be applied at all stages of the process of patient selection: the diagnostic criteria, the prognostic criteria, the distribution of patients in the treatment groups, and the decisions made about adverse effects. In some areas the patient population may have certain demographic, diagnostic, or therapeutic idiosyncrasies which would not deleteriously affect the outcome of the study but, if included, might improve the availability of patients.

These comments are an argument against many large, multicenter trials. The latter studies tend to be carried out by "data gatherers" instead of investigators. The necessary patient selection judgments are the province of the investigator on the scene. I believe that data gatherers have neither the time nor the training nor the inclination for these tasks.

VI. AVOIDING STUDIES THAT RESEMBLE FINE SCOTCH (AGED IN THE CASK)

Another suggestion for rapid completion and for statistically and clinically significant results is to start all patients in the study at the same time whenever possible. This avoids long drawn-out studies during which the quality of the data deteriorates as investigator interest wanes. Additionally, starting patients as a group decreases "improvement bias" noted in rheumatoid arthritis studies (Miller and Willner, 1974). Given the cyclic pattern of many chronic diseases (e.g., rheumatoid arthritis), patients often enter studies during an exacerbation. They then would be expected to get better with time, no matter what their treatment (regression toward the mean). Assigning treatment and starting all patients at one time, generally after a delay during which patients are selected for the study, diminishes improvement bias. Some patients will have passed the worst of the exacerbation and others will be at some middle point.

The procedures used in obtaining consent can also be improved. We allow patients to bow out gracefully for whatever reasons they advance. We do not attempt to convince a patient to enter a clinical trial and if a patient asks for more time to decide, we do not contact them again. They must contact us if they later decide to participate. Whenever possible in our studies, an investigator not associated with the daily care of the patient obtained the consent after discussing the pros and cons of entering the study. (This is a safeguard that cannot be used when physician/data gatherers conduct clinical trials.) Patients may feel constrained to participate when their own physicians are the ones obtaining consent. We have found that group discussions are an effective way of informing participants about a study. Potential participants gather together and are given information on the study and possible adverse effects. They ask questions and hear the concerns of others which might not have occurred to them. Video tapes, interactive computer programs, readability testing and other newer methods for improving communication have been applied to the consent process.

Discussion, worry, and thought about patient selection are often left completely to the investigators. Pharmaceutical industry monitors should continue involvement after the inclusion and exclusion criteria have been established. Once the design and protocol have been established, the patient selection process may be the single most important determinant of the outcome of a clinical trial. Proper monitoring of patient selection becomes increasingly important now that physicians in nonacademic centers are taking part in multicenter clinical trials with standardized protocols imposed upon them. They may have neither the expertise nor the experience to assess the influence of patient selection on the outcome of studies. They may fail to realize how their entering a patient into a clinical trial could affect the outcome because they see only a small portion of the patients in the study. Patient selection problems are less likely when investigators trained in clinical pharmacology or having wide experience in performing clinical trials are involved in the design and management of a study.

Precise or quantitative data of the impact of patient selection on the outcome and extrapolatability of a study do not now exist. Although the population in the study discussed above had an incomplete response to standard treatment and differed from the population of arthritis patients as a whole, we were able to demonstrate significant drug effects. How one uses the information from that study in making a therapeutic decision or regulatory decision is a difficult problem. More thinking and research is needed in this area.

Part B: The Inpatient

José F. Calimlim

I. INTRODUCTION

The term clinical trial covers a wide variety of different activities. To many people outside medicine it implies something exciting and dramatic and possibly dangerous involving the early administration of a new drug to man. There are clinical trials of that kind, but the majority are more mundane but no less important (Fleiss, 1986).

It is not easy to generalize about clinical trials in new drugs because applications vary so widely. On the one hand, the drug concerned might offer the first effective treatment for a hitherto untreatable form of cancer, and on the other hand it might be a new substance for the treatment of pain. Obviously the approach to these two problems would be very different. However, there are some basic principles which apply throughout this type of work.

Before a drug is offered for a clinical trial a great deal of work has been done on it and a lot of money has been spent. If the drug is reasonably safe in animal and Phase I clinical trials in man and has an action which might be useful in the treatment of pain, it will probably be accepted for study by a clinical investigator. If the secondary trials (Phases II and III) are successful and no serious toxicity is observed in man, analytical and descriptive papers on the drug then go back to the Food and Drug Administration for approval. If the proof of the efficacy and safety is acceptable, the pharmaceutical company will be given permission to market the drug.

We take the need for clinical trials to be self-evident. It is impossible to conceive of a modern civilized society without the benefits of modern drugs. Development and assessment of new analgesics is not possible without clinical trials. We do not think that many people would deny this general case as long as trials are carried on with utmost safety and efficiency.

In some clinical trials the use of placebo is essential (Dollery, 1967). The drug can only be assessed on the basis of what the patient tells the doctor or observer about the pain. The pain is often lessened somewhat by a tablet that does not contain any active ingredients, a "placebo effect"; therefore, a comparison of no treatment with the active tablet might give a falsely positive result because of this effect. Here it is necessary to compare the active tablet with a placebo. (See also Chapters 9 on analgesic trials and 13 on placebo effects.)

It is more difficult to generalize about the role of the patient in the trial of new drugs. If the drug is for the treatment of a serious condition, it is easy to find patients who are unresponsive to established drugs. A new drug is usually offered to patients who are in this position. But for new drugs with unproven efficacy to relieve pain, it is a little bit more difficult to obtain patient participation. In these circumstances a heavy responsibility falls upon the clinical investigator conducting a clinical trial for the patient's safety.

The first step in an analgesic clinical trial is to choose which kind of pain to study, a choice determined in part by the goal of the research (Murphee, 1966). Two kinds of pain have been studied in assaying analgesics: experimentally-induced and clinical pain. Experimental pain now has few protagonists, partly because with the institution of double-blind procedures it was found that many of the most famous experiments involved bias and cuing, and partly because assay of the therapeutic value of a drug is more appropriately done against clinical pain.

The next decision to be made is what kinds of subjects to utilize. In studies of clinical pain, the subjects will be patients of some sort. As with experimental pain, they also will be volunteers, although the factors which influence their participation and understanding are likely to be somewhat different.

If one works with outpatients, some special difficulties arise. One can never be certain that they take their doses when and as directed (Mainland, 1960). Their interest and cooperation cannot be actively and continuously engaged.

If inpatients are selected, the next choice is between acute and chronic patients. Acute postsurgical pain and its relief have been the subject of many reports (Calimlim et al., 1977). The meaningfulness of much of this work is evaluated subject to the diversity of etiologies and preoperative as well as postoperative surgical states. Postsurgical patients are also, to a varying and not altogether predictable extent, still recovering from the anesthesiologist's marvelous bag of tricks. Postoperative pain has been described as "the most frequent and neglected painful state in the hospital situation." Many others, including intelligent and informed patients, have echoed this sentiment, but postoperative pain relief is still too often left to the junior physician's "cautiously administered opiate and the balm that comes from time alone."

Most civilized men today surely concede that there is need for the relief of postoperative pain on humanitarian grounds alone. There are however, other obvious reasons for mitigating the discomfort of the patients. These include the need to promote deep breathing and cooperation with the physiotherapist and the desirability of early mobilization to avoid deep venous thrombosis.

A considerable increase in our knowledge of the efficacy of certain old and new analgesic compounds and in the elaboration of valid techniques for the evaluation of pain-relieving drugs has occurred (Lasagna and Meier, 1959). For clinical trials to be of any value, one must be able to extrapolate the results to the general population for whom the drug is designed. For extrapolation to be valid, there must be a relationship between the study sample and the population from which it was selected.

This discussion is an attempt to examine the degree of selection and attrition due to protocol and other factors that occurred in the course of obtaining 100 consenting volunteers completing a single-dose postsurgical analgesic study (Calimlim et al., 1977). An attempt is also made to compare the study population and the population from which it was selected.

II. DESCRIPTION OF THE STUDY

The protocol was written for a study intended to evaluate the efficacy of three analgesic treatments and a placebo administered in single-doses in double-blind fashion for postoperative pain. Subjects were postoperative surgical patients. Surgical procedures of potential participants were classified as general superficial surgery, gynecological surgery, plastic surgery, dental surgery, and superficial neurosurgical procedures. Cardiovascular, thoracic, and abdominal surgical procedures were excluded from the study.

There were several criteria for admission to the study. Patients must have been 21–65 years old, weighing 120–200 lbs. Patients whose medical or surgical history was consistent with a reasonable suspicion of gastrointestinal, liver, or urinary disease which might interfere with the absorption, metabolism, or excretion of medications were excluded.

Acceptable patients were those who had not participated in any other drug studies in the past 3 weeks, who had recovered from a surgical procedure sufficiently to request and receive oral analgesic medication during the first 3 postoperative days, who had at least a moderate degree of pain after surgery, and who did not have history of tolerance to analgesic medication.

Permission to visit patients for discussion of the study was obtained from the patients' physicians prior to the operation.

III. SELECTION PROCESS

A. Schedule Survey and Screening

A survey of the daily elective surgical schedule for the duration of the study showed a total of 8027 patients were potentially available. This number was reduced by 39% (3103) because of patients not screened due to unavailability during appointed hours of interview for various reasons (e.g., late admissions, being worked up by staff, referral to specialty clinics, or out on pass). This left 4924 patients available for screening. Preliminary exclusion eliminated 4254 (86%) of these. Only 670 patients were thus available for interview, that is, only 8% of the total number of patients originally available for the study.

Table 6.1 Reasons for Preliminary Exclusion

Below 21 or above 65 years	1921	45%
Insufficient postoperative pain	1155	27%
Excluded surgical procedures	718	17%
No physician consent	337	8%
Short-term admissions	123	3%
Total	4254	

B. Preliminary Exclusions

The reasons for preliminary screening exclusions are described in Table 6.1; 1921 (45%) were excluded because of age: 1232 were below 21 years and 689 were over 65 years. The lower age limit, 21 years, was the legal age of majority in New York State at the time of the study. The upper age limit was chosen arbitrarily. Age was thus a major factor for excluding 45% of the number available for screening.

Insufficiency of postoperative pain excluded an additional 1155 (27%). Some patients do not require postoperative pain relief even after major operations. The scheduled surgical procedures included diagnostic curettage and superficial gynecological surgery, superficial general surgery, and excisions of small lesions, gingivectomy, some simple eye and nose surgery, and endoscopies, procedures in which pain is often mild postoperatively.

Some of the surgical procedures that were excluded were gastrointestinal resections, open heart surgery, spine fusions, facial or mandibular surgery, where oral administration may be ineffective or inappropriate in the early postoperative period. These made up 718 exclusions (17%). The remaining were excluded for miscellaneous reasons (e.g., admitted in the morning for scheduled surgery and sent home later in the afternoon). Such patients numbered 123 (3%). A consider-

Table 6.2 Patients Available for Interview

Patients not screened		21	
Patients screened		649	
	Total	670	
Patients interviewed		391	60%
Patients rejected before interview		258	40%
	Total	649	

able number of patients (337 or 8%) were not included in the study because some attending surgeons had not consented to let us interview their patients for the study.

C. Chart Screening and Patient Interview

After these preliminary exclusions, only 670 (14% of the total initially available) were left for screening and interview (Table 6.2). This number was cut further by such factors as patients not being available for interview at designated time, surgery canceled or surgical procedure changed to a less painful one (e.g., laparotomy to laparoscopy) so that there was insufficient postoperative pain. Twenty-one patients were in this group. The charts of the remaining 649 patients were screened and reviewed in detail prior to interview with emphasis on the past and present history and physical findings together with the available laboratory reports. This brought about rejection of an additional 258 patients prior to interview. Thus 391 patients were left to be interviewed.

D. Reasons for Rejection Prior to Interview

There were several reasons why we rejected these 258 patients prior to interview after review of their charts. Many had multiple medical problems unrelated to the indication for surgery (Table 6.3). Some were underweight or overweight. Allergy or sensitivity to the drug was also reported but not observed. Others were too apprehensive, high-strung and agitated, or overly concerned about the loss of a particular organ such as breast, uterus, testis, etc. Active peptic ulcer

Table 6.3 Reasons for Rejection Prior to Interview

Multiple medical problems	82	31%
Overweight or underweight	60	23%
Sensitivity to study medication	28	11%
Emotional overlay	20	8%
Chronic analgesic intake	17	7%
Active peptic ulcer disease	16	6%
Psychiatric history or illness	10	4%
Language problems	8	3%
Multiple allergies	6	2%
Physical impairment (deaf, blind)	4	2%
Refused surgery	4	2%
Mental retardation	3	1%
Total	258	

Table 6.4 Patients Interviewed

Number of patients interviewed	391	
Number interviewed but rejected	53	
Reasons for rejection		
Multiple allergies		14
Medical problems		13
Very apprehensive		13
Overweight		7
No relief from study drug		3
Language problem		3
Patients asked for consent	338	
Patients consenting		246	73%
Patients not consenting		92	27%

disease, chronic intake of analgesics, and psychiatric illness or history thereof added to the exclusions. Language problems were also encountered, as were physical impairments (blindness, deafness) and mental retardation. Some patients refused surgery. A few reported severe multiple allergies to drugs and were excluded on that account.

This left 391 patients interviewed (Table 6.4), of which 53 were interviewed but rejected. Some of these patients who were interviewed and rejected at this stage were found to have had incomplete or absent workups. Old records were not available for evaluation prior to interview; therefore, in the process of patient interview other exclusion criteria were noted that were not known prior to interview. Such patients were therefore rejected from the study for various reasons. Some patients had multiple allergies and sensitivity to drugs; others had medical problems. Quite a few were excessively apprehensive. Problems

Table 6.5 Reasons for not consenting

Prefers intramuscular medication	65	71%
Cannot decide	10	11%
Study drugs not effective in past	9	10%
Family refuses	5	5%
No reason	3	3%
Total	92	

of overweight and language were also encountered. Still others claimed no relief with one of the study drugs on the basis of previous use. These patients were interviewed and rejected but the explanation of the purpose and process in conducting the drug study were not discussed.

Two hundred forty-six patients (73% of those not rejected) consented to participate in the study; 92 (27%) did not consent. There were various reasons given for refusing to consent (Table 6.5). The majority of patients about to undergo surgery develop varying degrees of anxiety and tension related to the extent of surgery and its attendant risks. For many patients the prospect of pain still remains a dreaded specter, so that relief or avoidance of pain is one of the primary concerns of patients after surgery.

A major reason for not consenting (71%) was the preference for a parenteral pain medication for fast, effective relief. The effectiveness of parenteral agents like morphine and meperidine has been assisted by the introduction of recovery room and intensive-care areas; it is now feasible for the anesthesiologist to routinely administer narcotics intravenously to achieve an immediate effect. Lowenstein et al. (1969) demonstrated that surprisingly large doses of narcotic can be given intravenously to pain-free individuals without dangerous cardiopulmonary depression, but in practice, adequate analgesia from parenteral agents frequently leads to impaired respiratory function and pulmonary sequelae. Some patients cannot decide whether to participate or not. Others graciously refuse consent after a member of the family present during the interview has commented or made a subtle indication of disagreement during the interview. Others refuse without offering any reason. In another study (Calimlin, unpublished) which did not include a placebo in the protocol, the nonconsenters were much less common (13% of 396 patients interviewed). Perhaps the presence of a placebo in a study may be factor that influences nonconsenting. In another ongoing clinical trial that includes a placebo, there is already a 17% nonconsenting rate among 123 patients interviewed.

Of the 338 patients ultimately interviewed in this study (which is only 4.2% of the total number of patients originally available), we thus had a group of 246 patients who consented to participate in the study, underwent surgery, and were followed closely up to the third postoperative day. One hundred were medicated; 146 were not. Data on 12 patients who were not medicated were not available, leaving 134 patients (Table 6.6) whose various reasons for nonmedication can be analyzed. About 35% of patients were pain-free during the study hours (8:00 A.M. to 6:00 P.M.). Some patients (26%) were kept off oral intake during the immediate postoperative period and were subsequently discharged on the first postoperative day. These were usually patients who had superficial or relatively simple gynecological procedures such as an abdominal tubal ligation. Some patients (22%) were kept off oral intake more than 72 hours postoperatively, thereby going beyond the time limits set in our protocol.

Table 6.6 Reasons for Nonmedication

Consenting patients		
Not medicated	146	
Data not available	12	
Difference:	134	
Insufficient pain when oral medication allowed	47	35%
Oral medication not permitted then discharged day 1	35	26%
Oral medication not permitted days, 0, 1, 2, 3	29	22%
Dropped from study by request of patient or surgeon	10	7%
Medical complications after surgery	9	7%
Surgery canceled	4	3%
Total	134	

Some patients (7%) had medical complications after surgery and had to be taken off the study. A similar number (7%) were dropped from the study by request of the patient or the surgeon. Some had surgery canceled (3%).

IV. COMPARISON OF CONSENTERS VERSUS NONCONSENTERS

We decided to compare the 246 consenting patients and the 92 who did not consent out of the total 338 interviewed. Age, sex, social class, and anticipated pain severity after surgery were the bases for comparison. This seemed important to do in view of the need to extrapolate the results of the analgesic study. The mean age of the consenting group was 35, while the mean age of the nonconsenting group was 41, a statistically significant difference (p < .001). There was a slight trend to a higher percentage of women in the consenters (76.3%) than in the nonconsenters (65.4%), but this difference was not statistically significant.

We coded Hollingshead's occupational categories as follows: high social class—higher executives and major professionals, proprietors of medium businesses ($35,000–100,000); middle social—lesser professionals to semi-professionals and farmers ($24,000–35,000); and lower class (< $24,000)—clerical and sales workers and unskilled employees. The Hollingshead's housewife category was removed to reduce sensitivity to the difference in gender distribution. There was no significant difference between the consenters and nonconsenters in the proportions of high, middle, and low social classes.

Table 6.7 Comparison Among Consenters

	Medicated	Not medicated	Significance
Age	—	—	NS
Sex: Female	67%	83%	p = 0.01
Social class (housewives excluded)	—	—	NS
Anticipated pain severity after surgery			
Moderate	37%	54%	p = 0.02
Severe	50%	41%	
Very severe	13%	5%	

NS: Not significant

V. COMPARISON AMONG CONSENTERS

We then compared the 100 consenters who were medicated and the 134 consenters who were not medicated (Table 6.7). Data was not available from the other 12 consenters. There was no significant difference between the two groups in the distribution of age and social class. However, there was a significant and unexpected difference in regard to sex. More females tended not to be medicated (p = 0.01). As expected, there was a difference in the pain severity of the operation, in that a significantly greater proportion of the patients with operations deemed prior to the surgery to be more painful were medicated than was the case for the other operations. The p-value of the χ^2 statistic was 0.02.

VI. SUMMARY

Initially, most exclusions are for administrative reasons such as nonavailability during times of interview due to late admissions, referrals to specialty clinics, or work-up by other members of the staff; this reduced the pool by almost 40% in this example. At the next stage, 86% were eliminated, almost three-fourths of which were due to age and insufficient severity of pain after operations; these were the two major reasons for preliminary exclusions. Most of the rejections prior to or at interview were due to concurrent medical problems, being over- or underweight, or having allergies or possible sensitivity to study medication.

At the consent stage, most of the patients who refused reported doing so because they preferred parenteral medication. After consenting, it was mainly administrative reasons, degree of pain, or denial of oral intake that resulted in failure to provide data. The only statistically significant difference between consenters and nonconsenters was the factor of age.

VII. CONCLUSIONS

The patients who end up in clinical trials may represent only a small percentage of the theoretical universe of available patients. Nevertheless, in this study the analysis did not suggest that extrapolation from the study sample was unjustified.

REFERENCES

Bloomfield, S. S. (1969). Conducting the clinical drug study. In *Proceedings of the Institute on Drug Literature Evaluation*. American Society of Hospital Pharmacists, Washington, D.C., pp. 147-154.

Calimlim, J. F., Wardell, W. M., Lasagna, L., and Gillies, A. J. (1977). Analgesic efficacy of an orally administered combination of pentazocine and aspirin, with observations on the use and statistical efficiency of "global" subjective efficacy ratings. *Clin. Pharmacol. Ther.*, **21**:34-43.

Dollery, C. T. (1967). Problems in clinical trials. *Adv. Sci.*, **23**:508-511.

Fleiss, J. L., *The Design and Analysis of Clinical Experiments*, John Wiley & Sons, New York, 1986.

Gorringe, J. A. L. (1970). Initial preparations for clinical trials. In *Principles and Practice of Clinical Trials* (Harris, E. L. and Fitzgerald, J. D., Eds.) Churchill-Livingston, Edinburgh, pp.41-46.

Hassar, M., Pocelinko, R., Weintraub, M., Nelson, D., Thomas, G., and Lasagna, L. (1977). The free-living volunteer's motivation and attitudes toward pharmacologic studies. *Clin. Pharmacol. Ther.*, **21**:515-519.

Hassar, M. and Weintraub, M., (1976). "Uninformed" consent and the healthy volunteer: An analysis of patient volunteers in a clinical trial of a new antiinflammatory drug. *Clin. Pharmacol. Ther.*, **20**:379-386.

Joubert, P., Rivera-Calimlim, L. and Lasagna, L. (1975). The normal volunteer in clinical investigation; How rigid should selection criteria be? *Clin. Pharmacol. Ther.*, **17**:235-257.

Lasagna, L. and Meier, P. (1959). Experimental design and statistical problems. In *Clinical Evaluation of New Drugs* (Waife, S. O. and Shapiro, A. P., Eds.). Hoeber, New York.

Lowenstein, E., Hallowell, P., Levine, E. H., Daggett, W. M., Austen, W. G., and Lauer, M. B. (1969). Cardiovascular response to large doses of intravenous morphine in man. *N. Engl. J. Med.*, **281**:1389.

Lung Heath Study Research Group (1993). *Control. Clin. Trials*, **14(suppl.)**:1S-79S.

Mainland, D. (1960). The clinical trial: Some difficulties and suggestions. *J. Chron. Dis.*, **11**:484-496.

Metropolitan Height and Weight Tables (1983). *Stat. Bull.* **64**:2-9.

Miller, R. and Willner, H. S. (1974). The two part consent form: A suggestion for promoting free and informed consent. *N. Engl. J. Med.*, **290**:964-966.

Morley, J. E., Vogel, K., and Solomon, D. H. (1990). Prevalence of geriatric articles in general medical journals. *J Amer. Geriatric Soc.*, **38**:173-176.

Murphee, H. B. (1966). Methodology for the clinical evaluation of analgesics. *J. New Drugs*, **15**:15-22.

Ropes, M. W., Bennett, G. A., and Cobb, S. (1965). Diagnostic criteria for rheumatoid arthritis. *Bull. Rheum. Dis.*, **9**:302-334.

Taves, D. R. (1974). Minimization: A new method for assigning patients to treatment and control groups. *Clin. Pharmacol. Ther.*, **15**:443-453.

Weintraub, M. (1976). Capricious compliance and intelligent noncompliance. In *Patient Compliance, vol. 10*, (Lasagna, L., Ed.), Futura, Mt. Kisco, NY, pp. 39-47.

Weintraub, M., Jacox, R. F., Angevine, C. D., and Atwater, E. C. (1977). Piroxicam (CP 16171) in rheumatoid arthritis: A controlled clinical trial with novel assessment techniques. *J. Rheum.*, **4**:393-404.

Weintraub, M., Sundaresan, P. R., Madan, M., Schuster, B., Balder, A., Lasagna, L., and Cox, C. (1992). Long-term weight control study I (weeks 0 to 34): The enhancement of behavior modification, caloric restriction, and exercise by fenfluramine plus phentermine versus placebo. *Clin. Pharmacol. Ther.*, **51**:886-94.

Yahr, M. D., Duvoisin, R. C., Schear, M. J., Barrett, R. E., and Hoehn, M. M. (1969). The treatment of Parkinson's disease with levodopa. *Arch. Neurol.*, **21**:343-354.

7

Statistical Aspects of Cancer Clinical Trials

T. Timothy Chen

National Cancer Institute, National Institutes of Health,
Bethesda, Maryland

I. CANCER TREATMENT PROGRESS

In the 1930s, less than 20% of cancer patients were alive 5 years after diagnosis. In the 1940s, the figure was about 25%, and in the 1960s it was about 33%. Today about 40% of cancer patients will be alive 5 years after diagnosis. If we compared with a similar control population, then the 5 year relative survival rate was 48.9% for patients diagnosed in 1974–76 and 49.8% for patients diagnosed during the period 1980–85 (Ries, Hankey, and Edwards, 1990). In the past three decades, good progress in treating cancer was made in acute lymphocytic leukemia in children, Hodgkin's disease, Burkitt's lymphoma, Ewing's sarcoma, Wilms' tumor, rhabdomyosarcoma, choriocarcinoma, testicular cancer, ovarian cancer and osteogenic sarcoma. However, for other common cancers, effective treatments have not been found.

Debates about whether we had really made progress in fighting against cancer since the passage of the National Cancer Act in 1971 were kindled several years ago (Bailar and Smith, 1986). The observation that the proportion of deaths due to cancer has increased progressively in the last 60 years has led to the conclusion that we are losing the fight against cancer. The progress against cancer is demonstrated clearly from examination of cohorts of men and women between 20 and 44 years of age (Doll, 1990). As a result of debates, several measures of progress against cancer were examined and many recommendations about the modification or expansion of the current information base were made (Extramural Committee to Assess Measures of Progress Against Cancer, 1990).

More than sixty anticancer drugs have received FDA approval for marketing in the United States. More than forty of these had their INDs (Investigational New Drug Application) sponsored by the National Cancer Institute. Currently, nearly one hundred new drugs and seventy new biologics are under active clinical investigation.

New improved methods of treating cancer are being actively pursued in all types of cancer. To establish the effectiveness of a new treatment, appropriate clinical trials have to be carried out. Statistical methods are used, in design, conduct, analysis, and reporting, to ensure the validity and efficiency of cancer clinical trials.

II. BENEFIT TO RISK RATIO

Since an antineoplastic drug usually produces toxicity to normal cells as well as killing cancer cells, we always have to consider the efficacy and toxicity together in obtaining a favorable benefit to risk ratio in cancer treatment. In other words, an oncologist would want to make sure that the new drug can produce a net benefit when compared to no treatment or current standard treatment. The net benefit can result from a large improvement in efficacy with a small worsening in toxicity or from a large reduction in toxicity with a small decrease in efficacy. The ideal situation will be both an improvement in efficacy and a decrease in toxicity. Of course, how much improvement in efficacy can balance out the harm of increasing toxicity is usually subjective and ambiguous.

When determining efficacy, the decision can be framed as accepting a treatment if it is good, or rejecting a treatment when it is not good. There are two types of error in this decision framework. The false negative error (Type I) is the error of misclassifying a good treatment as not good. The false positive error (Type II) is the error of misclassifying a bad or not so good treatment as good. There are costs or consequences associated with these two kinds of error. Usually in early clinical trials of a drug, we will tolerate a larger false positive error rather than a large false negative error.

In the process of developing a new treatment, the type and stage of cancer and the usefulness of the current standard treatment have to be considered in the estimation of the net benefit and the decision of whether to accept a new treatment (Kessler, 1989; O'Shaughnessy et al., 1991). Suppose the current treatment is not very effective for a certain type and stage of cancer, then the risk of using another ineffective drug is relatively not so high. In this situation, patients are willing to accept a larger false positive error or a larger variation for the estimation of net benefit. This kind of situation can be found in chronic lymphocytic leukemia in blast crisis, metastatic renal cell or germ cell cancer,

Hodgkin's disease refractory to MOPP/ABVD, postmenopausal hormone refractory metastatic breast cancer, or advanced stage lung cancer. Some AIDS clinical trials are in this class (Byar et al., 1990).

In other types and stages of cancer (e.g., previously untreated testicular cancer and Hodgkin's disease) the current treatment is very effective. Therefore, estimation of net benefit of the new treatment should have a higher precision, both false positive and false negative errors should be very small, and the new treatment should be compared to the current standard through a randomized controlled clinical trial.

In the above discussion, the risks of false positive and false negative errors are determined relatively according to the disease. This is reasonable from the treatment decision point of view. But from the perspective of scientific progress, the magnitude of errors should be small. Some large scale postmarketing studies could fulfill this purpose.

III. TRIAL ENDPOINTS

In cancer clinical trials, the efficacy endpoints include overall survival, quality of life, complete and partial response rate and duration, and time to progression. Overall survival is measured from the date of registration or randomization to the date of death or last follow-up. In the latter case, the observation is censored since the patient is still alive. Quality of life consists of many components including disease related symptoms and can be measured through a validated psychosocial instrument (Pocock, 1991; Aaronson, 1988). Complete response denotes the total disappearance of tumor lesions under clinical and diagnostic staging. The duration of complete response is measured only for complete responders from the date of response to the date of relapse or last check-up. In the latter case, the observation is censored since the patient is still in remission. Time to progression is measured for all eligible patients from the date of registration or randomization to the date of relapse (for responders) or the date of progression (for nonresponders) or the date of death (before relapse or progression). In early stage cancer after complete surgical removal of cancer (adjuvant setting), time to progression is called disease-free survival which is the time from study entry to relapse or death, with patients alive without relapse considered as censored.

The toxicity endpoints include the following major categories: blood or bone marrow toxicity, clinical hemorrhage, infection, gastrointestinal, liver, kidney, bladder, alopecia, pulmonary, heart, blood pressure, neurologic, skin, allergy, fever, metabolic, and coagulation. Within each major category, there are subcategories. For each subcategory, the grade of toxicity ranges from 0 to 4. Grade 0 means none or normal; grade 4 is life-threatening (National Cancer Institute, 1986).

In recent years many hematopoietic growth factors are being tested in cancer patients. The usefulness of these agents are measured by shortened duration of neutropenia or hospitalization, reduced episodes of febrile neutropenia, or delivery of more intensive chemotherapy. Other chemoprotectors are used to reduce incidence of nephrotoxicity and ototoxicity.

IV. PHASE I CLINICAL TRIAL

Drugs for treating cancer have to demonstrate effectiveness in tumor cell lines before actual testing in humans. The initial clinical evaluation of a new drug, biologic, or radiotherapy technique is called a Phase I trial which evaluates dose, schedule, toxicity, pharmacology, and early evidence of clinical activity. The goal of a Phase I trial is to arrive at a recommended dose with the minimum number of patients receiving either biologically inactive or toxic doses.

The patients selected for Phase I clinical trials are in those categories of cancer with no effective treatment at present. Patients should have good performance status and normal organ function. Hopefully, they might receive some benefit in using the investigative agent. The purpose of doing a Phase I trial is to find a dose which will cause dose-limiting toxicity in acceptable percentages of patients under a given method of administration for a fixed number of cycles.

The dose-limiting toxicity is usually defined as any grade 3 or 4 toxicity. The maximum tolerated dose (MTD) is usually defined as the dose which produces dose-limiting toxicity in 30% of patients. Many cytotoxic agents are observed to have a steep dose-response relation; therefore, the MTD estimated in a Phase I study will be the dose used in Phase II trials. Since a Phase I trial is a preliminary step for a Phase II trial, the requirement of statistical precision of dose determination is not high. Usually the dose can further be fine-tuned in a Phase II trial. The MTD is usually not disease-site specific. In general, pediatric patients have a different MTD from that of adults.

The usual starting dose in humans is one-tenth of the $MELD_{10}$ (mouse equivalent of the LD_{10}, dose with 10% drug-induced deaths) in mg/m^2 of body surface area, unless that dose is toxic in any species tested. The doses of drugs for human testing are usually selected by a modified Fibonacci method. The second dose level is twice the starting dose. The third dose level is 167% of the second, the fourth dose level is 150% of the third, the fifth dose level is 140% of the fourth, and each subsequent dose level is 133% of the preceding dose (Edler, 1990).

Since it is possible that the starting dose could be far away from the MTD and many patients will be exposed to a subtherapeutic dose, a pharmacokinetically guided dose selection scheme was proposed (Collins, Grieshaber, and Chabner, 1990). This approach builds upon a pharmacodynamic hypothesis that similar biological effects (e.g., toxicity) would happen at similar plasma levels in mice and man. For many agents, the area under the curve for plasma concentration versus time ($C \times T$) of the MTD for humans is found to be fairly close to the $C \times T$ for mice at the LD_{10} if calculated in mg/m^2 equivalents ($MELD_{10}$). Therefore, $C \times T$ of $MELD_{10}$ is considered as an upper limit, and the ratio F of $C \times T$ ($MELD_{10}$) to $C \times T$ (starting dose in men) is used to guide the dose selection. One method takes the second dose as $\sqrt{F}$ times the starting dose, and the third dose is twice the second dose, then follows the modified Fibonacci scheme. The other method continues to double the dose until 0.4F times the starting dose is exceeded and then the modified Fibonacci scheme is followed. Another use of preclinical toxicologic information is in choosing a higher entry dose than one-tenth of $MELD_{10}$. All these methods have a potential of reducing overall completion times by 25% (Collins, Grieshaber, and Chabner, 1990).

Drugs with high schedule-dependency in preclinical models will use the existing optimal schedule. For drugs without particular schedule dependency, two extremes of schedules (e.g., single bolus dose per course and 5-day continuous infusion) are generally examined (National Cancer Institute, 1986).

The usual dose-finding is carried out through a dose-escalation and de-escalation procedure. 1) Three new patients are studied at a dose level at the first stage. 2) If none experience dose-limiting toxicity, then the next higher dose is used for the subsequent group of 3 patients. 3) If 2 or more experience dose-limiting toxicity, then the MTD has been exceeded and 3 more patients are treated at the next lower dose (if only 3 patients were treated previously at this dose). 4) If $\frac{1}{3}$ experience dose-limiting toxicity at the current dose, then 3 more patients are accrued at the same dose at the second stage. If 0 of these three experience dose-limiting toxicity, then the dose is escalated. Otherwise the MTD has been exceeded and 3 more patients are treated at the next lower dose (if only 3 patients were treated previously). 5) The MTD is the dose level where $\frac{0}{6}$ or $\frac{1}{6}$ experience dose-limiting toxicity with the next higher dose having at least $\frac{2}{3}$ or $\frac{2}{6}$ experience dose-limiting toxicity.

In the above procedure, we require new patients at each dose level. Sometimes at very low doses, we could reenter a patient at a higher dose level and include this patient in the analyses of both dose levels. For higher doses, this kind of intrapatient escalation of doses could confound the result due to possible cumulative toxicity. The toxicity that occurs to these re-entered patients could either be due to the higher second dose or due to the cumulated total dose. Therefore, if intrapatient escalation is used in high doses, then the patient is only included in the analysis of the first dose level.

Storer (1989) proposed other single-stage and two-stage designs and the methods of estimating MTD. The different designs were compared through computer simulation by assuming logistic dose-toxicity curves. The results indicated that there is little difference among the two-stage designs due to the small sample sizes.

A Phase I study usually has a pharmacokinetic component to understand the absorption, distribution, metabolism, and excretion of the drug in humans. If the variation of pharmacokinetic behavior is too large for a drug among the patient population, then some kind of adaptive dosing can be used to control a patient's plasma concentration within a desirable range (Sheiner and Beal, 1982). This approach has the potential of maximizing response and minimizing toxicity. Its usefulness is under investigation.

V. PHASE II CLINICAL TRIAL

After a MTD is determined, the drug at that dose and schedule is carried forward to get a better estimate of antitumor activity in a Phase II trial which will evaluate drug, biologic, or radiotherapy techniques in single modality or combined modality regimens. The definition of a Phase II trial in cancer clinical trials is different from that of a Phase II clinical study defined by the FDA in drug development. The Phase II cancer clinical trial usually requires fewer than 100 patients, whereas FDA Phase II requires several hundred patients (Kessler, 1989).

From past experience, efficacy of a drug is disease-site specific; therefore, a Phase II trial is limited to a specific type of cancer. The kind of cancer to be tested is determined through preclinical animal data and the data collected in the Phase I clinical trials. Patients should have good performance status and normal organ function.

Since cancer drugs can be cross-resistant—that is, a drug can be less effective as a second-line treatment than as a first-line treatment—a new drug should be preferably used as a first-line treatment. The phenomenon of cross-resistance is due to similar drug actions; when some tumor cells are refractory to a certain drug, they are going to be resistant to a similar-action drug. Therefore, a Phase II trial is preferably done in patients who have not been previously treated if there is no effective treatment at the present for this particular type of cancer. If we use previously treated patients, the efficacy could be low, and we cannot differentiate it from background noise.

For certain categories of cancer at early stages, there are some very effective treatments. A Phase II trial will usually be first done in patients with the late stage of these kinds of cancer where the existing treatment is not so effective and then later in patients with the early stage cancer. A Phase II trial should not diminish a window of opportunity for patients to get effective treatment (Moore and Korn, 1992).

The response variable for a Phase II trial is usually tumor response rather than survival; the tumor response can be determined in the first few months of treatment and it usually has good correlation with survival within a specific type of cancer. In order to obtain precise evaluation of response, the patients should have measurable disease which can be measured through diagnostic tools.

For solid tumors, response includes both complete and partial response. For leukemia, the response includes only complete remission since it is known that only complete remission is related to long term survival (Cheson et al., 1990). In order to qualify as a response, the tumor reduction should be long-lasting, usually one month.

The design of a Phase II trial is based on one-sample binomial statistics with the probability of success being the probability of achieving a response. From the current treatment results, both desirable and undesirable response rates are specified. We would like to reject the drug as not promising, if it is unlikely that it has the desirable response rate (the false negative error is small); and to accept a drug as promising if it is unlikely that it has the undesirable response rate (the false positive error is small). Since we prefer not to miss any promising drug, the false negative error should not be greater than the false positive error. The trial is carried out in two stages (Simon, 1989) so that if a drug is not promising, the trial can be terminated early at the end of the first stage. This two-stage design and other multistage designs (Fleming, 1982) are examples of a broad class of drug screening procedures (Schultz et al., 1973). This approach to Phase II trial is identical to sampling inspection in the industrial quality control setting.

For example, a Phase II trial for a new agent in nonsmall cell lung cancer can have the following two-stage design. For the first stage, twelve patients are enrolled. If there is no response (complete or partial) in these twelve patients, then the study is terminated and the agent is rejected. If there is at least one response, then 25 more patients are enrolled in the second stage. If there are less than four responses in 37 patients, then the agent is rejected; otherwise, the new agent will be deemed promising. This design is based on the current treatment result for nonsmall cell lung cancer; the response rate of 20% is desirable and 5% is undesirable. The false negative and positive rates are limited to 0.10. This design has the minimum expected sample size (23.5) when the agent has a response rate of 5% (Simon, 1989).

Phase II trials usually are repeated in at least two different centers. Response rates from two trials could be different due to several reasons: patient selection, different evaluation and response criteria, intra- and interevaluator variability or bias, and protocol compliance. Whether a drug will enter into a Phase III trial depends not only on the tumor response observed in Phase II trials, but also on its toxicity, dose-response relationship, and cross-resistance with other active agents.

In the situation where there is more than one new agent, a randomized Phase II can be done. The sample size used here is about the size of an usual Phase II trial and the comparison will not be as precise as a Phase III trial. The advantage of a randomized Phase II trial is that the results for the several new agents can be compared within the same patient population and protocol procedure.

VI. PHASE III CLINICAL TRIAL

A. General Consideration

After a drug or treatment regimen has been shown to have promising antitumor activity, it may progress forward to a Phase III trial. A Phase III trial compares the experimental treatment(s) with a standard control treatment. The purpose of a Phase III trial is to demonstrate that the new treatment is either better than or equivalent to the standard control. The response variable for a Phase III trial is usually the overall survival or the quality of life.

Sometimes instead of concurrent control, historical control data are used in comparison. The validity and usefulness of this approach is very limited (Cox and McCullagh, 1982). The historical controls have to be the patients treated in the same institution in the past few years with the same enrollment criteria and evaluation procedure. The baseline comparability of historical control patients and current patients can never be demonstrated beyond reasonable doubt. The usefulness of statistical adjustment is based on the validity of the model assumptions and the inclusion of all major prognostic variables.

In a Phase III cancer clinical trial protocol, the purposes of the trial have to be very clearly defined. For each stated purpose, the data collection procedure should be thought through and described in the protocol. The statistical techniques of analyzing these data need to be planned and stated in the statistical consideration section of the protocol.

The data collected should be valid and reliable. To be certain about its validity and reliability, unequivocal documentation must be provided. Overall survival is very reliable if every patient is followed until death. The response, progression, and relapse status are less reliable due to the limitation of diagnostic techniques or clinical evaluation. To ensure that these data are reliable, a uniform and unbiased follow-up, standardized supportive care, secondary treatment, and method of evaluation for all treatment groups is very important.

Treatment regimens need to be defined specifically. For chemotherapy, the dose, schedule, route of administration, and dose modification due to toxicity have to be described clearly. For radiotherapy, the dose, schedule, and field size should be similarly specified. For surgery, the incision margin and number of nodes to be sampled need to be specified. The quality of treatment delivered should be monitored very closely to maintain the protocol compliance. The purpose of doing quality control is twofold: to insure that patients get optimal treatment and to minimize variation in the treatment outcome.

The eligibility criteria of patients must be determined very carefully. Here we need to strike a balance between more stringent and less stringent criteria. Usually good performance status and normal liver and kidney function patients can show maximal difference between the treatments in a trial. However, limiting the patient eligibility also limits the applicability of the trial result. Therefore, it usually is better to enroll all the patients who could benefit from the treatment and are healthy enough to receive the treatment.

Case report forms including prestudy, flow sheet, pathology, surgery, and radiotherapy should be designed with extra care so that only necessary and useful information is collected. If too much information is required, then the quality of data will deteriorate.

B. Randomization

In a Phase III trial, a randomized trial will provide the best design (Byar et al., 1976). The purpose of randomization is two-fold: 1) randomization will provide a theoretical foundation for the validity of the statistical analysis of the trial data; and 2) randomization will render the treatment groups comparable regarding unknown and known prognostic factors and reduce the bias in assigning patients to treatments. The actual process of randomization is usually done through a centralized statistical office.

The randomization scheme is usually not unrestricted randomization, but the sample sizes of the treatments are constrained to be equal. More desirable is block randomization with a block size of four to eight patients with blocks nested within each clinical center. With each block there is equal assignment of patients to each treatment. The purpose of doing this is to ensure that the final numbers of patients on the treatments are almost equal. The information about block size should not be revealed to avoid bias in the enrollment of patients.

The time of randomization should take place as close as possible to the time of beginning different treatments (Durrleman and Simon, 1991). For example, the protocol can have the same induction regimen, but two different intensification regimens. The patients should be registered twice: once before induction, the second time for randomization before intensification. This approach will minimize the possible bias in the eligibility determination for intensification and control the variability of number of patients on the two intensification arms. The statistical analysis to compare the two intensification arms can be restricted to those patients who were randomized.

C. Stratification

Usually some important prognostic factors can be identified before a trial. In this case, it is advisable to stratify patients by these prognostic factors, and then randomize within each stratum. The purpose of stratification is two-fold: 1) to make the result of the study more convincing, and 2) to increase the efficiency of statistical analysis.

Peto et al. (1976) stated that stratified randomization is not necessary for a large trial, since the probability is high that the balance among important prognostic factors can be achieved by unrestricted randomization. However, a stratified randomization is similar to an insurance policy to insure against the unlikely event of unbalanced distribution of patients among the important prognostic factors. If this event happens, an adjusted analysis may not alleviate the doubt because the statistical adjustment usually depends on model assumptions.

If one wants to balance patient assignment on many prognostic factors, some kind of dynamic allocation scheme can be considered (Pocock and Simon, 1975). For a multicenter clinical trial, it is always desirable to balance treatments at each institution since there is usually an institutional effect on the trial outcome (Simon, 1979).

D. Size of the Trial

The size of a trial depends on the degree of precision we would like to have about the estimate of the treatment difference. This is related to the width of the confidence interval for the treatment difference, or to the Type I and Type II errors of differentiating two hypotheses for possible values of the treatment difference. In a clinical trial, the sample size determination is usually done through the latter approach since the two types of error are usually not equally serious, and therefore not symmetric. However, there is a natural connection between the confidence interval approach and the hypothesis testing approach. In the analysis and report of the study, a confidence interval will provide more information than just a p-value. A p-value is the result of comparison of the observed difference with only one value of the hypothetical difference. A confidence interval or a standard error of the observed difference provides information about the whole range of values of the hypothetical difference. If the main variable is the overall survival, then the difference between treatments can be formulated in terms of the hazard ratio. If the main variable is the tumor response, then the difference between treatments can be formulated in terms of the odds ratio. If the main variable is a normal, continuous one, then the difference between two treatments is just the mean difference.

If the purpose of the trial is to show that a new treatment T_1 is better than the standard control treatment T_0, then we specify as the null hypothesis that the two treatments are the same, and try to use data to reject this null hypothesis. Since the likelihood of observing a treatment advance is not great (according to the past history of cancer clinical research), we want to control the Type I error α which is the probability of rejecting the null hypothesis when it is true. The Type I error is specified as 0.05. We also specify an alternative hypothesis which says the difference between the two treatments is a certain amount. We would like to control the Type II error β which is the probability of not rejecting the null hypothesis when the alternative hypothesis is true. The Type II error is usually specified between 0.1 and 0.2 (i.e., the power is between 0.9 and 0.8). We usually specify the treatment difference in the alternative as a clinically meaningful difference, or the minimum difference we would like to detect. This value is usually subjectively obtained and usually is a compromise between what is really important and what can be done. If this value is too large then the study will not have enough power to detect a smaller difference.

For a one-sided alternative and normally distributed data, assuming that the variance of both treatments are the same, the formula to obtain the required number of patients for each treatment is $n = 2\tau^2/\delta^2$ where $\tau = (z_{1-\alpha}) + (z_{1-\beta})$ and $\delta = (\mu_1 - \mu_0)/\sigma$. The z_P is the value of the normal deviate corresponding to the P point of the cumulative standard normal distribution. The value δ is the difference in means divided by the standard deviation. The value of τ is determined by the Type I and the Type II errors. For a two-sided alternative, α should be replaced by $\alpha/2$. Note that the total sample size required for the trial is $4\tau^2/\delta^2$.

For binomial response data, the formula for n is similar with $\delta = 2 \cdot (\arcsin \sqrt{p_1} - \arcsin \sqrt{p_0})$. The arcsine transformation of the square root of the observed proportion stabilized the standard deviation as $\frac{1}{2}\sqrt{n}$. A more accurate formula for the sample size to compare two binomials has been published (Casagrande, Pike, and Smith, 1978).

For the survival data, the formula for n is again similar with $\delta = \ln(\lambda_0/\lambda_1) \cdot \sqrt{2}/\sqrt{(1/p_1 + 1/p_0)}$. Here we make the assumption that the distribution of the survival data is exponential with hazard rates λ_0 and λ_1, and p_i $(i = 0, 1)$ is the proportion of actual events (deaths) for the i-th treatment at the time of data analysis. This expected proportion of events p_i is a function of total accrual time (M), the further follow-up time (L) after the termination of accrual, and the hazard rate λ_i. Assuming a Poisson patient arrival over M, and all patients are observed till the end of the further follow-up time L, then p_i is

$$1 - \frac{\exp(-\lambda_i M) \cdot \exp(-\lambda_i L)(\exp \lambda_i M - 1)}{\lambda_i M}$$

$$(7.1)$$

The reason that δ is more complicated is because the sufficient statistic is the number of events and not the number of total enrollments for exponential data with censoring. The formula for n can be rewritten as

$$\frac{\tau^2}{(\ln(\lambda_0/\lambda_1))^2} = \frac{np_0p_1}{p_0 + p_1}$$

$$(7.2)$$

In the design of a Phase III trial comparing overall survival, an appropriate follow-up period should be allowed after the closure of enrollment so that the number of expected events under the alternative hypothesis will be np_0 and np_1 at the time of the final analysis (Bernstein and Lagakos, 1978; Rubinstein, Gail, and Santner, 1981).

If the follow-up period is long enough, then p_0 is very close to p_1. The right hand side of the above formula would be very close to one-fourth of the total number of events in the trial. The total number of events in the trial is $4\tau^2/(\ln(\lambda_0/\lambda_1))^2$. If α (one-sided) is 0.025, β is 0.2, and the alternative of $1.5\lambda_1 = \lambda_0$ (a 50% improvement in median time to event) is to be detected, then the total number of events in the trial should be 192. If the follow-up period is not long enough, and p_0 is not close to p_1, then the total number of events should be greater, but not more than 110% of $4\tau^2/(\ln(\lambda_0/\lambda_1))^2$. For example, in a Phase III trial of stage IIIA and IIIB inoperable nonsmall cell lung cancer with vinblastine and cisplatinum followed by radiation therapy as the control, the experimental treatment could be vinblastine and cisplatinum followed by radiation therapy with concurrent carboplatin. If the accrual rate is 6.2 patients per month, and the control arm has a median survival of 15.5 months, then the study design should have an accrual period of three and a half years (260 patients) and a follow-up period of one and a half years in order to have 208 as the total number of events in the trial at the end of five years.

If the purpose of a Phase III trial is to show that the new treatment is equivalent to the standard control, then we have to define the term "equivalency." Some will define equivalence as within 10% of the control mean or an odds or hazard ratio between 0.9 and 1.1. Since the observed difference has variation, the definition will further require that the 95% confidence interval for mean ratio, odds ratio, or hazard ratio must be within the interval of 0.9 and 1.1. This kind of requirement is very stringent. If the true ratio is close to either 0.9 or 1.1, then the sample size would have to be very large to have the 95% confidence interval within the interval of (0.9, 1.1). Therefore, an equivalency trial usually requires more patients than a trial to prove superiority.

Sometimes a new treatment has less toxicity than the standard control and we are willing to "accept" the new treatment if it is not more than 10% inferior in efficacy. This kind of trial is also called an equivalency trial but it is different from the trial described in the previous paragraph. Here we only require that the lower limit of the confidence interval be greater than 0.9 and we set no bound for the upper limit. Again, the sample size will depend on how close the true efficacy ratio is to the value of 0.9. If we are willing to carry more risk, then the sample size will not be as large as the true equivalency trial in the previous paragraph. This trial is very similar to a superiority trial except the null value is shifted from 1 to 0.9 (Dunnett and Gent, 1977). To obtain a size for this kind of equivalency trial, one usually takes 0.9 as the null value and 1 as the alternative value, the Type I error associated with the null value at 5%, and the Type II error can be between 5 and 20%. This kind of design can be interpreted as a test which places more burden upon a new treatment to prove it is not more than 10% inferior in efficacy.

The expected accrual rate should be considered in the design. If a trial takes more than four years to complete, then the treatments being compared could become obsolete as the trial progresses and the investigator could lose interest in enrolling patients. Therefore, a Phase III trial is usually done by a cooperative oncology group. Sometimes for a rare cancer, the trial is done through an intergroup mechanism which pools several oncology groups together. This kind of pooling of resources makes sure that the trial can be finished in a reasonable amount of time. During the progress of the trial, accrual rate should be monitored to ensure it is not too far from the expected rate.

To compare survival or disease-free survival for two treatments, sometimes the comparison is done in terms of proportion of patients without event (survived, or survived and free of disease) at a specified time. If the sample size is calculated for comparison of two proportions, it is larger than that for the comparison of two exponential curves (Gail, 1985). Since the final test is usually a logrank test which compares the entire survival curves, the latter sample size is preferable unless the exponentiality assumption is grossly untrue.

If patient compliance or loss to follow-up could become major problems in the conduct of a trial, an allowance should be provided in the sample size calculation (Lachin and Foulkes, 1986). The sample size requirement based on the logrank statistic without the exponentiality assumption has been derived under very general conditions which include cure rate models (Lakatos, 1988; Halpern and Brown, 1987).

E. Data Analysis

Since randomization provides a theoretical basis to do statistical analysis, the analysis of the trial should include all the randomized patients. However, some ineligible patients could also get randomized by mistake. If they do not belong to the patient population for which the treatments question is being asked, they can be excluded from the analysis. In rare instances, patients may cancel their registration before the treatment begins and are not included in the analysis. Other than these, all patients should be included in the analysis. This approach is called "intent-to-treat" analysis since we intend to treat all eligible patients according to the randomly assigned treatments. Those patients who do not comply with the protocol are called inevaluable patients; they are included in the intent-to-treat analysis in the groups into which they were randomized. Since strict adherence to the protocol treatment is not easy when it is used in everyday practice, and we would like to find out in a Phase III trial what would happen if we use the treatments in a general patient population, doing intent-to-treat analysis will provide an answer closer to the general practice. Also treatment could affect early death and early dropout and compliance; therefore, excluding these inevaluable patients in the analysis will not provide a fair comparison of the total treatment effect. Including all randomized eligible patients also minimizes the possibility of biased exclusion since cancer trials usually cannot be blinded to the patients or to the investigators. Of course, if there is no bias involved, then the analysis excluding inevaluable patients could provide additional useful information.

There has been tremendous development in the statistical methods for censored survival data since 1960. For the estimation of survival distribution, the product-limit estimator (Kaplan and Meier, 1958) is widely used in clinical trials whereas the life table method is mostly used in epidemiology (Berkson and Gage, 1950; Cutler and Ederer, 1958).

For comparing two treatments, one method was an extension of the Wilcoxon rank statistics (Gehan, 1965; Prentice, 1978). Later, another method treating data as a series of 2×2 tables was an extension of Mantel-Haenszel test and is now called the logrank test (Mantel and Haenszel, 1959; Mantel, 1966). Both methods were shown to be in a class of weighted-sum statistics (Tarone and Ware, 1977; Harrington and Fleming, 1982). The logrank test has the benefit of being easily extended to cover the stratified analysis, and hence has been used extensively in the meta-analysis of cancer clinical trials (Early Breast Cancer Trialists' Collaborative Group, 1992). The logrank statistic and its variance estimate can be used to calculate the hazard ratio estimate and its standard error (Early Breast Cancer Trialists' Collaborative Group, 1988).

In analysis, the data of the major endpoints should be analyzed both unadjusted and adjusted by the stratifying factors. The adjusted analyses can be carried out through stratified analyses (Mantel and Haenszel, 1959; Cochran, 1954) or a regression model (Cox, 1958, 1972; Kalbfleisch and Prentice, 1980). If a prognostic factor is ordinal or numerical, and the relationship between the response and the prognostic factor can be approximated by the regression model, the regression approach will provide a better test for the treatment effect.

If there is a statistically significant difference between the treatments, then it is appropriate to show that there is no qualitative interaction between treatments and known prognostic factors. This can be tested by a statistical procedure (Gail and Simon, 1985) and by a tabulation of means or medians of the major endpoints by treatments within patient subgroup (Meier, 1983). This kind of tabulation to show there is no qualitative interaction will provide more credence for the conclusion. If there is no statistically significant difference between the treatments in the overall analysis, then subset analysis should not be done. Any intended subset analysis should be prespecified in the protocol and the sample size should be adequate within the subset. Any unprespecified subset analysis can only be used as hypothesis generating and any finding should be confirmed by another trial.

There is some controversy about using a one-sided or two-sided test. However, after using a two-sided test in a clinical trial, if we reject the null hypothesis of no difference, we will usually conclude that the new treatment is better than the control or the new treatment is worse than the control. Since we will not just conclude they are different, one-sided p-values are more relevant if we use the p-value as a strength of evidence to support our conclusion. Therefore, in order to be consistent in drawing conclusions after either a one-sided or two-sided test, a one-sided p-value of 0.025 should be seen as providing strong evidence.

Sometimes a better survival for responders than the survival for non-responders is used to argue for the efficacy of a new treatment in an uncontrolled single arm trial. This argument is not valid due to two reasons (Anderson et al., 1983). First, the better survival for the responders could be due to favorable prognostic status and not treatment. Second, the responders have to survive long enough to get a response. A randomized controlled trial is almost always needed to demonstrate the efficacy of a new agent beyond any reasonable doubt.

F. Interim Analyses

In the progress of a cancer clinical trial, if a new treatment has shown its superiority early, then a proper procedure should be in place to terminate the trial so that more patients will get the new effective treatment. Similarly, if the new treatment is worse than the standard control, the trial should be terminated early so that fewer patients will be exposed to the ineffective new treatment. However, if the interim analyses are carried out many times at the 5% significance level, the overall significance level is much larger than 5% (Armitage et al., 1969). Since the false positive result is quite common, interim analysis has to be carried out carefully so that the overall significance level (Type I error) stays at 5%. (See Chapter 14.)

For a large Phase III trial, the data monitoring procedure and stopping rule should be specified in the protocol and a data monitoring committee should convene periodically to decide whether to terminate the trial early. The members of this committee are privy to see the unblinded interim analysis results, whereas in semiannual oncology group meetings the reporting of interim analysis is blinded and the overall survival and toxicity results are usually pooled across treatments. The purpose of blinding the interim results is to safeguard the progress of the trial.

In doing interim analyses one controls the probability of making a wrong conclusion (Type I error) at 5% if the treatments are equally efficacious. Since the wrong conclusion can be reached at any interim analysis, the sum of (spending) probabilities of making wrong conclusions should total 5%. Lan and

DeMets (1983) proposed several use functions to spend this 5% in the process of the trial. Once a use function is specified as a function of the information fraction accrued in a trial, interim analysis can be carried out at any time with incremental error probability determined from the use function.

Originally interim analyses were proposed to be carried out at fixed time intervals with equal number of events (information time) in each interval. Pocock (1977) proposed a procedure which uses the same critical value for each interim standardized test statistic. O'Brien and Fleming (1979) proposed a procedure which uses different critical values for each interim standardized test statistic and that these critical values are inversely proportional to the square root of the number of events. This procedure is more conservative in the early looks; therefore, the power is reduced very little as compared with a fixed size trial. Since many cancer trials are analyzed before the semiannual oncology group meeting, it was shown that both procedures for repeated logrank analyses are quite robust if done at equal intervals of calendar time rather than information time (DeMets and Gail, 1985). Pocock (1982) showed that there is not much gain in efficiency or ethical benefit in doing more than five interim analyses.

Since any testing procedure can be converted into a confidence interval, interim tests can be converted into repeated confidence intervals (Jennison and Turnbull, 1989). The width of the interval is z_p times the standard error of the observed treatment efficacy ratio where z_p is the critical value for the interim standardized test statistic. The repeated confidence intervals can be used for early decision about any prespecified treatment efficacy ratio, not just the value of one.

Since interim analysis is making inference with incomplete data, the assumption of uninformative censoring is a critical one. Using early results to predict the long-term outcome, we assume that the early trend will continue. It is prudent to have sufficient follow-up before terminating a trial. A subsequent analysis for the long-term result is appropriate in many cases.

Sometimes early termination is done to conserve patient resources when the two treatments are quite similar. Lan, Simon, and Halperin (1982) proposed a stochastic curtailing procedure to terminate a trial in this situation. They compute a conditional probability of not rejecting the null hypothesis given the current observed data and the alternative hypothesis is true. If this probability is very high, then the trial can be terminated. The complement of this procedure can be used to terminate a trial when the two treatments are very different; however, it is more conservative than the group sequential approach.

VII. TRIAL REPORT

After the data set is analyzed, a paper should be written for peer review. Some guidelines have been suggested for writing up a report (Zelen, 1983). The following has been proposed by Simon and Wittes (1985).

1. The paper should discuss briefly the quality control methods used to ensure that the data are complete and accurate. A reliable procedure should be cited for ensuring that all patients entered in the study are actually reported upon. If no such procedures are in place, their absence should be noted. Any procedures employed to ensure that assessment of major endpoints is reliable should be mentioned (e.g., second-party review of response) or their absence noted.

2. All patients registered in the study should be accounted for. The report should specify for each treatment the number of patients who were not eligible, died, or withdrew before treatment began. The distribution of follow-up times should be described for each treatment, and the number of patients lost to follow-up should be given.

3. The study should not have an inevaluability rate greater than 15% for major endpoints due to early death, protocol violation, and missing information. Not more than 15% of eligible patients should be lost to follow-up.

4. In randomized studies, the report should include a comparison of survival and/or other major endpoints for all eligible patients as randomized, that is, with no exclusions other than those not meeting eligibility criteria.

5. The sample size should be sufficient to either establish or conclusively rule out the existence of effects of clinically meaningful magnitude. For "negative" results in therapeutic comparisons, the adequacy of sample size should be demonstrated by either presenting a confidence interval for the true treatment difference or calculating the statistical power for detecting differences. For uncontrolled Phase II studies, a procedure should be in place to prevent the accrual of an inappropriately large number of patients, when the study has shown the agent to be inactive.

6. Authors should state whether there was an initial target sample size and, if so, what it was. They should specify how frequently interim analyses were performed and how the decisions to stop accrual and report results were arrived at.

7. All claims of therapeutic efficacy should be based upon explicit comparisons with a specific control group, except in the special circumstances where each patient is his own control. If nonrandomized controls are used, the characteristics of the patients should be presented in detail and compared to those of the experimental group. Potential sources of bias should be adequately discussed. Comparison of survival between responders and nonresponders does not establish efficacy and should not be included. Reports of Phase II trials which draw conclusions about antitumor activity but not therapeutic efficacy do not require a control group.

8. The patients studied should be adequately described. Applicability of conclusions to other patients should be carefully dealt with. Claims of

subset-specific treatment differences must be carefully documented statistically as more than the random results of multiple-subset analyses.

9. The methods of statistical analysis should be described in sufficient detail that a knowledgeable reader could reproduce the analysis if the data were available.

ACKNOWLEDGMENT

The author would like to thank Drs. Edward L. Korn and K. K. Gordon Lan for their comments and suggestions.

REFERENCES

Aaronson, N. K. (1988). Quality of life: What is it? How should it be measured? *Oncology*, **2**:69-74.

Anderson, J. R., Cain, K. C., and Gelber, R.D. (1983). Analysis of survival by tumor response. *J. Clin. Oncol.*, **1**:710-719.

Armitage, P., McPherson, C. K., and Rowe, B.C. (1969). Repeated significance tests on accumulating data. *J. R. Statist. Soc.*, **A132**:235-244.

Bailar, J. C. III. and Smith, E. M. (1986). Progress against cancer? *N. Engl. J. Med.*, **314**:1226-1232; correspondence on Progress against cancer. *N. Engl. J. Med.*, **315**:963-968.

Berkson, J. and Gage, R. P. (1950). Calculation of survival rates for cancer. *Proc. Staff Meetings Mayo Clinic*, **25**:270-286.

Bernstein, D. and Lagakos, S. W. (1978). Sample size and power determination for stratified clinical trials. *J. Statistic. Comp. Simul.*, **8**:65-73.

Byar, D. P., Simon, R. M., Friedewald, W. T., et al. (1976). Randomized clinical trials: Perspectives on some recent ideas. *N. Engl. J. Med.*, **295**:74-80.

Byar, D. P., Schoenfeld, D. A., Green, S. B., et al., (1990). Design considerations for AIDS trials. *N. Engl. J. Med.*, **323**:1343-1348.

Casagrande, J. T., Pike, M. C., Smith, P. G. (1978). An improved formula for calculating sample sizes for comparing two binomial distributions. *Biometrics*, **34**:483-486.

Cheson, B. D., Cassileth, P. A., Head, D. A. et al., (1990). Report of the National Cancer Institute-sponsored workshop on definitions of diagnosis and response in acute myeloid leukemia. *J. Clin. Oncol.*, **8**:813-819.

Cochran, W. G. (1954). Some methods of strengthening the common χ^2 tests. *Biometrics*, **10**:417-451.

Collins, J. M., Grieshaber, C. K., and Chabner, B. A. (1990). Pharmacologically guided Phase I clinical trials based upon preclinical drug development. *J. Natl. Cancer Inst.*, **82**:1321-1326.

Cox, D. R. (1958). The regression analysis of binary sequences. *J. Roy. Statist. Soc.*, **B20**:215-242.

Cox, D. R. (1972). Regression models and life tables (with discussion). *J. Roy. Statist. Soc.*, **B74**:187-220.

Cox, D. R. and McCullagh, P. (1982). Some aspects of analysis of covariance. *Biometrics*, **38**:541-561.

Cutler, S. J. and Ederer, F. (1958). Maximum utilization of the life table method in analyzing survival. *J. Chronic Dis.*, **8**:699-712.

DeMets, D. L. and Gail, M. H. (1985). Use of logrank tests and group sequential methods at fixed calendar times. *Biometrics*, **41**:1039-1044.

Doll, R. (1990). Are we winning the fight against cancer? An epidemiological assessment. *European J. Cancer*, **26**:500-508.

Dunnett, C. W. and Gent, M. (1977). Significance testing to establish equivalence between treatments, with special reference to data in the form of 2 × 2 tables. *Biometrics*, **33**:593-602.

Durrleman, S. and Simon, R. (1991). When to randomize? *J. Clin. Oncol.*, **9**:116-122.

Early Breast Cancer Trialists' Collaborative Group (1988). Effects of adjuvant tamoxifen and of cytotoxic therapy on mortality in early breast cancer: An overview of 61 randomized trials among 28,896 women. *N. Engl. J. Med.*, **319**:1681-92.

Early Breast Cancer Trialists' Collaborative Group (1992). Systemic treatment of early breast cancer by hormonal, cytotoxic, or immune therapy: 133 randomized trials involving 31,000 recurrences and 24,000 deaths among 75,000 women. *Lancet*, **339**:1-15,71-85.

Edler, L. (1990). Statistical Requirements of Phase I Studies. *Onkologie*, **13**:90-95.

Extramural Committee To Assess Measures of Progress Against Cancer (1990). Measures of progress against cancer. *J. Natl. Cancer Inst.*, **82**:825-835.

Fleming, T. R. (1982). One sample multiple testing procedure for Phase II clinical trials. *Biometrics*, **38**:143-151.

Gail, M. H. (1985). Applicability of sample size calculation based on a comparison of proportions for use with the logrank test. *Controlled Clinical Trials*, **6**:112-119.

Gail, M. and Simon, R. (1985). Testing for qualitative interactions between treatment effects and patient subsets. *Biometrics*, **41**:361-372.

Gehan, E. A. (1965). A generalized Wilcoxon test for comparing arbitrarily singly-censored samples. *Biometrika*, **52**:203-223.

Halpern, J. and Brown, B. W. Jr. (1987). Cure rate models: Power of the logrank and generalized Wilcoxon tests. *Statistics in Medicine*, **6**:483-489.

Harrington, D. P. and Fleming, T. R. (1982). A Class of rank test procedures for censored survival data. *Biometrika*, **69**:553-566.

Jennison, C., and Turnbull, B. W. (1989). Interim analyses: The repeated confidence interval approach (with discussion). *J. R. Statist. Soc.*, **B51**:305-361.

Kalbfleisch, J. D. and Prentice, R. L. (1980). *The Statistical Analysis of Failure Time Data*, John Wiley and Sons, New York.

Kaplan, E. L. and Meier, P. (1958). Nonparametric estimation from incomplete observations. *J. Amer. Statist. Assoc.*, **53**:457-481.

Kessler, D. A. (1989). The regulation of investigational drugs. *N. Engl. J. Med.*, **320**:281-288.

Lachin, J. M. and Foulkes, M. A. (1986). Evaluation of sample size and power for analyses of survival with allowance for nonuniform patient entry, losses to follow-up, noncompliance, and stratification. *Biometrics*, **42**:507-519.

Lakatos, E. (1988). Sample size based on the log-rank statistics in complex clinical trials. *Biometrics*, **44**:229-241.

Lan, K. K. G. and DeMets, D. L. (1983). Discrete sequential boundaries for clinical trials. *Biometrika*, **70**:659-663.

Lan, K. K. G., Simon, R., and Halperin, M. (1982). Stochastically curtailed tests in long-term clinical trials. *Commun. Statist.-Sequential Analysis*, **1**:207-219.

Mantel, N. (1966). Evaluation of survival data and two new rank order statistics arising in its consideration. *Cancer Chemotherapy Rpt.*, **50**:163-170.

Mantel, N. and Haenszel, W. (1959). Statistical aspects of the analysis of data from retrospective studies of disease. *J. Natl. Cancer Inst.*, **22**:719-748.

Meier, P. (1983). Statistical analysis of clinical trials. In *Clinical Trials: Issues and Approaches*, (S. H. Shapiro, and T. A. Louis, Eds.), Marcel Dekker, New York and Basel, pp. 155-189.

Moore, T. D. and Korn, E. L. (1992). Phase II Trial design considerations for small-cell lung cancer. *J. Natl. Cancer Inst.*, **84**:150-154.

National Cancer Institute (1986). *Investigator's Handbook. A Manual for Participants in Clinical Trials of Investigational Agents Sponsored by the Division of Cancer Treatment National Cancer Institute*. National Cancer Institute, Bethesda, MD.

O'Brien, P. C. and Fleming, T. R. (1979). A multiple testing procedure for clinical trials. *Biometrics*, **35**:549-556.

O'Shaughnessy, J. A., Wittes, R. E., Burke, G., et al., (1991). Commentary concerning demonstration of safety and efficacy of investigational anticancer agents in clinical trials. *J. Clin. Oncol.*, **9**:2225-2232.

Peto, R., Pike, M. C., Armitage, P., et al., (1976). Design and analysis of randomized clinical trials requiring prolonged observation of each patient. I. Introduction and design. *Brit. J. Cancer*, **34**:585-612.

Pocock, S. J. (1977). Group sequential methods in the design and analysis of clinical trials. *Biometrika*, **64**:191-199.

Pocock, S. J. (1982). Interim analyses for randomized clinical trials: The group sequential approach. *Biometrics*, **38**:153-162.

Pocock, S. J. (1991). A perspective on the role of quality-of-life assessment in clinical trials. *Controlled Clinical Trials*, **12**:257S-265S.

Pocock, S. J. and Simon, R. (1975). Sequential treatment assignment with balancing for prognostic factors in the controled clinical trial. *Biometrics*, **31**:103-115.

Prentice, R. L. (1978). Linear rank tests with right censored data. *Biometrika*, **65**:167-179.

Ries, L. A., Hankey, B. F., and Edwards, B. K. (Eds.) (1990). *Cancer Statistics Review 1973-87*. NIH Publication No. 90-2789, National Institutes of Health, Bethesda, MD.

Rubinstein, L. V., Gail, M. H., and Santner, T. J. (1981). Planning the duration of a comparative clinical trial with loss to follow-up and a period of continued observation. J. Chron. Dis., **34**:469-479.

Schultz, J. R., Nichol, F. R., Elfring, G. L., and Weed, S.D. (1973). Multiple-stage procedures for drug screening. Biometrics, **29**:293-300.

Sheiner, L. B. and Beal, S. L. (1982). Bayesian individualization of pharmacokinetics: Simple implementation and comparison with non-Bayesian methods. *J. Pharmaceut. Sci.*, **71**:1344-1348.

Simon, R. (1979). Restricted randomization designs in clinical trials. *Biometrics*, **35**:503-512.

Simon, R. (1989). Optimal two-stage designs for Phase II clinical trials. *Controlled Clinical Trials*, **10**:1-10.

Simon, R. and Wittes, R. E. (1985). Methodologic guidelines for reports of clinical trials. *Cancer Treatment Reports*, **69**:1-3.

Storer, B. E. (1989). Design and analysis of phase I clinical trials. *Biometrics*, **45**:925-937.

Tarone, R. and Ware, J. (1977). On distribution-free tests for equality of survival distributions. *Biometrika*, **64**:156-160.

Zelen, M. (1983). Guidelines for publishing papers on cancer clinical trials: Responsibilities of editors and authors. *J. Clin. Oncol.*, **1**:164-169.

8

Antiepileptic Drug Development Program: A Collaboration of Government and Industry

Jia-Yeong Tsay[*]

*National Institutes of Health,
Bethesda, Maryland*

Gordon W. Pledger

*R. W. Johnson Pharmaceutical Research Institute,
Raritan, New Jersey*

I. INTRODUCTION

Epilepsy is a chronic neurological disorder characterized by recurrent seizures. It is the second most prevalent neurological disorder, preceded only by stroke. More than two million Americans, or approximately one percent of the U.S. population, suffer from this disorder (Porter, 1983). Epidemiologic studies indicate that 60 to 70 percent of individuals with a diagnosis of epilepsy become seizure free on antiepileptic medication (Hauser, 1990). The remaining 30 to 40 percent are not adequately treated with currently available antiepileptic drugs due to inadequate efficacy or unacceptable toxicity. More effective and less toxic antiepileptic drugs are needed. However, no major new antiepileptic drugs have been marketed in the U.S. since 1978.

[*] *Current affiliation*: Organon Inc., AKZO Pharma Group, West Orange, New Jersey

Generally speaking, drug development is a lengthy and costly process. On the average, it takes nearly 12 years and costs more than $230 million from synthesis of a new chemical entity (NCE) to Food and Drug Administration (FDA) approval in order to bring a new drug to the market (DiMasi et al., 1991). The pharmaceutical industry has not considered antiepileptic drug development a commercially attractive area for such large expenditures because of (i) the relatively small market; and (ii) the lack of proven clinical trial designs to establish safety and antiepileptic effectiveness for an investigational compound. In this situation, assistance from the U.S. Public Health Service is warranted.

The U.S. Government has been actively participating in antiepileptic drug development for more than two decades. The Epilepsy Branch of the National Institute of Neurological Disorders and Stroke (NINDS) was formed in 1966 to stimulate antiepileptic drug development and to develop relevant research methodology. For the first 10 years, all Epilepsy Branch-supported clinical trials were exclusively on antiepileptic drugs that were already available in a foreign market. To expand the spectrum of its involvement in the antiepileptic drug development process, the Branch established the Antiepileptic Drug Development (ADD) Program in 1975. The main goal of the ADD Program is to develop more efficacious and less toxic antiepileptic drugs than those currently available in the hope that a maximum number of patients with epilepsy can achieve seizure control.

The major components of the ADD Program are preclinical pharmacology and clinical trials. The former consists of two key projects: the Anticonvulsant Screening Project, which currently evaluates around 800 compounds a year, and the Toxicology Project. Compounds are submitted either from academia (about 30–35%) or from the pharmaceutical industry (about 65–70%). Compounds that successfully pass through these projects become candidates for ADD Program-sponsored clinical trials. The activities of these preclinical pharmacology projects have been discussed in detail by Kupferberg (1989). This chapter will focus on the clinical trials of the ADD Program and the statistical methodology developed for these trials.

II. MECHANISM FOR CONDUCTING ADD PROGRAM CLINICAL TRIALS

The National Institutes of Health (NIH) has supported clinical trials of new drugs or therapies through various funding mechanisms. Typically the NIH issues a Request for Proposals (RFP) for conducting a specific clinical trial of a new drug or therapy. Then through an open competitive process, the NIH selects the applicants best qualified to conduct the trial and contracts with a data coordinating center to process, monitor, and analyze the data, and write the project report.

The mechanism of the ADD Program clinical trials differs from this standard NIH model in several ways. First, the Epilepsy Branch solicits applications from medical research centers with interest in clinical testing of antiepileptic drugs for a Master Agreement which is an unfunded statement of prequalification spanning a 5-year period. Currently, more than 40 institutions hold ADD Program Master Agreements. Usually only an institution holding a Master Agreement may respond to an RFP for contracts to conduct ADD Program-supported clinical trials. The Master Agreement prequalification can save time in the application review process and thus expedite the development of antiepileptic drugs. When a compound has successfully passed preclinical testing (by the Epilepsy Branch or the drug sponsor) and an Investigational New Drug Application (IND) has been filed with the Food and Drug Administration (FDA), the compound becomes a candidate for clinical trials by the ADD Program. An Epilepsy Branch advisory committee will review and rank potential compounds for clinical trials. Once contracts are signed between the Epilepsy Branch and clinical research centers for a clinical drug trial, the Technical Information Section (the component of the Epilepsy Branch responsible for clinical trials) serves as the coordinating center and is responsible for clinical trial design, protocol development, trial monitoring, processing and analyzing the data, and writing the final report of the trial. Another unusual characteristic of the ADD Program is that for any compound to be tested by this Program, the sponsor (drug company) will hold the IND, not the Epilepsy Branch. Consequently, the ADD Program helps the pharmaceutical industry by sharing the financial and technical burden in the early stages of the antiepileptic drug development process.

III. CLINICAL TRIALS OF ANTIEPILEPTIC DRUGS

The demonstration that an investigational drug has antiepileptic activity is complicated by characteristics of epilepsy treatment and physicians' beliefs about epilepsy. Some of these complicating factors, and their implications for clinical trial design, are as follows:

1. Many neurologists who treat epilepsy and conduct epilepsy clinical research believe that seizures cause or at least may cause permanent damage. Thus, with few exceptions, preventable seizures must be prevented and patients at risk of recurrent seizures must receive some treatment that is potentially effective. Therefore, the traditional drug versus placebo monotherapy design is seldom used in clinical trials of antiepileptic drugs.

2. Generally, before a patient is considered a candidate for a new drug trial, one or more marketed antiepileptic drugs have been tried and have failed to control the patient's seizures. Thus, new drugs are often investigated initially in "therapy resistant" patients, making it difficult to detect antiepileptic activity.

3. Even though the traditional antiepileptic drugs have not controlled the patient's seizures, there is frequently a clinical impression that those drugs have exerted some effect. This impression leads to a reluctance to withdraw the drugs, resulting in the use of so-called add-on designs in clinical trials. That is, the test and control treatments are compared against a background of preexisting antiepileptic drug therapy.

4. Antiepileptic drug dosing is often guided by plasma concentrations. A variety of data sources have led to the calculation of a "therapeutic range" of plasma concentrations and although it is recognized that this range is not applicable to all patients, it is generally believed that for most patients concentrations below the therapeutic range tend to be ineffective and the concentrations above this range tend to be toxic. In clinical trials this translates into a desire to maintain stable plasma concentrations of concomitant antiepileptic drugs with the attendant emphasis on pharmacokinetic interactions between the test drug and the concomitant antiepileptic drugs.

IV. TRADITIONAL DESIGNS FOR ANTIEPILEPTIC DRUG TRIALS

The most commonly used design for clinical trials to evaluate the efficacy and safety of new antiepileptic drugs is the add-on, placebo-controlled, two-period crossover design. As discussed above, add-on trials ensure that patients are never completely untreated; some investigators consider this an ethical necessity.

Add-on trials, however, have clear disadvantages. First, such trials are likely to be insensitive to the effects that may reasonably be expected, because the investigational drug must reduce seizure rates in patients who have not responded completely to one or probably several other drugs. Thus, add-on trials may be prone to Type II errors (i.e., they may reject effective new drugs). Furthermore, the concomitant antiepileptic drugs, whether or not they have any efficacy for these refractory patients, do contribute adverse effects.

An additional complication of add-on trials is possible pharmacokinetic (pertaining to the actions of drugs in the body over a period of time) or pharmacodynamic (pertaining to the biochemical and physiological effects of drugs on living systems) interactions between the investigational drug and the concomitant antiepileptic drugs. This will cloud trial interpretation. Specific attribution of observed differences may be impossible.

Another commonly used traditional design for antiepileptic drug trials is the active control design. When antiepileptic drugs have been tested as monotherapy it has been in active control equivalence trials, that is, trials showing equivalence to a standard drug. Such clinical trials have very serious limitations as evidence of new drug efficacy. These limitations have been described extensively in the literature (e.g., Temple, 1982; Makuch et al., 1990). These authors discuss the type of historical data needed for evidential interpretation of active control equivalence trials, for example, previous placebo controlled trials showing that the activity of the active control drug can be reliably detected under conditions like those of the planned trial. For antiepileptic drugs, such data are not available.

From the above discussion, it is clear that new designs are needed for antiepileptic drug trials. In the following section we will discuss some alternative designs implemented in recent ADD Program trials. These new designs may allow earlier testing of new drugs as monotherapy, a clearer assessment of adverse experience profiles, and better information regarding appropriate doses; these points are elucidated in Tsay and Pledger (1991).

V. NEW ANTIEPILEPTIC DRUG TRIAL DESIGNS

A. Concentration Controlled Design

A randomized concentration controlled design was used in a clinical trial of flunarizine. Figure 8.1 shows the trial flow chart. Patients were randomly assigned to flunarizine or placebo at an individualized dosage aimed at achieving a prespecified plasma concentration.

This trial has some design features that are unique among drug trials in epilepsy. The most unusual aspect of the trial design is the way the test drug dosages were determined. This was a fixed dose trial with each patient receiving an individually determined fixed dosage aimed at achieving a specific plasma flunarizine concentration. This approach is based on the following rationale as discussed by Pledger and Treiman (1991):

1. It is reasonable to expect a stronger relationship between plasma drug concentration and response than between drug dose and response. On a theoretical basis, the quantity of drug actually available in the blood should be more informative than the amount ingested. This is considered to be the case with many of the marketed antiepileptic drugs. Sanathanan and Peck (1991) have discussed the relationship between plasma concentration and efficacy for other classes of drugs.
2. For flunarizine in particular, pilot study data support the plasma concentration approach over a more standard, dose-driven approach.

The design of the pilot study called for each of the first eight patients to be loaded to a plasma flunarizine concentration of 30 ng/ml and maintained at

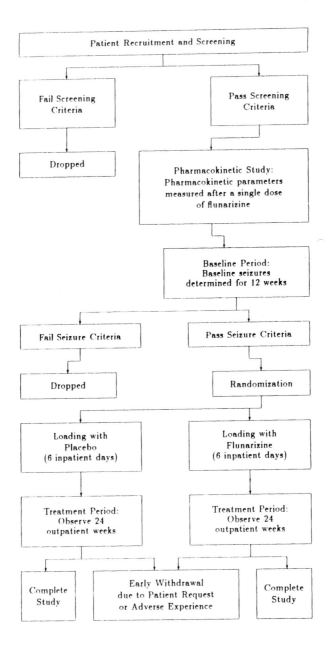

Figure 8.1 Flow chart of the flunarizine trial.

Table 8.1   Pilot Study of Flunarizine: Dosage and Plasma Concentration

Patient	Loading dose (mg)	Maintenance dosage (mg/day)	Mean concentration observed (ng/ml)	Mean concentration hypothetical (ng/ml)*
1	480	15	36	48
2	435	15	27	36
3	410	40	36	18
4	320	20	28	28
5	175	15	16	21
6	630	30	23	15
7	60	5	17	68
8	470	25	24	19
Mean:	372	21	25.9	31.6
S.D. =			7.5	18.4
Range =			(16,36)	(15,68)

* Based on the assumption that each patient received 370 mg loading dose and 20 mg/day maintenance dosage and that concentration is linearly related to dose.

that level for eight weeks. For each patient a loading dose and maintenance dosage were calculated with the goal of achieving and maintaining the 30 ng/ml plasma concentration. The actual flunarizine concentration was an outcome variable in the study and was not manipulated. The sole source of data for estimating the appropriate doses was a single-dose pharmacokinetic profile obtained for each patient. Upon study entry the patient received a 30 mg dose of flunarizine; then 25 blood samples were drawn over the next 34 days.

Table 8.1 summarizes the flunarizine dosages and observed plasma concentrations for the observation period aimed at the concentration of 30 ng/ml. One way of estimating the degree to which this approach decreases variability is as follows: if each patient had received the same dose of the test drug, say a 370 mg loading dose and a 20 mg/day maintenance dosage, then, assuming linearity, the observed mean concentrations would have been approximately those shown in the last column of Table 8.1. Thus it appears that this method of selecting doses tends to decrease variation in observed plasma concentrations.

Based on the pilot study data, the plasma flunarizine concentration of 60 ng/ml was chosen as the target concentration for the efficacy trial. As in the pilot study, a pharmacokinetic profile of plasma concentrations was determined for each patient upon trial entry. This profile was used to estimate the pharmacokinetic parameters, clearance and volume of distribution. These parameter esti-mates lead directly to a loading dose and a maintenance dosage for the individual patient by the following formulas:

Loading dose = (Volume of Distribution) × (Target plasma concentration)

Daily maintenance dosage = (Clearance) × (Target plasma concentration)

Typically, the plasma flunarizine concentration will be very low within a few days after the single dose, but will remain detectable for several weeks. Thus, the three-month baseline period began two weeks after the flunarizine single dose. Patients completing the baseline period are randomized to receive either flunarizine or placebo at the individually calculated dosage.

Another consequence of the emphasis on antiepileptic drug plasma con-centrations is the decision to blind those concentrations in the flunarizine trial. The plasma concentrations of the concomitant antiepileptic drugs, phenytoin and carbamazepine, should not be affected in a systematic way by administration of flunarizine based on previous studies, but there would likely be a strong temptation to react to any fluctuations outside the "therapeutic range" by altering dosages. Because these plasma concentrations are in a sense response variables, they should be simply observed and not manipulated so long as patient safety is protected. Protection of patient safety in the flunarizine trial was addressed in two ways:

1. An unblinded Epilepsy Branch monitor with no other connection to the trial monitored the plasma phenytoin and carbamazepine concentrations for alarming trends, (e.g., a concentration showing a clear downward trend over time, suggesting a serious compliance problem).
2. In reaction to a clinically observable event, the treating physician can obtain "statim" (immediate) phenytoin and carbamazepine concentrations and adjust dosages if those concentrations are out of range.

If the target plasma concentration is well-chosen and (approximately) achieved, the design should yield the following advantages: i) increased power to detect treatment effects in a population in which treatment effects are gen-erally small; and ii) fewer patients receiving subtherapeutic doses and fewer patients requiring dose reductions that can jeopardize the blinding.

The same methodology could be used to compare several plasma concen-trations. The so-called therapeutic ranges for the currently available antiepileptic drugs have resulted from sifting through the data of clinical studies, controlled and uncontrolled, and identifying the plasma concentrations achieved by the patients classified as responders. However, in those studies both the test drug plasma concentration and the seizure reduction were response variables. A more proper source for such therapeutic ranges would be controlled trials in which the plasma concentration is an independent variable (i.e., trials in which patients are randomized to several plasma concentrations of the test drug). This would allow calculation of an EC_{50} (the effective concentration for 50 percent of the patients) and a TC_{50} (the toxic concentration for 50 percent of the patients), which are reasonable operational definitions for the lower and upper limits of a therapeutic range respectively as mentioned by Dodson (1989).

The Epilepsy Branch has initiated a trial comparing three plasma concentrations of felbamate. In this trial, patients having at least two seizures per month while receiving antiepileptic drug therapy are converted to felbamate monotherapy at a dosage of 1200 mg t.i.d. (three times a day) over a 4-week period. This felbamate dosage is thought to be well-tolerated and effective. A 4-week baseline period follows the conversion phase; during the baseline, seizure type and frequency are recorded and weekly plasma felbamate concentrations are obtained. Patients are then randomly assigned to one of three parallel groups identified with three target felbamate concentrations. Each patient receives an individualized dosage based on the target concentration, the observed baseline concentrations, and a linearity assumption. The treatment period duration is six months or until "therapeutic failure" which is defined in the protocol by specific criteria similar to those described by Pledger and Kramer (1991). During the treatment period, plasma felbamate concentrations are monitored by a central laboratory and doses are adjusted if necessary. Blinding is maintained; each patient receives the same number of capsules, some of which may be placebo. This requires unit dose packaging of drug to ensure that the patient will not ingest all of the active drug at one dosing time and take only placebo at other times.

B. N-of-1 Design

The design for a clinical trial with only one patient is called an "N-of-1", "single case", "single patient", or "intensive" design. The N-of-1 design shares some common features with the crossover design (e.g., the patient serves as "his own control"). But in the crossover design, different sequences of treatments apply to different groups of patients. Hence, a sequence of treatments can be identified with an independent group of patients. On the other hand, in the N-of-1 design different sequences of treatments are randomly assigned to the same patient as the design dictates. Because the N-of-1 design can be considered a special kind of the crossover design, all precautions for the crossover design apply to the N-of-1 design. For instance, the N-of-1 design should not be used, if there is a carryover or delayed effect, or the disease is acute in nature, or the disease is self-limited.

The N-of-1 design has been extensively used in behavioral science research (Chassan, 1960, 1961, 1965; Kazdin, 1982; Barlow and Hersen, 1984).

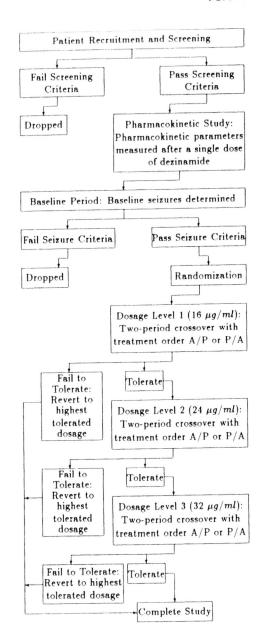

Figure 8.2 Flow chart of the dezinamide study.

Table 8.2. Randomization Plan of N-of-1 Designs for the Dezinamide Trial

Study center I		Study center II	
Patient	Randomization	Patient	Randomization
101	(P/A, P/A, P/A)	201	(A/P, A/P, P/A)
102	(A/P, P/A, P/A)	202	(A/P, P/A, P/A)
103	(P/A, A/P, A/P)	203	(P/A, A/P, A/P)
104	(A/P, P/A, A/P)	204	(P/A, A/P, P/A)
105	(P/A, P/A, A/P)	205	(A/P, P/A, A/P)
106	(P/A, A/P, P/A)	206	(P/A, P/A, A/P)
107	(A/P, A/P, A/P)	207	(A/P, A/P, A/P)
108	(A/P, A/P, P/A)	208	(P/A, P/A, P/A)

Guyatt and his colleagues (1986, 1988, 1990) have been advocating this methodology for clinical trials. They indicate that N-of-1 randomized clinical trials may play an important role in the early phases of drug development, particularly to gain information about rapidity of onset and termination of drug action, and the optimal dose.

The Epilepsy Branch has implemented the methodology of using multiple N-of-1 designs in a multicenter, randomized, double-blind, placebo-controlled clinical trial of dezinamide (a compound thought to have antiepileptic properties in an open-label trial). The trial flow chart is given in Figure 8.2. The objectives were to observe the effect of three different plasma concentrations of dezinamide on seizure frequency and safety variables in phenytoin comedicated patients, to demonstrate that target plasma concentrations of dezinamide can be achieved, and to obtain information on possible interaction between dezinamide and phenytoin. There were two trial centers, each with eight patients.

Two different initial sequences, A/P and P/A, were used. The design called for three sequences of treatments for each patient such that no two patients could have the same three sequences in the same order. Therefore, the treatment sequences A/P and P/A were randomly assigned three times with replacement to each patient with a restriction of no repetition of another patient's randomization sequences. There were eight possible outcomes and at each center there were eight patients. Therefore, the design was completely balanced in the treatment order. Table 8.2 displays the results of the randomization plan.

In Table 8.2 each patient has three sequences (or pairs) of treatments. The first sequence dosage was targeted at a plasma dezinamide concentration of 16 μg/ml, the second sequence at 24 μg/ml, and the third sequence at 32 μg/ml. The individualized dosage was determined by a single dose (400 mg) pharma-

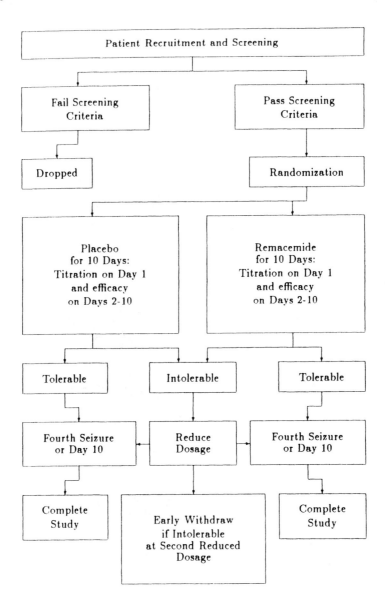

Figure 8.3 Flow chart of the remacemide study.

cokinetic profile as described in the concentration controlled design in Section V-A. It is interesting to note that each sequence aforementioned is a two-period crossover design to compare dezinamide at the underlying concentration with the placebo within the same patient. Thus, a patient serves as his or her own control.

C. Inpatient Seizure Evaluation Design

Commonly used endpoints for evaluating efficacy of an antiepileptic drug include seizure frequency in the treatment period, seizure frequency reduction from baseline to treatment period, and percent seizure frequency reduction from baseline to treatment period. However, if one wants to conduct a monotherapy trial, or a pseudomonotherapy (the concomitant antiepileptic drug is reduced to minimum), then the patient's exposure to an ineffective treatment (possibly placebo) must be limited.

Van Belle and Temkin (1983) proposed as an alternative endpoint time-to-k^{th} seizure. Shofer and Temkin (1986) did Monte Carlo simulations in the crossover setting with k equal to 3, 6, 9, and 12 to compare the endpoints of seizure frequency and time-to-k^{th} seizure. They found that for the time-to-k^{th} seizure as endpoint, power increased with increasing k, but in general, seizure frequency has higher power than time-to-k^{th} seizure. The advantages of time-to-k^{th} seizure design include: to reduce the chance of dropout for patients with high seizure frequency, to increase the patient recruiting rate by allowing patients with low seizure frequency to enter the trial, and to allow conduct of the trials under monotherapy or pseudomonotherapy conditions.

The time-to-4th seizure endpoint has been used by the Epilepsy Branch for a multicenter, randomized, double-blind, parallel-group, placebo-controlled trial of remacemide. There are three study centers with 20 patients to be enrolled at each center. Figure 8.3 shows the trial flow chart. This trial will evaluate the safety, efficacy, and the pharmacokinetics of remacemide in the treatment of patients with partial seizures who have had their concomitant antiepileptic drugs reduced or discontinued as part of an evaluation for epilepsy surgery.

At the completion of the presurgical evaluation and without further antiepileptic drug manipulation, patients will be randomly assigned to receive remacemide or matching placebo. Patients complete the trial after ten days or the fourth seizure, whichever is first. Efficacy will be assessed by comparing the seizure frequency (standardized to the same duration denominator) between the two randomized groups. Safety will be evaluated by comparing the nature and duration of the adverse experiences. The interpatient pharmacokinetic variability of remacemide will also be evaluated.

In this trial, patients may be evaluated on monotherapy or pseudomonotherapy which has been considered infeasible in more traditional antiepileptic drug trials. Using the time-to-4th seizure endpoint, patients will not be exposed to excessive risk even if they are receiving placebo treatment only. This type of design is very useful in the early stage of drug development. The caveat is that it can not provide proper evaluation of long-term drug effects.

VI. CONCLUSION

The designs discussed above will allow evaluation of investigational anti-epileptic drug efficacy and safety more efficiently than the commonly used add-on crossover design and active control equivalence design. Specifically, concentration controlled designs can define the therapeutic range of an anti-epileptic drug in a more scientific way than those commonly used in the past. N-of-1 designs can, with a small number of patients, gain preliminary information on efficacy, safety, optimal dose, and rapidity of onset and termination of drug action. Inpatient seizure evaluation designs with time-to-k[th] seizure end-points can enable the conduct of monotherapy, placebo controlled trials which should provide greater power to detect antiepileptic activity.

REFERENCES

Barlow, D. H. and Hersen, M. (1984). *Single Case Experimental Designs: Strategies for Studying Behavior Change, 2nd Ed.* Pergamon, New York.

Chassan, J. B. (1960). Statistical inference and the single case in clinical design. *Psychiatry*, **23**:173-184.

Chassan, J. B. (1961). Stochastic models of the single case as the basis of clinical research design. *Behav. Sci.*, **6(1)**:42-50.

Chassan, J. B. (1965). Intensive design in clinical research. *Psychosomatics*, **VI**: 289-294.

DiMasi, J. A., Hansen, R. W., Grabowski, H. G., and Lasagna, L. (1991). Cost of innovation in the pharmaceutical industry. *J. Health Economics*, **10**:107-142.

Dodson, W. E. (1989). Level off. *Neurology*, **39**:1009-1010.

Guyatt, G. H., Heyting, A., et al., (1990). N-of-1 randomized trials for investigating new drugs. *Controlled Clinical Trials*, **11**:88-100.

Guyatt, G. H., Sackett, D., et al., (1986). Determining optimal therapy: randomized trials in individual patients. *N. Engl. J. Med.*, **314**:889-892.

Guyatt, G. H., Sackett, D., et al., (1988). A clinician's guide for conducting randomized trials in individual patients. *Can. Med. Assoc. J.*, **139**:497-503.

Hauser, W. A. (1990). The natural history of drug-resistant epilepsy: epidemiologic considerations. Abstracts of NIH Consensus Development Conference on Surgery for Epilepsy, March 19-21, 1990. National Institutes of Health, Bethesda, Maryland.

Kazdin, A. E. (1982). *Single-Case Research Designs: Methods for Clinical and Applied Settings.* Oxford University Press, New York.

Kupferberg, H. J. (1989). Antiepileptic drug development program: a cooperative effort of government and industry. *Epilepsia*, **30(Suppl. 1)**:S51-S56.

Makuch, R., Pledger, G. W., Johnson, M., Herson, J., Hsu, J. P., and Hall, D. (1990). Active control equivalence studies. In: *Statistical Issues in Drug Research and Development*, (K. E. Peace, Ed.). Marcel Dekker, New York, pp. 225-262.

Pledger, G. W. and Kramer, L. D. (1991). Clinical trials of investigational antiepileptic drugs: monotherapy designs. *Epilepsia*, 32:716-721.

Pledger, G. W. and Treiman, D. T. (1991). Design of an individualized fixed-dose trial to test the antiepileptic efficacy of a plasma flunarizine concentration. *Controlled Clinical Trials*, 12(6):768-779.

Porter, R. J. (1983). Antiepileptic drug development program. In: *Orphan Drugs and Orphan Diseases: Clinical Realities and Public Policy*, Alan R. Liss, Inc., New York, pp. 53-66.

Sanathanan, L. P. and Peck, C. C. (1991). The randomized concentration controlled trial: a strategy for improving efficiency. *Controlled Clinical Trials*, 12(6):780-794.

Shofer, J. B. and Temkin, N. R. (1986). Comparison of alternative outcome measures for antiepileptic drug trials. *Arch. Neurol.*, 43:877-881.

Temple, R. (1982). Government viewpoint of clinical trials. *Drug Inform. J.*, 1:10-17.

Tsay, J.-Y. and Pledger, G. W. (1991). Innovative designs for antiepileptic drug trials. *Proceedings of the 1991 Biopharmaceutical Section of the American Statistical Association*. American Statistical Association.

Van Belle, G. and Temkin, N. R. (1983). Design strategies in the clinical evaluation of new antiepileptic drugs. In: *Recent Advances in Epilepsy* (Pedley, T. A., and Meldrum, B. S., Eds.). Churchill Livingston, Inc., New York, pp. 93-111.

9

Design and Analysis of Clinical Trials of Analgesic Drugs

Jack W. Green

G. H. Besselaar Associates,
Princeton, New Jersey

I. INTRODUCTION

Analgesic drugs were among the first drug classes for which well-controlled clinical trials were routinely used to compare the safety and efficacy of alternative medications. Consequently, a large body of results has been developed based on similar experimental techniques and evaluation criteria.

A positive response to standard drugs such as aspirin and morphine has been consistently demonstrated. This has proven to be very helpful to researchers and statisticians in the investigation of new analgesics. By including a standard drug as a positive control, researchers have an internal check on the sensitivity of their methodology for measuring analgesic activity. Likewise, the statistician can better determine whether the clinical trial meets its objectives than is possible in areas with less-well-defined criteria and with controls whose activity is not adequately characterized.

Unfortunately, there are many factors associated with analgesic trials which can distort the results of the study. Probably the most complicating factor is the subjective nature of pain, which varies with each individual. Even though all of us have experienced pain, the difficulty in deriving precise definitions and objective measurements means that analgesic trials must rely on subjective observations of complex phenomena. There is frequently a high placebo response and a large variability in the response to active drugs. This is primarily due to the many factors which influence the patient's perception of pain, such as the doctor-patient relationship, the patient's expectation based on response to previous therapy, and the impact that the patient's illness or injury has on his life situation.

For these reasons, a large number of patients must usually be studied before conclusive results can be derived. Fortunately, it is often possible to find investigators with a ready source of patients. Thus, a large sample size can usually be obtained without relying on multiclinic studies with their associated problems of protocol adherence and pooling of results.

Useful information concerning a drug's analgesic potential can often be derived by studying the effect of a single dose of the drug. This reduces the problems involved with changes in environment, missed doses, and study dropouts, all of which are major considerations in studies of longer duration. The remaining sections of this chapter will be restricted to a discussion of the design and analysis of single-dose studies. For readers interested in additional information on multiple-dose studies, Stambaugh and McAdams (1987) describe a method that allows single and multiple doses to be studied in the same subjects.

II. DESIGN OF ANALGESIC DRUG TRIALS

A. Parallel Group Designs

A variety of experimental designs have been employed in analgesic clinical trials. The design which is most straightforward and easiest to interpret is a parallel group design in which each patient is randomly assigned only one of the study medications. Frequently, the patients are stratified into homogeneous subgroups in order to improve the sensitivity of the study. In postpartum pain studies, for example, the random code may be generated in such a way that the treatment groups have comparable percentages with uterine cramping pain and with episiotomy pain. Baseline pain intensity level is also used as a blocking factor when randomizing the patients to treatment groups.

Most experts believe that patients with only mild pain at baseline should not be eligible to enter a study, since this would not allow enough discrimination to evaluate analgesic response.

B. Crossover Design

The complete crossover design, in which each patient receives all of the agents in the study, is an alternative design which has been used in analgesic trials. There are some types of pain, such as postpartum or dental surgery pain, which are of such short duration that a crossover design may not be feasible. For more chronic types of pain, the crossover design can theoretically increase experimental sensitivity by the comparison of treatments "within" rather than "between" patients.

However, the interpretation of data from crossover designs may be complicated by such factors as expectations based on pain relief due to preceding drugs, interest in the study, carryover effects, and differences in drop-out rates. In some instances corrections can be applied to derive valid estimates based on the complete crossover data, and in any case the first dose information can be analyzed even if the results of the following drug administrations are ambiguous. The potential advantages of the crossover design have to be weighted against the disadvantages of added administrative, time, and cost factors, and difficulties in interpretation of the results.

In a methodologic survey of a sample of investigators who were conducting clinical pain studies Sriwatanakul et al. (1983) report that only 2 of 13 respondents considered a crossover study to be the best experimental design for demonstrating the efficacy of a new analgesic. However, Forrest et al. (1988) analyzed 55 postoperative pain studies which were planned as four-period crossover designs for four treatments. These authors conclude that crossover designs are appropriate for the evaluation of selected parenteral analgesics because there is no evidence of significant residual analgesic effects and because of the efficiency of crossover designs.

More complex types of crossover designs allow the estimation of treatment effects in the presence of carryover effects. For example, if there are two treatments A and B, a design with sequences A–B–B and B–A–A allow the carryover effect of each treatment to be estimated within subjects. Chapter 10 contains additional information on crossover designs.

C. Incomplete Block Designs

The incomplete block design has been used by some analgesic researchers. In this design, each patient receives some (but not all) of the study medications according to a randomization schedule designed to extract the maximum amount of information about the study drugs. Different types of incomplete block designs can be utilized depending on the comparisons of most interest. In a balanced incomplete block design, each of the possible combinations of treatments is allocated to an equal number of patients.

Another type of incomplete block design, called a twin crossover design, has been utilized in studies comparing two dose levels of a test drug to two dose levels of a standard analgesic (Wallenstein et al., 1986). Each patient receives only two study medications on separate days: the lower dose of one study drug and the upper dose of the other. The order of administration of the pairs of doses are randomized, but in each block of four patients, drug, dose and order of administration are balanced (Table 9.1).

Table 9.1 Twin Crossover Design

	Standard drug		Test drug	
Sequence	Dose 1	Dose 2	Dose 1	Dose 2
1	–	1[*]	2	–
2	2	–	–	1
3	1	–	–	2
4	–	2	1	–

[*] Numbers denote order of administration for each sequence.

A weakness of this design is that there is no straightforward test of whether the two dose-response curves are parallel. Jones and Kenward (1989) list other types of incomplete block designs which may be more appropriate.

The potential ambiguities which were noted in the discussion of complete crossover designs also can occur with incomplete block designs. The assumptions required for valid estimates of efficacy and safety parameters should always be checked in clinical trials which use either type of design.

D. Selection of Appropriate Treatment Groups

Another part of the experimental design considerations is the selection of appropriate treatment groups. There are two prominent methods for testing the efficacy of a new medication. In one approach the test medication is compared with a placebo and a standard analgesic such as aspirin, ibuprofen, or morphine. This provides a check on the sensitivity of the experimental model based on the comparison of the standard to the placebo and also provides a comparison of the test medication to a positive control.

An alternative method of testing a new medication is the determination of a potency estimate of the test medication relative to a standard analgesic by establishing dose-response curves based on graded doses of both the test and standard medications. The demonstration of positive dose-response curves can be used as a measure of the sensitivity of the experimental model, thus eliminating the need for inclusion of a placebo in the study. Studies designed to estimate relative potency require special adherence to the protocol criteria since it is usually much more difficult to establish analgesic dose-response curves than to find significant differences between active analgesics and a placebo.

The 2 × 2 factorial design has been commonly used to evaluate analgesic combinations. The treatment groups consist of the two individual ingredients, the combination, and a placebo. Response surface designs can also be used to evaluate the relationship between the combination doses and the response variable over a wide range of doses (Goldstein and Brunelle, 1991).

III. ANALYSIS OF ANALGESIC DRUG TRIALS

A. Standard Techniques

The measurements of analgesic effectiveness are usually based on the patient's subjective evaluations of pain before and after administration of the study medication. Evaluations are made at specified time intervals and a global evaluation is recorded at the end of the study. Binary scales such as the presence or absence of pain are generally considered not as sensitive as four- or five-point scales such as pain intensity (none, mild, moderate, and severe) and pain relief (none, slight, moderate, and completely gone).

The conventional method of describing the results of a study is to assign numerical scores with unit increases to each of the scale categories (e.g., 0 = none, 1 = slight, 2 = moderate, 3 = completely gone). Time-action curves based on the treatment group mean scores at each observation time are then plotted. Figure 9.1 illustrates the pain relief time effect curves for a study comparing a test medication to aspirin and placebo.

Another approach is to ask the patient at each time point to mark where her current intensity of pain lies on a line of fixed length. The line is labeled at the ends with the extremes of pain scale, for instance:

No |————————————————————| Extreme
Pain | | Pain

The lengths of the line segments on this visual analog scale are then used to assess relative changes in pain over time.

Littman et al. (1985) compare three analgesic rating scales—visual pain analog, verbal pain intensity and verbal pain relief—and conclude that the scales are highly correlated with minimal differences in sensitivity.

Traditionally, two statistics derived from the pain rating scale have been given the most emphasis in the evaluation of analgesic response (Laska et al., 1967):

1. Sum of pain intensity differences (SPID): the sum across the observation points of the pain intensity differences from the initial pain score
2. Total pain relief (TOTPAR): the sum across the observation points of pain relief scores

These statistics are usually calculated by weighting the scores by the length of time between successive observations.

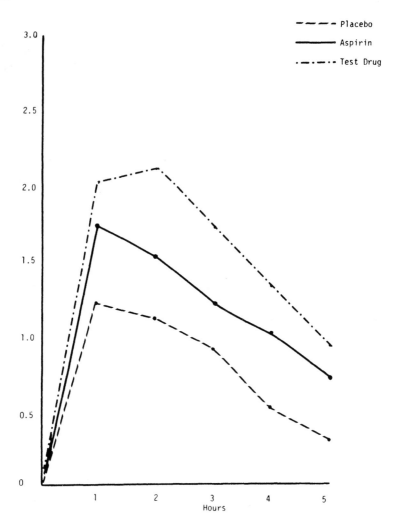

Figure 9.1 Mean pain relief scores.

The testing for statistically significant differences has been based on parametric analyses of SPID, TOTPAR, and other secondary statistics derived from the pain intensity and relief scores. Trials which are designed to compare a test drug to a standard analgesic and a placebo often use analysis of variance and multiple comparison procedures to test for differences between treatment groups (Sriwatanakul et al., 1983). Similar analyses using a nonparametric approach also have been frequently used (Bloomfield et al., 1986). Trials which are designed to determine the potency of a test agent relative to a standard analgesic have used statistical methods appropriate for bioassay experiments (Laska et al., 1967). Since bioassay methods are infrequently used in most areas of clinical trials research, these techniques will be described in greater detail in the next section.

B. Estimation of Relative Potency

The estimation of relative potency is based on the parallel line assay, which is applicable when the analgesic response is linearly related to log dose and when the response on log dose is parallel for the standard and test drugs. In designing the assay, the doses of the test and standard drugs should be chosen to give equivalent responses over the range in which the response is linear. Beaver and Feise (1977a, b) discuss an application of a sequential decision-making process which expedites choosing the doses of the test medication which are most closely equianalgesic with the standard.

The validity of the assumptions necessary for the estimation of relative potency can be examined statistically by an analysis of variance (ANOVA). Figure 9.2 illustrates the dose-response curves based on the TOTPAR scores calculated in a study comparing three dose levels of a test drug against three dose levels of a standard analgesic. The ANOVA table is given in Table 9.2.

For a satisfactory assay the separate regressions for the test and standard medications should not depart significantly from parallelism and the common slope should be significantly different from zero. The test for deviations from linearity indicates whether the dose-response curves can be adequately approximated by a straight line over the range of doses tested. If the test is significant, this may indicate that the dose range should be changed in future studies. A nonsignificant test for medication effect indicates that the test and standard medications yield approximate equianalgesic responses over the range of doses tested. It is desirable to choose the doses to give similar responses in order to avoid errors in the potency estimate due to extrapolation.

The logarithm of the relative potency of the two medications is given as the distance between the parallel dose-response lines measured on the log dose scale. The potency estimate is thus independent of dose. In the example, the relative potency estimate indicates that the test medication is approximately 7.3 times as potent as the standard. The 95% confidence limits for the potency estimate, based on Fieller's theorem (Finney, 1978) range from 3.3 to 20.7. Since the limits do not include unity, the clinical trial demonstrates that the test medication is significantly more potent than the standard at the 5% level of significance. Multivariate techniques can also be used to combine multiple responses such as SPID and TOTPAR (Laska and Meisner, 1987).

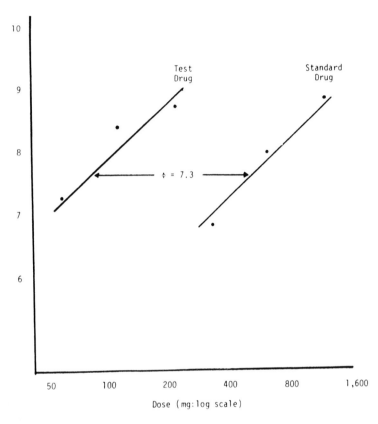

Figure 9.2 Relative potency study.

C.   Alternative Procedures

The SPID and TOTPAR scores provide measures which intuitively resemble "area under the curve" calculations. These statistics have proven to be useful for translating the results of statistical analyses into terms which are familiar to researchers working in the analgesic area and provide a convenient frame of reference for a comparison of new studies with results found in the literature. They are, however, arbitrary quantities based primarily on empirical justification.

Various alternative techniques have been used in place of or in parallel with the conventional analyses of SPID and TOTPAR. Rather than attempting to provide an exhaustive list of all possible methods of analyzing analgesic trials, a brief description will be given of some particularly useful or novel techniques which are more completely described in the references.

Table 9.2 Summary of Relative Potency Analysis of TOTPAR Scores

Source of variation	Analysis of variance			
	Degrees of freedom	Mean square	F	P
Slope	1	90.34	9.79	.01
Deviations from parallelism	1	2.79	0.30	NS
Medication effect	1	2.70	0.29	NS
Deviations from linearity	2	4.17	0.45	NS
Error	174	9.23		
Estimate of relative potency =		7.3		
Lower 95% confidence limit =		3.3		
Upper 95% confidence limit =		20.7		

Laska et al. (1991) describe methods to directly measure onset and duration of analgesia. The patients are given two stopwatches with instructions to stop the "onset" stopwatch when a specifically defined clinically significant amount of pain relief is first experienced and to stop the "offset" stopwatch when the clinically significant pain relief is no longer felt. Siegel et al. (1989) use a parametric bivariate cure model to analyze onset and duration in a clinical trial of meptazinol versus morphine in postoperative pain.

The conventional method of assigning numerical scores with unit increases to the subjective rating scales assumes that the intervals between the categories are equal, for example, a change in the pain intensity scale from "severe" to "moderate" (3 to 2) is equivalent to a change from "mild" to "none" (1 to 0). Criticism of this technique has led to the application of various transformations that more fully satisfy the distributional assumptions of parametric analyses and to the more frequent use of categorical and nonparametric techniques.

Laska et al. (1986) examine an approach for assigning values to the pain intensity categories so that the resulting scores maximize the simultaneous fit of the dose-response regression lines in a bioassay analysis. The authors compare the new method to the standard equal-step scoring scheme for the data from two analgesic studies and state that the improvement obtained in regression fit by the new method is relatively small. Wallenstein et al. (1990) report that the log conversion of an eight-point extended Pain Intensity Difference scale provides analgesic measures equal to or superior to other measures in terms of assay sensitivity in a relative potency assay.

A chi-square partitioning analysis and two logit analyses are applied by Cox and Chuang (1984) to model responses on pain from a postoperative analgesic trial. The chi-square partitioning technique uses continuation ratios and allows an examination of several subtables independently. The logit analyses use the multiplicative row and column effects association model considered by Goodman (1979) and the cumulative odds model proposed by McCullough (1980).

Chuang and Agresti (1986) propose a slightly different model which assumes monotone scores for ordered response categories. This model allows a stochastic ordering of the rating distributions of the drugs under comparison, which leads to simpler interpretation of the results.

There has been increased interest in the use of serum drug levels to better discriminate among treatments. Laska et al. (1986) demonstrate a possible bioequivalency problem with an experimental formulation of ibuprofen based on an evaluation of blood levels and pain relief data. Velagapudi et al. (1990) report on a pharmacokinetic/pharmacodynamic analysis of aspirin analgesia data.

McQuay (1991) describes recent work with single patient or "N-of-1" analgesic trials. This design, which tests within patient differences using multiple crossover treatment periods, would be most applicable in chronic pain settings where there is a rapid onset and short duration of treatment effects. There are many complexities involved in pooling single-patient results into a group analysis which would need to be worked out in the protocol design stage.

REFERENCES

Beaver, W. T. and Feise, G. A. (1977a). Twin crossover relative potency analgesic assays in man, I: Morphine vs. morphine. *J. Clin. Pharmacol.*, **21**:461-479.

Beaver, W. T. and Feise, G. A. (1977b). Twin crossover relative potency analgesic assays in man, II: Morphine vs. 8-methoxycyclacine. *J. Clin. Pharmacol.*, **21**:480-489.

Bloomfield, S. S., Mitchell, J., Cissell, J., and Garden, T. P. (1986). Analgesic sensitivity of two post-partum pain models. *Pain*, **27**:171-179.

Chuang, C., and Agresti, A. (1986). A new model for ordinal pain data from a pharmaceutical study. *Statistics in Medicine*, **5**:15-20.

Cox, C., and Chuang, C. (1984). A comparison of chi-square partitioning and two logit analyses of ordinal pain data from a pharmaceutical study. *Statistics in Medicine*, **3**:273-285.

Finney, D. J. (1978). *Statistical Method in Biological Assay*, Griffin, London.

Forrest, W. H., Jr., James, K. E., and Ho, T. Y. (1988). Residual analgesic effects of morphine in 55 four-period crossover analgesic studies. *Clin. Pharmacol. Ther.*, **44**:383-388.

Goldstein, D. J. and Brunelle, R. L. (1991). Dose-response model for combination analgesic drugs. In: *Advances in Pain Research and Therapy, Vol. 18*, (Max, M., Portenoy, R., and Laska, E., Ed.) New York: Raven Press.

Goodman, L.A. (1979). Simple models for the analysis of association in cross-classifications having ordered categories. *J. Amer. Stat. Assoc.*, **74**:537-552.

Jones, B. and Kenward, M.G. (1989). *Design and Analysis of Crossover Trials*, Chapman and Hall, London.

Laska, E. M., Gormley, M. Sunshine, A., Bellville, J. W., Kantor, T. G., Forrest, W. H., Siegel, C., and Meisner, M. (1967). A bioassay computer program for analgesic clinical trials. *Clin. Pharmacol. Ther.*, **8**:658-669.

Laska, E. M. and Meisner, M. J. (1987). Statistical methods and applications of bioassay. *Annu. Rev. Pharmacol. Toxicol.*, **27**:385-397.

Laska, E. M., Meisner, M., Takeuchi, K., Wanderling, J. A., Siegel, C., and Sunshine, A. (1986). An analytic approach to quantifying pain scores. *Pharmacotherapy*, **6**:276-282.

Laska, E. M., Siegel, G. and Sunshine, A. (1991). Onset and duration: measurement and analysis. *Clin. Pharmacol. Ther.*, **49**:1-5.

Littman, G. S., Walker, B. R., and Schneider, B. E. (1985). Reassessment of verbal and visual analog ratings in analgesic studies. *Clin. Pharmacol. Ther.*, **38**:16-23.

McCullough, P. (1980). Regression models for ordinal data (with discussion), *J. R. Stat. Soc., Ser. B*, **42**:109-142.

McQuay, H. J. (1991). N-of-1 trials. In: *Advances in Pain Research and Therapy, Vol. 18*, (Max, M., Portenoy, R., and Laska, E., Ed.), New York: Raven Press.

Siegel, C., Sunshine, A., Richman, H., et al. (1989). Meptazinol and morphine in postoperative pain assessed with a new method for onset and duration. *J. Clin. Pharmacol.*, **29**:1017-1025.

Sriwatanakul, K., Lasagna, L., and Cox, C. (1983). Evaluation of current clinical trial methodology in analgesimetry based on experts' opinions and analysis of several analgesic studies. *Clin. Pharmacol. Ther.*, **34**:277-283.

Stambaugh, J. E., Jr., and McAdams, J. (1987). Comparison of intramuscular dezocine with butorphanol and placebo in chronic cancer pain: A method to evaluate analgesia after both single and repeated doses. *Clin. Pharmacol. Ther.*, **42**:210-219.

Velagapudi, R., Brueckner, R., Harter, J.G., Peck, C.C., Viswanathan, C.T., and Sunshine, A., Laska, E.M., Mason, W.D., Byrd, W.G. (1990). Pharmacokinetics and pharmacodynamics of aspirin analgesia. *Clin. Pharmacol. Ther.*, **47**:179.

Wallenstein, S. L., Houde, R. W., Portenoy, R., Lapin, J., Rogers, A., Foley, and K. M., (1990) Clinical analgesic assay of repeated and single doses of heroin and hydromorphine. *Pain*, **41**:5-13.

Wallenstein, S. L., Kaiko, R. F., Rogers, A. G., and Houde, R. W. (1986). Crossover trials in clinical analgesic assays: Studies of buprenorphine and morphine. *Pharmacotherapy*, **6**:228-235.

10

Crossover Designs in Medical Research

Alan C. Fisher

Immunobiology Research Institute,
Annandale, New Jersey

Sylvan Wallenstein

The Mount Sinai Medical Center,
New York, New York

I. INTRODUCTION

Let us consider the following information which could be from a clinical trial designed to evaluate the effect of an investigational hypnotic medication: 12 insomniac outpatients are randomly assigned to two groups of 6 patients each. All patients will receive both a placebo (P) and an active investigational new drug (A). Patients in sequence group 1 are scheduled to take a placebo capsule (P) on Monday night (the first evaluation period) and the active investigational new drug (A) on Thursday night (the second evaluation period). Patients in sequence group 2 are scheduled to take A and then P. The medication is dispensed in a double-blind manner so that neither the patient nor the investigator is able to determine the order of the treatments. One variable evaluated in hypnotic trials is duration of sleep; typical data are displayed in Table 10.1.

The experimental design employed in the hypnotic trial described above is a two-period crossover design. The essential features of this design are patients crossed over to a second treatment, two treatments, and two periods of observation. In the statistical literature, the term *crossover* has been used interchange-

Table 10.1 Duration-of-Sleep Data (hr)

Period	Treatment	\ Patient\</div\>						Total	Mean
		11	12	13	14	15	16		
1	P	7.00	6.75	5.90	7.20	6.25	4.85	37.95	6.325
2	A	8.50	7.40	6.80	8.60	6.66	6.00	43.96	7.327
Totals:		15.50	14.15	12.70	15.80	12.91	10.85	81.91	

Period	Treatment	\ Patient\</div\>						Total	Mean
		11	12	13	14	15	16		
1	A	6.50	6.00	8.17	7.00	8.12	7.35	43.14	7.190
2	P	7.50	4.75	6.10	6.05	7.25	6.16	37.81	6.302
Totals:		14.00	10.75	14.27	13.05	15.37	13.51	80.95	

ably with the terms simple *reversal* (Lucas, 1950) and *changeover* (Grizzle, 1965). Changeover designs, in general, are experimental designs in which the patient receives more than one treatment during the course of the study. One type of changeover design is the Latin square changeover trial. In this type of clinical trial, the treatments under study are randomized among the patients in such a way that each patient receives each treatment once and each treatment appears in each medication period the same number of times. It should be noted that actual experiments using this design may consist of one or several squares, usually two or more. Other types of changeover designs include balanced incomplete block designs, switchback or double-reversal trials, extra period Latin square changeover trials, and incomplete Latin square changeover trials. The statistical analysis of the many types of changeover designs has been discussed by Lucas (1950), Federer (1955), and Cochran and Cox (1968). An extensive catalog of changeover designs can be found in the monograph by Paterson and Lucas (1962).

Many of these designs are complex and difficult to employ in actual clinical trials. They are also difficult to analyze if there are missing observations. For these reasons, an investigator is well advised to consult an experienced biostatistician before the initiation of this type of trial. However, the two-period crossover design, the simplest type of changeover design, has been commonly used in medical experiments. We will therefore limit our discussion to this design.

II. MEDICAL REASONS FOR USING THE CROSSOVER DESIGN

The crossover design has had great appeal among medical investigators. There are several reasons why this is so. The first is that it is often quite difficult to recruit patients into clinical trials who are both suitable according to the protocol criteria and are willing to participate in the study. It is therefore easy to understand why the medical investigator would want to use this select group of patients for both treatments in the study. Furthermore, if a major factor against completing a study is the difficulty in recruiting patients, the crossover design would allow an earlier completion of the study. In addition, it alleviates the ethical consideration that the study may not be helpful to patients randomized to only receive placebo therapy. Another reason why the crossover design is popular is the intuitive feeling that this allows the investigator to study the same subject under two different conditions. A person accustomed to medical-statistical jargon would translate this phrase to mean that "the patient serves as his/her own control." Somewhat related to this is the intuitive idea that a crossover design, as compared to a parallel group design, would require one-half the total number of patients.

These considerations appear to be perfectly valid and reasonable. However, as will be discussed later, the last reason is often invalid due to experimental factors and changes within the patient.

III. SIMPLE TWO-PERIOD CROSSOVER DESIGN: STATISTICAL CONSIDERATIONS

A. The Model

The mathematical properties of the two-period crossover design have been reported by Grizzle (1965). Interested readers are referred to that paper for a thorough discussion as well as to the paper by Hills and Armitage (1979) which gives a comprehensive account of the analysis and applications of such trials for both quantitative and binary data. Jones and Kenward (1989) and Wallenstein et al. (1990) provide additional comprehensive discussions. The purpose of this section is to acquaint the reader with some of the main properties of the design.

Following Grizzle (1965), we will assume for the data in Table 10.1 that an adequate model, denoted as model I, is

$$Y_{ijk} = \mu + \lambda_{l'} + \zeta_{ij} + \varphi_l + \varepsilon_{ijk}$$

(10.1)

$$j = 1, ..., n_i; \ i = 1, 2; \ k = 1, 2; \ l = 1, 2; \ l' = 1, 2$$

where

Y_{ijk} = the observation for the jth subject within the ith sequence at period k

μ = overall mean

ζ_{ij} = the effect of the jth subject within the ith sequence, which for the sake of testing hypotheses we assume to be a normally distributed random variable with mean 0 and variance σ_s^2

φ_l = the direct effect of the lth drug

$\lambda_{l'}$ = $\begin{cases} 0 \text{ for period 1} \\ \text{the residual effect of the } l'^{th} \text{ drug for period 2} \end{cases}$

e_{ijk} = the random fluctuation which is normally distributed with mean 0 and variance σ_e^2, and is independent of ζ_{ij}

The term *residual effect* is used interchangeably with the term *carryover effect*. The presence of a residual effect is often due to the effect of the second treatment conditioned in part by the preceding treatment (Lucas, 1950).

The general procedure for comparing direct treatment effects is to first perform a preliminary test for the equality of the carryover effects as indicated in Table 10.2. If the results of this test indicate that unequal carryover effects may be present, the analysis of treatment effects is calculated as indicated in Table 10.2. Brown (1980) and Hills and Armitage (1980) note however that this preliminary test has comparatively low power. The test for treatment effect in Table 10.2 is equivalent to a t test performed on period 1 data, and thus, at least for the analysis of direct treatment effects, the effort in collecting data for period 2 is wasted.

Table 10.2 Analysis of Variance for Crossover Design with Residual Effects (Model I)

Source of variation	df	Sum of squares	Expected mean square
Residual	1	$\dfrac{1}{2n_1n_2n}(n_2Y_{1..}-n_1Y_{2..})^2$	$\sigma_e^2 + 2\sigma_s^2 + \dfrac{n_1n_2}{2n}(\lambda_1-\lambda_2)^2$
Subject (seq)	$n-2$	$\dfrac{1}{2}\left\{\displaystyle\sum_{j=1}^{n_2}Y_{2j}^2 - \dfrac{Y_{1..}^2}{n_1} - \dfrac{Y_{2..}^2}{n_2}\right\}$	$\sigma_e^2 + 2\sigma_s^2$
Treatment	1	$\dfrac{1}{n_1n_2n}(n_2Y_{1.1}-n_1Y_{2.1})^2$	$\sigma_e^2 + \sigma_s^2 + \dfrac{n_1n_2}{n}(\varphi_1-\varphi_2)^2$
Error	$n-2$	$\displaystyle\sum_{i=1}^{2}\left\{\sum_{j=1}^{n_i}Y_{ij1}^2 - \dfrac{Y_{i.1}^2}{n_i}\right\}$	$\sigma_e^2 + \sigma_s^2$

Table 10.3 Analysis of Variance for Crossover Design without Residual Effects (Model II)

Source of variation	df	Sum of squares[*]	Expected mean square
Sequence	1	$\dfrac{1}{2n_1n_2n}(n_2Y_{1..} - n_1Y_{2..})^2$	$\sigma_e^2 + 2\sigma_s^2 + \dfrac{2n_1n_2}{n}(\psi_1 - \psi_2)^2$
Subject (seq)	n−2	$\dfrac{1}{2}\left\{\displaystyle\sum_{j=1}^{n_1}Y_{1j.}^2 + \sum_{j=1}^{n_2}Y_{2j.}^2 - \dfrac{Y_{1..}^2}{n_1} - \dfrac{Y_{2..}^2}{n_2}\right\}$	$\sigma_e^2 + 2\sigma_s^2$
Treatment	1	$\dfrac{1}{2n_1n_2n}(n_2G_1 - n_1G_2)^2$	$\sigma_e^2 + \dfrac{2n_1n_2}{n}(\varphi_1 - \varphi_2)^2$
Period	1	$\dfrac{1}{2n_1n_2n}(n_2G_1 + n_1G_2)^2$	$\sigma_e^2 + \dfrac{2n_1n_2}{n}(\pi_1 - \pi_2)^2$
Error	n−2	$\dfrac{1}{2}\displaystyle\sum_{i=1}^{2}\left\{\left[\sum_{j=1}^{n_i}(Y_{ij1} - Y_{ij2})^2\right] - \dfrac{(Y_{i.1} - Y_{i.2})^2}{n_i}\right\}$	σ_e^2

[*] $G_1 = Y_{1.1} - Y_{1.2}$; $G_2 = Y_{2.1} - Y_{2.2}$; $n = n_1 + n_2$

If the results of the preliminary test indicate that unequal carryover effects are not present, the analysis of direct treatment effects is as shown in Table 10.3. Grizzle suggests that failure to reject at the $\alpha = 0.10$ level would in most cases be an adequate indication of no carryover effects. Note that this test for treatment effects utilizes information from both periods.

Table 10.3 should also be used when it is known that carryover effects cannot occur. In this case both a sequence and period effect can be included in the model. The sequence effect is not estimable if the carryover effect is included in the model. The model without carryover effects, denoted as model II, may be written as

$$Y_{ijk} = \mu + \psi_i + \zeta_{ij} + \pi_k + \varphi_l + \varepsilon_{ijk} \tag{10.2}$$

where ψ_i = effect of the ith sequence group and π_k = effect of the kth period.

The sequence effect in this model could be used to test for homogeneity of the two groups or to evaluate experimental conditions that cannot interact

Table 10.4 Analysis of Variance for Crossover Design with Residual Effects (Model I)

Source	df	SS	MS	F	P
Residual	1	0.04	0.04	0.027	0.873
Subject (seq)	10	14.84	1.48		
Treatment	1	2.24	2.24	2.986	0.115
Error	10	7.54	0.75		

with treatment or period. For example, Albert et al. (1974) allocated subjects to sequence on the basis of body weight and related differences in sequence to the effect of volume of distribution on blood levels. The period effect could be used to test for different experimental conditions in the two evaluation periods. In a bioavailability trial, it could indicate different storage conditions, leading to lower assay results in one period than in the other. Alternatively, it could indicate lower pollen counts and thus fewer symptoms in evaluating an antihistamine, an approaching holiday in a psychotropic drug trial, or a small change in a rating scale or a laboratory assay procedure. Lastly, it could indicate an overall improvement or deterioration in the patients' conditions. None of the above situations would invalidate the analysis of treatment effects, although the last condition would suggest a look at possibly unequal carryover effects.

For the data shown in Table 10.1, a preliminary test for the equality of the carryover effects shows this source of variation to be nonsignificant. The analysis of model I as displayed in Table 10.4 also shows that the difference in treatment effects is not significant at the 10% level. (However, only the data from period 1 is used in performing this test for treatment effects.) Since the preliminary test indicates that unequal carryover effects are not present, the analysis of direct treatment effects should be analyzed by means of model II. The analysis of model II as shown in Table 10.5 indicates that the difference in treatment effects is significant at the 1% level.

B. A Reformulation of the Problem

To attain a better understanding of the construction and assumptions for the tests already described, as well as to plan alternative analyses of the data if the normality assumptions are not met, a reformulation of the problem is helpful. For each subject, the difference between the period 2 and period 1 values is a measure of the difference in treatment effects. This measure will be $(\pi_2 - \pi_1) + (\varphi_2 - \varphi_1) + (\varepsilon_{ij2} - \varepsilon_{ij1})$ for those subjects in sequence 1, and $(\pi_2 - \pi_1) - (\varphi_2 - \varphi_1) + (\varepsilon_{ij2} - \varepsilon_{ij1})$ for those subjects in sequence 2. If treatment effects are equal, (i.e., $\varphi_2 - \varphi_1 = 0$), both sets of differences have the same expected values.

Table 10.5 Analysis of Variance for Crossover Design without Residual Effects (Model II)

Source	df	SS	MS	F	P
Total (adj.)	23	23.31			
Sequence	1	0.04	0.04	0.027	0.873
Subject (seq)	10	14.84	1.48		
Treatment	1	5.36	5.36	17.87	0.0018
Period	1	0.02	0.02	0.07	0.796
Error	10	3.05	0.30		

The appropriate statistical procedure to test the equality of two sets of measurements is the two-sample t test or, if assumptions of normality are not met, the Wilcoxon rank sum test. Thus, the test for direct treatment effects in Table 10.3 is computationally equivalent to performing a two-sample t test to compare the differences between period 1 and period 2 in the two sequence groups. The test for period effects is computationally equivalent to comparing the (treatment 2 − treatment 1) means for the two sequence groups. In a similar manner, the analysis of both the sequence effect and the carryover effect can be based on a two-sample t test comparing the sum of both periods. The reader is referred to Chassan (1964) for a statement of this property and to Wallenstein and Fisher (1977) for further elaboration. Applications of the reformulation are given in the following section and in Section III.E for the repeated measurements designs.

C. Assumptions and Alternative Analyses

The above reformulation of the problem clearly indicates the assumptions for various statistical analyses. If carryover effects are not included in the model, it is only necessary to assume that differences in observations are normally distributed, but, as in the paired t test, it is not necessary to assume that the original data are normally distributed or that the variances for the two treatments are equal. If carryover effects are included, it would be necessary to both assume that the sums are normally distributed for the test of carryover effects and the differences are normally distributed for the test of direct treatment effects.

If the above assumptions are not satisfied, a nonparametric analysis may be used, as described by Koch (1972). Applying this procedure to the data displayed in Table 10.1 for model I shows that both the residual effects and treatment effects are not significant ($p > 0.10$). The analysis of model II shows the difference in treatment effects is significant at the 5% level. These analyses are shown in Table 10.6. Other nonparametric approaches are given by Fidler (1984) and Jones and Kenward (1989).

D. Intraclass Correlation

The concept of the "patient serving as his/her own control" in statistical terms implies that the variation among patients can be eliminated from the error term which is used to test the differences between treatments. According to this criterion, the crossover design is advantageous in those experimental situations in which the between-subject variation is large relative to the within-subject variation. Studying this criterion is equivalent to studying the intraclass correlation coefficient ρ, defined as

$$\rho = \frac{\sigma_s^2}{\sigma_e^2 + \sigma_s^2}$$

and estimated from

$$\frac{MS\ [subject(seq)] - MS\ [subject(error)]}{MS\ [subject(seq)]}$$

Grizzle (1965) notes that "when $\rho > 0$, even as small as .25, the two-period changeover design produces a worthwhile decrease in the amount of experimentation" for a specified power, "and when $\rho = 0$ it requires no more experimentation than the completely random assignment." However, it should be recognized that for a specified number of patients in a crossover design, twice as many patients are required in a completely randomized design to complete the same amount of experimentation.

E. Repeated Measurements Design

Let Y_{ijkm} be the observation for the jth subject in the ith sequence group at the kth period and m^{th} time point within the period. Schematically the data can be presented as follows:

Sequence	Period 1 time (1 ... t)	Period 2 time (1 ... t)
1	$Y_{1j11}, \ldots, Y_{1j1t}$	$Y_{1j21}, \ldots, Y_{1j2t}$
2	$Y_{2j11}, \ldots, Y_{2j1t}$	$Y_{2j21}, \ldots, Y_{2j2t}$

Subjects in sequence 1 ($j = 1, \ldots, n_1$) receive treatment 1 in period 1 and treatment 2 in period 2; subjects in sequence 2 ($j = 1, \ldots, n_2$; $n_1 + n_2 = n$) receive the treatments in reverse order.

Model II-R, the extension of model II to the repeated measurements case, may be written as

$$Y_{ijkm} = \mu + \psi_i + \zeta_{ij} + \pi_k + \varphi_l + e_{ij} + (\varphi\tau)_{im} + \omega_{ijm} + (\pi\tau)_{lm} + f_{ijkm}$$

where symbols i, j, k are defined in Equation 10.2, where τ_m is the effect of the mth time point, and where the interactions of time with treatment, period, and sequence are included in the model. The terms ζ_{ij}, e_{ijk}, ω_{ijm}, and f_{ijkm} are different expressions for error used in the analysis. A similar repeated measurements model can be written for model I.

As noted in Section III.B, the analysis of the basic crossover design can be done by analyzing the differences between period 1 and period 2. We extend this concept to the repeated measurements design by defining

$$d_{ijm} = Y_{ij1m} - Y_{ij2m} \, ,$$
$$s_{ijm} = Y_{ij1m} + Y_{ij2m} \, .$$

Canceling out the appropriate terms, we find that the preliminary tests for carryover effects and for carryover by time effects are computationally equivalent to tests performed in a repeated measurements, completely randomized, two-treatment design with dependent variable s_{ijm}. Similarly, the analysis of direct treatment effects and the treatment by time effects are equivalent to tests performed in a repeated measurements completely randomized design with d_{ijm} as the dependent variable. For further details, including an analysis of variance table, refer to Wallenstein and Fisher (1977).

For the complete analysis of model I-R, it is necessary to assume that each set of error terms $\{\zeta_{ij}\}$, $\{e_{ijk}\}$, $\{\omega_{ijm}\}$, and $\{f_{ijkm}\}$ be independent and identically normally distributed with variances σ_s^2, σ_e^2, σ_w^2, and σ_f^2, respectively. A sufficient condition for this assumption to be satisfied is that the variance-covariance matrix of $(Y_{ijkl}, \ldots, Y_{ijkt})$ is uniform, that is, the variance at each time point is some common value σ^2, and that the correlation between any two time points is ρ. This implies that the time interval between measurements does not affect the correlations. It is possible that the assumptions of this model would also be satisfied under less restrictive conditions.

For the analysis of treatment effects in model II-R using the above analysis, it is only necessary that the variance-covariance matrix of the differences over time be uniform. Note that this restriction does not require equal variability of the error terms e_{ijk} and f_{ijkm} for both treatments, nor does it necessarily require independence of the error terms f_{ijkm} and $f_{ijkm'}$. Note also that if these assumptions are not satisfied, multivariate or nonparametric tests may be performed based on s_{ijm} or d_{ijm}.

F.   Baseline Value

If there is a single baseline value X_{ij} followed by two treatment values Y_{ij1} and Y_{ij2} then an analysis of covariance utilizing X_{ij} as the covariate, and the differences and sums of the posttreatment values as outcomes will have slightly greater power than the analysis of differences and sums alone.

    If there are baseline values preceding each observation, then a possible approach is to subject the changes from baseline to Grizzle's analysis. Patel (1983) compared this approach with one that ignores the baseline values and with one that utilizes them as covariates. Fleiss, Wallenstein, and Rosenfeld (1985) noted that the analysis of changes from baseline can in some circumstances give a biased test of the carryover effects, and can lead to an inefficient analysis of the direct effects. Wallenstein and Fleiss (1988) indicate that an analysis based on changes from baseline would be preferred for an analysis of direct effects only for quite long washout periods and large correlations between observations. Under very restricted conditions, the baseline values could also be used alone to test for lack of carryover effects. Of course, even if there are two baselines, use of the first baseline alone will produce a modest increase in power. Other discussions of baseline values are given by Willan and Pater (1986) and Kenward and Jones (1987).

G.   Multiperiod Designs

The two-period two-sequence design is for many diseases not the design of choice for comparing two treatments. Extra period designs (EPDs) offer efficient within patient estimation of direct effects in the absence of certain types of carryover effects. The ideal application for these designs is one in which the effect of the first period persists to the second, but not subsequent periods.

    Kershner and Federer (1981) considered several two-treatment designs incorporating extra periods and/or sequences under a number of different models both with and without baseline observations. Among the designs that they considered, the EPD of Lucas (1950) repeating the last period of the two-period two-sequence design provided the most efficient test of direct effects. Optimal multiperiod designs have also been discussed by Laska, Meisner and Kushner (1983), Laska and Meisner (1985) and Matthews (1987). Some limitations of these types of designs are noted by Fleiss (1986).

IV.   USE OF CROSSOVER DESIGNS IN SPECIFIC CLINICAL AREAS

Although the crossover design has great intuitive appeal, there are specific areas of medical research in which it is generally accepted that this design is not recommended. In this section we discuss the advantages and disadvantages of this experimental design in certain areas of research.

The advantages of crossover designs in comparative bioavailability trials are generally acknowledged. The term *bioavailability* is defined as the rate and extent to which the active drug ingredient or therapeutic moiety is absorbed from a drug product and becomes available at the site of drug action. A comparative bioavailability trial is a trial in which the rate and extent of absorption of two drug formulations is compared (Bioequivalence Requirements, 1977). For example, this might involve obtaining blood samples from each subject at various times following administration so that the blood levels of the different formulations may be compared, or urine samples in a time interval so that the cumulative percentage of drug recovered in the urine may be compared. For the latter case, an example is given in Section V of Chapter 12. It has been shown that even in a group of subjects chosen to be homogeneous in regard to such background variables as age, height, sex, weight, and body frame, the observed concentration of drug in the blood is quite variable. Differences in blood levels in such a group could be due to differences in metabolism. Since the crossover design removes this patient variability (which is usually considerable) this design is often recommended (Westlake, 1974). The design of these trials should be based upon such considerations as the half-life of the drug so that duration of time in which blood samples are to be obtained and the proper washout time between administrations of dose can be determined. (See Chapter 12 for a general discussion of the design and analysis of bioavailability studies.)

The use of crossover designs in specific clinical areas has been addressed in the FDA clinical guidelines for 22 clinical areas. It has been observed by Dubey (1986) that the crossover design was recommended in five of the guidelines. The specific areas were for the study of analgesics, antianginals, anticonvulsants, antiinflammatories (ankylosing spondylitis), and hypnotics. For three additional areas, antianxiety agents, psychoactive drugs in children, and radiopharmaceuticals, the guidelines permitted the use of crossover designs. In the other 14 guidelines, crossover designs were permitted but discouraged in eight and not recommended in six. The investigator is strongly advised before utilizing a crossover design in the areas in which they are either recommended or permitted to ascertain the specific type of study for which they have been sanctioned.

The remaining portion of this section is a discussion of some possible pitfalls encountered in utilizing the crossover design. The examples are taken from the clinical evaluation of psychotropic agents.

In conducting clinical trials in schizophrenic populations, a requirement usually imposed is that patients have a drug-free "washout" period prior to receiving medication. This means that a washout period between active medication periods be imposed so that a patient could return to the status evident prior to the first period. Therefore, the lack of knowledge concerning the central nervous system metabolism of antipsychotic compounds in regard to rate of absorption, excretion, etc., can lead to confusing results, particularly since several studies have shown that some schizophrenic patients require a "no-drug" period of 90 days or more to reach their baseline mental status (Bishop and Gallant, 1966; Good et al., 1958). Although this problem can be overcome by designing the crossover trial to allow for a variable washout period between therapies, there are several reasons why this procedure should be avoided. One reason is that it could be difficult to conduct a double-blind study in which the duration of the treatment for each individual patient depends upon his results. Furthermore, it would be difficult to both set up proper decision rules and demonstrate that the status of each patient is the same prior to each of the treatment periods. For these and other reasons, Gallant et al. (1971) have suggested crossover designs be avoided in the evaluation of antipsychotic agents.

The crossover design should not be used if the nature of the disease, rather than the result of treatment, results in a change in the patient's condition. An example of this can be seen in patients with acute depression. Because most depressions are of short duration, the condition of the patients may have changed by the time they receive the second treatment. In addition, a significant number of patients may improve and drop out of the study. Because of these considerations, crossover designs are not recommended for evaluation of a treatment such as acute depression, although they may be useful in the treatment of chronic and/or recurring depression (Klerman, 1971).

As mentioned in the previous section, a significant carryover effect implies that only the data from the first medication period should be used. The carryover effect may be either physiological or psychological. The presence of psychological carryover effects precludes the use of crossover designs in anxious psychoneurotic patients, who tend to improve rapidly and maintain their early improvement (Frank et al., 1963). Another example of where crossover trials are not useful is in the study of mania since the disease is cyclic in nature (Gershon, 1971).

In conclusion, crossover designs can be very valuable to the investigator in those situations where the procedure is valid. Statisticians should be cautioned that the theoretical advantages are frequently not obtained because of nonlinear disease status from one period to the next. In certain types of trials, prior knowledge may indicate that the design could be used. In other cases, there are preliminary statistical tests which can be used to test the assumptions of the model, and certain biological variables which can be used to support the assumption of no carryover effects. However the investigator should be cautioned that lack of carryover effects cannot be proven either statistically or biologically (medically) and that in new areas of investigation use of the crossover design should be undertaken with caution.

REFERENCES

Albert, K. S., Sakman, E., Hallmark, H., Weidler, D., and Wagner, J. (1974). Bioavailability of diphenylhydantoin. *Clin. Pharmacol. Ther.*, **16**:727-735.

Bioequivalence requirements and in vivo bioavailability procedures (1977). *Federal Register*, **42**:1624-1653.

Bishop, M. P., and Gallant, D. M. (1966). Observations of placebo in chronic schizophrenic patients. *Arch. Gen. Psychiatry*, **14**:497-503.

Brown, B. W. (1980). The crossover experiment for clinical trials. *Biometrics*, **36**:69-79.

Chassan, J. B. (1964). On the analysis of simple crossovers with unequal number of replicates. *Biometrics*, **20**:205-208.

Cochran, W. G., and Cox, G. M. (1968). *Experimental Designs (2nd ed.)*, Wiley, New York.

Dubey, S. D. (1986). Current thoughts on crossover designs. *Clin. Res. Practices and Drug Reg. Affairs*, **4**:127-142.

Federer, W. T. (1955). *Experimental Designs*. MacMillan, New York.

Fidler, V. (1984). Changeover clinical trial with binary data: Mixed-model-based comparison of tests. *Biometrics*, **40**:1063-1070.

Fleiss, J. L. (1986a). Letter: On multiperiod crossover studies. *Biometrics*, **42**:449-450.

Fleiss, J. L., Wallenstein, S. and Rosenfeld, R. (1985). Adjusting for baseline measurements in the two-period crossover study: A cautionary note. *Controlled Clinical Trials*, **6**:192-197.

Frank, J. D., Nash, E. H., Stone, A. R., and Imber, S. D. (1963). Immediate and long-term symptomatic course of psychiatric outpatients. *Amer. J. Psychiatry*, **120**:429-439.

Gallant, D. M., Bishop, M. P., Free, S. M., Goldberg, W. T., and Simpson, G. M. (1971). Evaluating efficacy of psychotropic agents in schizophrenic populations. In *Principles and Problems in Establishing the Efficacy of Psychotropic Agents*, (Levine, J., Schiele, B. C., and Bouthilet, L., Eds.). Public Health Service Publication no. 2138. Washington, D.C., pp. 59-90.

Gershon, S. (1971). Methodology for drug evaluation in affective disorders: Mania. In *Principles and Problems in Establishing the Efficacy of Psychotropic Agents* (Levine, J., Schiele, B. D., and Bouthilet, L., Eds.). Public Health Service Publication no. 2138, Washington, D.C., pp. 123-236.

Good, W.W., Sterling, M., and Holtzman, W.H. (1958). Termination of chlorpromazine with schizophrenic patients. *Amer. J. Psychiatry*, **115**:443-448.

Grizzle, J. E. (1965). The two-period change-over design and its use in clinical trials. *Biometrics*, **21**:467-480 (correction note: *Biometrics*, **30**:727).

Hills, M., and Armitage, P. (1979). The two-period crossover clinical trial. *Br. J. Clin. Pharmacol.*, **8**:7-20.

Jones, B., and Kenward, M. G. (1989). *Design and Analysis of Cross-Over Trials.* Chapman and Hall.

Kenward, M. G. and Jones, B. (1987). The analysis of data from 2×2 crossover trials with baseline measurements. *Statistics in Medicine*, **6**:911-926.

Kershner, R. P. and Federer, W. T. (1981). Two-treatment crossover designs for estimating a variety of effects. *J. Amer. Stat. Assoc.*,**76**:612-619.

Klerman, G. L., (1971). Methodology for drug evaluation in effective disorders: Depression. In *Principles and Problems in Establishing the Efficacy of Psychotropic*

Agents (Levine, J., Schiele, B. C., and Bouthilet, L., Eds.). Public Health Service Publication no. 2138, Washington, D.C., pp. 91-122.

Koch, G. G. (1972). The use of nonparametric methods in statistical analysis of the two-period change-over design. *Biometrics*, **28**:577-584.

Laska, E. M. and Meisner, M. (1985). A variational approach to optimal two-treatment crossover designs: Applications to carryover effect models. *J. Amer. Stat. Assoc.*, **80**:704-710.

Laska, E. M., Meisner, M. and Kushner, H. B. (1983). Optimal crossover designs in the presence of carryover effect. *Biometrics*, **39**:1087-1091.

Lucas, H. L. (1950). *Design and Analysis of Feeding Experiments with Milking Dairy Cattle*. Mimeo Series no. 18. Institute of Statistics, North Carolina State University, Raleigh.

Matthews, J. N. S. (1987). Optimal crossover designs for the comparison of two-treatments in the presence of carryover effects and autocorrelated errors. *Biometrika*, **74**:311-320.

Patel, H. I. (1983). The use of baseline measurements in two-period crossover design in clinical trials. *Communications in Statistics - A*, **12**:2693-3712.

Paterson, H. D., and Lucas, H. L. (1962). *Changeover Designs*. Technical Bulletin no. 147, North Carolina Agricultural Experiment Station, Raleigh.

Wallenstein, S. (1986). Analysis of crossover design with baseline. *Proceedings of the 1986 Biopharmaceutical Section of the American Statistical Association*, American Statistical Association, pp.157-160.

Wallenstein, S., and Fisher, A. C. (1977). The analysis of the two-period repeated measurements crossover design with application to clinical trials. *Biometrics*, **33**: 261-269 (correction note: *Biometrics*, **37**:875).

Wallenstein, S., and Fleiss, J. L. (1988). The two period cross-over design with baseline measurements. *Communications in Statisistics*, **17**:333-3343.

Wallenstein, S., Patel, H. I., Fava, G. M., Polansky, M., Peace, K. E, Dubey, S. D., Kerschner, R. P., Lynch, G., and Koch, M. (1990). Two treatment crossover designs. In *Statistical Issues in Drug Research and Development* (K.E. Peace, Ed.) Marcel Dekker, Inc., New York, pp. 171-224.

Westlake, W. J. (1974). The use of balanced incomplete block designs in comparative bioavailability trials. *Biometrics*, **30**:319-327.

Willan, A. R. and Pater, J. L. (1986). Using baseline measurements in the two-period crossover clinical trial. *Controlled Clinical Trials*, **7**:282-289.

11

Statistical Considerations in Clinical Trials of AIDS Patients

C. Ralph Buncher

University of Cincinnati Medical Center,
Cincinnati, Ohio

I. THE DISEASE

In the 1980s, the disease AIDS broke out in epidemic and then pandemic [a worldwide epidemic] proportions. The disease is now known to be caused by the HIV or Human Immunodeficiency Virus and is referenced in that way. The earlier and less specific name of Acquired Immune Deficiency Syndrome (AIDS) is the more popular name and it will be called AIDS in this chapter. Statistical projections leading to doomsday predictions from a uniformly lethal disease have given way to the realization that AIDS is becoming another very serious chronic disease because of new pharmaceutical treatments and its innate biology.

In this disease a person becomes infected with the virus many years before the disease manifests itself clinically. Persons with antibodies to the AIDS virus are infected with HIV but we do not know when they will show symptoms of the disease. The median time from infection to symptoms is about ten years. Progression of the disease is followed by measuring the lymphocytes (white blood cells) known as CD4 and when the CD4+ count per microliter of these cells falls below 200, the person meets the U.S. Centers for Disease Control and Prevention (CDC) guidelines for AIDS. Alternatively, a person may be diagnosed for some other condition that turns out to be AIDS Related Complex (ARC), signifying that the disease can be diagnosed because of one of the effects of losing the normal immune status. Much of the pharmaceutical development in the general field of AIDS research has been to treat or prevent opportunistic infections or increase the immune response through immunomodulating agents. Clinical trials have been done with persons who are asymptomatic but virus-positive, persons who have ARC, and those who have a clinical diagnosis of AIDS.

The disease is important in its own right but it also has brought up problems that have forced statisticians to re-evaluate clinical trial methodology. The pharmaceutical situation for this disease has had many interesting differences from any prior situation. For example, at an early stage in drug development, a sizeable part of the patient population became involved with government sponsored clinical trials (specifically those of the National Institute of Allergy and Infectious Diseases) in clinical centers known as the AIDS Treatment Evaluation Units and collaboratively as the AIDS Clinical Trials Group. Thus, the interaction and sometimes competition of the private and federal sectors have produced a much different drug development setting than usual. This is another example of the studies described in Chapter 8.

II. COLLABORATIVE TRIALS

One of the reasons for organizing the patients into federally sponsored trials in the earlier years was because there were many candidate drugs and a limited supply of patients. An orderly research process based on scientific decisions was considered preferable to the alternative of drug sponsors competing for patients and recognition. An important by-product of this research was help in financing the care of these patients. The many candidate medications for this large market and numerous interested companies required a consensus to set priorities for testing. Others would state that politics also played a major role in decision making. One of the benefits of the trials from the standpoint of the AIDS patients has been help in paying for the costs of this disease, especially medication costs.

Unlike most other diseased populations, persons with AIDS have been in the prime of life when they have faced the disease and their impending death. They do not have the usual distribution of elderly or younger persons who are those most often ill. Children born to mothers with AIDS are another group. Moreover, the initial and defining concept was that the disease was uniformly fatal. The patients have been well-educated and very active in commenting on methods of care for themselves and their friends. The result has been a questioning of established biostatistical research methods. This chapter includes an effort to understand some of those questions, to provide the sides of the discussion when there are ongoing disagreements, and to point the way to research that might help settle these disputes.

Consider a simple example. Protocol 002 of the ACTG (AIDS Clinical Trials Group) was designed to study whether a lower or a higher dose regimen of AZT (Azidothymidine or more commonly zidovudine) was better for persons with AIDS who also were positive for *Pneumocystis carinii* pneumonia (PCP), one of the common sequela of the disease. When the clinical trial began in 1986, no prophylaxis for PCP was permitted in the trial but, in August, 1987, PCP prophylaxis for a second episode of PCP was authorized. In April, 1988 PCP prophylaxis was allowed to be used in any patient and by the time the trial was completed, prophylaxis for PCP was standard treatment. Thus some of these patients were given prophylaxis for their PCP during this early trial while others were not and hence this factor became an observational one rather than a controlled/randomized one. Therefore, the data accumulated on patients not receiving prophylaxis is not directly pertinent in the current medical climate. This is an example of doing research in the middle of a revolution which can be translated into the statement: when doing research in a rapidly changing environment, there is a high probability that the study will not conclude as originally designed. This possibility of a protocol change is always present in a clinical trial but is especially likely when the situation is changing rapidly as in many of the AIDS controlled trials.

III. PLACEBOS IN AIDS RESEARCH

One of the first questions in AIDS research concerned the use of placebos in the early trials. The statistical concept has been that only by giving a placebo concurrently and under blindfold conditions do we know what would have happened if the new medication was not given. Thus, the placebo group provides the baseline and differences from placebo are our measure of efficacy (see Chapter 13). AIDS patients said that the disease was uniformly fatal; therefore, giving some persons placebo was not ethical since placebos can not cure the patients. With a test medication, there is at least some chance that the patient's course can be improved substantially and it was this group that argued placebos should not be used in these trials.

Statisticians know that sometimes placebo is the better treatment in clinical trials. An example is "Mortality and morbidity in patients receiving encainide, flecainide, or placebo, the cardiac arrhythmia suppression trial" (Echt et al., 1991) in which there were statistically significantly fewer deaths on placebo than the other two drugs. Most patients seem to believe that placebo is inert in all ways and can never be the better treatment. If patients really believe the fiction that the new treatment can never be worse than placebo, then the research community has yet to adequately inform the volunteer community of the reality of clinical trials. Freedman (1992) provides additional insight into this situation.

An interesting anecdote appeared in a news account of AIDS testing (Navarro, 1992). "Four weeks after enrolling in a trial, John G. saw his CD4 cells [a marker for the progress of the disease] rise from 300 to 649. Thrilled, he called up other infected friends to urge them to get the drug. He started to take better care of himself—he ate three meals a day, he exercised—seeing a future for himself again. Then he found out he was on a placebo. .... 'I was totally shocked,' John G., a 31-year-old circulation manager for a magazine in Manhattan, said. 'I thought it was the miracle drug.'"

The elements in this anecdote are clear. Thinking you are doing better can be autocorrelated with actions that make you healthier. Patients do not believe that placebo, including the hope and special care involved, can cause them to improve. Patients may relapse after placebo therapy is halted and revealed.

Statisticians need to resolve these issues on the use of placebos in clinical trials. One thought along this pathway is that major changes in treatment, especially for lethal diseases, can be detected without placebos; placebos become of great importance in measuring smaller initial improvements and in evaluating side effects. Perhaps early trials for usually lethal diseases should proceed without placebos in an effort to make progress more rapidly (e.g., fewer patients required in the study) and then be followed by trials controlled with active medication that measure the effects more carefully. Alternatively, several "active" treatments could be used with the effect of each being measured against all of the others. In the case of AIDS, later trials have used the first effective medication, AZT, as the standard. The news account states "In AIDS trials, placebos are now used only when the drug being tested is the only promising treatment or when there is no immediate danger in withholding treatment temporarily" (Navarro, 1992).

IV.   INFORMATION AND TRIAL PARTICIPANTS

Another characteristic that makes AIDS patients unusual is Newsletters. Patients and other interested persons direct numerous newsletters to persons with AIDS and they routinely report on the latest findings in the clinical trials. Compared to the usual newsletter produced by a patient group, these newsletters are of quite high quality. They report results factually and correctly. What then can be the problem with patients being informed of the clinical trial results that pertain to them? One problem is that patients, who are battling over their own mortality, are frequently more responsive to the latest finding than is the research community, even if the finding is not yet replicated. Therefore patients may want to receive the medication that looked good in the last trial and may not wish to participate in a follow-up trial or other "research" to improve on the prior trial. Patients with a lethal disease do not feel they have the time to wait until the medication has been scientifically proven effective in confirmatory trials before they use it. In addition, these trials routinely use remote data entry to facilitate real time data analysis.

The important principle is that we all believe that information should be given to patients; the challenge is that most trials have not had to negotiate with such an informed patient group and some procedures of routine design in clinical trials may have to be changed. We in the statistical community must review our clinical trial methodology to make sure that it will work efficiently with very well-informed patients. "Intent-to-treat" could become an adjectival fact of life rather than just a methodology for analysis.

Information from newsletters and other information gathered by the patients may lead to other medication problems. Medication problems include patients who do not take the correct medicine. In most trials, the problems include some mistake by the patient or lack of adequate communication about the dosage schedule or the patient forgetting to take the medication. In the case of AIDS trials, there has been the problem of patients sharing medications. If the first patient is on one unknown medication and the second on another, it was reasoned by them, why not share medication. If we are both on the same medication, then no change has occurred. If we are on different medications, that is, one on "active medication" and the other on placebo, then we each get a half dose. Obviously the statistician trying to analyze the study may be unaware of these unintended modifications to the randomization code and errors will increase.

V.   DROPOUTS AND NONCOMPLIANCE

Another problem is dropout rates. In all trials, one must face the possibility that the dropouts are not a random sample of patients. If the rate of dropout is very different in the drug group than in the placebo group, there are many problems in interpretation. In some of the AIDS trials, there was a much greater dropout rate in the placebo group than in the drug group. Many explanations are possible. One unusual explanation in these trials was that the patients were having their blinded medication tested in a chemical laboratory and if it was placebo, they dropped out of the study. Statisticians now have to give greater concern to this possibility in any new clinical trial.

An old question is especially important in AIDS research. This is the issue of privacy and confidentiality. Social stigma for those diagnosed with this disease and economic considerations with respect to health insurance are important factors to consider in doing clinical trials with this patient group. Privacy and confidentiality thus must be achieved if a trial is to be successful. In one sense, these are not new issues but rather routine issues since patients in any trial must be assured of privacy and confidentiality. In some instances one must be concerned that the location of the treatment clinic may become an issue in the success of the trial. Some patients would rather go to another city for treatment rather than run the risk of being seen in an AIDS clinic in their home area. One possibility is to have patients cared for in a large facility in which many different types of patients are seen so that AIDS patients are not singled out. Some AIDS centers are separate from other patients which prevents that solution. Most patients wish to maintain their privacy and thus these issues will have to be considered in setting up any study.

In most clinical trials, persons who are not deemed likely to cooperate are eliminated from the pool of subjects to be included (see Chapter 6). A special group in this category is the intravenous (illegal) drug user. In general, patients known to be in this group are not expected to cooperate sufficiently in clinical trials. Once again, the study of persons with AIDS produces a dilemma. Intravenous drug users are such a large portion of the cases that the question must be asked whether the medications tested in others are as effective in this special group of patients. Moreover many in this group of patients want to participate in trials but the limited data available suggest that intravenous drug users are not as compliant as patients without this habit. The statistical issues include what degree of noncompliance is so great that there is a lack of value from including this group. Good information on the relation of valuable patient data to degree of noncompliance would be helpful in creating a cutoff between studies that might be cost effective and those studies that would be a waste of money and effort since no useful information would emerge from that study.

VI. BIOMARKERS

A final issue concerns the use of biomarkers rather than using a disease endpoint. In many clinical trials, waiting for the biological endpoint takes too long. For example, trials may take many years if one wishes to use survival as an endpoint and the improvement offered by the new treatment is likely to be small. This leads many scientists to substitute for the actual endpoint (e.g., survival), a marker of that endpoint such as the CD4 lymphocyte count. A valid biomarker will produce the same ultimate statistical differences but at an earlier timepoint. Thus if a decrease in CD4 counts predicts mortality, this provides a good biomarker that will shorten the time in clinical trials. Validating biomarkers and choosing the best from a series of potential summary statistics is a challenging task (Dawson and Lagakos, 1991).

More research on biomarkers in clinical trials would be helpful as there are unsolved statistical problems and biomarkers are now being used more commonly in clinical trials. Clearly in some instances using biomarkers in clinical trials will produce different answers than using the ultimate biologic measure such as survival. Statisticians need to do more work modeling the savings in time and cost that will result from using biomarkers in clinical trials in comparison to the increase in Type I or Type II errors which will result from using a surrogate measure. Under what circumstances are biomarkers more efficient and under what circumstances are they less efficient as the endpoint in clinical trials?

An excellent discussion of the "Statistical issues arising in AIDS clinical trials" is given by Ellenberg, Finkelstein, and Schoenfeld (1992) followed by an extensive discussion by other experts in clinical trials. They discuss some of these same points and raise others. For example, if AZT is given to all patients as the standard therapy instead of placebo, how do you identify drugs that are inferior to AZT but better than no treatment? A particularly fascinating problem is caused by patients who wish to be in more than one study simultaneously; this multiple enrollment or "co-enrollment" creates problems in understanding toxicity or side effects.

The process of developing the pharmaceutical treatments for AIDS has created new problems for statisticians and reminded us of some old ones. The solution of these problems will create progress in clinical trials methodology.

REFERENCES

Dawson, J. D. and Lagakos S. W. (1991) Analyzing laboratory marker changes in AIDS clinical trials, *J of Acquired Immune Deficiency Syndromes* 4:667-676.

Echt, D. S., Liebson, P. R., Mitchell, L. B., et al.. (1991) Mortality and morbidity in patients receiving encainide, flecainide, or placebo; the cardiac arrhythmia suppression trial. *N. Engl. J. Med.*, **324**:781-788.

Ellenberg S. S., Finkelstein D. M, Schoenfeld D. A., (1992) Statistical issues arising in AIDS clinical trials. *J. Amer. Stat. Assoc.*, **87**:562-569 and discussion **87**:569-583.

Freedman, B., (1992) Suspended judgment, AIDS and the ethics of clinical trials: Learning the right lessons. *Controlled Clinical Trials*, **13**:1-5.

Navarro, M., Into the unknown: AIDS patients test drugs, *New York Times*, 29 February 1992, pp. 1, 10.

12

Bioavailability Studies

Daniel L. Weiner

Syntex Development Research,
Palo Alto, California

Lianng Yuh

Parke-Davis Pharmaceutical Research Division,
Warner-Lambert Company,
Ann Arbor, Michigan

I. INTRODUCTION

One of the many steps necessary for characterizing a new drug dosage form is to conduct an in vivo bioavailability study. The Food and Drug Administration (FDA) requires that pharmaceutical companies perform these studies in order to procure an investigational new drug application (IND), new drug application (NDA), or an abbreviated new drug application (ANDA). Specific details related to requirements for an NDA are presented in the Federal Register (January 7, 1977) while Cabana and Douglas (1976) give bioavailability guidelines for IND development.

Bioavailability, as defined by Metzler (1974), "includes the study of the factors which influence and determine the amount of active drug which gets from the administered dose to the site of pharmacologic action as well as the rate at which it gets there." However, since it is often difficult or impossible to measure drug at its site of action, drug concentrations in the systemic circulation (blood, plasma, or serum) are often used as a surrogate. Thus, to put it another way, the object of a bioavailability study is to quantify the relative amount and rate of absorption of the administered drug which reaches the

215

general circulation intact. Note that the FDA has defined bioavailability as "the rate and extent to which the ingredient is absorbed."

There are a number of types of biopharmaceutic studies which involve the measurement of bioavailability of one or more dosage forms. These include the following:

Absolute bioavailability: These studies involve a comparison of a nonparenteral dosage form (such as an oral form) relative to an IV solution. If an IV solution cannot be administered, then an oral solution or suspension is often used.

Comparative bioavailability or bioequivalence: These studies are used to determine if the rate and extent of a test formulation is equivalent to that of a reference formulation. These studies are discussed in detail in this chapter.

ADME (Absorption, Distribution, Metabolism, Excretion): These studies are run to determine the pharmacokinetics and metabolic profile of a dosage form.

Dose proportionality: These studies are run to determine the relationship of increasing dosage strengths of a drug to its bioavailability.

In the course of development of a drug, many other special studies are often run. These include studies to determine the effects of renal impairment, food, and concomitant medications on the bioavailability of the drug.

Factors influencing the bioavailability of a dosage form are numerous but essentially fall into two categories. The first is made up of physical characteristics of the dosage form itself and includes such items as tablet compression force, particle size, and dissolution rate. The second category contains certain design factors which can affect the drug level profile of a drug. For example, the choice and spacing of the sampling times must be carefully chosen so that the drug concentration can be accurately determined. The choice of sampling times also affects the precision of the estimates of the pharmacokinetic parameters. Other design factors such as the amount of physical activity permitted during a study and whether or not a subject fasted prior to administration of medication can also have a marked effect on drug levels. However, even under controlled conditions blood and urine concentrations often differ markedly, even for subjects who are similar in age, height, weight, general health, and other demographic factors (Gibaldi, 1977, Chapter 7).

Because of the complexity of the problems related to establishing the bioavailability of a dosage form, a multidisciplinary approach to the design and analysis of these studies seems mandatory. A statistician's input into the planning of these studies relates to determining which factors need to be controlled, whether patients or subjects should be studied, and what statistical design will yield the needed information. The purpose of this chapter is to provide some

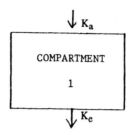

Figure 12.1 One-compartment model.

insight into the choice of design and method of statistical analysis of single-dose in vivo bioavailability studies.

II. BIOAVAILABILITY OF A SINGLE FORMULATION

A. Compartmental Models

In order to determine the pharmacokinetic properties of a drug it is frequently useful to represent the body as a system of interconnected compartments. The compartments often have only limited physiological meaning but serve as a conceptual device useful for analyzing drug concentration data. The most commonly used compartmental models are the one-compartment model (Figure 12.1) and the two-compartment model (Figure 12.2).

In Figure 12.1 the entire body is viewed as a single compartment with K_a and K_e, representing, respectively, the rate of absorption of drug into the system and the rate of elimination of drug from the system. Gibaldi and Perrier

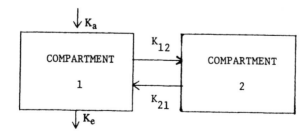

Figure 12.2 Two-compartment model.

(1982) point out that viewing the body as a one-compartment model does not mean that the concentration of drug throughout the body tissues at any given time is the same, but the model does assume that changes of drug concentrations in the plasma quantitatively reflect drug concentration changes in the tissues. A system of two compartments is shown in Figure 12.2. Typically the first compartment is taken to represent blood plasma and the second compartment, peripheral tissue. The constants K_{12} and K_{21} represent transfer rates of drug between plasma and tissue, and K_e denotes the elimination rate of the drug.

Input of drug into the model is usually assumed to be zero- or first-order, depending on the route of administration, while elimination from the compartments and transfers between compartments are assumed to be first-order. By zero-order input we mean that the input rate of drug is constant; first-order input and elimination implies that the input and elimination rates are proportional, respectively, to the amount of drug at the absorption site and to the amount of drug in the compartment through which the drug is eliminated. The above assumptions regarding rates of input, transfer, and elimination require that the equation relating the concentration of drug to time can be written as a sum of exponentials. For example, the equation corresponding to the concentration of drug in the single-compartment model (Figure 12.1) is as follows:

$$C(T_i) = \frac{FD}{V} \frac{K_a}{K_a - K_e} (e^{-K_e t_i} - e^{-K_a t_i}) \tag{12.1}$$

where

$C(t_i)$ = concentration of drug in the compartment at time t_i
F = fraction of the dose that is absorbed
D = dose of drug
V = apparent volume of distribution of the drug
K_a = absorption rate constant
K_e = elimination rate constant

Note that for suitable choices of a_j and λ_j, Equation 12.1 can be written as

$$\sum_{j=1}^{2} a_j e^{-\lambda_j t_i}$$

Equation 12.1 is appropriate when the drug is administered orally. If the drug is administered through IV injection or some other route of administration and the body is viewed as a single compartment, then Equation 12.1 would no longer be appropriate. Wagner (1975) discusses the one-compartment and various

Table 12.1 Concentration of Drug A in Whole Blood

Sampling time (hr)	Concentration (ng/mL)
0.5	0.70
1.0	1.11
2.0	1.36
3.0	1.17
4.0	0.99
6.0	0.71
8.0	0.50
10.0	0.31
14.0	0.14
18.0	0.06
24.0	0.20

multicompartmental models and provides equations for the concentration of drug in the various compartments following several different routes of administration.

B. Parameter Estimation

1. Individual Modeling

Table 12.1 displays concentrations of drug A in whole blood for one subject following a single oral solution of 260 mg. The first step in the analysis of such data is the identification of the underlying model, which is usually based on information obtained from a preliminary study or from data reported in the literature. If such additional data is not available, then one should begin by fitting the sequence of concentrations to the equations generated by both the one- and two-compartment systems. Computer programs utilizing nonlinear least-squares algorithms are usually employed to obtain best-fitting estimates of the pharmacokinetic parameters (e.g., K_a, K_e) for a particular model. The criterion of goodness of fit is usually taken to be minimization of the quantity,

$$\sum_{i=1}^{n}(Y_i - \hat{Y}_i)^2$$

(12.2)

where Y_i is the observed concentration at time t_i, $\hat{Y}_i$ is the concentration predicted by the fitted model at time t_i and n is the number of samples collected from the subject. Hence we attempt to find the set of estimates of the pharmacokinetic parameters that minimizes the residual sum of squares.

Table 12.2 PCNONLIN Nonlinear Estimation Program

Parameter	Estimate	Standard error	95% Confidence limits		
Volume/F	51.843825	1.355671	48.717613	54.970038	Univariate
			46.986464	56.701187	Planar
K_a	1.017178	.053985	.892687	1.141670	Univariate
			.823749	1.210608	Planar
K_e	.202892	.005700	.189748	.216036	Univariate
			.182470	.223315	Planar

Several problems make the fitting procedure difficult. First, since the algorithms employ iterative techniques, they usually require initial estimates of the parameters. In addition, if one considers the residual sum of squares (Eq. 12.2) as a p + 1 dimensional surface (where p = number of pharmacokinetic parameters), it is easy to envision other problems. This surface may have many peaks and valleys causing the program to converge to a local and not a global minimum. Additionally, the minimum may occur at a nearly flat spot on the surface, thus implying that there is a wide range of parameter values that give approximately the same value of Equation 12.2. Fortunately, these difficulties can often be overcome if one has available initial estimates of the parameters which are fairly close to the true values. One method of obtaining initial estimates is discussed in the Appendix.

Example. As an illustration of the model-fitting technique the data presented in Table 12.1 were fit to Equation 12.1 using the program PCNONLIN (1992). In this particular example, F was unknown, so the parameters to be estimated were K_a, K_e, and V/F. A portion of the output from the program is presented in Tables 12.2 and 12.3. The model fits the data fairly well, at least as evidenced by the residuals.

There are several factors which must be considered in assessing the adequacy of the model. In particular, one should always make a careful comparison of the observed and predicted concentrations. Examination of the concentration time profile may reveal a delay in the start of absorption as zero or near-zero concentrations at the first sampling time(s). This indicates the need for a lag time in the model. Equation 12.1 with a lag time incorporated into the model is as follows:

$$C(t_i) = \frac{FD}{V} \frac{K_a}{K_a - K_e} \left[e^{-K_e(t_i - L)} - e^{-K_a(t_i - L)} \right]$$

(12.3)

Table 12.3 PCNONLIN Nonlinear Estimation Program. Summary of Nonlinear Estimation, Function 1

X	Observed Y	Calculated Y	Residual	Weight*	SD-$\hat{Y}$	Standardized residual
.5000	.7000	.7281	−.2811E−01	.1841	.8463E−02	−3.414
1.000	1.110	1.096	.1428E−01	.1161	.7344E−02	1.545
2.000	1.360	1.291	.6927E−01	.0947	.5604E−02	6.664
3.000	1.170	1.197	−.2699E−01	.1101	.6294E−02	−2.701
4.000	.9900	1.029	−.3899E−01	.1301	.6523E−02	−3.962
6.000	.7100	.7078	.2152E−02	.1815	.5474E−02	.2057
8.000	.5000	.4746	.2537E−01	.2577	.4882E−02	2.359
10.00	.3100	.3167	−.6699E−02	.4156	.5406E−02	−.6381
14.00	.1400	.1407	−.7047E−03	.9203	.6101E−02	−.0697
18.00	.6000E−01	.6250E−01	−.2496E−02	2.147	.6088E−02	−.2467
24.00	.2000E−01	.1850E−01	.1500E−02	6.442	.4675E−02	.1384

Corrected sum of squared observations = 2.25042
Weighted corrected sum of squared observations = .728326
Sum of squared residuals = .87432E−2
Sum of weighted squared residuals = .111548E−2
S = .118083^{-1} with 8 degrees of freedom
Correlation (Y, $\hat{Y}$) = .998
AIC criteria = −68.78317
SC criteria = −71.18632
AUC computed by trapezoidal rule = 9.29250

---

* For this example, the weight for each observation was taken as WT = 1/y. The weights were then normalized so that the sum of the weights equals the number of observations, 11.

A lag time does not improve the fit with this data set. However, a lag time may considerably improve the fit with other data sets.

In assessing the goodness of the parameter estimates, one must consider the widths of the confidence limits. As can be seen in Table 12.2, there are two types of limits available. The univariate limits are calculated in the usual way as the parameter estimate ± the appropriate t statistic multiplied by the standard error of the parameter estimate. The planar limits are derived from an estimated 95% joint confidence region for the parameters and will always be wider than the univariate limits. From the width of the confidence limits reported in Table 12.2, we can infer that there are a variety of estimated parameters that yield equivalently good fits of Equation 12.1 to the data.

In addition, one should always check to see if there are systematic deviations from the fitted model. This is easily done by inspection of the residual values in Table 12.3. For example, if the model overestimates larger concentrations and underestimates smaller concentrations, then one should try weighting the observed concentrations, the Y values, by $1/E(Y)$ or $1/\sqrt{E(Y)}$. Other weighting schemes may help remove other biases. If weights are used, the criteria for "best" fit would be to use those estimated parameters that minimize

$$\sum_{i=1}^{n} W_i(Y_1 - \hat{Y}_i)^2$$

where W_i is the weight assigned to the ith concentration. An excellent review of this topic was presented by Peck et al. (1984).

An examination of the residual values can also help identify outliers or aberrant observations. Least-squares procedures are very sensitive to outliers and the presence of one in the data set can cause sizable bias in the parameter estimates. To combat this problem Rodda et al. (1975) have developed a non-parametric procedure which lessens the effect that an aberrant observation has upon the estimates of the parameters. Their procedure, known as OSEP, performed quite well relative to least-squares procedures on simulated data sets containing several different types of outliers. The OSEP procedure is recommended by those authors when the true model is known and outliers are known to exist.

Giltinan and Ruppert (1989) have demonstrated the use of the generalized least squares estimator (GLS) in pharmacokinetic modeling. Sheiner et al. (1972) applied extended least squares (ELS) to obtain the regression and variance parameters simultaneously. ELS is equivalent to the maximum likelihood estimator (MLS) using the normal likelihood function. The advantages and disadvantages of the ELS method have been discussed extensively by Jobson and Fuller (1980), Carroll and Ruppert (1982), and van Houwelingen (1988) in the statistical literature, and by Finney (1985) in the pharmacokinetic literature. Beal and Sheiner (1988) have compared GLS and the modified extended iteratively reweighted least squares estimators (MEIRLS). Their results indicated that these two estimators have similar efficiencies and both are superior to ELS. Finally, the approach using Bayes' theorem was discussed by Katz et al. (1981), and Racine-Poon (1985).

2. Population Modeling

Thus far we have discussed how one derives estimates of the pharmacokinetic parameters based on data obtained from a single subject. In a multisubject trial it is usually desired to estimate the distribution of the parameter values in the population. As Steimer et al. (1985) indicated, there are two types of data from the multisubject studies: experimental data and observational data. Generally, experimental studies involve health volunteers. They are well designed and the data are balanced and less variable. The results of the studies provide the estimates of basic pharmacokinetic parameters and the information for adjusting the individual dosage regimens.

Population studies involve the collection of observational data from the treated patients in the efficacy trials. These studies are unbalanced and more variable than the experimental studies. The results of the observational studies can be used for pharmacokinetic screening and post market surveillance. We refer to Whiting et al. (1986) and Grasela et al. (1987) for more discussion of the applications of these data.

Many procedures have been developed for population modeling. The basic assumption is that the subjects have a common pharmacokinetic model. Certainly, the Naive-Pooled-Data and Naive-Averaging-of Data approach are not efficient and are biased. (Rodda et al., 1972; Steimer et al., 1985; Katz, 1988). Another approach is called the "Two Stage Method". In the first stage of this method, individual data are fitted to obtain the individual parameter estimate using the methods suggested in Section II.B. The final population parameter estimate is obtained using the combined individual estimates. To apply this method, each subject needs sufficient data to derive a reasonable individual estimate. In addition, the simple two stage method does not take the variability associated with each individual estimate into account. Thus, it often underestimates the variabilities associated with the population parameters. A modified two stage method was proposed by Prevost (1977). This revised procedure incorporated both inter- and intrasubject variabilities in the model. The two stage method approach using Bayes' theorem was discussed by Katz and D'Argenio (1984), Racine-Poon (1985), and Racine-Poon and Smith (1989).

Weiner and Jordan (1978) have investigated a procedure which fit all subjects simultaneously and letting the pharmacokinetic parameters be functions of known physiological factors. Particularly, V, the apparent volume of distribution, is frequently related to body weight or body surface area. This can be of particular importance if one is using the estimates of the pharmacokinetic parameters to predict drug levels at steady state. Sheiner and Beal (1980, 1981, 1983) proposed a nonlinear mixed effects model (NONMEM) using ELS with the first order approximation to pool the individual data to obtain a population estimate and the individual estimates simultaneously. A computationally intensive procedure using the Nonparametric Maximum Likelihood Estimator was derived by Mallet (1986). Steimer et al. (1985) and Katz (1988) have reviewed these procedures. In general, the nonlinear mixed affects model approach is preferred for the observational data because the lack of the quality of the individual data. Linstrom and Bates (1990) have developed a new computing algorithm using the nonlinear mixed effects model approach while Gelfand et al. (1990) proposed a full Bayesian procedure using Gibb's sampler. No one has determined which of the above methods is best in any given situation because of the difficulty of defining the criteria for evaluating different procedures. However, ELS is the only procedure with a commercial computer package (NONMEM). We recommend that the analyst analyze the data using several of the above methods before drawing any definitive conclusions.

III.   COMPARATIVE BIOAVAILABILITY STUDIES

A.   Introduction

Thus far we have been concerned with assessing the bioavailability of a single formulation. Frequently, though, it is necessary to conduct comparative bioavailability studies. For example, one of the requirements for procuring an abbreviated new drug application (ANDA) is to compare the bioavailability of the new generic formulation to a standard or reference formulation. In addition, changes in manufacturing practices may necessitate conducting a comparative bioavailability study.

In this section we will provide guidelines regarding the choice of criteria for comparison of two dosage forms and discuss the designs and analyses most commonly used in relative or bioavailability trials. Other pharmacokinetic studies such as food effect, age effect and dose proportionality studies can be assessed using similar principles. The FDA held a Bioequivalence Public Hearing in 1986. Based on the results of this meeting, the statistical decision criteria for solid dosage formulation bioequivalence studies were clearly defined (Food and Drug Administration, 1988).

B.   Choice of the Criteria for Comparison

Before turning to a discussion of the design and analysis of relative bioavailability studies we will first discuss several choices of bioavailability criteria for comparing the two dosage forms. In practice it is customary to choose several criteria, each related to specific aspects of the drug concentration profiles, with which to compare the formulations. The choices usually depend on the type of bioavailability study being run. For example, the criteria for comparing a timed-release capsule with an immediate release tablet may differ from those for comparing a single formulation of drug manufactured at different locations. The following are the criteria most commonly used as a basis for comparing two formulations. For a given study, the investigator chooses one or more of these variables as a basis for comparing the two dosage formulations.

1. Area Under the Concentration Curve

Area under the concentration curve (AUC), the integral of concentration of the drug over time, measures the total amount of drug absorbed. AUC is probably the most commonly used variable for comparing two formulations. For a given profile, this quantity is usually calculated from the sequence of plasma concentrations by the trapezoidal rule as follows:

$$AUC = \sum_{i=2}^{N} \frac{[C(t_i) + C(t_{i-1})]}{2} (t_i - t_{i-1})$$

$$(12.4)$$

where $t_0 = 0$ and N is the number of samples that were taken after administration of the dosage form. Equation 12.4 should be used as an estimate of total absorption only if $C(t_N)$ is zero or near zero. Otherwise, one could seriously underestimate the bioavailability of the formulation. If $C(t_N)$ is somewhat greater than zero, then one should extrapolate the AUC to infinite time by adding the term $C(t_N)/\lambda$ to the value of Equation 12.4. In the additional term, λ represents the terminal rate constant. However, it is preferred to report AUC (0: last quantifiable concentration) if λ can not be precisely estimated.

2. Peak Concentration (C_{max})

Peak concentration is usually calculated as the maximum observed $C(t_i)$. Although in actuality the peak concentration will most likely be higher and occur at some time point other than one of the sampling times, the approximation is usually adequate for comparative purposes.

3. Time to Peak Concentration (T_{max})

Time to peak concentration is usually taken to be the time at which the maximum concentration was observed. T_{max} and C_{max} are used jointly to measure the rate of absorption.

4. Cumulative Percentage Recovered of Drug ($A_c\%$)

Applicable to urine data, the cumulative percentage of drug recovered is calculated as the cumulative amount recovered in some time interval (e.g., 24 hours) divided by the initial dose. As with AUC, $A_c\%$ can be extrapolated to infinity as well (Wagner, 1975).

5. Estimated Absorption Rate (K_a)

It is sometimes desirable to compare two formulations on the basis of their estimated absorption rates. For example, suppose that in order to lengthen the period of time that a tablet retains its potency, a pharmaceutical firm begins to manufacture the tablets with a new protective coating. It would then be of interest to compare the absorption rate of the coated tablets with that of the uncoated tablets. As discussed in Section II, it is necessary that one first be able to identify the underlying pharmacokinetic model in order to estimate K_a.

6. Elimination Half-Life

The elimination half-life of a drug, $t_{1/2}$, can be estimated by dividing 0.693 by the absolute value of the slope of the terminal linear phase of the concentration profile when plotted on semi-log scale. For the model displayed in Figure 12.1, $t_{1/2} = 0.693/K_e$.

7. Concentration Profiles

Instead of comparing formulations on the basis of univariate values derived from the sequence of blood levels, one can instead compare the entire concentration profiles using a multivariate method such as profile analysis (Morrison, 1976, pp. 205–216). This can help distinguish between two formulations which differ in onset and duration but are equivalent in terms of AUC.

One inherent problem in the statistical evaluation of these data is that the parameters (1–6) are estimated rather than measured directly. Often, the standard errors of the estimates are quite heterogeneous across subjects which may invalidate the assumptions underlying an analysis of variance.

Many of the variables discussed above can be estimated in an alternative manner by taking an appropriate function of the estimated pharmacokinetic parameters. For example, by integrating Equation 12.1 from 0 to ∞ one can show that an estimate of AUC is $FD/\hat{V}\hat{K}$. Other expressions can be derived for peak concentration and time to peak concentration. This method of estimation is not to be recommended unless the underlying pharmacokinetic model can be identified and the parameters accurately estimated. Regulatory agencies prefer calculation of parameters by the above methods rather than from fitted parameters due to problems associated with model misspecification.

For bioequivalence studies, AUC and C_{max} are the primary response variables. T_{max} is not considered a primary response variable due to its highly discrete nature (Chen and Jackson, 1991). Pharmacokinetic constants, elimination half-life, and the concentration-time profile are also considered as secondary variables. For urine samples, the cumulative percentages of drug recovered are often used as the primary response variables. If a multiple dose study is conducted, the minimum plasma concentration (C_{min}) and the fluctuation index $[(C_{max} - C_{min})/C_{avg}]$, both of which are measured under steady-state conditions, are also parameters of interest (Food and Drug Administration, 1988).

C. Designing a Comparative Bioavailability Study

It is well known that there is tremendous intersubject variability in the bioavailability of most drug formulations. As a result of this phenomenon, designs that enable each subject to serve as his own control (see Chapter 7) are most

Table 12.4 A Two-Period Crossover Design

Sequence	Period	
	1	2
1	A	B
2	B	A

frequently employed in comparative bioavailability studies except for drugs with an extremely long half-life or toxicity problems. Such designs have an advantage over other designs in that the comparison of formulations is free from subject-to-subject variation. One example of such a design, the two-period crossover design, is presented in Table 12.4 (Grizzle, 1965; Jones and Kenwald, 1989). Note that each subject received each of the two formulations according to one of two predetermined sequences. Between each dose period there is a washout period of sufficient length to ensure that the first formulation has been eliminated from the subject's system. The washout period should be at least 5 times the terminal half-life of the drug.

The design presented in Table 12.4 can easily be extended to compare more than two formulations. These larger crossover designs are usually constructed so that the allocation of formulations to periods forms a Latin square design. One such design for comparing four formulations is presented in Table 12.5.

In a 2×2 crossover design, the carry over effect is confounded with the formulation, sequence, and period effects unless one utilizes the Balaam design (crossover design with extra periods, Table 12.6). Using this type of design, one can not only estimate the within subject variability but also assess the

Table 12.5 A Four-Period Crossover Design

Sequence	Period			
	1	2	3	4
1	A	B	C	D
2	B	D	A	C
3	C	A	D	B
4	D	C	B	A

Table 12.6 An Extra-Period Crossover Design

Sequence	Period		
	1	2	3
1	A	B	A
2	B	A	B

formulation by subject interaction (Hwang et al., 1978; Ekbohm and Melander, 1989; Rodda, 1989).

Westlake (1974) has discussed the use of balanced incomplete block designs for comparative bioavailability studies involving a large number of treatments. Yuh and Ruberg (1990) have presented several balanced and unbalanced crossover designs using an oversampling technique for a drug to drug interaction study (Table 12.7).

D. Choosing the Sample Size

Once the appropriate design has been determined for the bioavailability study the next step is to determine the number of subjects to be included in the study. Traditionally, using the results of the analysis of variance (ANOVA) which includes sequence, period and treatment effects, the sample size was determined by the power of the test statistics that can detect a 20% difference between the reference and test means. However, because current bioequivalence criteria is based on the two-sided test procedure (Food and Drug Administration, 1988),

Table 12.7 Incomplete Partially Balanced Crossover Design

Sequence	Period			
	1	2	3	4
1	A	E	C	D
2	B	F	D	E
3	C	G	E	F
4	D	A	F	G
5	E	B	G	A
6	F	C	A	B
7	G	D	B	C

the sample size should be derived based on this procedure which is equivalent to the confidence interval approach. For a given coefficient of variation, the difference between mean parameters, and acceptable limits, Phillips (1990) provided a procedure which yields the sample size estimate based on the power of the two-sided tests procedure. In general, the sample sizes derived by the usual ANOVA method are somewhat smaller than those based on the power of the two one-sided test procedure. For example, if the coefficient of variation is 20% and the mean difference 5%, a study with 24 subjects will be required to assure the 90% confidence interval values are within 80–120% range with a power of .8.

E. Sampling Times

The final step in designing the study is to determine the number and spacing of collection times for the plasma or urine samples. There are several guidelines which can be helpful in selecting the times for drawing samples.

A control urine or plasma sample should always be taken just prior to administration of drug. This ensures that the subject has no residual amounts of drug in his or her system from previous study days and that the subject has not inadvertently ingested the drug with some concurrent medication or food. The remaining samples should be taken at times sufficient to determine the profiles of the concentration curves. For orally administered drugs, samples should be taken most frequently during the absorption phase when the profile is most rapidly changing and less frequently during the elimination phase.

If pharmacokinetic parameters are to be estimated from the data, then the total number of samples should be sufficiently large so that there are enough degrees of freedom in estimating σ^2 (e.g., ≥ 6) to allow for a reasonable test of the adequacy of the compartmental model. In addition, the plasma or urine concentrations should have returned to near-zero levels by the time the last sample is taken. For more discussion about optimal sampling designs, we refer to the papers by D'Argenio (1981), DiStefano (1981, 1982), Landaw (1982), Katz and D'Argenio (1983), and Drusano et al. (1988).

Having discussed various considerations in designing comparative trials and the choice of the criteria for comparison, we now turn to a discussion of the analysis of such data.

V. ANALYSIS OF COMPARATIVE BIOAVAILABILITY STUDIES

A. Analysis of Variance Model

Before performing an analysis of the data one should first check to see if the concentrations of drug in the predosing samples are at or near the lower limit

Table 12.8 Area Under the Curve (0–24 hours)

	Period 1		Period 2	
Subject	Formulation 1	AUC (μg hr /mL) (0–24 hrs)	Formulation 2	AUC (μg hr /mL) (0–24 hrs)
1	2	1697.45	1	1636.21
2	1	800.54	2	636.29
3	2	1050.98	1	615.61
4	1	1049.72	2	113.16
5	2	936.53	1	113.20
6	1	1504.35	2	1094.96
7	2	1277.87	1	1603.63
8	2	1548.28	1	1411.44
9	2	964.41	1	1129.55
10	1	1458.60	2	1341.08
11	1	1602.51	2	1470.85
12	1	688.35	2	576.33
13	1	1841.53	2	1312.68
14	1	1259.93	2	1037.75
15	1	1227.14	2	931.89
16	2	1339.08	1	1241.25
17	2	931.91	1	613.14
18	2	1136.23	1	547.66
19	2	1534.56	1	823.46
20	2	794.44	1	952.24
21	2	1538.59	2	1131.16
22	1	874.06	2	1133.22
23	2	1478.62	1	1385.12
24	1	1328.58	2	1131.90

of quantitation. If other than trace amounts are measured in these control samples, then it is recommended that only the first-period data be analyzed. The analysis would then proceed along the lines of a one-way analysis of variance. For the remainder of the discussion we will now assume that at most trace amounts of drug are measured in the control samples.

The analysis of data obtained in a crossover trial comparing two formulations is presented in Chapter 10. Note that it is not necessary to fit the data to model 1 of that chapter since the control samples serve as a check on the existence of residual effects. Thus one can proceed directly to their model 2 (Eq. 12.5), which allows for testing of equality of sequence, treatment, and period effects:

$$Y_{ijk} = \mu + \psi_i + \zeta_{ij} + \pi_k + \varphi_1 + \varepsilon_{ijk}$$

$$(12.5)$$

where

Y_{ijk} = univariate response [(AUC), peak concentration, etc.] for the jth subject within the ith sequence at period k

μ = overall mean

ψ_i = effect of the ith sequence

ζ_{ij} = effect of the jth subject within the ith sequence

π_k = effect of the kth period

φ_1 = effect of the lth formulation

ε_{ijk} = random error

Example. A study was undertaken to compare the relative bioavailabilities of two batches of drug B in terms of the 0 to 24 hour AUC (Table 12.8). This study involved a two-period crossover design with twelve subjects randomly allocated to each of the two sequences. The analysis of the data was carried out according to the methods outlined in Chapter 7 and is reported in Table 12.9. Note that none of the F tests yielded significant results: thus there is no reason to reject the hypothesis that the two formulations are comparable in terms of extent of absorption.

The analysis of variance for more complicated crossover designs is discussed in Winer (1971). In addition, Westlake (1974) as well as Yuh and Ruberg (1990) have discussed the analysis of the balanced and partially balanced incomplete crossover designs.

Table 12.9 Analysis of AUC Data

Source	df	Sum of squares	F
Sequences	1	1878.82	0.01
Subjects (sequences)	22	3869649.05	
Formulations	1	310807.54	8.18
Periods	1	9150.34	0.24
Error	22	836364.11	
Summary of Results			
Estimate of μ_T/μ_R		97.7%	
90% Confidence interval		(89.4%, 105.9%)	
90% Westlake interval		(90.9%, 109.1%)	
Power ($\alpha = 0.05$)		0.977	

B.   The Power Approach

The analysis of dose proportionality studies was produced by Yuh et al. (1991). Because the objective of the bioavailability studies is to show that the test formulation is equivalent to the reference formulation, the FDA used a decision criterion that is based on the power of the test to detect a 20% difference between the reference and test treatment means. This rule stated that the two formulations are bioequivalent if the p-value for the formulation effect is greater than a prespecified level of significance and the power of the test to detect a 20% difference between the reference and test means is greater than or equal to 80%. This rule is no longer used by the FDA as the primary criteria for assessment of bioequivalence.

C.   Confidence Interval Approach

The development of the bioequivalency of two formulations discussed in Section VI.A was based on the classical theory of hypothesis testing. Traditionally, this methodology has been applied in those situations where it is desired to show that a significant difference exists in the effects of two treatments. Usually, though, this is not the objective of a comparative bioavailability study. Instead, it is usually desired to assess the difference in the relative bioavailabilities of the two formulations in terms of some univariate measure such as AUC, and determine if the difference is within acceptable limits. It would thus seem that a confidence interval approach might be more appropriate in some instances. Westlake (1972), Metzler (1974), Shirley (1976), and O'Quigley and Baudoin (1988) have discussed this approach in the context of comparative bioavailability trials.

In general, the 90% confidence interval for the ratio in formulation means is computed as follows:

$$\left\{ \left[\frac{\overline{X}_T - \overline{X}_R}{\overline{X}_R} \pm \frac{t_{v,0975} S_d}{\overline{X}_R} \right] + 1 \right\} \times 100\% \qquad (12.6)$$

where

$\overline{X}_i$ = mean observed response with formulation i, i = T,R (where T and R denote the test and reference formulations)

S_d = standard error of the difference in formulation means using the error mean square obtained from the analysis of variance

v = degrees of freedom for error

Substituting the appropriate values into Equation 12.6 we obtain (89.4%, 105.9%) as the 90% confidence interval for the ratio in formulation means.

Thus, these two formulations are considered bioequivalent if we use ± 20% as the acceptable limits. Note that the confidence interval on the difference in formulation means contains 100%, as is expected, since the test of equality of formulations effects was not significant at the 0.05 level (Table 12.9). Also, the end points of this percent scaled confidence interval are obtained by treating the reference mean as a constant. The correct (exact) confidence interval for the ratio of the means can be obtained using Fieller's theorem (Locke, 1984; Schuirmann, 1990). Note that this asymmetric exact confidence interval is usually somewhat wider than the approximate confidence interval.

Westlake (1976) has also modified the confidence interval approach and developed a confidence interval symmetric about the origin. Mantel (1977) indicated that these symmetric confidence intervals do not show the location of the sample value and they are always longer than the traditional confidence interval. However, as Westlake mentioned, the symmetric interval is used merely as a decision making tool. It is noted that the confidence coefficient for the Westlake interval is always greater than $1 - \alpha$. Again, since the sample mean is used as a constant, Westlake's symmetric confidence interval is only an approximation. The exact symmetric confidence interval has been derived by Mandallaz and Mau (1981). We refer to Steinijans and Diletti (1983), Kirkwood (1981), and Metzler (1988) for additional discussion of the use of the symmetric confidence interval. For comparison purposes, the symmetric confidence interval is also presented in Table 12.9.

D. Bayesian Approach

Rodda and Davis (1980) first suggested that bioequivalence can be assessed using Bayesian analysis. They computed the posterior probability that the observed relative bioavailability is within the acceptable limits. Selwyn et al. (1981, 1984) have also proposed a Bayesian procedure using a more complex statistical model. According to Wijnand and Timmer (1983), Rodda and Davis' rule is equivalent to Westlake's while the decision rule based on Mandallaz and Mau's confidence interval is equivalent to that using the Bayesian procedure proposed by Selwyn et al. Gelfand et al. (1990) suggested a Bayes procedure using Gibb's sampler to evaluate bioavailability/bioequivalent studies.

E. Anderson and Hauck's Procedure

Because of the nature of equivalence trials, one has to set up an alternative hypothesis that the difference of the two formulation means will fall within a prespecified interval. This concept was discussed originally by Lehmann (1959) and Bondy (1969). Anderson and Hauck (1983) have adopted this idea using the following pair of hypotheses.

$H_0 : \mu_T - \mu_R \leq \log(.8)$ or $\mu_T - \mu_R \geq \log(1.2)$

$H_1 : \log(.8) < \mu_T - \mu_R < \log(1.2)$

Note that the null hypothesis states that the two formulation means are not equivalent and the alternative hypothesis states that they are equivalent. The test statistic they consider is

$$T = \frac{\overline{X}_T - \overline{X}_R - 1/2\,[\log(.8) + \log(1.2)]}{S(1/n_T + 1/n_R)}$$

where the X's and n's are the sample means (in the logarithmic scale) and sample sizes, respectively, and S is the standard deviation (square root of MSE) obtained from the analysis of variance mode. As they discussed, the distribution of T is generally unknown. Thus, the exact critical value cannot be found. However, the Student's t distribution can be used as an adequate approximation. A similar approach was used by Rocke (1984). It is noted that there is a small probability that the null hypothesis will be rejected even when the difference of the two means is not within the acceptable limits.

F.   Two One-Sided Tests Procedure

Schuirmann (1987) proposed a procedure which consists of two pairs of testing hypotheses

$H_{01}: \mu_T - \mu_R \leq \log(.8)$

$H_{11}: \mu_T - \mu_R > \log(.8)$

and

$H_{02}: \mu_T - \mu_R \leq \log(1.2)$

$H_{12}: \mu_T - \mu_R > \log(1.2)$

This procedure decomposes the interval hypotheses H_0 and H_1 into two sets of one-sided hypotheses. Note this procedure has also been developed based on the log scaled values. This two one-sided tests procedure will conclude equivalence of two formulation means if and only if both H_{01} and H_{02} are rejected at a chosen nominal level of significance (e.g., .05). The design rule based on the two one-sided tests procedure ($\alpha = .05$) is equivalent to the rule

Table 12.10 Listing of Individual Bioequivalence (%)

Subject	AUC reference (2)	AUC test (1)	Ratio (formulation 1/ formulation 2)%
1	1735	3340	193%
2	2594	2613	101%
3	2526	1138	45%
4	2344	2738	117%
5	938	1287	137%
6	1022	1284	126%
7	1339	1930	144%
8	2463	2120	86%
9	2779	1613	58%
10	2256	3052	135%
11	1438	2549	177%
12	1833	1310	71%
13	3852	2254	59%
14	1262	1964	156%
15	4108	1755	43%
16	1864	2302	124%
17	1829	1682	92%
18	2059	1851	90%
Median			109%
(Min/Max)			(43%, 193%)

based on the traditional 90% confidence interval. In the same paper, Schuirmann also compared the rejection regions and the probability characteristics of his procedure to the power approach and showed that the two one-sided tests procedure is superior to the power approach in general. Metzler (1988) performed a simulation study to evaluate the power of different decision rules—two one-sided tests procedure (traditional confidence interval), Westlake's symmetric confidence interval (Rodda and Davis' procedure), Mandallaz and Mau's approach, and Anderson and Hauck's procedure. He defined the power of a decision rule for bioequivalence to be the probability of rejecting bioequivalence of a test formulation, given the true relative bioavailability. Based on the probability of rejection curve, Westlake's rule is very similar to the Anderson and Hauck's procedure if the coefficient of variation is less than 20%. The probability of rejection is close to .90 for all rules except the one based on the two one-sided tests procedure. To make this rule more like the others, a larger value (namely two times α) should be used.   If the coefficient of variation is greater than or equal to 30%, only the rule by Mandallaz and Mau is different from the others.

F. Individual Bioequivalence

Another useful technique in assessing the relative bioavailabilities of two dosage forms is to compare the dosage forms on a by-subject basis. In particular, the proposed rules for bioequivalence studies involving tricyclic antidepressants (Federal Register, February 17, 1978) require such by-subject comparisons in addition to the standard analyses outlined in previous sections.

One method of making the by-subject comparison is to express the response obtained with formulation 1 as a percentage of the response obtained with formulation 2 (or vice versa). As an example, consider the data reported in Table 12.10. Note that 6 out of 18 (33%) of the subjects had a bioavailability on formulation 1 within 25% of that of formulation 2. In the late 1970s, the FDA proposed a rule using the individual ratios. This rule states that two formulations are equivalent if and only if at least 75% of these individual ratios are between 75% and 125% limits. Clearly, these two formulations are not bioequivalent if the 75/75 rule is applied. As Westlake (1979) pointed out, the underlying principle of this 75/75 rule is similar to the construction of the tolerance interval. Haynes (1981) has performed a simulation study to investigate the 75/75 rule and shown that there is a greater probability to accept the test formulation if the coefficients of variations for the test and reference formulations, respectively, are identical and smaller. In addition, a test formulation compared to a reference formulation with equal variability had less chance of acceptance than the case when the reference formulation has smaller variability.

However, if the individual ratio is the parameter of clinical interest, one should construct a confidence interval for the central tendency of these individual ratios (Yuh, 1990a, 1990b). Yuh also indicated that individual bioequivalence is similar to the percentage change from baseline in the conventional efficacy trials. This parameter is more difficult to analyze because it is a ratio of two random variables. That is, the ratios are not normally distributed even if the original random variables are normally distributed. Thus, the sample median or a robust estimator should be used to assess the central tendency of these ratios (Yuh, 1988, 1990b). Anderson and Hauck (1990) also discussed a procedure treating the individual ratio as a binary response.

Finally, Yuh (1992) has proposed to assess bioequivalence using the concordance correlation. The concordance correlation measures a combination of the formulation mean and variability differences and the linear correlation between the formulations. Note that the concordance correlation is equal to 1 if two formulations have identical means and variabilities and the linear correlation is equal to 1.

G. Other Topics

Other approaches include the transformation and nonparametric methods. There are several reasons one may perform a logarithmic transformation of the pharmacokinetic parameters. Both Westlake (1988) and Rodda (1989) gave a good discussion of the rationale for performing the transformation. An Advisory Committee to the FDA recommended a logarithmic transformation of pharmacokinetic data. However, our experiences show that the distribution of the difference in formulations is usually symmetric if the subject effect is removed (as in the analysis of a crossover study). It should be noted that the acceptable lower (upper) limit under an antilogarithmic transformation is 80% (125%) and the power calculation is different from that using the raw data.

If the underlying distribution is not normally distributed or there are potential outliers, nonparametric methods may be utilized to assess bioequivalence. Several nonparametric methods have been proposed for the crossover studies (Koch, 1972; Hauschke et al., 1990; Pabst and Jaeger, 1990). Generally, these methods are useful for testing hypotheses; however, that they may not be appropriate for estimation because, clinically, it is difficult to interpret the confidence interval on rank scale.

Testing the homogeneity of variance for bioavailability is also a topic of interest. Haynes (1981) discussed a procedure to test the equality of variances for crossover studies. This test statistic was originally proposed by Pitman (1939) and Morgan (1939) and used for testing the variances of two correlated data sets. Assuming there is neither period nor sequence effects, the test statistic F is computed as follows: $F = [(g - 1)^2 \times (n - 2)]/[4 \times g \times (1 - r^2)]$, where n is equal to the number of subjects, g is equal to the ratio of the formulation variances and r^2 is equal to the correlation of two formulations. Using the data presented in Table 12.8, F = .562 and the corresponding p-value is equal to .16. This indicates that there is insufficient evidence that the variabilities of these two formulations are different.

Bioavailability studies are conducted to evaluate the rate and extent of absorption of a drug product. Mean AUC and C_{max} values are used to assess the extent and rate, respectively. According to the current criterion, two formulations are bioequivalent if the end points of the percent scale 90% confidence interval for the ratio of the treatment means are within the 80–120% range for both AUC and C_{max} (FDA, 1988) or 80–125% for the log scaled data. Because the sampling time points are usually not optimal, T_{max} data are more variable than C_{max} and both parameters are more variable than AUC. In general, to

Table 12.11 Simulated Data

Time	Concentration	Extrapolated concentration using $T(t_i)$	Residual concentration $R(t_i)$
0.5	35.43	80.60	45.12
1.0	45.47	66.29	20.55
1.5	45.33	54.52	9.19
2.0	40.85	44.84	3.99
2.5	35.25		
3.0	29.68		
5.0	13.78		
7.0	6.19		
9.0	2.78		
11.0	1.27		

obtain a meaningful confidence interval for AUC, C_{max}, and T_{max}, there is a need to develop an optimal sampling time strategy to reduce the variability of these parameters.

If studies involve more than two treatments, multiple confidence intervals for each pharmacokinetic parameter will be constructed and adjustments for these multiple comparisons should be made. However, most classical procedures are developed to adjust Type I error which may widen the confidence interval. Little work has been done to address the issue of adjustment of multiple confidence intervals.

If a crossover study design is used, a logical approach will utilize the within-subject comparisons. Thus, the individual ratio is also a parameter of interest. Note that one is analyzing the individual ratios implicitly if the logarithmic transformation is performed. Finally, replicated designs can be important because they can provide a good estimate for inter- and intrasubject variability, respectively. In addition, these designs can also adjust for carry over effects. Clearly, there are many unanswered questions and more research is needed to be done in the future.

APPENDIX: PEELING TECHNIQUE FOR OBTAINING STARTING VALUES

We have previously discussed the importance of good starting values in obtaining the best estimates of the parameters. In this section we discuss one method

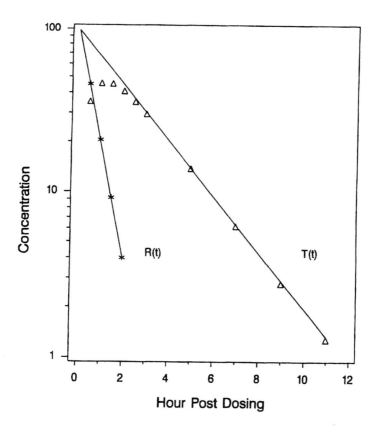

Figure 12.3 Application of the Peeling technique for estimating the pharmacokinetic parameters in a one-compartment model.

of obtaining initial estimates. The method will be illustrated by estimating the parameters in Equation 12.1 for the simulated data presented in Table 12.11. The first step is to plot the data on semilogarithmic paper, as in Figure 12.3. Note that in Equation 12.1 if we assume $K_a > K_e$ (which is true for most drugs), then for large enough t, the second exponential term is essentially zero. Let $T(t_i)$ denote the equation of this terminal phase. Then

$$T(t_i) = \frac{FD}{V} \frac{K_a}{K_a - K_e} e^{-K_e t_i}$$

$$(12.7)$$

Taking logarithms to base 10 of both sides we obtain

$$\log T(t_i) = \log\left(\frac{FD}{V}\frac{K_a}{K_a - K_e}\right) - \frac{K_e t_i}{2.303}$$

(12.8)

Thus, if we were to draw a line through those points lying on the terminal linear portion of the curve on semilog scale, it would have slope $-K_e/2.303$ and intercept

$$\log T(t_i) = \log\left(\frac{FD}{V}\frac{K_a}{K_a - K_e}\right)$$

The constant 2.303 is the reciprocal of the log to base 10 of e. Although working with logarithms to base 10 may seem somewhat unusual since the underlying pharmacokinetic models can be written as sums of exponentials, it is the conventional methodology of practitioners in this area and its use will be adopted here.

In Figure 12.3 we see that the last six points fall nearly on a straight line. Selecting the first and last of these six values, the slope can be estimated by

$$\frac{\log T(11) - \log T(2.5)}{11 - 2.5} = 0.17$$

Thus $K_e = -0.17(-2.303) = 0.39$ and the estimated intercept from Figure 12.3 is log (98). We have then that

$$T(t_i) = \log (98) - 0.17t_i$$

(12.9)

If we subtract Equation 12.1 from Equation 12.7 and call the result $R(t_i)$, we have

$$R(t_i) = \frac{FD}{V}\frac{K_a}{K_a - K_e} e^{-K_a t_i}$$

(12.10)

Again, taking logs of both sides of Equation 12.10, we note that log $R(t_i)$ is linear with slope $-K_a/2.303$ and intercept

$$\log\left(\frac{FD}{V}\frac{K_a}{K_a - K_e}\right)$$

To obtain an estimate of K_a we first use Equation 12.9 to obtain extrapolated concentrations for those four points not in the terminal linear portion of the curve. These values are recorded in the third column of Table 12.11. Then $R(t_i)$, i = 1, 2, 3, 4 are obtained by subtracting column 2 from column 3. The slope of log $R(t_i)$ can easily be found to be −0.69; thus $K_a = -0.69 \times (-2.303) = 1.59$. Using the intercept of either log $T(t_i)$ or log $R(t_i)$ (they should be equal), one can estimate either V or V/F depending on whether or not F is a known quantity. If the two intercepts did not coincide, the implication is that a lag time is needed in the model.

The true values used in simulating the data were $K_a = 1.50$ and $K_e = 0.40$ with $\sigma = 0.01$. Although the estimated values were very close to the true values of the parameters in this example, this will not always be the case. It is sometimes difficult to determine which points to include as lying on the terminal linear portion of the curve. This is especially the case when K_a and K_e are nearly equal. Wagner (1975, pp. 59–63) illustrates the use of the peeling technique for more complicated models.

As is evidenced by the discussions in Sections II and III, model identification can be quite difficult. Even if the model can be assumed to be known, it may be difficult to obtain precise estimates of the pharmacokinetic parameters. If estimates of the pharmacokinetic parameters are not needed, there are alternative models available. For example, splines and polynomial models often give good fits to drug concentration data. Usually, though, estimates of the pharmacokinetic parameters are needed to predict blood levels after multiple dosing (Wagner, 1975, pp. 144–147).

ACKNOWLEDGEMENT

We are grateful to J. Cook for his helpful comments.

REFERENCES

Anderson, S. and Hauck, W. W. (1980). Consideration of individual bioequivalence. *J. Pharmacokinet. Biopharmaceutics*, **18**:259-271.

Anderson, S. and Hauck, W. W. (1983). A new procedure for testing equivalence in comparative bioavailability and other clinical trials. *Commun. Statis.*, **A12**:2663-2692.

Beal, S. L. and Sheiner, L. B. (1988). Heteroscedastic nonlinear regression. *Technometrics*, **30**:327-337.

Bondy, W. H. (1969). A test of an experimental hypothesis of negligible difference between means. *Amer. Statistician*, **23**:28-30.

Cabana, B. E. and Douglas, J. F. (1976). Bioavailability: Pharmacokinetic guidelines for IND development. Paper presented at 15th Ann. Internatl. Pharm. Conf., Austin, TX. February 23-27, 1976.

Carroll, R. J. and Ruppert, D. (1982). A comparison between maximum likelihood and generalized least squares in a heteroscedastic linear model. *J. Amer. Statistic. Assoc.*, **77**:878-882.

Chen, M. and Jackson, J. (1991). The role of metabolites in bioequivalency assessment. I. Linear pharmacokinetics without first-pass effect. *Pharmaceut. Res.*, **8**:25-32

D'Argenio, D. Z. (1981). Optimal sampling times for pharmacokinetic experiments. *J. Pharmacokinet. Biopharmaceutics* 9:739-755.

DiStefano, J. J. III (1981). Optimized blood sampling protocols and sequential design of kinetic experiments. *Amer. J. Physiol.*, **240**:259-265.

DiStefano, J. J. III (1982). Algorithms, software and sequential optimal sampling schedule designs for pharmacokinetics and physiologic experiments. *Mathematics and Computers in Simulation*, **XXIV**:531-534.

Drusano, G. L. Forrest, A., Snyder, M. J., Reed, M. D., and Blumer, J. L. (1988). An evaluation of optimal sampling strategy and adaptive study design. *Clin. Pharmacol. Ther.*, **44**:232-238.

Ekbohm, G., Melander, H. (1989). The subject-by-formulation interaction as a criterion of interchangeability of drugs. *Biometrics*, **45**:1249-1254.

Federal Register, Vol. 42. No. 5, January 7, 1977, Book I, pp. 1624-1653.

Finney, D. J. (1985). Models, formulations and statistics. In *Variability in Drug Therapy.* (Rowland, M., Sheiner, L. B., and Steiner, J-L., Eds.). Raven Press, New York, pp. 11-123.

Food and Drug Administration (1988). *Report by the bioequivalence task force on recommendations from the bioequivalence hearing conducted by the FDA.* Bioequivalence Task Force, FDA, Washington, D.C.

Gelfand, A. E., Hills, S. E., Racine-Poon, A., and Smith, A. F. M. (1990). Illustration of Bayesian inference in normal data models using Gibbs sampling. J. Amer. Statistic. Assoc., **85**:972-985.

Gibaldi, M. and Perrier, D. (1982). Pharmacokinetics, Marcel Dekker, New York.

Gibaldi, M. (1977). *Biopharmaceutics and Clinical Pharmacokinetics*, Lea & Febriger, Philadelphia.

Giltinan, D. M. and Ruppert, D. (1989). Fitting heteroscedastic regression models to individual pharmacokinetic data using standard statistical software. *J. Pharmacokinet. Biopharmaceutics*, **17**:601-614.

Grasela, T. H., Antal, E. J., Ereshefsky, L., Wells, B. G., Evans, R. L., and Smith, R. B. (1987). An evaluation of population pharmacokinetics in therapeutic trials Part II. Detection of a drug-drug interaction. *Clin. Pharmacol. Ther.*, **42**:433-441.

Grizzle, J. E. (1965). The two period changeover design and its use in clinical trials. *Biometrics*, **21**:467-480.

Hauschke, D., Steinjans, V. W., and Diletti E. (1990). A distribution-free procedure for the statistical analysis of bioequivalence studies. *Int. J. Clin. Pharmacol. Ther. Toxicol.*, **28**:72-78.

Haynes, J. D. (1981). Statistical simulation study of new proposed uniformity requirement for bioavailability studies. *J. Pharm. Sci.*, **70**:673-675.

Hwang, S., Huber, P. B., Hesney, M., and Kwan, K. C. (1978). Bioequivalence and interchangeability. *J. Pharm. Sci.*, **67**:IV.

Jobson, J. D. and Fuller, W. A. (1980). Least squares estimation when the covariance matrix and parameter vector are functionally related. *J. Amer. Statistic. Assoc.*, **75**:176-181.

Jones, B., Kenward, M. G. (1989). *Design and analysis of crossover trials.* Chapman and Hall, New York.

Katz, D. (1988). Population density estimation. *Prog. Food Nutr. Sci.*, **1**:325-338.

Katz, D., Azen, S. P., and Schumitzky, A. (1981). Bayesian approach to the analysis of nonlinear models. *Biometrics*, **37**:137-142.

Katz, D. and D'Argenio, D. (1983). Experimental design for estimating integrals by numeric quadrature, with applications to pharmacokinetic studies. *Biometrics*, **39**:621-628.

Kirkwood, T. B. L. (1981). Bioequivalence testing - a need to rethink. *Biometrics*, **37**:589.

Koch, G. G. (1972). The use of nonparametric methods in the statistical analysis of two-period changeover design. *Biometrics*, **28**:577-584.

Landaw, E. M. (1982) Optimal multicompartmental sampling designs for parameter estimation: practical aspects of the identification problem. *Mathematics and Computers in Simulations*, **XXIV**:525-530.

Lehmann, E. L. (1959). *Testing Statistical Hypotheses*, John Wiley & Sons: New York.

Linstrom, M. J. and Bates, D. M. (1990). Nonlinear mixed effects models for repeated measures data. *Biometrics*, **46**:673-687.

Locke, C. S. (1984). An exact confidence interval from untransformed data for the ratio of two formulation means. *J. Pharmacokinet. Biopharmaceutics*, **12**:649-655.

Mallet, A. (1986). A maximum likelihood estimation method for random coefficient regression models. *Biometrika*, **73**:645-656.

Mandallaz, D. and Mau, J. (1981). Comparison of different methods for decision making in bioequivalence assessment. *Biometrics*, **37**:213-222.

Mantel, N. (1977). Do we want confidence intervals symmetrical abut the null value? *Biometrics*, **33**:759-760.

Metzler, C. M. (1974). Bioavailability: a problem in equivalence. *Biometrics*, **30**:309-317.

Metzler, C. M. and Huang, D. C. (1983). Statistical methods for bioavailability. *Clin. Res. Pract. Drug. Reg. Affairs*, **1**:109-132.

Metzler, C. M. (1988) Statistical methods for deciding bioequivalence of formulations. In *Oral Sustained Release Formulation: Design and Evaluation.* (Yacobi, A. and Halphin-Walega, E., Eds.) Pergamon Press: New York.

Morgan, W. A. (1939). A test for the significance of the difference between the two variables in a sample from a normal bivariate population. *Biometrika*, **31**:13-19.

Morrison, D. F. (1976). *Multivariate Statistical Methods.* McGraw-Hill, New York.

O'Quigley, J. and Baudoin, C. (1988). General approaches to the problems of equivalence. *The Statistician*, **37**:51-58.

Pabst, G. and Jaeger, H. (1990). Review of methods and criteria for the evaluation of bioequivalence studies. *Eur. J. Clin. Pharmacol.*, **38**:5-10.

PCNONLIN (1992). *User's Manual*, SCI Software, Lexington, Ky.

Peck, C. P., Beal, S. L., Sheiner, L. B., and Nichols, A. I. (1984). Extended least squares nonlinear regression: a possible solution to the choice of weights problem in analysis of individual pharmacokinetic data. *J. Pharmacokinet. Biopharmaceutics*, **12**:545-558.

Phillips, K. F. (1990). Power of the two one-sided test procedure in bioequivalence. *J. Pharmacokinet. Biopharmaceutics*, **18**:137-144.

Pitman, E. J. G. (1939). A note on normal correlation. *Biometrika*, **31**:9-12.

Prevost, G. (1977). Estimation of normal probability density function from samples measured with nonnegligible and nonconstant dispersion. Internal report 6-77, Adersa-Gerbios, 2 avenue de ler Mai, F-91120 Palaiseau.

Racine-Poon, A. (1985). Bayesian approach to nonlinear random effects models. *Biometrics*, **41**:1015-1023.

Racine-Poon, A. and Smith, A. F. M. (1989). *Population models in Statistical Methodology in the Pharmaceutical Sciences*, (Berry, D. A., Ed.), Marcel Dekker, New York, pp. 139-162.

Rocke, D. M. (1984). On testing for bioequivalence. *Biometrics*, **40**:225-230.

Rodda, B. E. (1989). *Bioavailability: Design and Analysis in Statistical Methodology in the Pharmaceutical Sciences*, (Berry, D. A., Ed.), Marcel Dekker, New York, pp. 57-81.

Rodda, B. E. and Davis, R. L. (1980). Determine the probability of an important difference in Bioavailability. *Clin Pharmacol. Ther.*, **28**:247-252.

Rodda, B. E., Sampson, C. B., and Smith, D. W. (1972). The one-compartment open model: Some statistical aspects of parameter estimation. Abstract 2057. *Biometrics*, **28**:1180.

Rodda, B. E., Sampson, C. B., and Smith, D. W. (1975). The one-compartment open model: Some statistical aspects of parameter estimation. *Appl. Stat.*, **24**: 309-318.

Schuirmann, D. J. (1987). A comparison of two one-sided tests procedure and the power approach for assessing the equivalence of average bioavailability. *J. Pharmacokinet. Biopharmaceutics*, **15**:657-680.

Schuirmann, D. J. (1990). Confidence intervals for the ratio of two means from a crossover study. *Proc. 1989 Biopharmaceutical Sect. Amer. Statistic. Assoc.*, American Statistical Association, pp. 121-126.

Selwyn, M. R., Dempster, A. P., and Hall, N. R. (1981). A Bayesian approach to bioequivalence for the 2x2 changeover design. *Biometrics*, **37**:11-21.

Selwyn, M. R. and Hall, N. R. (1984). On Bayesian methods for bioequivalence. *Biometrics*, **40**:1103-1108.

Sheiner, L. B., Rosenberg, B., and Melmon, K. L. (1972). Modeling of individual pharmacokinetics for computer-aided drug dosage. *Computers and Biomedical Research*, **5**:441-459.

Sheiner, L. B. and Beal, S. L. (1980). Evaluation of methods of estimating pharmacokinetic parameters: I. Michaelis-Menten model: routine clinical pharmacokinetic data. *J. Pharmacokinet. Biopharmaceutics*, **8**:553-571.

Sheiner, L.B. and Beal, S. L. (1981). Evaluation of methods of estimating pharmacokinetic parameters: II. Bioexponential model and experimental pharmacokinetic data. *J. Pharmacokinet. Biopharmaceutics*, **9**:635-651.

Sheiner, L. B. and Beal, S. L. (1983). Evaluation of methods for estimating population pharmacokinetic parameters. III. Monoexponential model: Routine clinical pharmacokinetic data. *J. Pharmacokinet. Biopharmaceutics*, **11**:303-319.

Shirley, E. (1976). The use of confidence intervals in biopharmaceutics. *J. Pharm. Pharmacol.*, **28**:312-313.

Steimer, J., Mallet, A., and Mentre, F. (1985). Estimating individual pharmacokinetic variability. In *Variability in Drug Therapy*. (Rowland, M., Sheiner, L. B., and Steiner, J., Eds.). Raven Press, NY. pp. 65-111.

Steinijans, V. W. and Diletti, E. (1983). Statistical analysis of bioavailability studies: Parametric and Nonparametric Confidence Intervals. *Eur. J. Clin. Pharmacol.*, **24**:127-136.

van Houwelingen, J. C. (1988). Use and abuse of variance models in regression. *Biometrics*, **43**:1073-1081.

Wagner, J. G. (1975). *Fundamentals of Clinical Pharmacokinetics*, Drug Intelligence, Hamilton, IL.

Weiner, D. L. and Jordan, D. C. (1978). Incorporating subject variability into pharmacokinetics models. Paper presented at Midwest Biopharmaceut. Stat. Workshop, Muncie, IN., May 23-24, 1978.

Westlake, W. J. (1972). Use of confidence intervals in analysis of comparative bioavailability trials. *J. Pharm. Sci.*, **61(8)**:1340-1341.

Westlake, W. J. (1974). The use of balanced incomplete block designs in comparative bioavailability trials. *Biometrics*, **30**:319-329.

Westlake, W. J. (1976). Symmetric confidence intervals for bioequivalence trials. *Biometrics*, **32**:741-744.

Westlake, W. J. (1979). Statistical aspects of comparative bioavailability trials. *Biometrics*, **35**:273-280.

Westlake, W. J. (1988). Bioavailability and bioequivalence of pharmaceutical formulations. In *Biopharmaceutical Statistics for Drug Development*, (Peace, K. E., Ed.), Marcel Dekker, New York, pp. 329-352.

Whiting, B., Kelman, A. W., and Grevel, J. (1986). Population pharmacokinetics-theory on clinical application. *Clin. Pharmacokinetics*, **11**:387-401.

Wijnand, H. P. and Timmer, C. J. (1983). Minicomputer programs for bioequivalence testing of pharmaceutical drug formulations in two-way crossover studies. *Comput. Prog. Biomed.*, **17**:73-78.

Winer, B. J. (1971). *Statistical Principles in Experimental Design*, McGraw-Hill, New York.

Yuh, L. (1990a). On robust estimation in bioavailability/bioequivalence studies. *Proc. Bio International 1989*, Toronto, Canada. pp. 162-165.

Yuh, L. (1990b). Robust procedures in comparative bioavailability studies. *Drug Information Journal*, **24**:741-751.

Yuh, L. Ruberg, S. J., and Eller, M. G. (1991). A stepwise procedure for dose proportionality studies. *Proc. 1990 Biopharmaceutical Sect. Amer. Statistic. Assoc.*, American Statistical Association, pp. 47-50.

Yuh, L. (1992). Recent developments of statistical criteria to BA/BE studies. *Proc. 1992 Bio International*. Bad Hamburg, Germany. In press.

Yuh, L. and Ruberg, S. J. (1990). Latin square designs in a comparative bioavailability study. *Drug Information Journal*, **24**:289-297.

13

The Wonders of Placebo

Roy L. Sanford

Baxter Healthcare Corporation,
Deerfield, Illinois

I. INTRODUCTION

Concern about the appropriateness, safety, and efficacy of medical treatments has been expressed by many people over the past 200 years. According to Bourne (1971), Voltaire is said to have remarked, "Doctors pour drugs, of which they know little, for diseases, of which they know less, into patients—of whom they know nothing." One reason for this pejorative view of medical practice is that throughout much of the history of medicine until the twentieth century, for many diseases the *placebo effect* was the best a physician was able to offer his patients. Experience has shown that most of the historically available medications were pharmacologically inert. Brodeur (1965) indicates that some of the earliest doubt concerning the efficacy of prescription medication surfaced in the late eighteenth century. During that period, the word placebo was derived from the Latin verb placere meaning "to please." A definition of placebo as "a commonplace method or medicine" appeared in the 1787 edition of Quincy's Lexicon. In the 1811 edition of Harper's Medical Dictionary the word placebo was defined as "an epithet given to any medium adopted more to please than to benefit the patient." This definition has carried through to the twentieth century with recent revisions.

The role played by the placebo in contemporary medical practice as well as in medical history cannot be overlooked. Current medical literature indicates that reactions to placebos may involve practically any organ system in the body. Placebos are known to cause undesirable side effects. Response to the administration of placebos tends to be varied and broad in extent throughout the population. To date, no single accepted explanation of the placebo effect has been advanced. Shapiro (1970) sees the placebo response as an adaptive mechanism which has helped mankind to survive. Uncertainty about how to harness this effect for the benefit of people remains and gives rise to several ethical considerations. Because of this, the role placebos play in medicine will continue to receive attention in coming years as the field of medicine evolves.

During the twentieth century much progress has been made in basing the practice of medicine on science and technology. As Thomas (1977) points out, the discovery that restoration of absent biochemical agents would control pellagra, pernicious anemia, and diabetes, and the discovery of the sulfanamides and penicillin, demonstrated the importance of the scientific method. The pursuit of new knowledge through human experimentation has raised many ethical considerations. Standards of clinical research have developed and the methodology of clinical experimentation continues to be studied.

Placebos play an important role in this work by providing a standard against which to compare the effects of a potential new therapeutic agent in a valid, controlled clinical trial. In certain cases it would be difficult, if not impossible, to determine whether a new medication was pharmacologically active and effective without the use of a placebo in a controlled trial. This use of placebos in clinical research brings us full circle in clinical practice. In those cases where a therapeutic agent is not found to be superior to a placebo used as a standard in a controlled trial, if patient benefit is observed, one might argue that the placebo remains a potential therapy. Buncher (1972) states, "Placebo is the best drug therapy available until some other drug has been proven superior in a valid controlled clinical trial." Such a viewpoint is not without its controversy. The intelligent therapeutic use of placebos remains as much an issue today as it was 150 years ago.

In order to provide an exposition of the uses and wonder of placebos, this chapter is divided into four basic parts. First, current thoughts about the basis of placebo response are discussed. Second, various attitudes concerning placebo use are explored. Third, consideration is given to the role of the placebo in double-blind trials. Fourth, an interesting clinical trial in which a placebo was used is reviewed. A list of supplementary reading is supplied at the end of the chapter (following the references) as the literature on this subject is much too broad to cover in a single chapter. It is hoped that the reader will be encouraged to explore this literature.

II. PLACEBO RESPONSE

As the understanding of the general effects of medication has expanded, modern-day definitions of placebo have broadened. Shapiro (1970) has proposed a definition that he feels fulfills historic and heuristic criteria. He proposes that

> A placebo is any therapy (or component of therapy) that is deliberately or knowingly used for its nonspecific, psychologic, or psychophysiologic effect, or that is used unknowingly for its presumed or believed effect on a patient, symptom, or illness, but which, unknown to patient and therapist, is without specific activity for the condition being treated.

This definition covers the case when a placebo is used in experimental studies as a control without specific activity for the condition being evaluated, as well as the use of placebos in medical practice. Shapiro's definition does not require a beneficial treatment outcome. When a placebo is administered, there may or may not be a resulting response. If a response occurs, it may be favorable or unfavorable. Shapiro's definition also erases the distinction between the placebo and the therapy. The existence of placebos as a separate class of treatments is very questionable (Horvath, 1987). Shapiro's definition implicates the presence of a psychological mechanism. It is currently believed that a placebo response is the result of an interaction among a combination of factors involving the placebo, the patient, the physician, and the process of treatment. In fact, the mechanism leading to a placebo response may be a component of all treatments, in addition to specific pharmacological effects (Lundh, 1987).

A. Examples of Placebo Response

Current medical literature indicates that reactions to placebos may involve practically any organ system in the body. Such reactions include relief of headache, cough, angina pectoris, the pain of rheumatoid and degenerative arthritis, the symptoms of hay fever, the pain of peptic ulcer, and at least transient reduction of hypertension. The high frequency of placebo response in angina pectoris has been well documented for more than three decades and has encompassed a broad range of therapeutic approaches. For example, at one time xanthines, khellin, vitamin E, ligation of the internal mammary artery, and implantation of the mammary artery were believed to be effective and were used extensively. In each case improvement rates of 70% and above were reported (Chong, 1987). These treatments are now known to have no specific physiologic effect and have been abandoned. Similar discoveries continue to be reported in the literature. For instance, ultrasound therapy is used to reduce pain and inflammation and to accelerate healing after soft tissue injury. In a placebo-controlled, double-blind clinical trial examining the contribution of placebo and massage effects in ultrasound therapy following bilateral surgical extraction of lower third molars, it was determined that ultrasound therapy can significantly reduce postoperative morbidity but by placebo-mediated mechanisms which are unrelated to ultrasound itself (Hashish et al., 1988). Another example is the response to placebo observed during a double-blind trial of recombinant alpha-2 interferon in multiple sclerosis patients (Hirsch, 1988). Cousins (1977) has summarized a variety of cases reported in the medical literature over 25 years that point toward the efficacy of various placebos. On the other hand, Buncher (1972) points out that

> Side effects from placebo therapy are even more extensive than the list of conditions that are aided by placebo. Headaches, nausea, vomiting, dizziness, diarrhea, pain, dermatitis, drowsiness, anxiety-nervousness, weakness-fatigue, dry mouth, abdominal pain, insomnia, urinary frequency, urticaria, loss of libido, tinnitus, and so forth, have all been caused by the administration of placebos.

Both the extent and variety of the placebo effect are so great that it has been remarked that a study showing no response to a placebo might well be suspected of lack of objectivity on that ground alone (Brodeur, 1965). While the existence and frequency of occurrence of the placebo effect in the treatment of a variety of diseases is well substantiated, the mechanism and nature of the effect is not.

B. Factors Influencing Placebo Response

According to Bush (1974), the factors influencing placebo response may be divided roughly into two groups—bound and manipulable. Bound factors are those that not cannot be modified and are predisposed, dependent on the history and personality of the patients, health personnel, and investigators. Manipulable factors include treatment settings, communications to the patient, type of treatment and type of drugs, including characteristics of drugs that can be detected by the patients (i.e., taste and route of administration). To the extent that optimism, enthusiasm, empathy, and faith in treatment can be controlled, these factors are also manipulable. Perhaps the most important factor is the doctor himself and the atmosphere surrounding therapy that is under this direct control. Insofar as the placebo effect seems to be derived from a combination of the factors involving the patient, the physician, and the relationship between the two, it represents a social-psychological phenomenon.

Benson and Epstein (1975) have recognized a parallel between the placebo effect and other social-psychological phenomena. Examples of such related phenomena include Milgram's experiment, which demonstrated the compliance of people resulting from the commands of a scientific authority figure in a clinical setting, as well as the "Hawthorne effect" where the efficiency of factory workers was found to improve as a direct result of the increased attention received during investigation. What is perhaps unique about placebo response in a clinical setting, however, is that there must be congruence between the physician's approach to therapy and the patient's attitudes toward illness and expectations of treatment.

Lundh (1987) argues, "an important part of the placebo effect is due to the development of placebo beliefs (beliefs of the form 'This treatment is going to cure me'), which may counteract the kind of cognitions that produce anxiety and depression; placebo beliefs produce emotional responses (hope, calm, etc.), which are antagonistic to depression and anxiety. . . . placebo beliefs may be expected to have a positive influence on physical health." Senger (1987) observes that for maximum effect the patient must believe in three things: "I can be helped. This therapist can help me. This treatment is right for me." Correspondingly, the therapist is more believable to the degree he or she shows in the above beliefs: "I am a helper. This patient can be helped. My method is right for this patient." The implications are that a placebo effect will be absent or negative if a patient cannot or does not favorably interpret cultural symbols represented by physicians and hospitals. The consensus that Christian Scientists are not placebo reactors is a case in point as recognized by Bush (1974).

C. Current Pharmacological Research into Placebo Response

Research reported at the Second World Congress on Pain in 1978 and cited by Heylin (1978) suggests ingestion of a placebo triggers a series of neural events that result in the release of endorphins, naturally occurring, opiate-like peptides produced in the brain. These studies involved administration of placebos to dental surgery patients with selective administration of the narcotic antagonist naloxone. Naloxone produced no additional increase in pain in those who failed to respond to placebos, but it did increase the pain levels of placebo responders. The mechanism triggering production of endorphins in this situation is unknown.

In a prior publication Snyder (1977) indicates that

> The chemical identification of endorphins has major relevance both for practical medical practice and a fundamental understanding of brain functions. Since opiates such as morphine cause euphoria and relieve pain, clarification of how these opiate-like peptides function will greatly advance our understanding of how the brain regulates emotions and the perception of pain. Already, it is clear that neurons

containing these compounds are highly localized in "emotion" and "pain sensing" parts of the brain.

Snyder (1977) also reports on research involving structural modifications of natural occurring enkephalins to produce more potent analgesics with the promise of use as practical therapeutic agents. Costa and Trabucchi (1978) provide an overview of what is currently known about endorphins. Levine et al. (1978) have presented evidence that naloxone may block placebo analgesia, which may indicate that endorphins play a role in this kind of placebo analgesia; this conclusion is disputed by Gracely et al. (1983). More research is needed on the role of endorphins in producing placebo responses.

D. Identification of Placebo Responders

Who, then, is a placebo responder? According to Gelbman (1967), it is extremely doubtful if a "placebo reactor" personality exists. It is not unreasonable to expect the placebo effect to be present to some degree in almost any therapy, whether or not the therapy involved is known to have a specific effect for the condition being treated, as long as it is not known or believed to be a placebo. Reactions to a placebo are highly variable from person to person and from time to time. A group of people who respond to a placebo today may not respond next week and vice versa. Those who respond for one indication may not respond for another. Susceptibility to placebo therapy can be expected to vary depending upon the ailment involved, the total doctor-patient relationship, and contingent social-psychological circumstances. Senger (1987) sums it up, "One person's placebo is another's aggravation." As a consequence, it would be difficult, if not impossible, to develop a screening test for placebo responsiveness. No one is likely to respond to a placebo if he knows it is one; however, virtually everyone can react favorably to some therapeutic remedy that is later found to be a placebo. In summary, the concept of placebo reactor is now generally considered not viable (Shapiro and Morris, 1978).

III. PLACEBO USE: ATTITUDES AND ALTERNATIVES

Adler and Hammett (1973) indicate that 60 to 80% of the case load carried by a general practitioner falls into the category of functional illness. In view of this, it is not surprising that some writers (Brodeur, 1965; Benson and Epstein, 1975; Bush, 1974) suggest that the placebo response needs to be better understood so that its benefits can be used out of insight rather than ignorance. The question of the proper clinical use of placebos has been and is currently a subject of discussion. Recent publications have suggested that, while all health professionals have a general obligation to benefit their patients, nursing has a special obligation to enhance the placebo effect—maximizing its positive effects and minimizing its negative effects (Connelly, 1991; Seeley, 1990). Despite the fact that the placebo response is involved in many forms of therapy today, this discussion will continue as long as we are dealing with a complex phenomenon whose mechanism is not understood. Physicians as a community are very much concerned with the ethical issues involved with the use of placebos. In general, there is great public concern for human rights in the provision of medical care and in the conduct of clinical research. Mike and Good (1977) note that there are now institutes and scholarly journals devoted to the exploration of ethical problems in medicine and biology.

A. Physicians' Attitudes About Placebo Use

Before the physician tries any treatment on a patient, he must weigh, as best he can, the potential assets and liabilities of alternative courses of action and consider these not only as a scientist, but also from the point of view of the patient and of society. The focus is on benefiting the patient. Depending on the ailment, a scientifically established efficacious treatment may not exist and only nonspecific treatment is possible. While the placebo effect may be therapeutic, it is interesting to note that Shapiro and Struening (1973) report many physicians are defensive about the scientific basis for their treatment of patients when the possibility of a placebo effect is considered. Not all physicians are defensive regarding the potential therapeutic value of treatment with placebos. The defensiveness that exists seems to stem from the fact that physicians attempt to fulfill the expectations of patients with treatment based upon scientific principles. Based on a survey, Shapiro and Struening (1973) found a tendency for physicians to deny personal use of placebos and ascribe use to other physicians and specialties. They found that physicians tend to define placebo in such a way that a connection with their therapies was excluded. This tendency is one reflection of the ethical concerns felt and expressed by many people about the use of placebos.

B. Alternatives to Placebo Use

One school of thought says that the use of placebos should be carefully considered and restricted. The argument expressed by some in favor of this position (Bok, 1974; Goshen, 1966) is that when a physician knowingly prescribes a placebo without informing the patient, the patient has been deceived and has been encouraged to continue to believe that formally prescribed placebos are necessary for relief. Some authors argue that such deception can result in a decline in trust in the physician that, in turn, results in potential injury to the institution of medicine. The cost of such therapy, as well as the possibility of uncomfortable side effects, represent additional complicating factors.

As an alternative to the knowing use of placebo, some authors have suggested psychotherapy and others have suggested communication. An interesting study illustrating these possibilities has been reported by Bourne (1971). Anesthesiologists assigned each of 97 preoperative patients to one of two randomly selected groups. One group received the routine preoperative visit and preoperative medications. The other group had more intensive contact with the anesthesiologist before and after the operation discussing the nature, causes, and course of normal postoperative pain. Both groups underwent elective intra-abdominal operations of equal risk. The surgeons, who were unaware of the study, prescribed almost 50% less analgesic medication to the treated group and, on the average, sent these patients home 2.7 days earlier. Whether such medical tactics would benefit patients at other times with other ailments is not known.

Cousins (1977) reports that an increasing number of physicians believe they should not encourage their patients to expect prescriptions. This is one step in stopping the perpetuation of false ideas about disease and its cure. However, in spite of the concerns expressed regarding the pharmacological properties of placebos, patients have received therapeutic benefit with placebos. A number of drugs have been prescribed for years because they were thought to be effective. These same drugs were later found to be equivalent to placebos as a result of controlled clinical trials. Once a drug is shown to be equivalent to a placebo in a controlled study, the physician is faced with a dilemma. Does he prescribe the drug or the placebo? Should he search for another therapy that may, under similar controlled circumstances, be shown to be no more effective than placebo and in some cases more harmful? What is in the best interests of the patient? Under these circumstances a placebo, in conjunction with a positive physician/patient relationship, may be the best therapy available until another method of treatment has been found to be superior in a controlled clinical trial.

Weighing the risks and benefits of different therapies, including placebo, will never be easy. Medical experience has shown, however, that the placebo plays the role of a catalyst in connecting the will to live with a tangible entity such that the mind can carry out its functional relationship to the body. Therapy using a placebo may be necessary to help the patient help himself. This is not to say alternatives should not be considered to help the patient. Dr. Albert Schweitzer's viewpoint gives a valuable perspective. According to Cousins (1977), Dr. Schweitzer is said to have remarked, "We are at our best when we give the doctor who resides within each patient a chance to go to work."

IV.   CLINICAL EVALUATION OF DRUGS

Since 1946 controlled clinical trials have been conducted with increasing frequency, and the design and analysis of clinical trials have come to constitute an important area of statistical research. Before a drug can be considered to remedy or mitigate a disease ailment, it must be evaluated and found to perform acceptably in a clinical setting. In developing a protocol for a drug trial, the statistician will frequently find it necessary to trade off various considerations to arrive at an ethical study that is workable by clinicians and that will yield reliable data upon which a sound determination can be made.

A. Ethical Considerations

Sissela Bok (1974) questioned the ethics of placebo use in a major paper. Three major concerns were raised. The first ethical issue was that of the planned deception involved in placebo administration. The withholding of an active treatment and the lack of understanding by research participants of what is a placebo-control trial were the other concerns. Each point requires consideration by the statistician in protocol development.

Informed consent is a necessary part of the clinical research process. Deception becomes much less of an issue in placebo-control trials when knowledge of the study design is shared with the patients so that they become partners in the deception. Problems can still arise when patients respond either favorably or adversely to the placebo and learn after the fact they received an inactive substance. Counseling by the physician may be necessary in such situations.

Withholding an active treatment is the second issue to address. This most often arises when a new drug is proposed for a condition in which there is a known treatment. Stanley (1988) discusses a number of considerations that pertain. Is the illness a life threatening condition? Does the illness have a relatively predictable and stable course not masked by spontaneous remissions or exacerbations? Is the standard treatment of established efficacy? If one or more of these questions are answered yes, Stanley (1988) suggests the use of a placebo becomes difficult to reconcile. On the other hand, a placebo may be appropriate despite the above if the standard treatment is high in toxicity or has serious side effects. Such considerations must be deliberated during protocol development to reach a suitable conclusion.

The third ethical issue relates to what patients are capable of understanding about placebo-controlled trials. Do they really understand what a placebo is? Do they understand what the process of research is? Stanley (1988) points out that patients may have difficulty appreciating that the entire purpose of participation is not solely to help or benefit the patient. Some patients will not believe what is told to them no matter how they are informed. In designing a study, careful attention should be given to clear and frank communication of that information needed by a patient to make an intelligent decision about participation.

Concern over ethical norms in the field of biomedical research has resulted in several declarations. Haegerstam et al. (1982) notes that the 1975 Tokyo revision to the Helsinki declaration of 1964 states "concern for the interest of the subject must prevail over the interest of science and society." The Hawaii declaration for therapy and research in psychiatry states that "every patient must be offered the best available treatment." Such declarations fail to provide unequivocal answers to resolve ethical problems presented by clinical trials. Moreover, the emphasis placed in these guidelines is on immediate patient benefit in contrast to assuring that valid inferences can be drawn for the benefit of future patients. Guidelines issued by various regulatory bodies, such as the U.S. Food and Drug Administration (FDA) (1979), for the clinical evaluation of drugs place more weight on the achievement of the best possible scientific documentation of drug efficacy. The contrast in emphasis between international declarations and regulatory guidelines serves to reinforce the critical importance of careful and comprehensive study design. Both ethical and regulatory concerns must be served by the same study. The statistician must share in the responsibility for reaching an acceptable course of action.

B. Regulatory Considerations

Experience has indicated that certain standards of research are necessary to insure the validity of a study. A set of guidelines has been published by the U.S. Food and Drug Administration (1979). The following six standards of a good clinical drug trial have been recognized by Buncher (1972):

1. Controlled comparisons
2. Blind evaluation
3. Randomized allocation
4. Orthogonal contrasts
5. Replicated study
6. Proper administration

Much has been written about experimental designs that have these characteristics. Prien (1988) and Haegerstam et al. (1982) discuss some of the advantages and disadvantages of different clinical trial designs. One issue to bear in mind when selecting the choice of design is the demonstrated fact that the order in which active agents and the placebo are administered can present a problem with crossover designs. Several analgesic studies have been reported (Batterman and Lower, 1968; Laska and Sunshine, 1973; Moertel et al., 1976) that indicate that patients treated initially with placebo have a lower response rate than those who are exposed to placebo after taking one or more pharmacologically active agents. Such a situation can give rise to a treatment by period interaction that, as O'Neill (1977) has indicated, reduces the validity of a randomized two-period changeover experiment. This example illustrates why the existence of a placebo response needs to be carefully considered in designing clinical trials. After reviewing the literature the reader will find in practice that there is no one perfect clinical trial and that designing a workable, ethical trial free from obvious defect is an accomplishment.

C. The Role of a Placebo as a Control for Experimentation

In evaluating a drug's performance, a standard or control for comparison is needed to determine the relative effects of the drug. Without a suitable standard for control in a drug trial, the evaluation could be rendered invalid due either to symptomatic relief as a result of a placebo response, or due to the natural reversal of the ailment on its own accord. Some of the complications that can occur have already been discussed. Placebos are frequently used as a standard or control for comparison in clinical drug trials against which relative effects of the test substance can be estimated. Other controls are possible depending on the situation. The May 8, 1970 Code of Federal Regulations states, "Uncontrolled studies or partially controlled studies are not acceptable as the sole basis for the approval of claims of effectiveness." These regulations recognize four types of control groups: 1) no treatment; 2) placebo control; 3) active treatment control; and 4) historical control (see Chapter 5 for details).

The selection of the control to be used depends on the specific situation being investigated. The historical control and no-treatment cases represent alternatives useful only in special situations. When the historical control is employed, one is searching for a radical improvement in therapy. When the no-treatment case is considered, fairly persuasive evidence is needed to support the objectivity of the measurements and the lack of a placebo response. Without evidence to the contrary, the statistician needs to expect that some patients will experience therapeutic relief that may not be the result of the test substance. On the basis of these considerations, the more frequently chosen controls are the placebo and active treatment cases. It is important to recognize that these two controls provide for different measurements of treatment effect. The placebo control allows for a comparison that evaluates whether or not the test substance has effects greater than those due to the placebo, with the implication that the difference is due to pharmacological properties of the test substance. The active treatment control results in a direct comparison between the therapeutic properties of the control and those of the test substance because one would assume the placebo effect would be common to both treatment groups.

A decision of which of these two controls to use depends on the availability of an active treatment control, the ethics of using a placebo in place of an active treatment control, and the goals of the research program. Frequently, placebo controls are selected and used in double-blind clinical trials because the simplicity and practicality of such experiments allow for determination of the pharmacological effect of the test substance. Sometimes both controls are used, for example, in analgesic trials, because in the case of single control, insignificant differences between the test drug and placebo or between the test drug and active drug may be due to the insensitivity of the clinical trial rather than the ineffectiveness of the test drug. This situation can be best understood when both placebo and an active drug are in the trial.

D. Some Comments Regarding Double-Blind Clinical Trials and Placebo Controls

A typical double-blind clinical trial involves randomization of patients into control and treatment groups upon entry to the study. Random assignment follows a predetermined plan that is usually devised with the aid of a table of random numbers. Randomization ensures that the personal judgment and prejudices of both the investigator and the patient do not influence treatment allocation. In the long run, randomization achieves a balance of patient characteristics between groups. However, with a small series of patients, randomization may not always produce groups that are alike in every respect, and the analysis of the data may have to take discrepancies into account.

The actual allocation of treatments may be *single-blind* in that only the patient is unaware of whether he is in a treatment or control group, or *double-blind* in that both the patient and physician are unaware of which patient is assigned to which group. The previous discussion of placebo indicated that many factors, such as suggestibility of the patient, suggestibility of the investigator, their mutual interaction, and other bound or manipulable factors can influence placebo responses. The intent of the double-blind clinical trial is to minimize the impact of these and other biases on the outcome of the study in order to make the investigation and the comparison between treatment groups as objective as possible. Depending upon setting, blinding can be very difficult to achieve. Consider the difficulty in blinding the surgeon about an injection administered during surgery when the active agent imparts a definite color to the solution being injected, and this color cannot be reproduced in the placebo injection. At times established medical procedures must be changed to achieve a blinded study and this may necessitate careful negotiations with the participating physicians to win their concurrence.

Informed consent is a necessary part of clinical trial research such that patients know they may receive a placebo instead of the test substance. It is important to recognize that informed consent and its written documentation should be considered possible sources of bias in the design of a clinical trial.

Levine (1987) presents the conclusion that behaviors are likely to be influenced in subjects who are informed of the expected therapeutic and adverse effects of pharmacologically active agents in placebo-controlled trials. For example, armed with such information, patients may be able to guess correctly their treatment assignment. As a remedy, Levine suggests that disclosure be amended and proposes adding to the list of potential side effects a list of irrelevant side effects that are of the same order of importance as those expected. Levine believes this will reduce the probability of correct guesses about treatment assignment, although the number of subjects who refuse to participate may increase. This proposal does nothing to minimize the probability that investigators and study coordinators may also guess correctly the treatment assignment. Levine notes the frequency and importance of this phenomenon needs to be assessed.

A placebo control allows estimation of the comparative therapeutic benefit of the pharmacological activity of the test substance by contrasting this to the symptomatic relief provided by the placebo, plus the natural reversal rate of many illnesses. Such a contrast is often necessary to prove that a proposed medical procedure is safe and efficacious. Whereas a well-designed placebo-controlled study is intended to give a valid comparison of the test procedure versus the control, such a study should not be expected to predict the observed efficacy rates in clinical practice. Physicians commonly seek to maximize the placebo effect since the goal of treatment in the clinical setting is to improve the quality of the patient's life. This is not the case in a research study (Packer, 1990). As Lundh (1987) notes, "in double-blind placebo-controlled studies, there is usually no attempt to maximize the placebo effect by means of instructions, enthusiasm, etc." The same is true for the test substance since both groups are to be contrasted. Kirsch and Weixel (1988) report on a study comparing double-blind versus deceptive administration of a placebo, which illustrates how the pattern of response can change depending on study design. The extrapolation of study results into a broader clinical setting with different patients, physicians, and institutions requires expert medical insight. In practice a statistician will be confronted in certain situations with the attitude that the use of a placebo in a double-blind clinical trial is unnecessary or unethical and immoral. Examples of such situations include the following:

1. Evaluations of toxicity
2. Pilot studies preceding controlled comparative studies
3. Studies of drugs that are highly effective for the treatment of specific diseases
4. Situations when the researcher is convinced a priori that the medication is superior to a standard drug as reference

5.   Situations when the phenomenon being monitored is said not to be subject to psychological influences
6.   Situations where the disease is extremely destructive or life-threatening.

    The last situation mentioned above deserves particular attention. Over the last few years, the public has become much more aware of people dying of AIDS, Alzheimer's disease, cancer, and other scourges that are destructive or life-threatening. This has led to a passionate debate involving the public, the scientific community, government, and business. The discussion centers around the early release of promising drugs to patients with extremely destructive or life-threatening diseases without completion of the traditional three-stage FDA approval process which requires controlled trials that are capable of providing definitive data on a drug's safety and efficacy. Some argue that for certain diseases, proof of safety, accompanied by indications of efficacy, would suffice. The rationale is that quality of life is supremely important to the victims; so, why shouldn't a drug that offers only symptomatic benefit be put on the market or made available if no other effective treatment is available? For many such patients, such a gamble is the only hope for them and their family. Hence, one view is that denying treatments from which such patients might benefit is not moral and should not be sustained. So, who should make the decision—the regulators or the physician, patient, and family?

    On the other hand, informed individuals argue that such a course of action is irresponsible to patients and their families and could lead to the uncontrolled use of unproven chemical agents on vulnerable, desperately ill people. Such drugs would cost money and could have side effects. Worst, such a step disregards what has been learned during the past generation about the importance of clinical trials when it comes to establishing drug safety and efficacy. For example, in the 1989 Cardiac Arrhythmia Suppression Trial, 1,455 patients participated, with 730 receiving the drug and 725 a placebo at random. Of those who received the experimental drug, 7.7% died, while only 3.0% of the placebo group died.

    This controversy has sparked review of the old drug-approval process by top federal health officials. Under the old drug-approval system, companies could make experimental drugs available, free of charge, on a compassionate basis to gravely ill patients. But drugs couldn't be sold until they had completed all phases of drug testing. A couple of alternatives have been proposed to this system. For example, there is the treatment Investigational New Drug application (IND). Under this procedure, companies testing a product for serious conditions, such as Alzheimer's disease, can release it for treatment to doctors or sell it if they have some preliminary evidence of safety and effectiveness and a go-ahead from the FDA. Alternatively, there is the parallel-track distribution plan (Bacon, 1989). Parallel-track distribution would take place while normal clinical tests are continuing. It would begin after drugs had passed initial safety tests but before they complete the Phase II effectiveness trials. The objective of this initiative would be to make promising investigational agents available to persons for whom there are no satisfactory alternative therapies and who are not eligible or not able to participate in a controlled clinical trial. Other initiatives may be taken in the future.

It should be clear that this recent debate is far from over. The perspectives are evolving and the considerations are difficult involving ethical, moral, and scientific issues. The question of how much evidence is enough to justify release of a drug for treatment of a specific condition will always confront the statistician involved with clinical trials.

When one is faced with one of these situations or other comparable situations not mentioned, it is important to recognize the basic reasons for using the placebo and to consider each case as it arises on its own merits relative to the overall research plan.

At some point in the evaluation of a new therapy, a controlled study using a placebo is usually necessary, if only from a regulatory standpoint. Both Prien (1988) and Haegerstam et al. (1982) note in their review papers that placebo control is an irreplaceable tool that plays a critical role in new drug development. To a large extent, as Gilbert et al. (1977) mention, arguments against placebo-controlled, double- blind studies imply that investigators know in advance which is the favorable therapy. Experience has shown presumably efficacious treatments to be equivalent to placebo on enough occasions to provide ample grounds on which to reserve judgment. Tukey (1977) indicates the statistician must insist on obtaining reliable evidence on which to make a decision, and that the double-blind methodology is one very effective, practical, and efficient means of obtaining such evidence.

V. A CASE STUDY OF A CLINICAL TRIAL OF THE DRUG CHYMOPAPAIN

Published results of studies with chymopapain serve to illustrate the revealing nature of placebo-controlled, double-blind trials and the importance of not underestimating the magnitude of the placebo effect. A brief historical review is in order. Chymopapain is an enzyme which dissolves the protein in injured vertebral discs. According to Smith, the dissolution of this protein may alleviate the pain experienced by slipped-disc sufferers. In 1963 Smith (1964) injected the enzyme intradiscally into patients with symptomatic lumbar disc disease and termed his new treatment, chemonucleolysis. Prior to 1975, almost 17,000 patients had undergone chemonucleolysis by neurosurgeons and orthopedists. These studies were uncontrolled, and success rates ranged from 50% to 80% from investigator to investigator. FDA approval of chemonucleolysis seemed a formality. In 1974 the American Academy of Orthopedic Surgeons endorsed chymopapain as safe and effective. However, prior to approval, it was determined that a placebo-controlled, double-blind study was necessary in the United States. Based on the successful uncontrolled clinical experience since 1963, this study was considered by many to be unethical. Unfortunately, the only therapy available other than chemonucleolysis was surgery and could not be accommodated in a double-blind format.

This situation prompted the election of a placebo-controlled study. Four hospitals were selected to participate in a double-blind clinical study that was conducted for approximately one year and was completed on December 31, 1975. A total of 106 patients were admitted with 56 patients receiving a placebo injection and 50 receiving chymopapain. The placebo injection consisted of the vehicle without the enzyme but with an inert substance for bulk. The vehicle consisted of cysteine hydrochloride and ethylenediamine tetraacetic acid with sodium iothalamate, and was considered to be pharmacologically inert when included as part of the injection procedure. All other aspects of the treatment program were identical for both the control and treatment groups. At the end of one year, out of 50 patients receiving chymopapain, 20 patients were determined to be treatment failures, and out of 56 treated in the control group, 28 were determined to be treatment failures. Determination of treatment failure was made jointly by the physician and patient. The overall success rates were not found to be significantly different between control and treatment groups, nor did any additional evaluation of all the data collected demonstrate a significant benefit in favor of chymopapain compared to the placebo control. Long term follow-up did not alter this situation. Further details and results were reported by Schwetschenau et al. (1976).

This one chymopapain study raised numerous questions. Was the placebo response rate due to true pharmacological activity on the part of the vehicle, or was the true placebo response rate under these circumstances comparable to those of chemonucleolysis and surgery? Should measurement of patient improvement and the reduction of pain from lower back problems have been done differently to result in a more efficient estimate of the contrast in success rates between chymopapain and placebo? How long should patients be followed before success or failure is determined? Should a larger sample size have been used? Lower back problems tend to reverse themselves and then recur. Should different patient entry criteria be used? Based on the findings of this study, the new drug application (NDA) was not approved and further use of chymopapain in the United States was discontinued. Several articles appeared in the press (e.g., Steinmetz, 1976) questioning the nonapproval of the NDA and the conduct of the study; two congressional investigations were launched, documentaries appeared on television, and scientific papers were published in different journals discussing chemonucleolysis.

McCulloch (1977) published results of a 7-year, unblinded, single-treatment study of 480 patients who underwent enzymatic dissolution of the nucleus pulposus with chymopapain. He reported that 70% of patients with the clinical criteria for a disc herniation had a favorable response to chemonucleolysis. Those patients with spinal stenosis or psychogenic components or those having had a previous operation were found to have poor results. In 1976 Smith, the discoverer of chemonucleolysis, indicated, "In comparison with usage of the drug over 12 years, this one study [referring to the double-blind study] is relatively insignificant." The controversy continued for years. Final approval of chymopapain was received from the FDA in 1982 after two additional placebo-controlled studies were completed. One was conducted in Australia and is reported by Fraser (1982). The other was conducted by Smith Laboratories and is reported by Haines (1985) in a review of the three published randomized clinical trials of chymopapain. The three published studies were double-blind placebo-controlled studies utilizing a total of 234 patients. Haines (1985) pooled the results of these studies on the basis that study design, selection criteria, technique, and outcome assessment were very similar. He concluded that the odds of successful outcome were 2.6 times as great with chymopapain than with placebo, or that chemonucleolysis provided a 23% increase in the number of successfully treated patients compared with placebo. The pooled success rate for chymopapain was 70% and for placebo was 47%. Haines demonstrated that in these three studies the estimated powers for finding a 50% increase in success rate for chymopapain relative to placebo ranged from 0.51 to 0.61, and he concluded that "the failure of the original double-blind study . . . probably resulted from small sample size." Of course, in the design of the first double-blind study, no one anticipated a placebo response rate of around 47%, particularly for patients that met the selection criteria. These people were in chronic pain not alleviated by bed rest. Such is the wonder of placebo. Hopefully, this example will provoke the reader to contemplation of, if not commitment to, carefully conducted controlled trials.

VI. CONCLUSION

Any statistician who participates in clinical trials will have to deal with many of the considerations raised in this chapter. This discussion gives the reader, who will become a statistician or clinical monitor, an informed viewpoint from which to address the many intriguing medical research problems that continue to be investigated. The suggested further readings following the references add to what has been presented here. So much has been written on the subject of placebos that a complete list of references has not been given. This chapter provides an introduction to the wonders of placebo, and illustrates why anyone who enters the field of medical research will gain considerable respect for placebo.

ACKNOWLEDGMENT

The author would like to extend sincere appreciation and thanks to Marcia Parkinson for typing and organizing several versions of this chapter.

REFERENCES

Adler, H. M. and Hammett, V. O. (1973). Doctor-patient relationship revisited: An analysis of the placebo effect. *Ann. Intern. Med.*, **78**:595-598.

Bacon, K. H. (1989). Plan to speed availability of AIDS drugs is endorsed by top U.S. health officials. *Wall Street Journal*, 7-21-89, pg. B2.

Batterman, R. C. and Lower, W. R. (1968). Placebo responsiveness: Influence of previous therapy. *Curr. Ther. Res.*, **10**:136-143.

Benson, H. and Epstein, M. D. (1975). The placebo effect: A neglected asset in the care of patients. *J. Amer. Med. Assoc.*, **232**:1225-1227.

Bok, S. (1974). The ethics of giving placebos. *Sci. Amer.*, **231**:17-23.

Bourne, H. R. (1971). The placebo: A poorly understood and neglected therapeutic agent. *Ration Drug Ther.*, **5**:1-6.

Brodeur, D. W. (1965). A short history of placebos. *J. Amer. Pharm. Assoc.*, **5**:642-662.

Buncher, C. R. (1972). Principles of experimental design for clinical drug studies. In *Perspectives in Clinical Pharmacy*, (Francke, D. E. and Whitney, H. A. K. Jr., Eds.). Drug Intelligence, Hamilton, IL, pp. 504-525.

Bush, P. J. (1974). The placebo effect. *J. Amer. Pharm. Assoc.*, **14**:671-674.

Chong, T. M. (1987). The placebo in the practice of medicine. *Int. J. Psychosomatics*, **34(2)**:25-30.

Connelly, R. J. (1991). Nursing responsibility for the placebo effect. *J. Med. Philos.*, **16(3)**:325-341.

Costa, E. and Trabucchi, M. (1978). *The Endorphins*, Raven Press, New York. Advances in Biochemical Psychopharmacology, Vol. 18.

Cousins, N. (1977). The mysterious placebo: How mind helps medicine work. *Saturday Rev.*, October 1, 1977, pp. 9-16.

Fraser, R. D. (1982). Chymopapain for the treatment of intervertebral disc herniation. A preliminary report of a double-blind study. *Spine*, November-December, **7(6)**:608-12.

Gelbman, F. (1967). The physician, the placebo, and the placebo effect. *Ohio State Med. J.*, **63**:1459-1461.

Gilbert, J. P., McPeek, B., and Mosteller, F. (1977). Statistics and ethics in surgery and anesthesia. *Science*, **198**:684-689.

Goshen, C. E. (1966). The placebo effect: For whom? *Amer. J. Nurs.*, **66**:293-294.

Gracely, R. H., Dubner, R., Wolskee, P. J., and Deeter, W. R. (1983). Placebo and naloxone can alter post-surgical pain by separate mechanisms. *Nature*, 306:264-265.

Haegerstam, G, Huitfeldt, B, Nilsson, B. S., Sjovall, J., Syvalahti, E, and Wahlen, A. (1982). Placebo in clinical drug trials—a multi-disciplinary review. *Methods Find Exp. Clin. Pharmacol.*, 4(4):261-278.

Haines, S. J. (1985). The chymopapain clinical trials. *Neurosurgery*, 17(1):107-110.

Hashish, I., Hai, H. K., Harvey, W., Feinmann, C., and Harris, M. (1988). Reduction of postoperative pain and swelling by ultrasound treatment: a placebo effect. *Pain*, 33(3):303-311.

Heylin, M. (Ed.) (1978). Placebo action determined. *Chem. Eng. News*, 56:24.

Hirsch, R. L., Johnson, K. P., and Camenga, D. L. (1988). The placebo effect during a double-blind trial of recombinant alpha 2 interferon in multiple sclerosis patients: immunological and clinical findings. *Int. J. Neurol.*, 39(3-4):189-196.

Horvath, P. (1987). Demonstrating therapeutic validity versus the false placebo-therapy distinction. *Psychotherapy*, 24(1):47-51.

Kirsch, I. and Weixel, L. J. (1988). Double-blind versus deceptive administration of a placebo. *Behavioral Neurosci.* 102(2):319-323.

Laska, E. and Sunshine, A. (1973). Anticipation of analgesic: A placebo effect. *Headache*, 13:1-11.

Levine, J. D., Gordon, N. E., and Fields, H. L. (1978). The mechanism of placebo analgesia. *Lancet*, 2:654.

Levine, R. J. (1987). The apparent incompatibility between informed consent and placebo-controlled clinical trials. *Clin. Pharmacol. Ther.*, 42(3):247-249.

Lundh, L. G. (1987). Placebo, belief, and health: A cognitive-emotional model. *Scand. J. Psychol.*, 28(2):128-143.

Martins, A. N., Ramirez, A., Johnston, J., and Schwetschenau, P. R. (1978). Double-blind evaluation of chemonucleolysis for herniated lumbar discs. Late results. *J. Neurosurg.*, 49(6):816-827.

McCulloch, J. A. (1977). Chemonucleolysis. *J. Bone Joint Surg., Ser. B*, 59:45-52.

Mike, V. and Good, R. (1977). Old problems, new challenges. *Science*, 198:677-678.

Moertel, C. G., Taylor, W. F., Roth, A., and Tyce, F. A. J. (1976). Who responds to sugar pills? *Mayo Clin. Proc.*, 51:96-100.

O'Neill, R. (1977). Current status of crossover designs. Paper presented at Pharm. Manuf. Assoc. Conf., Statisticians Session, Arlington, VA.

Packer, M. (1990). The placebo effect in heart failure. *Amer. Heart. J.*, 120(6(Pt. 2)): 1579-1582.

Prien, R. F. (1988). Methods and models for placebo use in pharmacotherapeutic trials. *Psychopharmacol. Bull.*, 24(1):4-8.

Schwetschenau, P. R., Ramirez, A., Johnston, J., Barnes, E., Wiggs, C., and Martins, A. N. (1976). Double-blind evaluation of intradiscal chymopapain for herniated lumbar discs. *J. Neurosurg.*, 45:622-627.

Seeley, D. (1990). Selected nonpharmacological therapies for chronic pain: the therapeutic use of the placebo effect. *J. Amer. Acad. Nurse. Pract.*, 2(1):10-16.

Senger, H. L. (1987). The "placebo" effect of psychotherapy: A moose in the rabbit stew. *Amer. J. of Psychother.*, **41(1)**:68-81.

Shapiro, A. K. (1970). Placebo effects in psychotherapy and psychoanalysis. *J. Clin. Pharmacol.*, **Mar/Apr**:73-78.

Shapiro, A. K. and Struening, E. (1973). Defensiveness in the definition of placebo. *Comp. Psychiatry*, **14**:107-120.

Shapiro, A. K. and Morris, L. A. (1978). The placebo effect in medical and psychological therapies. In *Handbook of Psychotherapy and Behavior Change, An Empirical Analysis, 2nd ed.*, Wiley & Sons, New York, pp. 369-411.

Smith, L. (1964). Enzyme dissolution of the nucleus pulposus in humans. *J. Amer. Med. Assoc.*, **187**:137-140.

Snyder, S. H. (1977). The brain's own opiates. *Chem. Eng. News*, **56**:26-34.

Stanley, B. (1988). An integration of ethical and clinical considerations in the use of placebos. *Psychopharmol. Bull.*, **24(1)**:18-20.

Steinmetz, J. (1976). Dr. Lyman Smith tangles with the FDA over his papaya enzyme to treat bad backs. *People*, **5**:58-59.

Thomas, L. (1977). Biostatistics in medicine. *Science*, **198**:675.

Tukey, J. (1977). Some thoughts on clinical trials, especially problems of multiplicity. *Science*, **198**:679-684.

U.S. Food and Drug Administration (1979). *General considerations for the clinical evaluation of drugs.* U.S. Department of Health, Education and Welfare (FDA), pp. 77-3040.

ADDITIONAL SUGGESTED READING

Amery, W. and Dony, J. (1975). A clinical trial design avoiding undue placebo treatment. *J. Clin. Pharmacol.*, **Oct**:674-679.

Byar, D. P., Simon, R. M., Friedewald, W. T., Schlesselman, J. J., DeMets, D. L., Ellenberg, J. H., Gail, M. H., and Ware, J. H. (1976). Randomized clinical trials. *N. Engl. J. Med.*, **295**:74-80.

Kirkendall, W. M. (1967). The placebo and clinical investigation. *J. Clin. Pharmacol.*, **Sept/Oct**:245-247.

Koch, G. G., Amara, I. A., Brown, B. W., Colton, T., and Gillings, D. B. (1989). A two-period crossover design for the comparison of two active treatments and placebo. *Statistics in Medicine*, **8**:487-504.

Meier, P. (1975). Statistics and medical experimentation. *Biometrics*, **31**:511-529.

Park, L. C., Covi, L., and Uhlenhuth, W. H. (1967). Effects of informed consent on research patients and study results. *J. Nerv. Ment. Dis.*, **145**:349-357.

Tetreault, L. and Bordeleau, J. (1971). On the usefulness of the placebo and of the double-blind technique in the evaluation of psychotropic drugs. *Psychopharmacol. Bull.*, **7**:44-64.

Zaroslinski, J. F., Browne, R. K., and Almassy, A. (1969). Placebo response in the evaluation of hypnotic drugs. *J. Clin. Pharmacol.*, **9**:91-98.

14A

Interim Analysis

Interim Analyses in Clinical Trials

Robert L. Davis[*] and Irving K. Hwang

Merck Research Laboratories, Inc.
West Point, Pennsylvania

I. INTRODUCTION

The double-blind, randomized, controlled clinical trial is the predominant method of evaluating new treatments and therapeutic procedures in clinical research. Since clinical trials may last for months or years, there are incentives for investigators and sponsors (e.g., the pharmaceutical companies) to review and analyze the data periodically for evidence of efficacy and safety in the ongoing trials. Most importantly, the investigators and sponsors want to minimize risk to patients and to stop the study if ethically necessary. Also, since the sponsors may have a financial stake in the final outcome of the trial, assessing the study results in midstream is more than mere curiosity. These or any other examinations of the data prior to study completion are referred to as *interim analyses.* The Food and Drug Administration (FDA) has continually prodded the industry to document the planned and unplanned interim analyses performed for a study and has emphasized that "the need for statistical adjustment because of such analyses should be addressed." (FDA Guideline, 1988)

---

[*]*Current affiliation:* Astra/Merck Group of Merck & Company, Wayne, Pennsylvania

Interim analyses in the pharmaceutical industry generally fall into two categories: formal and administrative. Formal interim analyses usually are applied to the Phase III pivotal studies or large scale mortality trials with long-term patient follow-up. The Phase III trials are primarily run by the pharmaceutical companies for confirmation of efficacy and safety before the New Drug Application (NDA) submission. The large-scale mortality trials are usually sponsored by companies, but run by the National Institutes of Health (NIH), academic institutions, or other similar bodies, with a multidisciplinary policy board, steering committee, and data and safety monitoring committees. For mortality trials, any interim analysis which demonstrates overwhelming efficacy may lead to early stopping of the trial. To maintain the overall Type I error probability, early stopping rules and some adjustment of the p-value may become necessary. In Section II we provide a review of the methods for these types of interim analyses and give examples of a few large-scale mortality trials in which they have been employed.

Equally important, however, are the interim analyses of an administrative nature, in which the interim data are reviewed and analyzed, and which may not require p-value adjustment and detailed documentation. Administrative interim analyses are employed primarily for making crucial internal project management decisions during the early stages (i.e., Phases I–II) of drug development such as: 1) whether to stop the drug project entirely as a failure; 2) whether to change the dose(s) used; 3) whether to study a different patient population; 4) whether to run additional studies; 5) how many patients to include in additional studies; 6) whether to add patients to the existing study; 7) whether to begin new production facilities; and 8) whether to let existing patients stay in the study longer. Industry statisticians feel that detailed documentation and p-value adjustment need take place only for the formal interim analyses, but not for the administrative ones at the early stages of drug development (see PMA Biostatistics and Medical Ad Hoc Committee on Interim Analysis, 1993). A PMA/FDA workshop (1992) on Clinical Trial Monitoring and Interim Analysis in the Pharmaceutical Industry was held in Washington, DC. Consensus and recommendations reached at the workshop were very much in agreement with this view. Sections III.A to III.C review methods, problems, and possible resolutions with regard to interim analyses appropriate for internal project management decisions during the early stage of drug development.

II. FORMAL INTERIM ANALYSES

A. Methodology Review

When the accumulating data from a fixed sample size trial are analyzed repeatedly, the true Type I and Type II error probabilities associated with the testing of hypotheses will be inflated above the prespecified levels. For example, if we chose the conventional $\alpha = 0.05$ level for the Type I error probability (two-sided) and performed repeated tests from a normal distribution with known variance, then the true Type I error probability would escalate considerably with the number of tests, that is, it would be 0.05 for testing once; 0.08, twice; 0.11, three times; 0.14, five times; 0.19, ten times; and to 0.25 for testing twenty times, which is five times the true value of 0.05. (See Table 2, Armitage, McPherson, and Rowe, 1969.)

To protect the Type I error probability some ad hoc rules have been proposed. Haybittle (1971) and Peto et al. (1976) suggested using a conservative critical value of $z = 3.00$ at each interim analysis, but keeping the conventional $z = 1.96$ at the final analysis. This rule is simple to use and is conservative at each interim analysis, but it does not precisely guarantee the prespecified Type I error probability.

Wald (1947) introduced the classical sequential probability ratio test (SPRT) to control the Type I and Type II error probabilities. There are, however, two major drawbacks to the SPRT which severely limit its applications: the continuation region extends indefinitely so that there exists no upper bound on the number of observations and the SPRT unrealistically requires repeated analyses of data after each observation. Because of these limitations, few pharmaceutical trials use SPRT. To minimize the maximum number of observations, Armitage (1957) and Anderson (1960) developed closed or restricted sequential designs. Later, Armitage (1975) proposed repeated significance tests (RST) for paired data. Although the closed sequential design and RST are superior to the SPRT, again their use has not been widespread due to the requirement of patient pairing or repeated data analysis after each outcome.

Pocock (1977) extended the RST of Armitage and developed the group sequential method (GSM). Instead of pairing patients, the GSM performs repeated significance tests on larger, equal-sized groups of patients, so that the difficulty of continuous assessment of the response of every pair of patients is overcome. Pocock obtained results for discrete group sequential boundaries for a normal response with known variance and showed that the normal results are readily adapted to other types of response data. He suggested performing no more than five interim analyses with the GSM. Independently, O'Brien and Fleming (1979) introduced different group sequential boundaries for comparing two treatments when the response is both dichotomous and immediate. Whereas Pocock boundaries are constant throughout the repeated tests, O'Brien-Fleming boundaries decrease over time. Consequently, the α_i associated with each repeated test, $i = 1, \ldots, N$, increases with i such that α_N, the final nominal α-level, is only slightly smaller than α, the prespecified level. This is probably the most appealing reason for using O'Brien-Fleming-like boundaries in practice.

DeMets and Ware (1980, 1982) considered one-sided as well as asymmetric group sequential boundaries. Jennison and Turnbull (1984, 1989) and others have proposed using group sequential confidence intervals. Siegmund (1978), Tsiatis, Rosner, and Mehta (1984) and many others have approached the estimation problems using confidence intervals following sequential tests. Slud and Wei (1982) suggested the construction of group sequential boundaries, by choosing a steadily increasing sequence of error probabilities, α_i, $i = 1, ..., N$, so that the sum of these probabilities is equal to the prespecified level, α. Wang and Tsiatis (1987) proposed a class of group sequential boundaries in terms of minimizing expected sample size. Armitage, Stratton, and Worthington (1985), Geary (1988), Lee and DeMets (1991), and Wu and Lan (1992) investigated repeated significance tests with repeated measurements. In addition, O'Brien (1984) and Pocock, Geller, and Tsiatis (1987) used a generalized least squares approach to sequential analysis of multiple endpoints.

Group sequential methods such as Pocock's and O'Brien and Fleming's require that the maximum number of interim analyses be prespecified. Also, the analyses must be equally-spaced with respect to the information time of the trial. The total information of a trial is represented by either the total number of patients, the total number of endpoint events, or surrogates such as patient-weeks, patient-months, etc. The information time (ranging between 0 and 1) can be calendar time or can be rescaled in terms of the total information. During the course of a trial, various factors may influence the frequency and interval of interim analyses. If the frequency or interval of the analyses changes, then the above described group sequential methods in theory would be inapplicable. The work by Lan and DeMets (1983), overcomes these problems through the use of an a spending (use) function, which lets the experimenter determine in advance how to "spend" his overall Type I error probabilities over the information times. Hwang (1988) and Hwang, Shih, and deCani (1990) extended this approach to a general family of spending functions to construct customized group sequential boundaries for interim analyses at arbitrarily or unequally spaced information times. For example, if in a mortality trial the total expected number of deaths is 200, and we perform an interim analysis after 120 deaths, the corresponding information time is

$$t_1 = \frac{120}{200} = 0.6.$$

A parallel methodology to the group sequential methods is the curtailing procedure. Whereas the sequential methods focus on data already on hand, the curtailing procedure focuses on both the existing and future data. Lan, Simon, and Halperin (1982) proposed the method of stochastic curtailment for early termination of a trial when the probability of a trend being reversed is small. Using stochastic curtailing, one calculates the conditional probabilities given the interim information accumulated under either the null or alternative hypothesis, to project the outcome of the trial at the planned end. The curtailing procedure can be useful in justifying early termination of a fixed-sample trial and as a means for evaluating the performance of a trial when formal interim analyses were not planned in advance. A natural way to convert a nonsequential design to a sequential one is curtailing. Curtailing of a nonsequential trial is extremely conservative in terms of early stopping, and therefore, it is usually not advisable to employ the procedure to stop a trial until at least two-thirds of the total information of a trial is available.

Another development in interim analysis methodology is the Bayesian approach as described by Berry (1985, 1987, 1989) and by Freedman and Spiegelhalter (1989, 1992). In a similar vein, a procedure called predictive probabilities by Choi, Smith, and Becker (1985) used a noninformative prior to estimate the probability of a significant difference at study completion. Spiegelhalter, Freedman, and Blackburn (1986) suggested calculating the predictive power derived by averaging the conditional power with respect to the current belief about the unknown parameters. Snapinn (1992) proposed a similar procedure with rejection and acceptance boundaries carefully balanced in order to maintain the overall significance level of the trial.

A useful set of guidelines for the content of interim analyses for practitioners can be found in Geller and Pocock (1987). Another valuable review paper is that by Enas et al. (1989), which describes interim analyses in the pharmaceutical industry. Other reviews on the development of sequential methods can be found in DeMets and Lan (1984), and Hwang (1992)

The statistical methodology selected in performing an interim analysis is crucial. Regardless of the procedure chosen, the objective should always be to minimize bias and control error probabilities, and in turn, preserve the credibility and integrity of the trial. The next section gives examples of a few large mortality trials which used some of the methods described above.

B. Examples of Interim Analyses Leading to Early Stopping for Overwhelming Efficacy

Although most of the large scale studies which use the methods referred to in the previous section are run by organizations such as the NIH, the pharmaceutical industry is beginning to support such trials also. The following well-publicized clinical trials, all sponsored by pharmaceutical companies, undertook planned or unplanned interim analyses during the course of the trial.

1. The Beta-Blocker Heart Attack Trial (BHAT)

BHAT was a randomized, double-blind, placebo-controlled, multicenter trial designed to evaluate the effect on mortality of propranolol, a beta-blocker, in 3,837 post myocardial infarction (MI) patients (BHAT Research Group, 1982; DeMets et al., 1984). The BHAT was originally planned to have a total of 4,200 patients to compare an average 3-year survival in patients randomized to propranolol or placebo. The sample size was determined based on an estimated 3-year mortality rate of 18% on placebo, a 28% reduction by propranolol to 13% with power 90% at $\alpha = 0.05$, two-sided. Recruitment began in June, 1978 and continued until October, 1980. The trial was planned to terminate in June, 1982 with an average patient follow-up of approximately three years. An independent Policy and Data Monitoring Board received the unblinded results of the trial every six months. The results reported in October, 1981 at the sixth interim analysis showed 9.5% (183) deaths and 7% (135) deaths in the placebo and propranolol groups, respectively. The normalized logrank test statistic was 2.82 (p = .005), a value considerably greater than the critical O'Brien-Fleming boundary value of 2.23 (p = .0257). After a thorough evaluation of many issues including the strong statistical evidence, the Board recommended that the BHAT be terminated nine months before its scheduled end.

Comments: The BHAT was originally designed as a fixed-sample trial with built-in interim analyses. The group sequential boundary and early stopping rule of O'Brien and Fleming with adjusted p-values was later adopted while the trial was ongoing. The stochastic curtailing procedure of Lan, Simon, and Halperin was also employed at the termination of the trial to strengthen the belief of the small likelihood of trend reversal. Although the post hoc application of the O'Brien-Fleming boundary in this mortality trial at equally-spaced calendar times (every six months) is questionable, the use of stochastic curtailing remains valid.

2. CONSENSUS

The Cooperative North Scandinavian Enalapril Survival Study (CONSENSUS) was a randomized, double-blind, placebo-controlled trial to evaluate the influence of the angiotensin-converting-enzyme inhibitor, enalapril, in addition to conventional therapy, on the prognosis of patients with severe congestive heart failure (CHF) (CONSENSUS Group, 1987; Lubsen, 1988). The target sample size was 400 patients based on the estimated six-month mortality rates of 40% on placebo versus 24% on enalapril with 90% power at $\alpha = 0.05$, two-sided. Two-hundred fifty-three patients entered the trial and were randomized to treatment with either enalapril (127) or placebo (126). Patient follow-up averaged six months with a maximum of 20 months. The crude mortality rates at six months were respectively 44% (55 deaths) on placebo and 25% (33 deaths) for enalapril, a reduction of 40% for enalapril (p = 0.002). The study concluded that the addition of enalapril to conventional therapy in patients with severe CHF can reduce mortality and improve symptoms.

The data monitoring was done under the auspices of an independent Ethical Review Committee (ERC), which unblindedly reviewed the accumulating data including the mortality information every three months. A formal interim analysis was performed at 200 patients with 99 and 101 patients randomized, respectively, in the placebo and enalapril groups. Six-month life-table mortality rates were 49% (52 deaths) for placebo and 24% (28 deaths) for enalapril (p = 0.0002). Despite the substantial difference, the ERC decided not to terminate the trial until further information became available three months later, at which time the ERC concluded that further continuation was no longer justifiable on ethical grounds and would be of limited scientific value.

Comments: There was a formal interim analysis performed when one-half of the target sample (200 patients) was reached. There were also frequent informal looks (every three months) by the independent ERC chairman. Although many group sequential methods and early stopping rules were reviewed and evaluated at the ERC meetings, no formal rule was adopted governing the termination of the trial for the purpose of interim analyses by the ERC, and consequently, the p-values were not adjusted. At an Advisory Committee Meeting to review this study, the FDA raised concerns about the frequent unplanned interim looks without a well-documented stopping rule.

3. Helsinki Heart Study

This was a randomized, double-blind, 5-year primary-prevention trial to evaluate the efficacy of the lipid-lowering agent gemfibrozil versus placebo in reducing the risk of coronary heart disease (CHD) in asymptomatic middle-aged men with dyslipidemia (Frick et al., 1987). A total of 4,081 men entered the trial and were randomized to treatment with either gemfibrozil (2,051) or placebo (2,030). The study began in 1980 and was completed in 1987. The cumulative rate of cardiac endpoints at five years was 2.73% in the gemfibrozil group and 4.14% in the placebo group, a reduction of 34% in the incidence of CHD for gemfibrozil (p < 0.02). There was no difference between the groups in the total mortality endpoint. The study concluded that the modification of lipoprotein levels with gemfibrozil reduces the incidence of CHD in men with dyslipidemia.

An early stopping rule was adopted while the trial was ongoing. The rule employed a flexible boundary based on the α spending function of Lan and DeMets.

Comments: An unscheduled interim analysis was performed and the trial was terminated early. The use of the elapsed calendar time instead of the information time (i.e., time rescaled in terms of the total expected cardiac endpoints) was not justified.

4. 4S

The Scandinavian Simvastatin Survival Study (4S) is a multicenter, randomized, double-blind, placebo-controlled trial to investigate the efficacy of simvastatin, a potent cholesterol-lowering drug, in reducing mortality in patients with ischemic heart disease and hypercholesterolemia (Scandinavian Simvastatin Survival Study Group, 1992). Patients were recruited throughout Scandinavia and randomized to treatment with either simvastatin or placebo. The study planned a one-year recruitment period and a follow-up period of at least three years from the randomization of the last patient. It was estimated that patients with stabilized angina and uncomplicated MI would have 4-year cumulative mortality rates of 10% and 12% respectively on placebo. It was further assumed that simvastatin would reduce mortality (the primary end-point) by 30% during this period. The total sample size required to provide 95% power at $\alpha = 0.05$ (two-sided), was approximately 4,000 patients, which was obtained using an exponential survival model, as described by Lachin and Foulkes (1986). The estimate of the total expected deaths, which represent the total information, was 380.

A group sequential design with interim analyses and early stopping rule has been planned for this trial. The analyses were originally planned (based on the rescaled information time) at 190 deaths ($t_1 = 0.5$), 300 deaths ($t_2 = 0.8$) and 380 deaths ($t_3 = 1.0$). The corresponding group sequential boundary ($\alpha = .05$, two-sided) for the scheduled analyses was 2.753, 2.343 and 2.020. The boundary was constructed using the a spending function, $\alpha(-4,t)$, of Hwang, Shih, and deCani.

The first patient entered and began treatment in May, 1988 and patient recruitment was completed in August, 1989. A total of 4,444 patients who satisfied the inclusion and exclusion criteria was randomized to treatment with either simvastatin or placebo. The total information of the trial was reestimated to be approximately 440 deaths. The interim analysis plan was further revised by the independent Data and Safety Monitoring Committee (DSMC) with the use of the prespecified a spending function $\alpha(-4,t)$. Instead of performing a total of three analyses, the DSMC recommended an additional interim analysis early at 100 deaths as a safety valve to guard against any unexpected safety problems (e.g. simvastatin causing greater mortality than placebo). Therefore, the revised analyses would be performed at 100 deaths ($t_1 = 0.23$), 200 deaths ($t_2 = 0.46$), 350 deaths ($t_3 = 0.80$), and 440 deaths ($t_4 = 1.00$). The corresponding revised group sequential boundary would be 3.200, 2.885, 2.341, and 2.022. The first interim analysis took place in August, 1990 and the second interim analysis in April, 1992.

Comments: The 4S provides a clear specification of a group sequential design with planned early stopping rule for performing formal interim analyses.

Making changes to the group sequential design and interim analysis plan in 4S further demonstrates the flexibility of using the a spending function approach.

III. ADMINISTRATIVE INTERIM ANALYSES

Although the above types of interim analyses are performed in the pharmaceutical industry, there are probably many more situations in which the interim data are reviewed or analyzed without intending to stop the study early. For most NDAs there are regulatory requirements for having long-term safety data on a certain number of patients for a certain length of time so there may be little advantage to terminating studies early.

On the other hand, drug development is highly competitive; shortening the development time is critical to a drug's success in the market place. For most drug development projects, sponsors conduct a number of studies before nailing down every aspect of the drug's dosage and patient population profile. In these situations interim analyses can be quite important as an aid in making crucial internal project management decisions such as:

1. Whether to stop the program entirely as a failure.
2. Whether to change the dose(s) used.
3. Whether to study a different patient population.
4. Whether to run additional studies.
5. How many patients to include in additional studies.
6. Whether to add patients to the existing study.
7. Whether to begin new production facilities.
8. Whether to let existing patients stay in the study longer.

For such decisions, the managements of most companies may want to look at data before the completion of studies, especially because they have to balance budget and resources among projects. It is best that these looks at the data and any adjustments to clinical programs be done in a formal way. This is especially important when adjustments are made in late Phase II trials which are theoretically nonconfirmatory studies, but which, because of the competitive nature of drug development may become the so-called "pivotal studies." The regulatory authority (e.g., the FDA) usually requires two of these pivotal trials for each claim. Since the FDA might take the position that any interim analysis would require some adjustment in the final reported p-value, the statistician must walk a fine line. She/he needs to guide the company into striking a delicate balance between gaining early information from the study and maintaining a credible study which could possibly be pivotal in the future. In such situations nobody wants to have to make any adjustments to p-values which might turn an otherwise potentially pivotal trial into one that is not pivotal.

The following sections review methods appropriate for making internal project management decisions. The statistician should keep in mind, however, that early results can be misleading because of the variability associated with small samples. Also, any change in the patient entry exclusion criteria or dosing schedule of an ongoing study as the result of an interim analysis will make it difficult to determine what conclusions can be drawn about which patient population when the study is over.

A.   Stopping a Trial by Abandoning a Lost Cause

A method by Gould (1983) called "Abandoning Lost Causes" can be used to handle the first three management decisions given above: that is, stopping the program as a failure, changing the dose, or changing the target patient population. It reflects the philosophy that management needs to know how the studies are going but that studies should be stopped early only if they are unlikely to yield positive results at completion. If an interim analysis that is built into the protocol suggests that positive results are likely, the study should continue to completion to obtain complete information on safety and tolerability, and perhaps some information on subgroups for initiating new studies. Otherwise, the study should be stopped early and abandoned as a lost cause.

An illustration of the method is a Phase II study of an analgesic exploring three possible doses versus placebo. Response was defined to be improvement of two or more categories from baseline score. The sample size of 100 per group was determined for 80% power at $\alpha = .05$, two-sided, to detect a difference in response rates of 70% for any active group versus 50% for placebo. The study design included the following rules:

1.   Analyze the data halfway through the study (50 patients per group).
2.   If the high dose versus placebo comparison has a p-value of 0.3 or greater, abandon the study as a lost cause.
3.   If the study continues to completion, run the final test at the 0.057 level to compensate for the fact that there may be some early decisions to abandon a lost cause which was not really lost.

Table 14.1   Interim Results of Analgesic Dose Study

	Patients with marked or moderate improvement
High dose	36 / 62 (58%)
Middle dose	41 / 59 (69%)
Low dose	39 / 61 (64%)
Placebo	37 / 62 (60%)

The results of the interim analysis are shown in Table 14.1. The p-value from a χ^2 test comparing the high dose versus placebo was greater than 0.3, so either the doses needed to be re-evaluated or the program stopped entirely.

It is of interest that the sample size for each treatment group was about 60 when the plan had been to stop at 50, and by the time the study was stopped completely the sample sizes had reached 80. This phenomenon, called "over-running", can be a problem with any interim analysis. Some resolutions of this problem have been addressed by Whitehead (1990).

Note that the method used for "Abandoning Lost Causes" is useful for abandoning the program entirely or restarting using a different dose or patient population. However, changing the dose or patient population in mid-study is not useful for a confirmatory study.

B. Interim Analysis Without p-Value Adjustment

Another method to make a possible alteration to a clinical program is a so-called no-penalty interim analysis. An example of a no-penalty design is another Phase II analgesic study to evaluate a new formulation (fast action) versus the marketed formulation. The sample size of 50 patients per group was determined for 80% power at $\alpha = 0.05$, two-sided, to detect a difference in one-hour response rate of 60% for the new formulation versus 30% for placebo.

Management needed to know whether the new formulation was effective. If not, another formulation would have to be studied. Otherwise, more studies using the current formulation could be started to gain additional information. Therefore, the plan was to look at the first half of the data purely to provide background information for the planning of additional studies. The trial was to continue to completion with the originally estimated sample size, regardless of the interim results.

The results of the interim analysis are shown in Table 14.2. On the basis of this assessment, the project team concluded that the new formulation was probably satisfactory. In this manner, not only some early information vital to the project development was obtained, but also the integrity of this study was maintained.

Table 14.2 Interim Results of Analgesic Formulation Study

Variable	Probability[*]
1 hour pain intensity difference	.66
1 hour pain relief	.90
1 hour 50% pain relief	.99

[*] Conditional probability that the new formulation will be better than the marketed formulation at study completion, given the interim results.

Experiences such as these suggest the following rules for making a no-penalty interim analysis credible to the FDA reviewers:

1. The interim evaluation should be done by a third party not involved in the actual conduct of the study, such as a statistician not associated with the design or conduct of the study.

2. Only the minimum amount of data required for planning further studies should be included in the interim evaluation.

3. Individuals involved in the conduct of the trial should not be informed of the interim results to avoid the temptation of altering the study design and introducing bias. Some sponsors have set up internal data monitoring committees (Kershner, 1987) composed of a statistician, a clinical monitor, and a regulatory specialist to review the data from such studies. Unfortunately, it is not clear whether these committees are truly independent.

4. Each center participating in the trial should enroll a fairly large number of patients and be encouraged to complete the planned enrollment.

5. The studies should continue to completion, with the number of patients originally planned, unless the project is abandoned altogether.

At the PMA/FDA Workshop on Clinical Trial Monitoring and Interim Analysis, Robert O'Neill of the FDA recommended that the study protocol might state that an interim analysis would be conducted at some infinitesimally small α. Obviously the penalty to be paid would be negligible, and the interim analysis would, in effect, be a no-penalty analysis.

C.   Adding Patients to an Ongoing Study

Another important possible change to a clinical program is whether to add additional patients to an ongoing study. Sample sizes depend on assumptions about within-treatment variance and between-treatment mean differences. If the variance is underestimated or the mean difference is overestimated, the sample size would be underestimated. There are several ways of adjusting the sample size, none of which is entirely satisfactory.

1. A monitoring committee could be established, either externally or internally, simply to estimate the variance based on interim data, or this committee could estimate both variance and treatment difference in an unblinded manner. In the latter case, some adjustment of p-values may need to be made.

2. A slightly more conservative method is for the data monitoring committee to look only at the placebo patients. If the placebo failure rate was lower than expected, then more patients than originally estimated would be required. Wittes and Brittain (1990) proposed the use of an internal pilot study. At the end of the pilot study the control (e.g., placebo)

group variance is examined and the required sample size for the main study is recalculated.

3. The above methods (1 and 2) require breaking the blind to some degree. Gould (1992) proposed a procedure which does not require unblinding the data. It uses the fact that the noncentrality parameter and hence the power for a binomial test is a function of the overall response rate and the ratio (or odds ratio) of the response rates. A peek at the overall response rate alone, not broken by treatment groups, can provide very useful information as to the adequacy of the original sample size. Gould showed that looking at the overall response rate at some convenient point during the study (e.g., halfway), with an appropriate adjustment of the sample size if the estimate of the overall rate was too large, does not materially increase the overall Type I error probability while keeping power reasonably high.

For example, suppose we wanted to compare the test drug and placebo with estimated response rates of 20% and 10%, respectively. For 90% power, $\alpha = .05$, two-sided, 268 patients per group would be required (normal approximation to the binomial test). With equal allocation the expected overall response rates would be 15%. Halfway through the study we would look at the overall response rate. If the overall rate was 10%, we would increase the sample size accordingly, that is about 50% to 402 patients per group.

Gould and Shih (1992) developed an analogous method for normally distributed data. The advantage of these methods is that blind is maintained. The disadvantage is that without knowing the true treatment difference the sample size may be increased unnecessarily.

IV. CONCLUSION

Interim analyses for clinical trials are scientifically, ethically, economically, and administratively necessary for drug development in the pharmaceutical industry. These analyses can be formal or administrative in nature.

Formal interim analyses are primarily performed in pivotal or confirmatory studies and large-scale mortality/morbidity trials with long-term patient follow-up. The objective is to provide definitive proof of efficacy and safety. Since performing interim analyses requires unblinding accrued data, it is important to control the dissemination of interim results to minimize the potential for introducing bias. Another issue is to protect the overall Type I error probability due to repeated tests. In addition, there may be a possibility of early termination due to overwhelming efficacy or untoward harm. In these instances, group sequential methods (e.g., a spending function) may be employed to establish group sequential boundaries and early stopping decision rules, and to make appropriate p-value adjustments so that the overall significance level is maintained.

Administrative interim analyses, on the other hand, are conducted in studies during the early stages of drug development. The intent is not to terminate the studies early for compelling benefit. Rather, they are used to gather information for making crucial project management decisions such as detecting study design flaws, revealing unexpected toxicity, and assisting in planning future studies. Generally, p-value adjustment is not an important issue.

Regardless of what type of analysis is done or statistical method is chosen, it is important to recognize that interim analyses should be planned and done in a logical manner. That is, to satisfy the ethical, statistical, and regulatory concerns, the dissemination of interim results and selection of a specific decision rule must be carefully considered and carried out. In addition, there should be adequate documentation to indicate when the interim analyses were done and what their consequences were. The primary goal should always be to maintain the credibility and integrity of the trial.

ACKNOWLEDGEMENTS

The authors would like to acknowledge the secretarial assistance of Ms. Cynthia Gilchrist and Ms. Laurie Rittle. We also thank Ms. Cynthia Johnson and Dr. Albert Getson for helpful comments.

REFERENCES

Anderson, T. W. (1960). A modification of the sequential probability ratio test to reduce sample size. *Ann. Math. Stat.* **31**:165-197.

Armitage, P. (1957). Restricted sequential procedures. *Biometrika*, **44**:9-26.

Armitage, P. (1975). *Sequential Medical Trials, 2nd Ed.*, Oxford: Blackwell.

Armitage, P., McPherson, C. K., and Rowe, B. C. (1969). Repeated significance tests on accumulating data. *J. Roy. Stat. Soc. A*, **132**:235-244.

Armitage, P., Stratton, I. M., and Worthington, H. V. (1985). Repeated significance tests for clinical trials with a fixed number of patients and variable follow-up. *Biometrics*, **41**:353-359.

Berry, D. A. (1985). Interim analysis in clinical trials: classical vs. Bayesian approaches. *Stat. Med.*, **4**:521-526.

Berry, D. A. (1987). Interim analysis in clinical trials: the role of the likelihood principle. *Amer. Stat.*, **41**:117-122.

Berry, D. A. (1989). Monitoring accumulating data in a clinical trial. *Biometrics*, **45**:1197-1211.

Beta-Blocker Heart Attack Trial Research Group (1982). A randomized trial of propranolol in patients with acute myocardial infarction. I. Mortality results. *J. Amer. Med. Assoc.*, **247**:1707-1714.

Choi, S. C., Smith, P. J., and Becker, D. P. (1985). Early decision in clinical trials when the treatment differences are small. *Controlled Clinical Trials*, **6**:280-288.

CONSENSUS Trial Study Group (1987). Effects of enalapril on mortality in severe congestive heart failure. Results of the Cooperative North Scandinavian Enalapril Survival Study (CONSENSUS). *N. Engl. J. Med.*, **316**:1429-1435.

DeMets, D. L. and Lan, G. K. K. (1984). An overview of sequential methods and their application in clinical trials. *Commun. Stat. Theory and Meth.* **13**:2315-2338.

DeMets, D. L., Hardy, R., Friedman, L. M., and Lan, K. K. G. (1984). Statistical aspects of early termination in the Beta-Blocker Heart Attack Trial. *Controlled Clinical Trials*, **5**:362-372.

DeMets, D. L. and Ware, J. H. (1980). Group sequential methods for clinical trials with a one-sided hypothesis. *Biometrika*, **67**:651-660.

DeMets, D. L. and Ware, J. H. (1982). Asymmetric group sequential boundaries for monitoring clinical trials. *Biometrika*, **69**:661-663.

Enas, G. G., Dornseif, B. E., Sampson, C. B., Rockhold, R. W., and Wuu, J. (1989). Monitoring versus interim analysis of clinical trials: a perspective from the pharmaceutical industry. *Controlled Clinical Trials*, **10**:57-70.

FDA Guideline (1988). Guideline for the format and content of the clinical and statistical sections of an application. Center for Drug Evaluation and Research, Food and Drug Administration, 64.

Freedman, L. S. and Spiegelhalter, D. J. (1989). Comparison of Bayesian with group sequential methods for monitoring clinical trials. *Controlled Clinical Trials*, **10**:357-367.

Freedman, L. S. and Spiegelhalter, D. J. (1992). Application of Bayesian statistics to decision making during a clinical trial. *Stat. Med.*, **11**:23-35.

Frick, M. H., et al. (1987). Helsinki Heart Study: primary prevention trial with gemfibrozil in middle-aged men with dyslipidemia. Safety of treatment, changes in risk factors, and incidence of coronary heart disease. *N. Engl. J. Med.*, **317**:1237-1245.

Geary, D. N. (1988). Sequential testing in clinical trials with repeated measurements. *Biometrika*, **75**:311-318.

Geller, N. L. and Pocock, S. J. (1987). Interim analysis in randomized clinical trials: ramifications and guidelines for practitioners. *Biometrics*, **43**:213-223.

Gould, A. L. (1983). Abandoning lost causes (early termination of unproductive clinical trials). *Proc. Biopharm. Sec., Amer. Stat. Assoc.*, American Statistical Association, Toronto, pp. 31-34.

Gould, A. L. (1992). Interim analyses for monitoring clinical trials that do not materially affect the Type I error rate. *Stat. Med.*, **11**, 55-66.

Gould, A. L. and Shih, W. J. (1992). Sample size reestimation without unblinding for normally distributed outcomes with known variance. *Commun. Stat. Theory and Meth.* **21**:2833-2853.

Haybittle, J. L. (1971). Repeated assessment of results in clinical trials of cancer treatment. *Brit. J. Radiology*, **44**:793-797.

Hwang, I. K. (1988). *Group sequential significance tests for clinical trials*. Ph.D. dissertation, Department of Statistics, The Wharton School, University of Pennsylvania.

Hwang, I. K. (1992). Overview of the development of sequential procedures. In *Biopharmaceutical Sequential Statistical Applications*, (Peace, K. Ed.) Marcel Dekker, New York, pp. 3-17.

Hwang, I. K., Shih, W. J., and de Cani, J. S., (1990). Group sequential designs using a family of Type I error probability spending functions. *Stat. Med.*, **9**:1439-1445.

Jennison, C. and Turnbull, B. W. (1984). Repeated confidence intervals for group sequential trials. *Controlled Clinical Trials*, **5**:33-45.

Jennison, C. and Turnbull, B. W. (1989). Interim analyses: the repeated confidence interval approach. *J. Roy. Stat. Soc., Ser. B*, **51**:305-361.

Kershner, R. P. (1987). A company's experience with interim analysis and data monitoring. Presented at 1987 Annual Pharmaceutical Manufacturers Assoc. Biostatistics Subsection Meeting. San Diego, CA.

Lachin, J. M. and Foulkes, M. A. (1986). Evaluation of sample size and power for analysis of survival with allowance for nonuniform patient entry, losses to follow-up, noncompliance, and stratification. *Biometrics*, **42**:507-519.

Lan, K. K. G. and DeMets, D. L. (1983). Discrete sequential boundaries for clinical trials. *Biometrika*, **70**:659-663.

Lan, K. K. G., Simon, R., and Halperin, M. (1982). Stochastically curtailed tests in long-term clinical trials. *Commun. Stat.-Sequential Analysis*, **1**:207-219.

Lee, J. W. and DeMets, D.L. (1991). Sequential comparison of changes with repeated measurements data. *J. Amer. Stat. Assoc.*, **86**:757-762.

Lubsen, J. (1988). Monitoring methods, considerations, and statement of the Cooperative North Scandinavian Enalapril Survival Study (CONSENSUS) Ethical Review Committee. *Amer. J. Card.*, **62**:73A-74A.

O'Brien, P. C. (1984). Procedures for comparing samples with multiple endpoints. *Biometrics*, **40**:1079-1087.

O'Brien, P. C. and Fleming, T.R. (1979). A multiple testing procedure for clinical trials. *Biometrics*, **35**:549-556.

Peto, R., et al. (1976). Design and analysis of randomized clinical trials requiring prolonged observation of each patient. I. *Brit. J. Cancer*, **34**:585-612.

PMA Biostatistics and Medical Ad Hoc Committee on Interim Analysis. (1993). Interim analysis in the pharmaceutical industry. *Controlled Clinical Trials*, **14**:160-173.

PMA/FDA Workshop (1992). The PMA/FDA Workshop on clinical trial monitoring and interim analysis in the pharmaceutical industry. Washington, D.C., February 24-25.

Pocock, S. J. (1977). Group sequential methods in the design and analysis of clinical trials. *Biometrika*, **64**:191-199.

Pocock, S. J., Geller, N. L., and Tsiatis, A. A. (1987). The analysis of multiple endpoints in clinical trials. *Biometrics*, **43**:487-498.

Scandinavian Simvastatin Survival Study Group (1993). Design and baseline results of the Scandinavian Simvastatin Survival Study of patients with stable angina and/or previous myocardial infarction. *Amer. J. Cardiol.*, **71**:393-400.

Siegmund, D. (1978). Estimation following sequential tests. *Biometrika*, **65**:341-349.

Slud, E., Wei, L.J. (1982). Two-sample repeated significance tests based on the modified Wilcoxon statistic. *J. Am. Stat. Assoc.*, **77**:862-868.

Snapinn, S.M. (1992). Monitoring clinical trials with a conditional probability sequential stopping rule. *Stat. Med.*, **11**:659-672.

Spiegelhalter, D.J., Freedman, L.S., Blackburn, P.R. (1986). Monitoring clinical trials: conditional or predictive power? *Controlled Clinical Trials*, **7**:8-17.

Tsiatis, A. A., Rosner, G. L., Mehta, C. R. (1984). Exact confidence intervals following a group sequential test. *Biometrics*, **40**:797-803.

Wald, A. (1947). Sequential Analysis. New York: Wiley.

Wang, S.K., Tsiatis, A.A. (1987). Approximately optimal one-parameter boundaries for group sequential trials. *Biometrics*, **43**:193-199.

Whitehead, J. (1992). Overrunning and underrunning in sequential clinical trials. *Controlled Clinical Trials*, **13**:106-121.

Wittes, J. and Brittain, E. The role of internal pilot studies in increasing efficiency of clinical trials. *Stat. Med.*, **9**:65-72.

Wu, M.C., Lan, K.K.G. (1992). Sequential monitoring for comparison of changes in a response variable in clinical studies. *Biometrics*, **48**:65-72.

14B

Interim Analysis

A Regulatory Perspective on Data Monitoring and Interim Analysis

Robert T. O'Neill

*United States Food and Drug Administration,
Rockville, Maryland*

I. INTRODUCTION

The Food and Drug Administration (FDA) is responsible for evaluating the data, statistical analyses, reliability, and validity of conclusions of clinical studies performed by the pharmaceutical industry in support of the efficacy and safety of new drugs. The FDA reviews important protocols submitted by industry and evaluates completed studies submitted in New Drug Applications and often must either provide advice at the planning stage of a clinical trial or pass judgement on the appropriateness of an analysis where data monitoring and interim analysis has occurred.

In addition to the ethical issues involved, several factors are making the early access to and analysis of accruing data a reality. One is the computerization of clinical data, its timely entry into a computer data base, and its timely auditing and cleaning through call backs and visits to the site. Another is that virtually all trials for life threatening diseases are carried out with some type of monitoring and interim analysis, usually planned in the protocol. And perhaps as importantly, the current availability of a variety of statistical methodologies and the continuing development and refinement of statistical methodologies to guide decision making for trial planning, monitoring and early termination itself is responsible for increased use as well as increased consideration of such methods in clinical studies. While there has been an increasing acceptance of the use of these statistical methods for clinical trials in recent years, many of these methods are subject to misuse in the hands of inexperienced practitioners.

From a regulatory perspective, the statistical experience with interim analyses appear best explored in the large scale government sponsored clinical trials with mortality endpoints and it is not the intention of this article to describe this model or these statistical methods in detail. Less well-explored is the application of methodology to non-life threatening diseases in which a sponsor of a clinical trial might consider monitoring and unplanned analysis with possible termination of a trial or submission to the FDA of interim results, not necessarily for ethical reasons but perhaps for drug development reasons. Sometimes decisions are made by study investigators or other involved parties to terminate a trial when naive, poor, or inappropriate monitoring and analysis methods have been employed in a study. In some situations, novel approaches are used or suggested by a sponsor without much theoretical justification of the strategy or characteristics of the plan provided. The FDA must consider the appropriate conclusions that can be drawn from such trials and analyses, often after the fact.

Recognizing that documentation of what is planned and what is actually carried out in a study is an important part of a study write-up, the FDA included in its 1988 *Guideline for the Format and Content of the Clinical and Statistical Sections of a New Drug Application* a paragraph regarding reporting of clinical trials in which interim analyses have been performed. This part of the guideline (page 67) states:

> The process of examining and analyzing data accumulating in a clinical trial, either formally or informally, can introduce bias. Therefore, all interim analysis, formal or informal, by any study participant, sponsor staff member, or data monitoring group should be described in full, even if the treatment groups were not identified. The need for statistical adjustment because of such analyses should be addressed. Minutes of meetings of a data monitoring group may be useful (and may be requested by the review division).

This paragraph is brief on explanation and intent and does not distinguish between early exploratory trials or confirmatory trials in the latter stages of drug development. Consequently, a considerable response to this portion of the Guideline developed from those responsible for documenting pharmaceutical research submitted to the FDA, mostly by industry statisticians, regarding the circumstances to which it applies. The guideline paragraph was originally intended to address trials that were terminated earlier than planned and to serve notice to sponsors regarding the use of appropriate statistical procedures for interim analysis, especially focusing on methods that should be planned for in the protocol.

However, it is now clear the statistical methods and strategies for interim analysis cannot be divorced from the data monitoring strategies and various approaches and models followed by the industry in carrying out clinical studies. In contrast to the statistical literature on interim analysis methodology which has been evolving for at least ten years particularly within the context of large scale government sponsored clinical trials with external data monitoring committees, little published literature exists on its application in drug development and specifically by the pharmaceutical industry in more routine settings.

The dialogue is just beginning on the industry use of interim analysis and data monitoring and there is likely to be an evolutionary period to the interim analysis and data monitoring strategies as followed by the pharmaceutical industry.

II. TWO SOURCES OF BIAS OF CONCERN

From a statistical perspective there are two sources of bias, which may be of more or less concern depending upon the independence of the group charged with monitoring a clinical trial. The first source relates to the process of monitoring a trial that may unblind the trial in subtle or partial ways to participants or investigators or possibly to management of the trial's sponsor. This may have potential for influencing biased allocation schemes for future patients entered into a trial, changing the outcome criteria or assessment criteria during the trial in a manner to optimize observed effects for a treatment, dropping centers or sites that may be experiencing less favorable relative treatment benefits, or changing the protocol in some way that is not taken account of in the ultimate analysis. All these issues impact on the relative treatment comparisons in ways that may produce estimates or inferences that do not reflect the true effect of a test drug in the appropriate patient population, especially when not discussed, analyzed or documented in a trial report.

The second source of bias relates to the appropriate statistical quantification of uncertainty, usually captured in calculation of p-values and more specifically in estimates of treatment effects, confidence intervals for the treatment effects and other measures of statistical uncertainty. Most clinical trial questions are posed in terms of a hypothesis and the statistical research on repeated significance testing of accumulating data in clinical studies has articulated well the implications on Type I error of excessive statistical tests of hypotheses, the probability of concluding that a drug produces an effect when in fact it does not.

The specific attention to interim analysis in the FDA Guideline is not intended to encourage routine use of interim analysis in all clinical studies but rather to call attention to the issues that must be addressed if such analyses are actually done. The issue of data monitoring is part of interim analysis, particularly with respect to the committee or body which has access to the unblinded data, and with regard to who has advisory versus ultimate authority for stopping a trial. The FDA guidelines are silent on that matter, either with respect to which practices and procedures a data monitoring group should follow or with respect to what a sponsor should document regarding their composition, operation and decision making. In this regard, naive or unknowledgeable clinical trialists can potentially adversely impact on the credibility of a trial. The FDA would like to minimize this occurring for trials submitted for regulatory purposes.

The population of controlled clinical trials submitted to the FDA generally fall into three classes:

1. Trials with independent (or external to the sponsor) Data Safety Monitoring Board's (DSMBs), most all of which are in life threatening disease or which use mortality endpoints. These trials should have and almost always do have protocols which use planned interim analysis strategies employing some form of group sequential methods with stopping rules specified in various levels of detail. The model followed is along the lines of the large government sponsored by the National Institutes of Health (NIH).

2. Trials in non-life threatening diseases which do not have independent (e.g., internal to the sponsor) DSMBs which are monitored and sometimes have unplanned analysis or unusual proposals for termination (usually with no published methodology), or questions regarding termination which were not planned in the protocol. This is the population of trials where more a priori planning should be focused.

3. Controlled trials for which the trial sponsor has no expressed intention of terminating earlier than planned completion (assuming this is well-stated in a protocol) but which are being monitored for safety but not formally for efficacy outcomes.

Because of a variety of experiences, the FDA is concerned about situations where analyses that are unplanned in the protocol are carried out on trials and which are not being properly reported or discussed in study documentation and write-ups. Related to this concern is how trials which do not have independent external data monitoring committees operate. Even in situations where a sponsor includes plans in the protocol, these plans may not be well-thought out and, in fact, may create more difficulties than had they not been in the protocol because they give a false sense of license to stop a trial. The FDA advice to sponsors in this regard is described in the next sections.

A. Protocol

The monitoring of interim results should be planned in advance, preferably with a limited number of interim analyses focused on key endpoints. A protocol should describe:

Sample size planning assumptions, duration or follow-up and degree of certainty in these planning estimates (e.g., target event rates and minimal difference between treatment and control worth detecting), perhaps the degree of skepticism regarding the expected treatment effects, etc.

Strategies or contingency plans for stopping the trial earlier than planned in the case of efficacy monitoring and toxicity monitoring (each may require separate decision criteria and boundaries that can be asymmetric). If group sequential methods are used, there should be some discussion of the timing and number of looks, at least the class or shape of the Type I spending function that is planned to be used and not changed as a result of data-driven analyses.

There are a number of routine situations of interest that must be dealt with:

Terminating a trial for better than expected efficacy

Terminating a trial for lack of expected efficacy

Terminating a trial for unexpected toxicity

Modifying a trial design on the basis of comparative results observed during the monitoring of the trial

Dropping one arm of a trial in a multi-arm trial

Adjusting the sample size of a trial upwards to maintain planned statistical power because of lower event rates or higher variability than hypothesized.

Statistical methodology is emerging that appropriately deals with each of these situations, or at least provides a sensible strategy to follow; however, practitioners may not either be aware of the need for it or ignore it.

B. Administrative Looks

The pharmaceutical industry has introduced the concept of administrative looks which is intended, among other things, to allow for access not only to summaries of patient entry characteristics, accrual patterns, and other administrative data of interest, but also to relative treatment differences on primary and secondary outcomes during the trial but with no expectation to change, modify, or terminate the trial.

The concept of an "administrative" look cannot be separated from the data monitoring responsibility. Particularly important is the issue of access to unblinded summarized group results of efficacy and safety outcomes and the potential, regardless of intention, of possible early stopping, possible up-sizing of the trial, possible downsizing of the trial, or other variations. As a result of such an administrative look, it is natural to ask questions regarding the practices and procedures as well as reporting and documentation requirements regarding unblinding of trial results, who has access to the data in the decision-making chain, what safeguards there are for maintaining the integrity of the trial, which trials deserve internal versus external monitoring groups or when is it advisable to use external monitoring committees versus internal monitoring committees, etc. Any "administrative look" at accruing study results that is not intended to stop a trial early but which allows for unblinded relative treatment efficacy comparisons, should be done cautiously in a manner that does not allow early termination of a trial for rejection of the null hypothesis. This can operationally be accomplished by use of a very conservative constant spending function during the entire duration of the trial which essentially leaves one with the same statistical criteria at the completion of a trial that a fixed trial concept would have achieved.

These issues speak to the need for planned standard operating procedures (SOPs) to be in place prior to a trial to insure that unanticipated decisions are made in the context of some planned structure and that all responsible parties are aware of the issues beforehand.

In a broader context, the FDA is concerned about its proper role in the interaction with sponsors, and with DSMBs external to the sponsor and the mechanisms for flow of information, particularly in life threatening disease areas where special regulations focus on expediting therapies to patients. An evolving consensus is that the FDA does not need to be or think it wise to be in any decision making role for a study or a DSMB.

NOTICE

The views expressed in this paper are those of the author and are not necessarily those of the Food and Drug Administration.

15

Postmarketing Studies and Adverse Drug Experiences: The Role of Epidemiology

Kathy Karpenter Wille[*]

*Merrell Dow Research Institute,
Cincinnati, Ohio*

I. INTRODUCTION

Three female referral patients were diagnosed with sclerosing peritonitis, as reported in Brown et al. (1974), a case series article. One factor common to their medical histories was treatment with a beta-adrenergic-blocking drug. The treatment had been continuous for at least 15 months in each of these cases. How can the pharmaceutical company employees assess the relationship of these events to their drug? There are many questions to answer. Biostatistics can provide a partial answer. In addition, the application of epidemiologic methods will help with this assessment.

A. Definition

Epidemiology is the study of the ways in which factors influence the patterns of disease occurrence in human populations (Lilienfeld and Lilienfeld, 1980). Its application in the pharmaceutical industry can be pictured from two views. First, when applied in the classical sense, descriptive epidemiology can be used during the drug development phase to clearly define the natural history of the disease to be treated. Second, drugs are factors that influence the patterns of disease occurrence in human populations; epidemiologic methods can be used to evaluate the benefits and risks of drugs in the treatment of diseases.

---

[*]*Current affiliation:* The Procter & Gamble Company, Cincinnati, Ohio

The development of a drug typically represents an investment of more than 200 million dollars. Prior to initiating an expensive clinical trial program, it is only prudent to evaluate the impact of the indicated disease in the population and to ascertain whether intervention will benefit public health and be commercially attractive. It is important to know the frequency of occurrence, the demographics of the population likely to be affected, the signs and symptoms of the disease, and what quantitative measures of morbidity are available. Descriptive epidemiology can provide those answers.

In planning clinical trials, one needs to know about the population of patients, particularly, their age, sex, and race. Descriptive epidemiology focuses on characterizing person, place, and time with respect to the disease (Mausner and Kramer, 1985). Knowledge about the natural history of the disease is helpful in determining the time when intervention may be most successful. Further, knowing about the disease progression may help the person evaluating the trial results to differentiate between adverse drug effects and disease effects. (Guess et al., 1988).

There are three main types of descriptive studies: correlational studies, case reports and case series, and cross-sectional surveys. These types of studies are valuable in raising hypotheses but they are of limited value in testing hypotheses. To test hypotheses, either an observational or an interventional study is required. In observational studies, the investigator cannot allocate patients to exposure or any factors affecting disease status. Differences between groups can only be observed, not created experimentally. In an interventional study, the investigator randomizes the patient to the exposure (Hennekens and Buring, 1987). In the pharmaceutical industry, observational studies are known as epidemiologic studies and interventional studies are known as clinical trials. Clinical trials are the gold standard in establishing the efficacy of drugs. However, epidemiology is a tool that is useful in overcoming some limitations of clinical trials. Thus, observational trials complement interventional trials.

B. Limitations of Clinical Trials

The randomized controlled clinical trial is the scientist's most powerful tool in establishing efficacy; however, the clinical trial is an imperfect tool. We can only obtain information in a limited number of patients within a limited spectrum of the disease state (see Chapter 6).

As information is accumulated prior to approval (Phases I, II, and III), the number of patients studied is only a small fraction of the number of patients that will be treated subsequent to approval. During Phase I trials, perhaps 20 to 40 normal, healthy volunteers will be studied. Phase II studies may involve 100 to 200 patients with the disease of interest. In Phase III studies, the total number of patients studied rarely exceeds 3000, and this number may be much

Table 15.1 Number of Persons Required to Observe at Least One Occurrence of an Adverse Event (AE)

Frequency of AE	Probability of observing at least one AE			
	95%	90%	85%	80%
1/100	300	231	190	161
1/500	1,498	1,151	949	805
1/1,000	2,996	2,303	1,898	1,610
1/5,000	14,979	11,513	9,486	8,047
1/10,000	29,958	23,026	18,972	16,095
1/20,000	59,915	46,052	37,943	32,189
1/30,000	89,872	69,078	56,914	48,284
1/ 50,000	149,787	115,130	94,856	80,472
1/100,000	299,574	230,259	189,712	160,944
1/500,000	1,497,867	1,151,293	948,560	804,719

smaller (see Chapter 1). Therefore, relative to those who will be using the drug after approval, the number of patients tested is typically small.

Clinical trials are conducted under strictly defined conditions on a carefully demarcated group of patients who are chosen to be as homogeneous as possible. The drug will be used under much broader conditions and in a variety of patients in general use (Porta and Hartzema, 1988). Because of the limited number of patients studied, it is unlikely that adverse events that occur with a low frequency will be detected.

In order to have a 95% probability of observing at least one adverse event that has a true occurrence rate of 1 in 10,000, you would have to observe nearly 30,000 people. This follows from the Poisson probability law with parameter np, where n is the number of people observed and p is the incidence rate. Table 15.1 gives the study sizes for several combinations of rate and probability of observing at least one event. Note that when the probability of observing at least one event is 95%, the resulting sample size is generally three times the inverse of the rate. This is sometimes known as the "rule of three" (Sackett et al., 1986). The magnitude of these numbers indicates that it would be a logistic nightmare to plan a clinical trial to detect or compare rare adverse events.

C. Strengths of Clinical Trials

Randomization is the key to the strength of clinical trials. A primary role is to prevent bias in the allocation of treatments. This is the predominant way of controlling for potential confounding variables, particularly, confounding by indication (Porta and Hartzema, 1988). In practice, a physician treats a patient based on the symptoms presented. Most believe that the baseline characteristics of a patient affect the prognosis of that patient. If all patients presenting with similar baseline symptoms are treated with the same drug, and those patients presenting with different symptoms are treated with a different drug, then the association of drug and outcome is confounded by the baseline symptoms. Randomization ensures that patients have an equal probability of receiving the treatments being evaluated.

Compared to observational epidemiologic trials, intervention in the disease process by allocating patients to a treatment makes it easier to evaluate the effect. The value of the epidemiologic trial is realized after the drug has been approved. The epidemiologic trial is used to further study the safety of a drug by examining the occurrence of rare adverse events. In addition, the epidemiologic trial can be used to study the economic benefits, health status, and quality of life related to drug treatment.

II. POSTMARKETING STUDIES

A. Introduction

Phase IV clinical trials are those that are conducted after a drug has been approved and marketed. Some Phase IV studies are requested by the U.S. Food and Drug Administration (FDA) as a condition for approval; others are initiated voluntarily by the manufacturer to further investigate the drug. The typical Phase IV study uses a more heterogeneous population and is designed to more closely recreate the conditions found in general usage. The randomized controlled trial may be used to establish efficacy relative to a competitor or to broaden labeling claims. In order to distinguish the postmarketing clinical trial (interventional) from the epidemiologic trial (observational), the usage of Phases IV and V is growing in the vernacular. Phase IV refers to interventional postmarketing studies, and Phase V refers to observational postmarketing studies.

As the drug becomes more widely distributed through marketing, previously unreported adverse events are likely to be described. Epidemiologic studies provide a methodology for evaluating the risks of adverse events that were not detected prior to marketing. Not only should the natural history of the disease be understood, the pharmacologic action of a drug needs consideration in the interpretation of the data from such studies (Lawson, 1984). The application of epidemiologic methods to the study of drug effects has emerged as a specialized field of epidemiology, known as pharmacoepidemiology (Porta and Hartzema, 1988).

Pharmacoepidemiology joins together epidemiology and pharmacology, the study of the properties and reactions of drugs with respect to their therapeutic value. Most research referred to as pharmacoepidemiology occurs once a drug has been approved for marketing (Spitzer, 1991). The discipline of pharmacoepidemiology is growing rapidly, with scientific meetings and journals dedicated to this topic. The application of pharmacoepidemiology will continue to grow as requests by the FDA for postmarketing studies increase and as the FDA comes to a decision about the interpretation of what constitutes the "adequate and well controlled investigations" required for drug approval (Faich, 1991).

B. Observational Cohort Studies

In observational cohort studies, whether prospective or retrospective, the patients or subjects are classified based on the presence or absence of exposure (to the drug). In a prospective study, patients are followed to a specified endpoint or until the occurrence of the outcome of interest (an adverse event or disease). In a retrospective study, patients or subjects are still classified according to exposure; however, enough time has elapsed so the event of interest will have had the opportunity to occur. The controlled cohort study most closely resembles the controlled clinical trial and it shares many of the same limitations of the clinical trial (Edlavitch, 1988). To detect differences in the rates of rare adverse events, the sample size may be so large as to make this study design impractical. Further, because they are nonrandomized, cohort studies are subject to confounding by indication.

C. Case-Control Studies

A case-control study is an observational study in which cases (those with the disease or outcome of interest) and controls (those without the disease or outcome of interest) are selected. Patients are interviewed or medical records may be reviewed to determine the presence or absence of exposure prior to the development of disease or some other outcome. The exposure in the cases is then compared to the exposure among the controls, and inferences are drawn. (Hennekens and Buring, 1987) Since the participants for case-control studies are selected on the basis of disease status, the design allows for selection of adequate numbers of diseased (and nondiseased) individuals to detect a significant difference. This design is particularly valuable when the disease being studied is rare.

Case-control studies provided the first clear evidence that oral contraceptives do increase the risk of thromboembolic and thrombotic disease (Stadel, 1981). Based on the results of the case-control studies, which were further substantiated with the results from cohort studies, the product labeling for several oral contraceptives warns of the increased risk of thromboembolic and thrombotic disease in users of oral-contraceptives (Zurich, 1993).

Major problems encountered with the case-control study are selection bias, resulting from differential selection of cases and controls based on exposure status, and differential recording or reporting of exposure information based on disease status (Hennekens and Buring, 1987). Ideally, the cases would be all those occurring in a specific population (for example, a case registry, hospital records, or a Health Maintenance Organization) over a well-defined period of time, and the controls should be a sample of the population from which the cases developed (Sartwell, 1974).

When studying the role of drugs in relation to disease status, it is important to remember that drug exposure is usually related to some underlying illness. In the more traditional case-control study, the exposure (diet, occupation, chemical exposure, smoking history) tends to predate any medical problem. Thus, when evaluating the data from a case-control study in which a drug is hypothesized to be related to a disease, it is important to consider whether the underlying medical condition is related to the illness currently under investigation (Jick and Vessey, 1978).

D. Evaluation of Epidemiologic Studies

The interpretation of results and the conduct of epidemiological studies, particularly case-control studies, are often subjected to criticism and debate (Feinstein, 1988; Savitz et al., 1990; Weiss, 1990; Feinstein, 1990). The following points should be considered in the evaluation of epidemiologic studies.

1. The research hypothesis should be stated prior to collecting the data. If a relationship was not a part of the research hypothesis, then associations found subsequent to collecting the data should be viewed as hypothesis generating. "Data dredging" brings up all the statistical issues associated with multiple comparisons. Some results from case-control trials found through data dredging have been contradicted or the results cannot be confirmed with a cohort study.

2. In clinical trials, a great deal of effort is expended to ascertain the eligibility of patients; similar effort should take place in epidemiologic trials. For example, the researcher needs to ensure that the disease or adverse event, does not precede exposure. Exposure and disease status should be clearly defined and verified.

3. The data need to be obtained as objectively as possible. Relying on the memory of individuals to obtain exposure information can be misleading because of "recall bias". Cases may spend more effort searching their distant memory than controls. Studies should involve efforts to verify the

patients' reports. For example, if a patient says a medication was taken, can this be verified through prescription records?

4. To avoid selection bias, diagnosis of disease must be sought with equal rigor in the exposed group and the unexposed group. Preferably, the interviewer will be blinded to the exposure status.

E. Automated Data Bases

Data sources that link drug histories with medical care records can be used by pharmacoepidemiologists to investigate drugs and their relationship to adverse drug reactions (ADRs) in a specific population. If there are regulatory decisions to be made, particularly if there is some question about the safety of a drug, the study should be performed quickly. Automated data bases provide a rapid means to identify large numbers of individuals who were exposed to a drug or who developed a disease.

The data base should provide information on drug utilization, diagnosis, and demographics. There are many sources available for such data and it is important to understand the circumstances under which the data were collected if the results are to be interpreted correctly. For example, if outpatient diagnostic codes are related to health insurance reimbursement, the incidence of this code selection may not reflect that of the general population.

There are several automated data bases available. The Group Health Cooperative of Puget Sound in conjunction with the Boston Collaborative Drug Surveillance Program contains information on outpatient prescriptions and on hospital discharge diagnoses. The Saskatchewan Health Prescription Drug Plan captures prescription and in-patient and outpatient diagnoses for all the residents of the province of Saskatchewan. Kaiser Permanente's outpatient pharmacy records can be linked with Kaiser Permanente hospital discharge diagnoses. Refer to Table 15.2 for a list of these data bases that was compiled by Serradell, Bjornson, and Hartzema (1988).

The primary advantage of a large linked data base is that of speed and flexibility (Walker, 1989). The recorded exposures will include a vast array of drugs, and there will be many kinds of outcomes. Thus, the study possibilities are virtually limitless. Unfortunately, the large automated data base cannot be used to study drugs that are not yet being prescribed, and even these data bases cannot be used to detect very rare events (less than one in a thousand) with accuracy. However, the automated data base is perhaps one of the most useful resources for epidemiologic investigations available to the pharmaceutical industry. If there is some indication of a safety issue, such as the sponsor getting reports of a previously unreported adverse event, the large linked data base provides a means for testing a hypothesis in a comparatively short period of time.

Table 15.2 National Databases for Pharmacoepidemiological Research

Group Health Cooperative of Puget Sound	
Population:	HMO in Seattle area 280,000-300,000 members
Characteristics:	Boston Collaborative Drug Surveillance Program uses the databases for cohort studies; prepaid group HMO
Advantages:	Drug costs covered by membership fee; assumes complete drug use data in members
	Pharmacies fully automated since July 1976
	Complete record of all prescriptions filled by 28,000 people for a period of eight years
	Hospital discharge diagnosis available dating back to 1972
	95 percent of patients with a prescription will have it filled at an HMO pharmacy
	Validity of data high (reported to be more than 99 percent through interviews and record searches)
Disadvantages:	Population served is primarily young, employed; excludes elderly

Medicaid	
Population:	Primarily needy, indigent, disabled
Characteristics:	Programs designed for states to provide needed medical care to defined population
Advantages:	Temporal associations between drug and diagnostic events can be examined
	Longitudinal medical histories
	Data collection is continuous for several years in most states
	Fairly rapid retrieval and relatively inexpensive
	Reliable denominator data
	Absence of classic biases (e.g., interview, recall, reporting)
Disadvantages:	Data validity
	Database is a medical billing database rather than medical records
	Loss to follow-up because of eligibility changes
	Skewed population with respect to demographics, income, and social factors
	State-to-state variations in eligibility and treatments covered
	Potential confounders not documented (e.g., smoking, alcohol consumption, parity)

Table 15.2 National Databases for Pharmacoepidemiological Research (cont.)

Saskatchewan Data Base	
Population:	Approximately 1 million people in the province of Saskatchewan
Characteristics:	Drug plan database has accumulated 37 million prescriptions since 1975
	1600 formulary products covered
Advantages:	Drug database linkable to separation data from hospitals, physician services, psychiatric services, cancer foundation, and vital statistics

Boston University School of Medicine Drug Epidemiology Unit	
Population:	Over 50,000 cases have been studied, including 11,000 patients with cancer and 5500 children with congenital malformations
Characteristics:	Goal is to quantify the serious side effects of drugs
	Chief strategy is case-control approach
	22 collaborating hospitals
Advantages:	Information is collected by personal interview
	Information gathered on both prescription and nonprescription medications
Disadvantages:	Normal biases when using interview techniques (e.g., recall, interviewer)

Kaiser Foundation Health Plan: Southern California	
Population:	Approximately 1.8 million
Characteristics:	Prepaid group HMO
Advantages:	Age and gender distribution is similar to general population
	All areas have computerized pharmacies effective 1984
	Basic descriptive and demographic data on hospital discharges

National Databases for Pharmacoepidemiological Research	
Advantages:	Are stored on computer files
Disadvantages:	Outpatient encounters and details of hospital episodes available only by manual review

Table 15.2 National Databases for Pharmacoepidemiological Research (cont.)

| COMPASS (Computerized On-Line Monitoring) |

Population:	Compilation of billing information submitted by Medicaid in approximately 10 states
Characteristics:	Database run by Health Information Designs, Arlington, VA
Advantages:	Dates back to 1980
	Very large, comprehensive data collection
	Temporal associations between drug and diagnostic events can be examined
	Longitudinal medical histories
	Data collection is continuous for several years in most states
	Fairly rapid retrieval and relatively inexpensive
	Reliable denominator data
	Absence of classic biases (e.g., interview, recall, reporting)
Disadvantages:	Data validity
	Database is a medical billing database rather than medical records
	Loss to follow-up because of eligibility changes
	Skewed population with respect to demographics, income, and social factors
	State-to-state variations in eligibility and treatments covered
	Potential confounders not documented (e.g., smoking, alcohol consumption, parity)

| ARAMIS (American Rheumatism Association Medical Information System) |

Population:	Tracks 22,000 patients from 17 centers
Characteristics:	National arthritis data resource
	Prospective protocol assesses routine clinical data, mortality, disability, discomfort, drug side effects, and economic impact
Advantages:	Can investigate comparative toxicity of drugs, long-term drug effects, effects of drug combinations, effects of comorbidity, generalizability of premarketing estimates, clinical characteristics of those with side effects
Disadvantages:	Focused population

III. SURVEILLANCE/EPIDEMIOLOGIC INTELLIGENCE

A. Introduction

As discussed earlier, during Phase I through Phase III of drug development, the drug is studied in a limited population, the conditions under which a drug is studied are closely circumscribed, and the duration of administration is limited. Thus, adverse reactions that are unlikely to be detected prior to marketing are: a) those that occur with a rare incidence (less than one in ten thousand), b) those resulting from a specific interaction with concomitant drug therapies or a concurrent disease, or c) those requiring a long latency. One tool for detecting previously unreported adverse reactions is spontaneous reporting.

B. Spontaneous Reporting System

Faich (1986) has defined surveillance as the "systematic detection of drug-induced reactions by practical, uniform methods". One important aspect of this surveillance is the maintenance of a system for the reporting of adverse drug reactions. The FDA has a program to monitor adverse drug reactions of marketed drugs. The FDA monitoring program is based on reports that arise from the usual practice of medicine. This is experience during the marketed phase, in contrast to adverse events that arise from clinical trials prior to approval of the New Drug Application (NDA). Reports can be made directly to the FDA, but more frequently, reports by health professionals and consumers are made directly to the drug manufacturer. The manufacturer is required by law to submit all reports to the FDA. If the reaction is serious and not already listed in the product labeling, it must be submitted to the FDA within 15 working days of initial receipt of the information. "Serious" means an adverse drug experience is life threatening, permanently disabling, or requires patient hospitalization. In addition, the following outcomes are always considered serious: death, congenital anomaly, cancer, or overdose. All other spontaneous reports are submitted to the FDA periodically. When the FDA receives these reports, they are entered into a computer data base. These data are reviewed and analyzed by the Office of Epidemiology and Biostatistics at the FDA, and they are available upon request through the Freedom of Information Act.

C. Interpreting and Summarizing Spontaneous Data

Spontaneous reporting can provide a timely signal of risk. This early warning system was instrumental in establishing the association between flank pain syndrome (flank pain and transient liver failure) and suprofen, a nonsteroidal antiinflammatory drug (Rossi et al., 1988). Marketing of suprofen began in the United States in January of 1986, and by mid-March, five or six cases had been reported. In late-April a "Dear Doctor letter" was set out to more than 170,000 physicians. By the end of June, the FDA had received 117 reports of flank pain syndrome. Eventually, 366 cases were reported. In 291 of the 366 cases, the date of onset of flank pain syndrome was available and it was possible to demonstrate a correlation between the number of cases of flank pain syndrome and suprofen usage.

The early warning system worked well in this context because the event was unusual in the population not treated with suprofen and the onset of flank pain syndrome occurred very shortly after taking one or two doses of suprofen. In addition, the individuals taking suprofen were healthy, so there were relatively few underlying diseases or concomitant exposures to confound the results.

The spontaneous reporting system is best used for signaling. One must remember that calculated rates are reporting rates not true incidence rates. They usually have to be expressed in terms of sales or prescriptions written. Under-reporting is a common problem with this system. Physicians may not be aware of this system or they may be concerned about possible litigation (Faich, 1986). The surveillance system was designed to detect possible safety problems with drugs because they cannot all be known at the time of marketing, so physicians are encouraged to report all suspected adverse reactions. For any given report, there is no certainty that the suspected drug caused the reaction. Comparisons of drugs should not be made from this system; drugs are approved at different times and they are monitored under different circumstances. The length of time a drug has been marketed, the drug class, recent publicity, or any number of other factors influence reporting.

Manufacturers are required to report an increase in serious, labeled adverse drug reactions. An increase in reports could signal a change in awareness or in patterns of use of the medication. An increase could indicate increasing rates of ADRs associated with longer administration of the drug, or an increase in reporting rates could reflect an increase in the occurrence of that disease in the population that is independent of drug administration. An increase in reports might also be a reaction to a media report. The reasons for an increase in the reports of serious, labeled adverse drug reactions are many and probably cannot be discerned from the surveillance system. The surveillance system is a signalling system; an analytic study is frequently required to test the hypotheses it generates.

The reports for labeled adverse drug reactions should be reviewed at least as frequently as the cycle for submitting periodic reports. When a significant increase in frequency is found, a narrative summary must be submitted within 15 working days of its detection. Two approaches for detecting increased frequencies have been suggested in a FDA draft *Guideline for Postmarketing of Adverse Drug Reactions* (Food and Drug Administration, 1985), an arithmetic approach and a statistically based approach. Briefly, the arithmetic approach calls for reporting a doubling of reports from the comparative reporting period to the current reporting period after an appropriate adjustment for changes in drug usage has been made. Using the statistically based approach, the manufacturer would be required to report an increase in excess of the upper 95% confidence limit for the comparative reporting period after adjustment for drug usage.

The statistical approach uses a large sample approximation that is valid when the number of adverse drug reaction reports is large. Norwood and Sampson (1988) have developed an exact procedure based on a Poisson distribution to monitor adverse drug reactions that occur with a low frequency.

IV. SUMMARY

One important role of epidemiology in the pharmaceutical industry is to evaluate the safety of a drug. More specifically, epidemiology is used to evaluate the relationship of a drug to adverse events. Ideally, a signal would be detected through the spontaneous reporting system. Then, an epidemiologic study might be conducted to test the hypothesis and the data for the study could be drawn from an automated data base. An inference would be drawn from the study results based on a scientific rationale. If the conclusion is that there is indeed a risk related to drug treatment, evaluation of the risk then requires an assessment of the benefits of the treatment. Benefit can be measured in terms of quality of life and economic measurements.

Unfortunately, the safety of a drug can easily become an emotional and sensational issue. There are situations in which the data are inadequate for drawing a scientifically based conclusion, but they are of sufficient interest to attract comments by the press and the public. Once the media has mobilized the public against a drug, the damage is frequently irrevocable, even if the data are shown to be inadequate or subsequent studies exonerate the drug (Melmon, 1989). When there is a safety issue, the pharmaceutical company must act promptly to evaluate the relationship of the adverse drug reports to the drug.

REFERENCES

Brown, P., Baddeley, H., Read, A. E., Davies, J. D., and McGarry, J, (1974). Sclerosing peritonitis, an unusual reaction to a beta-adrenergic-blocking drug (Practolol), *Lancet*, 2:1477-1481.

Edlavitch, S. A. (1988). Postmarketing surveillance methodologies. In *Pharmacoepidemiology: An Introduction*, (Hartzema, A. G., Porta, M. S., and Tilson, H. H., Eds.), Harvey Whitney Books Company, Cincinnati, pp. 27-36.

Faich, G. A. (1986). Special report: Adverse-drug-reaction monitoring, *N. Engl. J. Med.*, 314:1589-1592.

Faich, G. A. (1991). Pharmacoepidemiology and clinical research. *J. Clin. Epidemiol.*, **44(8)**:821-822.

Feinstein, A. R. (1988). Scientific standards in epidemiologic studies of the menace of daily life, *Science*, **242**:1257-1263.

Feinstein, A. R. (1990). Scientific news and epidemiologic editorials: A reply to the critics, *Epidemiology*, **1(2)**:170-180.

Guess, H. A., Stephenson, W. P., Sacks, S. T., and Gardner, J.S. (1988). Beyond pharmacoepidemiology: The larger role of epidemiology in drug development, *J. Clin. Epidemiol.*, **41(10)**:995-996.

Hennekens, C. H. and Buring, J. E. (1987). Design strategies in epidemiologic research. In *Epidemiology in Medicine*, (Mayrent, S. L., Ed.), Little, Brown and Company, Boston, pp. 16-27, 132-149.

Jick, H. and Vessey, M. P. (1978). Case-control studies in the evaluation of drug-induced illness, *Amer. J. Epidemiol.*, **107(1)**:1-7.

Lawson, D. H. (1984). Pharmacoepidemiology: a new discipline, *Brit. Med. J.*, **289**:940-941.

Lilienfeld, A. M. and Lilienfeld, D. E. (1980). *Foundations of Epidemiology*, 2nd ed., Oxford University Press, New York, p. 3.

Mausner, J. S. and Kramer, S. (1985). *Epidemiology: An Introductory Text*, 2nd ed., W. B. Saunders Company, Philadelphia, p. 119.

Melmon, K. L. (1989). Second thoughts: Adverse effects of drug banning, *J. Clin. Epidemiol.*, **42(9)**:921-923.

Norwood, P. K. and Sampson, A. R. (1988). A statistical methodology for postmarketing surveillance of adverse drug reaction reports, *Statistics in Medicine*, **7**:1023-1030.

Porta, M. S. and Hartzema, A. G. (1988). The contribution of epidemiology to the study of drugs. In *Pharmacoepidemiology: An Introduction*, (Hartzema, A. G., Porta, M. S., and Tilson, H.H., (Eds.), Harvey Whitney Books Company, Cincinnati, pp. 1-6.

Rossi. A. C., Bosco, L., Faich, G. A., Tanner, A., and Temple, R. (1988). The importance of adverse reaction reporting by physicians: suprofen and the flank pain syndrome, *JAMA*, **259(8)**:1203-1204.

Sackett, D. L., Haynes, R. B., Gent, M., and Taylor, D. W. (1986). Compliance. In *Monitoring for Drug Safety*, 2nd ed., (Inman, W. H. W, Ed.), MTP Press, Lancaster, UK, pp. 471-483.

Sartwell, P. E. (1974). Retrospective studies: A review for the clinician, *Ann. Intern. Med.*, **81**:381-386.

Savitz, D. A., Greenland, S., Stolley, P. D., and Kelsey, J. L. (1990). Scientific standards of criticism: A reaction to "Scientific standards in epidemiologic studies of the menace of daily life," by A. R. Feinstein, *Epidemiology*, **1(1)**:78-83.

Serradell, J., Bjornson, D. C., and Hartzema, A.G. (1988). Drug utilization study methodologies: national and international perspectives. In *Pharmacoepidemiology: An Introduction*, (Hartzema, A. G., Porta, M. S., and Tilson, H. H., (Eds.), Harvey Whitney Books Company, Cincinnati, pp. 19-26.

Spitzer, W. O. (1991). Drugs as determinants of health and disease in the population an orientation to the bridge science of pharmacoepidemiology, *J. Clin. Epidemiol.*, **44(8)**:823-830.

Stadel, B. V. (1981). Oral contraceptives and cardiovascular disease, *N. Engl. J. Med.*, **305**:612-618.

Walker, A. M. (1989). Large linked data resources, *J. Clin. Res. Drug Devel.*, 1-5.

Weiss, N. S. (1990). Scientific standards in epidemiologic studies, *Epidemiology*, **1**(1):85-86.

Zurich, D. B., (Ed.) (1993). *Physicians' Desk Reference*, 47th ed., Medical Economics Data, Montvale, New Jersey, pp. 1723, 1779, 2253.

16

The Role of Contract Research Organizations in Clinical Research in the Pharmaceutical Industry

Roger E. Flora

Pharmaceutical Research Associates, Inc.,
Charlottesville, Virginia

I. INTRODUCTION

Contract Research Organizations (CROs) have played an increasingly important role in pharmaceutical research over the past few years. The role of these organizations within the pharmaceutical industry and their interactions with pharmaceutical companies in the development of new drugs are discussed in this chapter. First, CROs are defined in terms of the services they provide and their structure. Some reasons why pharmaceutical companies use CROs are then presented. The importance of CROs is assessed in terms of the amount of work CROs do for pharmaceutical companies in support of their clinical research activities. Finally, in keeping with the theme of statistics in the pharmaceutical industry, the role of statisticians in CRO activities is discussed. This is approached from two perspectives: the industry statistician's role in interfacing with a CRO, and the CRO statistician's role in performing necessary work within the CRO and in interfacing with the client company.

II. WHAT IS A CRO?

A "contract research organization" is defined in the *Code of Federal Regulations* (21 CFR Ch. 1, Sect. 312.3) as follows:

"Contract research organization" means a person that assumes, as an independent contractor with the sponsor, one or more of the obligations of the sponsor, e.g., design of a protocol, selection or monitoring of investigations, evaluation of reports, and preparation of materials to be submitted to the Food and Drug Administration.

This definition encompasses a broad range of activities or services. The primary services provided by CROs, however, are those associated with the planning and managing of clinical trials and with the data management, analysis, and reporting of the resultant data. More specifically, these services are:

Project management and planning
Clinical trial design, including protocol and case report form development
Management of clinical trials, including site monitoring of investigations
Clinical data management
Statistical analysis, reporting, and consultation
Preparation of final clinical study reports and regulatory submissions and support
        at regulatory meetings
Regulatory affairs services

The first of these, project management and planning, involves formulating a detailed clinical development plan for a compound, device, or biologic, and managing the execution of this plan. The remaining activities are key elements in the execution of the clinical development plan. All of these activities may be performed by the pharmaceutical company that is developing the compound, or any subset of these activities may be contracted to one or more CROs.

A full-service CRO is one that can provide the full range of activities associated with clinical development of a compound. Only a very few CROs, however, can claim to have this capability. The majority offer some subset of the complete list of services, or are particularly capable only in certain areas of the clinical development process. Thus CROs range from very large organizations that mirror the clinical research capabilities of a major pharmaceutical company to a single individual who provides a specific service for the sponsor company. Similarly, the structure of a CRO depends on the number of employees and the number of services provided. In general, however, even a full-service CRO will have fewer levels of decision-making authority than are typically encountered within pharmaceutical companies.

III.   IMPORTANCE OF CROs IN PHARMACEUTICAL CLINICAL RESEARCH

Definitive information about the amount of pharmaceutical clinical research that is accomplished by CROs is not readily available. Most CROs are privately held, so that annual reports are not available to the public. Also, there is no association or organization for the certification or registration of CROs from which information can be obtained. The only reliable source of information is the companies who use, or intend to use, the services of CROs. Some information on this point is available from a survey conducted in 1989 by Barnett Associates, Inc. for the Associates of Clinical Pharmacology (Barnett et al., 1990). The purpose of this survey was to determine key trends in pharmaceutical research over the next decade. A survey instrument was sent to "opinion leaders" in 26 companies, and responses were received from 17. One of the 8 trends reported was:

> Contract Research Organizations (CROs) will be increasingly involved in all aspects of the clinical research process, including managing clinical research projects and monitoring study sites for the industry.

Participants were asked to describe their past, current, and anticipated use of CROs for handling of complete development programs, as well as for components of the programs. Only 6 of the 17 responding companies contracted full clinical development programs to CROs, and this was for only 5% to 15% of their develop- ment programs. Some of the smaller companies reported much more extensive use of CROs for this purpose. The sample, however, was heavily skewed toward the larger companies (14 of the 17 responding companies were in the largest 30 pharmaceutical companies); therefore, this could not necessarily be cited as a trend among smaller companies.

Of the 17 responding companies, 10 reported using CROs for the selection of investigators through writing of the final study reports for about 8% of their studies. Only 6 of these companies, however, reported using CROs for this range of services 3 years ago, and then for only about 5% of their studies. A definite trend toward increased use of CROs for monitoring of individual studies was also indi- cated. Twelve of the 17 companies who responded to the survey said they currently use CROs to monitor 5–10% of their studies, and 15 projected that they would use CROs to monitor up to 17% of their studies in 3 years and up to 25% in 5–7 years.

Unfortunately, information about the amount of work in clinical data management and statistical analysis that is contracted separately from study monitoring services was not elicited by this survey. Thus, overall use of data management and analysis services is not documented. It seems likely, however, that the use of CROs for these services will also increase. The information is sketchy, but there is good reason to believe that CROs will play an increasingly important role in the conduct of pharmaceutical clinical research over the next decade.

IV. WHY DO PHARMACEUTICAL COMPANIES USE CROs?

An often asked question is, "Why do pharmaceutical companies contract services from CRO, rather than obtain the necessary resources to complete the work internally?" The most obvious answer is timing. The time required to bring a new product to market is quite long relative to its patent life. This, along with competition from other companies with similar products under development, make it extremely important to get a new product to market as quickly as possible. Saving a few months or even a few days can mean millions of dollars in additional revenue. If a company does not have trained staff available when a new pharmaceutical entity reaches the clinical development stage, valuable time could be lost before the necessary resources can be acquired. Furthermore, if additional resources were added, the flow of potential products through the company's research pipeline might not be sufficient in the future to justify the increased capacity for clinical development.

The use of CROs, who already have experienced staff available, is a natural way to implement development without delaying until additional resources can be recruited, or until development has been completed for another product. This strategy also avoids a commitment to resources for which the need may soon disappear. Used in this way, CROs become a shared resource among many companies. This helps smooth out fluctuations in the workload caused by fluctuations in the numbers of compounds at the development stage for individual companies.

Another reason why pharmaceutical companies may turn to CROs is that CROs are often able to focus resources more quickly on a new project than pharmaceutical companies can. This can be attributed, at least in part, to the fewer levels of decision-making authority within CROs.

For new companies or start-up companies that have compounds at the development stage, but have no drug development capabilities, CROs can play an even more important role. CROs provide these companies with a way to begin development of new pharmaceutical entities immediately without having to wait until they can develop the capabilities internally. Such companies would find it even more difficult to build the resources internally than companies that already have some clinical development staff but who would require additional personnel to undertake a new project. The start-up companies lack both the base structure and the expertise necessary to assemble such resources quickly. This is the situation, for example, with many of the so-called biotech companies.

A pharmaceutical company may also use a CRO because of a special expertise the CRO has acquired. The CRO may have recent valuable experience in an area in which the pharmaceutical company lacks experience. In addition, CROs can call upon the experience gained from having worked with different clients on similar projects. When addressing a new project, the CRO can select the best aspects of the different approaches it has encountered.

The trend toward internationalization of drug development provides a final reason why pharmaceutical companies turn to CROs for assistance. Both the high cost and the lengthy time required to develop a new drug are compelling reasons to use data from studies done in other countries when applying for registration within a given country. More uniform regulatory requirements and increased acceptance of data from other countries by regulatory authorities have made this much more feasible than in the past. Companies with international operations have moved, or are moving, to standardize procedures from country to country in order to meet the goal of international drug development. For companies without an international presence, CROs provide a means for accomplishing this goal. These companies can pursue development in different countries through CROs that have established operations in the target countries. Many CROs have established international operations specifically to meet the increased demand for international drug development.

V. THE ROLE OF THE STATISTICIAN IN CRO ACTIVITIES

The role of the statistician in CRO activities depends on whether the statistician works for the CRO or the sponsoring company. It also depends on the type of company contracting the services of the CRO and the amount of responsibility they are willing to transfer to the CRO. If the sponsor has no statistician, as might be the case for a start-up company, the statistician for the CRO would assume the full responsibility associated with the analysis and reporting function for a clinical trial or clinical development program. In this case, the CRO statistician's function would be essentially the same as that of a statistician working for the sponsoring company. Various aspects of this function are addressed in other chapters of this book and will not be dwelled on here.

The above would also hold if the sponsoring company assigned full responsibility for a program to the CRO. Such a complete transfer of responsibility, however, seldom occurs. Attention must then be given to the issue of how the company statistician and the CRO statistician are to interact to accomplish the objectives of the program effectively and efficiently. This issue, along with related aspects of the interaction between the sponsor and the CRO, will be the focus of the remainder of this chapter.

When a study or program is contracted to a CRO, the freedom given to the CRO's statistician(s) to control the statistical aspects of the program must be made clear. This is especially important if only a portion of the studies that constitute a development program are contracted to the CRO. In such cases, care must be taken to ensure that the statistical approach is consistent among studies. For example, the choice between a parametric or nonparametric analysis may appear arbitrary in one study, while in another the choice seems clear cut. Similarly, the choice between the Cochran-Mantel-Haenszel procedure or some other categorical data technique such as a log-linear-model analysis may be primarily a matter of preference. Consistency among studies should then dictate which analytic approach is used in the primary analysis. The primary project statistician for the sponsoring company must, in such cases, assume a major role in coordinating the analyses between the sponsor and the CRO. The statistical approach, conventions to be followed, and formats to be used must be determined and communicated to everyone involved. This does not mean, however, that the approach should necessarily be dictated by the statistician(s) of the sponsoring company. Selection of the best approach should be a joint effort between the sponsor and CRO statisticians who are involved in the analysis. The key to the success of the project is good communication by both parties throughout the process.

Good communication regarding analysis and reporting issues is especially important early in a project. It is, in fact, desirable to have as much information as possible during the resource planning phase of the project. This will allow for a more realistic estimate of the work to be done. The sponsoring company's statistician can help facilitate this by making sure that the specifications for the project clearly delineate any special analysis and reporting considerations. Although the statistician may not be responsible for developing the project description, he or she should still be involved in preparing the specifications for the analysis and reporting. Sample reports and examples of tabular displays in the desired format can be quite helpful in communicating such requirements. Additional background information, including findings and data problems from similar studies, can also help prevent time from being spent ineffectively during the analysis process. Failure to establish realistic expectations and clear specifications for the services required can lead to a poor working relationship and unfulfilled objectives.

In the situation discussed above, it is clear that the primary project statistician for the sponsoring company must use strong project management skills in coordinating the analysis and reporting. Although this is also true when the project is handled entirely within the company, additional considerations are necessary when a CRO is used for some studies. It must be recognized that the CRO statisticians are not a part of the "statistical culture" of the sponsoring company. Thus, they may not be accustomed to the same conventions and ways of doing things as those within the company. Failure to communicate this kind of information early in the process can lead to misunderstandings and to needless iterations in the analysis and reporting process. These, in turn, can lead to unwanted delays in completing the project.

The statistician for the CRO can also help facilitate the kind of interaction between the sponsor and the CRO that is essential to a successful joint development program. Preparing detailed analysis plans and submitting them to the sponsor for approval prior to performing the analyses and writing the reports for a project will ensure mutual understanding of the approach to be followed and the content of the reports. The plans should indicate specific tabular summaries to be included and the formats in which they will be presented. Such measures can substantially increase the efficiency of the analysis and reporting process by providing a focused plan of attack for the statistician and programmers to follow.

When a complete development program is contracted to the CRO, the intent may be to transfer full responsibility for coordinating the analysis and reporting function to the CRO. If, however, the sponsoring company has a strong internal statistical group, some interaction between the sponsor and CRO statisticians is still important. As noted earlier a certain "statistical culture" is likely to exist within the company that may have an impact on the statistical approach that is used and the conventions that are followed in presenting the data. Reviewers of the report will have certain expectations and may wish to retain a certain amount of control over what is done. It is incumbent upon the CRO statistician to be aware of these issues and to address them early in the program. As in the case of a joint development program, detailed analysis plans approved by the sponsor can enhance understanding and reduce the number of iterations required to complete the analyses and reports.

Whether a statistician works for a pharmaceutical company or a CRO, the purely statistical aspects of the work are the same and require the same basic knowledge and training. Clearly, however, it is very beneficial for a CRO statistician to have previous experience working for a pharmaceutical company. Such experience provides a greater understanding of the needs of the pharmaceutical companies and the analysis and reporting process within them. Additional skills required by statisticians involved in joint clinical development programs between CROs and pharmaceutical companies are strong communication and project management skills. These skills are required on both sides if the analysis and reporting function is to be effectively and efficiently accomplished through the joint efforts of two independent organizations.

REFERENCES

Barnett, S. T., Hilsinger, R., Harwood, F., Ballard, R. (1990). Clinical research practices for the 1990s: Results of a survey conducted for the Associates of Clinical Pharmacology. *J. Clin. Res. Pharmacoepidemiol.* 4:7-23.

Code of Federal Regulations, (1991). Title 21: Food and drugs. Part 312. Washington, DC: U.S. Government Printing Office, p. 63.

17

Documenting the Results
of a Clinical Study

**John R. Schultz, Jeffrey B. Aldrich, Arthur E. Hearron,
Karen E. Woodin, and Joseph R. Assenzo**

*The Upjohn Company,
Kalamazoo, Michigan*

I. INTRODUCTION

Although documentation of study results is the last step in the clinical trial process, documentation preparation actually begins in the planning and execution stages. Good planning and execution with appropriate documentation generated on an ongoing basis yields a scientifically valid, medically relevant report which can be produced in a timely manner. With this perspective, the following discussion will begin with the design phase of a clinical trial followed by the execution phase and finally by the reporting phase. The frame of reference will be a large multicenter trial. An organization may have several such trials underway. Although they may be in different therapeutic areas, the general approach and many of the operational procedures could be essentially identical. Documentation of other types of studies may be less complex but the objectives are the same. At a minimum, documentation must contain sufficient detail to permit the knowledgeable reader to understand how the study was conducted and to evaluate the results. The sponsor may also require detailed documentation related to study conduct such as procedures for investigator selection and training, preparation of drug supplies, drug accountability, informed consent, study monitoring, and data management. Documentation such as this will also be required by regulatory agencies if the objective of the clinical trial is to support

315

drug registration. These issues are considered in the context of the authors' experience with the clinical trials process rather than a summary of the literature.

II. DESIGN PHASE

The primary documents for the design phase are the protocol which includes the analysis plan and the data management plan. The protocol provides the plan for conducting the trial. It contains a clear statement of the primary objective and any secondary objectives, the study design and the patient population as defined by the inclusion and exclusion criteria. The protocol will also include a description of the treatment groups, treatment schedules, primary and secondary variables and when they are to be recorded. The analysis plan may be included in the body of the protocol or attached as an appendix. It describes how the data will be summarized and analyzed in order to meet study objectives. The data management plan is a document separate from the protocol and describes in detail how the analysis plan will be implemented. It also describes the case report form (CRF) modules which will be used, data flow, editing procedures, output tables, database validation methods and specific data analysis software packages and procedures. Further details related to the preparation and contents of the protocol, analysis plan, and data management plan are given in the following sections.

A. Protocol

Preparation of the protocol is a collaborative effort of the medical monitor and the biostatistician. Input may also be obtained from medical specialists with expertise in the therapeutic area, clinical investigators and field monitors. The protocol provides information needed to carry out the study and summarize the results. The Food and Drug Administration (FDA) has described elements which should appear in a protocol (Food and Drug Administration, 1988). A summary of these elements is as follows.

1. A statement of the objectives.
2. A clear statement of the study population and all patient selection criteria.
3. A statement of the basic experimental design.
4. A description of the control groups.
5. A description of the type of blinding and the specific procedures to carry it out.
6. A description of the method of assigning patients to treatment groups.
7. A description of the efficacy and safety variables to be recorded, the times they will be recorded, and the methods for measuring them.
8. A description of the steps taken to ensure accurate, consistent, and reliable data (e.g., training sessions, instruction manuals, data validation, or audits).

9. A description of any planned interim analysis of the data, including the monitoring procedures.

10. A description of the circumstances under which the study would be terminated before the planned number of patients has been enrolled and of circumstances in which individual patients would discontinue the study medication before planned completion.

11. A description of the statistical methods to be applied to the data.

12. A description of the statistical considerations which were used to determine the number of patients in the study.

These elements may be addressed in various sections of the protocol depending upon the conventions followed by the sponsor. An example protocol outline is given in Table 17.1. Items 9, 11, and 12 of the protocol requirements will usually be included in the Analysis Plan and will be considered in more detail later. There are three other elements of the protocol which are of particular relevance to this discussion. These are study objectives, study design and data quality assurance.

A clear, unambiguous objective statement is the key element of the protocol. The objective drives the definition of patient population, study design, patient treatment, variable selection, sample size, and analysis procedures. Some examples of good objective statements are:

1. The objective of this study is to compare the efficacy and safety of orally administered Drug A and Drug B in the treatment of acute pneumonia

Table 17.1 Example of Protocol Outline

A. Introduction
B. Study Objectives
C. Study Design
D. Subject Selection
E. Treatment
F. Study Activities and Observations
G. Data Quality Assurance
H. Analysis Plan
I. References

Appendices

Study activities schedule
List of required laboratory assays
Detailed description of assessment scales
Sample Informed Consent Form

    caused by pathogens susceptible to these two antimicrobials in geriatric patients.

2.   The objective of the study is to investigate and compare the effects of Drug A with those of Drug B on serum lipids, lipoproteins and apolipoproteins.

    In some cases, a protocol may also have secondary objectives. They should be clearly indicated as such and must be compatible with the main objective.

    An example of a protocol with primary and secondary objectives is as follows:

Primary Objective: To compare, using primary and secondary response variables, the efficacy of an antiinflammatory drug with analgesic properties (Drug A) versus that of a strictly analgesic drug (Drug B) in the management of osteoarthritis during a 12 week period.

Secondary Objective: To correlate findings of inflammation observed at arthroscopy/synovial biopsy of the left knee with the changes in clinical assessments of the right knee during a 12 week period.

In the examples above, the primary objective is stated in a manner which allows formulation of a clear, testable hypothesis. It is desirable to specify the variable(s) which will be used to evaluate the objective. This might be cure rate, a laboratory assay, or time to a specific event. As was done in the examples above, these criteria should be described in detail in the Study Activities and Observations section. Further details regarding the study population are given in the Subject Selection section.

    Documentation issues related to study design include statistical design, type of control groups, screening periods, baseline periods, and treatment periods. In situations where there is an upward titration of dose, a maintenance period, and a discontinuation period with dose reduction, the sequence must be described in detail to avoid confusion. In more complex cases a graphical representation may be useful. Consideration should also be given to the need for justifying choice of experimental design or type of control. Particular concerns related to carry over effects and changes in disease condition are associated with crossover designs (see Chapter 10). In the ideal situation, a placebo control is the comparison of choice; however, in many situations an effective treatment already exists. In these cases the standard treatment may be an appropriate control but efficacy of the standard in the specific experimental situation should not be accepted without question. A review of the literature with appropriate citations can provide useful documentation in this regard. Additional design considerations are discussed in other chapters of this volume.

    Quality assurance processes ensure that the data resulting from clinical studies fulfill the requirements of the protocol, are available when needed, and can be processed efficiently. Two important components are data editing and process validation. Both components involve standards, measurement, correction, and feedback. Measurement against standards identifies elements which need to be corrected, and feedback is the mechanism for process improvement. Standards are defined in policies, procedures, and operating documents such as the data management plan. Quality cannot be assured with undocumented standards. In all cases, success is dependent on measurement and correction without delay, and feedback to those whose actions can improve the process.

Although most of the documentation related to data quality will be provided in the Data Management Plan, it may be desirable to include some of the requirements in the protocol. Topics addressed in the Data Quality Assurance section of the protocol might include compliance requirements, training provided to personnel in the clinic that would contribute to data quality, and steps taken to standardize measurement procedures between clinics. Requirements for retention of source documents such as hospital charts might also be specified in this section.

B. Analysis Plan

The analysis plan gives the statistical model(s) to be used in data analysis, procedures for validation of model assumptions, any planned subgroup analysis, interim analysis procedures and stopping rules (if interim analysis is planned) and the α level(s) to be used in declaring a comparison to be statistically significant. It may also include a discussion of how missing data and dropouts will be handled, rationale for selecting the number of patients (sample size), including power of the test $(1 - \beta)$, and specific data listings and summaries needed. Examples of typical data listings and summaries are given in the Data Management Plan section.

Details of the statistical model should address analysis of primary and secondary efficacy variable(s). For data measured on at least an ordinal scale, analysis of variance and regression models are particularly useful. In the case of multiclinic trials, the difference between the analytical model for the analysis of primary and secondary variables may be the inclusion of a treatment by an investigator interaction term for the former but not the latter. These discussions should refer to the exact variables to be analyzed by each model and when they are observed. The procedures for handling missing values and dropouts should also be included here (i.e., ignore, estimate, use the last value carried forward, or use the worst value carried forward).

Analysis of safety variables may include a number of techniques based upon the type of study, duration, and likelihood of withdrawal effects. Analysis of safety variables may be as simple as the listing and flagging of aberrant laboratory assay values, and the tabulation and listing of medical events.

Additionally, the analysis may include statistical techniques ranging from simple χ^2 tests to categorical data modeling, clustering, and life table procedures.

The traditional a level used in determining statistical significance is 0.05. In most studies, a two-sided test of significance is used (e.g., $\alpha/2 = 0.025$). In testing for a treatment by investigator interaction effect, α is quite often set at 0.10, particularly in the case of small studies. In the case of interim analyses (multiple looks at the data) or multiple comparisons, other a levels may be appropriate.

To estimate sample size, one must decide upon a medically desirable difference to be detected between treatment groups with respect to the primary endpoint (if in fact such a difference does exist), decide upon a probability level $(1 - \beta)$ for detecting the desired difference (usually ≥ 0.80), obtain an estimate of the variance from prior studies and/or the literature, and specify the statistical model to be used. The sample size can then be determined from the appropriate equation (e.g., the noncentral F distribution or normal approximation to the binomial). In addition to including a power estimate $(1 - \beta)$ based upon the considerations above, it is informative to include power estimates over a range of alternatives. In the case of continuous data one can conveniently use standardized differences between treatment means $\delta = |(\mu_1 - \mu_2)|/\sigma$ versus $1 - \beta$ for fixed sample size. For binomial data one can use $\delta = |(p_1 - p_2)|$ for fixed sample size and $(p_1+p_2)/2$ constant.

C. Data Management Plan

The data management plan (DMP) defines the data processing necessary to meet the objectives of the protocol and analysis plan. It is prepared by the data manager (or data coordinator), statistician, output programmer and appropriate systems personnel. A great deal of interaction with the medical group is also required. To be effective, the DMP must be prepared before the study starts. Data management plans are prepared for individual studies and for projects involving multiple studies. The sections of a typical plan are described below.

1. Case Report Forms

Data collection forms implicitly define the study and consequently have the greatest effect on study result documentation. Their importance to successful data management and the fact that they cannot be changed after the study starts, warrant a great deal of attention and planning during the design phase. Although the paper form is the most tangible representation of study data, case report form design begins with definition of the underlying data which will form the database necessary to meet the objectives of the protocol and analysis plan. The database content and format must also satisfy the need for consistency and continuity required to merge data across studies for integrated summaries and other project level processing.

It is helpful to define the database in terms of sets of data items, or data modules, which naturally occur together. These modules have the properties that they are consistent across studies (enabling mergability), have purpose (they jointly describe something of interest), and they are commonly collected and retrieved as an entity. Examples are vital signs or the set of data items which describe an adverse reaction. Data modules are usually based on simple concepts which lend themselves to standardization and provide the foundation for an effective and efficient clinical database.

As the instrument for collecting data modules, case report forms are more complex and variable in nature because they must also satisfy the needs of the investigators who complete them and the data management persons who will handle the forms and enter the data. For example, a report form containing a vital signs module may exist in several languages and may contain different amounts of instructional information to support different clinical cultures. A rigorous understanding of the requirements of the collection instrument and the data to be collected promotes optimization of both database and case report form.

Where database modules are usually simple to understand and maintain, case report forms must satisfy a much broader audience requiring extensive review and agreement. There are few other activities in the study implementation process where it is so easy to misinterpret the requirements of others. Review of report form design includes investigators, clinical sponsors, statisticians, and data management personnel. As with data modules, standardization of associated report form modules is essential for effective and efficient data collection instruments. A chart identifying the modules to be collected at each time point is very useful. In the simplified example shown in Table 17.2, the lower case

Table 17.2 Example Case Report Form Modules

		Initial	Week 1	Week 2	Final
Physical exam	F	X			
EKG	F	X			
Vital signs	F	X	X	X	X
Lab chemistry	E	X	X	X	
Lab urinalysis	E	X	X	X	
Medical event	F		X	X	X
Efficacy 1	F		X	X	
Final report	F				X

See text for explanation

module name identifies a module which requires development. The character to the right of the module name indicates whether the data collection media is a form (F) or an electronic file (E). Initial refers to the visit before treatment is initiated, Week 1 and Week 2 are evaluation visits. Final refers to post-treatment follow up.

With this guideline, the study database is defined using the specified data modules and case report forms are designed using report form modules associated with their respective data modules.

2. Data Flow

This section describes the way data will be handled from the source to inclusion in the database to archival storage. Whether the collection media is a paper form or an electronic file, the routing, transfer mechanism, and custodial responsibility for each step is documented. Operations performed at each step are specified including methods for making and documenting corrections. Each person who handles the data from creation to archiving should know where they are getting the data, what the format is, what they will do with it, and where it goes when they have finished their part of the operation.

3. Edit Standards

A primary function of the data management process is assuring that the results being documented are accurate. This function is fulfilled by examining (or editing) the data against standards defined in this section of the data management plan. Each variable should have a set of rules which can be used to test its validity. In practice, the rules are the same for many variables and lend themselves to defaults so that the task is not as formidable as it might appear. This section of the DMP specifies the rules to be used for the study. From this information, operations personnel can implement the processing necessary to perform the editing process. Common data edits are ranges, domains, existence, dependencies, and logical relations which may involve several data items.

Successful editing requires feedback and control. The desire for an error free database can lead to excessive testing if the process is not controlled. Although testing of a single data item is usually straightforward with a few error modes, the number of permutations of relationships between data items can rapidly exceed the ability to efficiently test them. Although computerized data editing is a powerful tool, the editing process should be designed with attention to utilizing the right tool for the job. Much time can be wasted writing programs to examine a complex data relationship when a simple display or listing will allow a person to easily understand and verify the correctness of the data. Case report form design can also have a significant effect on the ability to recognize and correct errors when examining the completed form.

Process validation is measurement and documentation that data processing systems, human and electronic, are functioning as designed and maintaining the integrity of the data being processed. Where data editing seeks to identify errors in data, validation identifies errors in processes. Because editing and validation both use similar techniques of measurement, correction, and feedback, they can be confused without a well-defined operating model and understanding of purpose. When quality assurance processes are working properly, validation is simply documentation of their success. A validation process which is producing significant negative results is functioning as a pseudoeditor indicating gross failure of the other processes. Electronic and human processes are validated differently due to the differences in reproducibility. Electronic systems are reproducible and the acceptable error rate is zero. For human systems, there will be a finite error rate. The acceptable error rate is dependent on the nature of the data and the relative effect of errors on analyses. Final error rates of a few data items in ten thousand are achievable under routine operating conditions.

There are two major human processes which require periodic validation; data collection and data entry. Validation uses auditing procedures to produce error rates with statistically calculated confidence intervals from random samples supplemented with one hundred percent comparisons for critical variables. The choice of when to validate is based on experience and balances the cost of validation against the probability of a process going astray. With proven editing and monitoring processes, validation at the end of short studies is sufficient. Long studies spanning a year or more require validation during the course of the study or on demand if problems are suspected. An effective practice is to perform a validation early in the study to assure proper use of case report forms, database definition, and data entry systems.

Compliance edits are also defined here. These are the tests which determine if the rules of the protocol are being followed such as meeting eligibility criteria, adequate enrollment rates and completed case reports. As with data edits, the DMP should document the standards which need to be met, not the method of testing.

4. Data Summaries and Listings

The structure and content of data from a clinical trial is complex. Decisions on what and how to display results should be made in collaboration with the medical group. Certain types of data listings and summaries are almost always needed. They include the following:

1. **Listings**
   a. Lists of patients included in the summaries and analyses as well as patients not included and reason for exclusion.

b. Patient data listings. For each patient, this listing gives all observations on all key variables recorded during the trial. This is also useful as a final check on the quality of the data base.

c. Patient listings of observations used in summary and analysis tables. These listings will differ from those of (b) above if, for example, average values during specified time periods were used in the analyses. The accompanying report should clarify how listings (b) and (c) differ.

d. Listings of all data reported for key variables. These listings are arranged by treatment group, investigator, patient and time of the observation. For a cholesterol-lowering study, one such listing would include all total cholesterol as well as HDL and LDL values. If change and percentage change from baseline are to be used in the summaries and analyses, these values should be similarly listed.

e. Listings of medical events. These are presented in one or both of the following ways:

   i. By patient. All medical events reported (experienced) by a patient are listed with other detailed data such as demographic characteristics (e.g., age, sex, weight), treatment group, dose, time of occurrence, severity, outcome, and opinion about relationship to investigational medication.

   ii. By medical event. For each specified medical event, a listing of patients experiencing that medical event along with specified data such as listed above is presented.

f. Listings of dropouts. These data are listed by treatment group, investigator, patient and reason for dropout. Specific patient data may also be useful particularly in a study with many dropouts due to medical events.

g. Mortality listing. Detailed tables summarizing deaths are prepared. The tables include such information as cause of death, documents used to determine cause (e.g., autopsy results), history, results of physical examination and laboratory studies before and during the study, age at entrance to the study and at the time of death, race, gender, and other pertinent demographic data, duration of therapy and dose of study drug and concomitant medication.

h. Laboratory assay listings. All laboratory assays values should be listed by patient and time with abnormal values flagged.

2. **Summaries.** Data should be summarized over all investigators, and separately by investigator for primary endpoint variables and for important background variables. Data summaries by diagnostic subgroups and/or other prognosis variables may also be needed.

a. Patient inventory. This shows by treatment group the number of patients starting the study and the number for whom data are available at each planned analysis time point.

b. Summaries of characteristics noted immediately before the study began (baseline characteristics). Frequency distributions of baseline characteristic values by treatment group are often useful. Usually included in the display are summary statistics such as means, standard deviations, key percentile values, minimums, and maximums. These help describe the study population, determine if the treatment groups were homogeneous when the study began, and indicate which baseline variables might be useful as covariates or in adjustment procedures.

c. Summaries of dropouts. In addition to the listing described previously in Listings (f), the reason for discontinuation of study medication should be displayed in tabular form by treatment group and reason, and by treatment group and time.

d. Frequency distribution of dose of study drugs by treatment group and time. Often these are constructed for both dosage rate and total dose.

e. Summaries of efficacy variables. For efficacy variables, it may be desirable to display the data in frequency distributions using actual values and transformed values (e.g., absolute change or percentage change from baseline). Graphic displays of time versus means and proportions by treatment group (e.g., proportion improved, proportion of patients with a 15% or greater reduction in serum cholesterol, proportion of patients whose supine diastolic pressure is 90 mm Hg or less) provide an easy-to-evaluate assessment of important results from a study.

f. Safety Variables. Summaries of safety variables will involve a number of techniques based upon the type of study, duration, and likelihood of withdrawal effects. For medical events it is useful to present the number and percent of patients experiencing each specific event by treatment group and classified by body system for all events reported during the study. Another tabular representation of the data involves the display of treatment emergent signs and symptoms (TESS) (i.e., medical events which first occur after the start of study medication, plus those which increase in intensity after the start of study medication). For studies involving drugs with a potential for withdrawal effects, it is useful to tabulate discontinuation emergent events (DESS) (i.e., events with onset during and after the taper-withdrawal of study medication, plus those which occur with greater intensity during taper-withdrawal than at any prior time in the study). In long-term studies, the onset of certain medical events may be time dependent; in such cases, graphic or tabular presentations of incidence rates versus time are useful.

5. Analysis

This section provides details regarding the analyses described in the Analysis Plan. The set of patients included in each analysis, specific variables and time(s) observed, and the analysis model must be described. In some cases, it may be convenient to list this information in tabular form. Mathematical formulas used to create derived variables must be described and all analytical software which will be used should be specified (SAS, BMDP, etc.). Validation procedures for software which is created for any specialized analysis should also be provided.

6. Data and Program Archiving

The data and programs used to document study results must be stored in a manner which assures reproducibility of the results. The mechanism, location, and identification associated with archiving the results documented under the plan is specified in this section.

Many of the activities and specifications discussed in the data management plan can be defaulted to documented standards. A data management plan in a mature organization with a rich set of operating standards will be reduced to a discussion of those things which are exceptions to standard or new procedures and specifications requiring development.

III. EXECUTION PHASE

Potential study sites are identified and evaluated while the study is still in the planning phase. Several factors are considered in site selection. These include access to required study population, expertise in the therapeutic area, ability to meet protocol requirements, qualifications of support personnel, and total clinical load. After preliminary discussions and general review of protocol requirements, the final selection of study sites is made. At this point a meeting is held with the site investigators and key sponsor personnel.

A. Investigator Meetings

The primary purpose of an investigator (study startup) meeting is the dissemination of consistent information pertaining to study practices and procedures. This meeting is best held immediately prior to study initiation; if held too far in advance, information may be forgotten, while if held after the study is underway, the benefits of consistent training are lost.

Attendees at the startup meeting usually include the primary investigator from each study site as well as the study coordinator. Occasionally subinvestigators are included, especially if they will play a substantial role in conducting the study. The attendance of study coordinators is essential, as they are responsible for case report form completion and many other study duties. Often one session is held for all attendees to discuss and explain the protocol, after which the investigators may leave; following this an additional session is held for study coordinators to focus on CRF completion and study administrative issues. Attendees from the sponsoring company include the medical monitor and his or her staff, field monitors, the biostatistician, and other personnel as appropriate.

B. Study Monitoring

Study monitoring consists of three main phases: study initiation, routine monitoring, and study closure. Study monitoring is usually done by a Clinical Research Associate (CRA), who is often based in the field.

1. Study Initiation Visit

Study administration and operations issues are addressed at the site startup meeting. These issues include drug storage, dispensing, and accountability procedures, data management and correction procedures, data flow, time lines and expected accrual rates, plus program specific issues. Any conceivably confusing areas of the protocol or case report forms should be clarified, as well as potential problems (red flags) that may arise during the course of the study.

The Investigator is responsible for keeping all study documentation and materials in readiness in case of an audit by the sponsor or the FDA. A notebook is provided in which this documentation can be filed. These documents include:

Signed copy of the protocol
Signed copy of the FDA 1572 form (Statement of Investigator)
CVs for the Investigator, subinvestigators and other personnel
Listing of ancillary personnel involved in the study
Institutional Review Board (IRB) approval letter and other IRB correspondence
Approved Informed Consent Form(s)
Laboratory certification numbers and normal ranges
Drug shipment invoices
Drug accountability information
Drug return information
Clinical brochure

The CRA will help site personnel in accumulating the necessary documents, and should check periodically to ensure that these documents are current and readily available. If there are changes in personnel during the study, for example, the ancillary personnel list must be updated and the 1572 Form revised if required. Any amendments to the protocol must be sent to the IRB and the signed amendment as well as the approval letter for the amendment must be added to the notebook. Rarely is the documentation for a study stable throughout the entire time period, so vigilance must be used to assure everything is in order. At the conclusion of the trial, this documentation, along with the case report forms for every subject entered into the trial, must be retained in accordance with FDA regulations.

2. Routine Monitoring

Regular study monitoring visits occur throughout the study, usually on a monthly basis. At each visit, completed case report forms are reviewed for completeness, legibility, and accuracy. Most sponsors also do source document review on all, or at least a certain percentage, of the completed case report forms. Source document review consists of cross-checking the case report forms against any available source documents, such as patient charts and laboratory reports. These items include, but are not limited to, adverse medical events, concomitant medications, especially if contraindicated, and preexisting medical conditions. Special attention is paid to inclusion and exclusion criteria for the study, as well as significant, predetermined, study-specific items.

Before entering any study, the subject/patient must sign an Informed Consent Form. The CRA checks each informed consent to be sure that it is properly signed, dated, and witnessed, and that it pre-dates study participation. The consent must match the official consent form approved by the Institutional Review Board that sanctioned the study.

Ongoing drug accountability procedures take place at regular monitoring visits. These may consist of any of the following: checking distribution records against case report forms, perusal of pharmacy dispensing records, inventory of drug, request for the shipment of additional supplies, checking of drug expiry dates, and return of unused drug for patients who have completed their participation in the study. Other study supplies are also checked to ensure that critical materials are not lacking to meet patient accrual.

The CRA interacts with the study coordinator and investigator during each visit to discuss the progress of the study and to assist in solving any problems that may have occurred. Any specific problems related to data collection or forms completion which have been detected through the data editing process will also be reviewed. This feedback at the early stage of the study can contribute in a major way to subsequent data quality. The CRA is a main conduit of information between site personnel and the sponsor, and is charged with keeping the study running smoothly. If necessary, the CRA may contact the medical monitor in the sponsoring company during the monitoring visit to clarify any issues that have surfaced since the last visit.

At the conclusion of the monitoring visit, the CRA makes a written report, the Contact Report, discussing the visit. A listing of the case report forms being submitted to the sponsor, as well as any additional items being submitted, such as an Institutional Review Board reapproval, are included in this report. Any case report forms which have been retrieved are attached to the Contact Report and submitted to the sponsor.

3. *Final Study Visit*

At the completion of a site's involvement in a study, the CRA makes a final, or close-out, monitoring visit. By the time of this visit, all subject/patient participation is finished and all completed case report forms have been submitted to the monitor. The only potential case report form duty remaining is the correction of any outstanding errors, although these should be minor and few in number at this point. If additional corrections are apt to be forthcoming, it is best to close the study at a later date.

Any remaining drug should be inventoried for return to the sponsor, including partial containers left from study patients, as well as completely unused medications, both active and placebo. Some sponsors prefer to have all study drug containers returned, including those which are empty, while others will allow the disposal of empty containers at the investigative site. The CRA completes a Return of Drug inventory form, leaves a copy for the investigator's files, and encloses the others with the drug to be returned. The CRA also arranges for the shipping of the returned drug. Any other study supplies, such as specimen collection tubes, case report forms, etc. are either returned, disposed of, or left for the use of the investigator, as appropriate.

The investigator is required to notify the IRB that the study has been terminated and the CRA usually requests a copy of this letter for the sponsor files. The CRA reminds site personnel to add this letter to the study document notebook and rechecks the notebook to ensure that all other study documentation is included and complete. This material remains in the investigator's files and must be retained for two years after FDA approval of the drug or until the sponsor notifies the investigator that they can be destroyed. The CRA takes care of any other loose ends that may remain, then documents the closing activities on a final contact report.

A successful study is one in which the appropriate number of patients was accrued within the given time and where all procedures were followed carefully and completely. At the end of a successful study, both the investigator and the sponsoring company should be looking forward to additional opportunities to collaborate in the future.

IV. REPORTING PHASE

A. Analysis and Interpretation

After all the data have been collected and all the database validation procedures completed, the analysis phase can begin. The treatment code is broken, the listings, summaries, and analyses specified in the Data Management Plan are generated and the results are reviewed by the statistician and the medical monitor. Ideally, the study has been conducted exactly as planned and analysis can proceed in a straightforward manner. Unfortunately, this is seldom the case. There may be a number of issues which need to be resolved before planned analysis can get underway.

Inevitably there will be some patients enrolled who fail to meet the entrance requirements. A common solution in the past was to exclude them from all but safety analyses. The current trend is to do "intent-to-treat analyses". That is, all patients who are randomized to treatment are included in the analysis regardless of eligibility. Additional analyses might use only the subset of patients who have met all inclusion/exclusion criteria and who have complied with and completed the study in accordance with the protocol. Intent-to-treat analyses tend to carry the most weight if the two methods differ. For documentation purposes, one should compare the proportion of patients not meeting entrance requirements among treatment groups. If the difference is statistically significant, one might suspect that blindness and randomization procedures had been violated. A listing of those patients not meeting the entrance requirements along with the reason should be included in the report.

Patients sometimes return for follow-up evaluations within a few days of the evaluation time specified by the protocol. Such deviations occur due to vacations, illnesses, and other unexpected reasons. If the time discrepancy between a scheduled and actual visit date is short, it may be reasonable to include the data from such a visit in the analysis. The time interval within which a patient evaluation is to be considered acceptable will vary with the nature of the study and the study medications. Procedurally, it may be wise to decide upon an allowable tolerance for each scheduled visit during protocol development and to include it in the protocol. For example, in a short-term study requiring biweekly visits and recognizing that not every patient will return in exactly 2 weeks, one might choose to accept observations made within ± 4 days of each scheduled visit. Sometimes a mistimed visit results in more than one visit during a defined interval. The question then is, for example, in reporting the 3-month observations for a patient in a study that allowed a variation of ± 2 weeks, does one report the values 1 week prior to 3 months or 10 days after 3 months? Some might choose to use the first visit in the interval, some the last, some the visit closest to 3 months, some the mean of the scores from each visit (in the case of variables for which a mean can be calculated), and still others a more complicated procedure.

The handling of medical events poses an additional problem. That is, what should be done when the same medical event is reported more than once for mis- timed multiple visits. A reasonable choice is to report the most severe report of the medical event. For example, if there were three visits in the interval of 3 months ± 2 weeks, one could select the most severe report of drowsiness from the data base. Drowsiness would then be reported as absent only if it were absent at all three visits.

After provisions are made to accommodate deviations, results are generated within the framework of the Data Management Plan. It should be emphasized that evaluation of the listings, summaries, and analysis may raise questions which in turn require additional analysis. This requires close interaction between the medical monitor and the statistician. Some elements of the analysis and evaluation process are discussed below.

1. Initial Analysis

Analysis is frequently equated with hypothesis testing, but it is much more complex. Reviews of the data and quality checks are all a part of statistical analysis. However, assuming that the data have been reviewed and the data base to be used for further analysis has been defined, the next step is to consider patient characteristics. That is, how has selectivity influenced the sample of patients studied versus the target population? This analysis might also include a comparison of the evaluable population versus patients who were originally enrolled but were deemed nonevaluable, if the latter are several in number. The summaries over all patients and by investigator of baseline characteristics provide the necessary information for this analysis.

Having described the patient population, the next step is to check for differences that might affect the analysis. An important check is for differences among treatment groups overall and for differences among investigators. Different subgroups of patients may respond differently, leading to investigator by treatment interactions which may preclude the combining of data across investigators.

2. Model Assumptions

Analysis of variance is usually employed for continuous variables and is often used for all variables measured on at least an ordinal scale. A potential problem with this approach is inequality of variance, although the data should also be checked for other patterns which might invalidate the model. Close scrutiny of the data is the best safeguard. Also, if there are sufficient data, a frequency distribution may be worthwhile. A study of the minimum and maximum, as well as a comparison of means and medians is useful. In addition, after the analysis has been performed, a check of the assumptions can be made by reviewing residual plots.

3. Baseline Comparability

If the assumptions for the model appear to be met, statistical analysis of baseline comparability of the various treatment groups can proceed. If the groups are not comparable, covariance analysis or difference scores may solve the problem.

Alternatively, an analysis on poststratified data may be tried. In a sense, the comparison of baseline data is to get a feel for the data and to highlight areas in which special care will need to be taken in analysis of improvement. Specifically, if the drugs perform differently depending on the patient baseline characteristics, any baseline differences between treatment groups may lead to mistaken conclusions about treatment differences.

4. Hypothesis Tests and Estimation

The first step in analyzing data for a treatment effect is a review of summary statistics. Generally, a summary of scores among each investigator's patients and for key variables will determine if the responses fit a pattern. Scatter diagrams of change scores related to the pretrial scores may point out interesting relationships.

Assuming that patient groups do not have to be subdivided because of unusual patterns, the next step is to review the model. The model used should be the one specified *a priori* in the protocol analysis plan, unless the model appears to be inadequate or otherwise inappropriate at this point in the analysis. If it is necessary to deviate from the model specified in the protocol, it is important to document why additional models were considered and why the final one was chosen.

In setting up a model one of the questions which has to be addressed is how to treat the investigator variable—is it a fixed or random effect? Since the purpose of most drug studies is the extension of the results to imply how patients treated by other investigators will respond, then one would like to consider the investigator effect to be random. At the same time, the clinical investigators are not a random sample of all clinicians, just as the study participants are not a random selection of all patients (see Chapter 6). With this in mind, most statisticians consider investigator effects to be fixed.

A model should also include interaction terms. The preliminary analysis should include at least all first-order interactions involving treatment. If the treatment interacts with time, which is likely if one of the treatments is a placebo, data should be analyzed separately for different time periods. A treatment interaction with a covariate can probably be best handled by dividing the patients into subgroups. If the interaction is not large relative to the observed treatment effect and if the interaction is of size and not direction, the overall conclusion as to which treatment is preferred can still be determined based on the analysis of the combined groups. However, in these cases estimating the size of the differences within the subgroups would probably be important.

5. Analysis of Dropouts

It is important to evaluate dropouts, since a patient's decision to leave the study may be due to medical events. When one of the treatment groups is receiving placebo, dropouts may occur because of lack of efficacy leaving only placebo responders. Finally, patients being treated for acute illness may drop out when they believe they have improved sufficiently. Ignoring dropouts can obviously lead to mistaken conclusions.

In long-term studies, the problem of dropouts is equally involved. Patients have a higher likelihood of discontinuing the study because of problems unrelated to the drug or the disease. If the dropout rates differ in the treatment groups, then the treatment groups may not be comparable by the end of the study. For this reason, a repeat of the analysis of baseline characteristics should be done with only those patients who finish the study included in the data set. As before, any population mix difference which cannot be adequately handled by covariates or other modifications in the model will require analysis of patient subgroups.

6. Analysis to Show Equality

When comparing a test drug to placebo, the study is designed to show superiority of the test drug. The anticipated conclusion of the hypothesis test, then, is a statistically significant difference. In this case, a next step would be to form confidence intervals for the size of the difference and to evaluate the difference among subgroups of patients. But the primary task has been accomplished when it has been shown that the groups differ and that the difference favors the drug.

Frequently, the drug comparison is not with placebo but with the current standard for treating that disease. Then the expected result may not be statistical significance. The test drug may be about equivalent in efficacy but is thought to have fewer side effects. The same situation might also occur in a placebo-drug comparison, where although the efficacy is expected to be different the side effects are hoped to be the same.

When the intention is to demonstrate equality, a careful evaluation must be made at the time of study design to insure that the power of statistical tests to be performed is considered sufficient. The statement then can be made that the lack of statistical significance, together with the acknowledged power of the procedure to pick up differences felt to be medically important, is sufficient to support the conclusion that the drugs do not differ meaningfully. A better approach to the reporting of the results would be to create confidence intervals for the differences. If the power of the procedure was high and if the statistical test shows no significant differences, the confidence intervals should show that the range of possible values does not include medically important differences.

B. Report Preparation

In many organizations, the final report is prepared by the medical monitor, the statistician, and a medical writer. The statistician prepares the statistical method-

Table 17.3 The Full Integrated Clinical and Statistical Report of a Controlled Clinical Study

I.	Title page	
II.	Table of contents for the study	
III.	Identity of the test materials, lot numbers, etc.	
IV.	Introduction	
V.	Study objectives	
VI.	The investigational plan	
	A.	Overall design and plan of the study
	B.	Description and discussion of the design and choice of control group(s)
	C.	Study population
	D.	Method of assigning patients to treatment
	E.	Dose selection
	F.	Blinding
	G.	Effectiveness and safety variables recorded; data quality assurance
	H.	Compliance with dosing regimens
	I.	Appropriateness and consistency of measurements
	J.	Criteria for effectiveness
	K.	Concomitant therapy
	L.	Removal of patients from the study
VII.	Statistical methods planned in the protocol	
	A.	Statistical and analytical plans
	B.	Statistical determination of sample size
VIII.	Disposition of patients entered	
IX.	Effectiveness results	
	A.	Data sets analyzed
	B.	Demographic and baseline features of individual patients and comparability of treatment groups
	C.	Analysis of each effectiveness measure and tabulation of individual patient data
		1. Analysis of measures of effectiveness
		2. Statistical/analytical issues
		a. Adjustments for covariates
		b. Handling of dropouts or missing data
		c. Interim analyses and data monitoring
		d. Multicenter studies
		e. Multiple endpoints
		f. Use of "efficacy subset" of patients
		g. Active-control studies

Table 17.3 (continued)

3. Examination of subgroups
4. Tabulation of individual response data
- D. Documentation of statistical methods
    1. Statistical considerations
    2. Format and specifications for data requested from sponsor
- E. Analysis of doses administered and, if possible, dose-response and blood level-response relationships
- F. Analysis of drug-drug and drug-disease interactions
- G. Special by-patient displays

X. Safety results
- A. Extent of exposure
- B. Adverse experiences
    1. Overall experience
    2. Display of all adverse events and occurrence rates
    3. Grouping adverse event terms that probably represent the same event
    4. Analysis of adverse events
    5. Listing of each patient's adverse events(s)
    6. Display and analysis of deaths and dropouts due to adverse events (adverse dropouts) and other serious or potentially serious adverse events
- C. Clinical laboratory evaluation
    1. Listing of individual laboratory measurements by patient
    2. Listing of each abnormal laboratory value
    3. Evaluation of each laboratory parameter
        a. Mean (median) values over time
        b. Individual patient changes
        c. Individual marked abnormalities

XI. Summary and conclusions
XII. References
XIII. Appendices
- A. Cross-references of all pertinent materials
- B. Protocol, sample case report form, and amendments
- C. Publications based on part or all of the results of the study
- D. List of investigators
- E. Randomization scheme and codes
- F. Documentation of statistical methods
- G. Patient data listings

ology section and will usually prepare the Results section. The statistician must ensure that the results of the study are adequately documented to support the interpretations and inferences made. Most of the material describing statistical methodology can be taken from the Analysis section of the protocol. In cases where extensive discussion is needed to describe the methodology, it may be desirable to place this material in an appendix of the report. This will ensure appropriate level of detail is documented without disturbing the flow of the report for the nonstatistical reader.

The statistician and medical monitor review the summaries and listings which have been generated and select those which should be included in the report. As a general rule, key summary tables and displays are included in the body of the report and supporting tables may be included in an appendix. Analysis of variance tables would be included in an appendix as well. A complete report which is intended to support drug registration may contain over 1000 pages including all detailed tables and appendices. It is therefore important to organize the report in a manner which facilitates review and evaluation. The FDA (1988) has published guidelines which include the format and content of an integrated clinical and statistical report. The outline given in the guidelines is presented in Table 17.3. While some of the items in the outline may not be relevant to a particular study, it provides a useful overall guide regarding material which should be included in the report.

The final component of report preparation is quality verification or review. Although this process is ongoing throughout the report preparation phase, it is the last one completed. Since the report may contain hundreds of tables, it is essential that they be reviewed for consistency and accuracy. Ideally, the review team should include people who have not been involved in report preparation. Individuals with a fresh perspective can often spot flaws which have been overlooked by those deeply involved in the trial. At a minimum, consistency of patient counts across all the tables should be examined. Numerical values given in the text of the report should be consistent with values presented in tables and graphs. Finally, the body of the document should be reviewed to ensure all the conclusions are consistent with results presented in the tables.

The level of review related to accuracy depends to a great extent on the procedures in place while the report was being prepared. If the quality assurance procedures described earlier are in place and if tables are not rekeyed, the level of accuracy should be acceptable. Even with this degree of control, however, it may be prudent to conduct an audit back to the source documents for critical variables such as mortality or a medical event of particular concern.

In some cases, the report may be reviewed in draft form by others involved in development of the drug but not in that particular trial. This review may focus more on external consistency or how findings from the current study compare with results from other clinical trials or studies in animals. It should be emphasized again that report preparation and review is a dynamic process. Some components of the report may be in the preparation stage while others are undergoing review. Conducting these activities interactively contributes to timely completion and at the same time enhances overall quality.

In summary, a massive amount of information is generated during planning, execution, analysis, and reporting of a large multiclinic trial. Early planning and preparation are essential to generate appropriate documentation and organize it effectively. Sound design, execution and analysis are also critical components of the process. Effective integration of the documentation will yield a high-quality, scientifically valid, and medically relevant report which can be produced in a timely manner.

REFERENCE

Food and Drug Administration, (1988). *Guideline for the Format and Content of the Clinical and Statistical Sections of an Application*, U.S. Food and Drug Administration, Center for Drug Evaluation and Research., Rockville, MD.

18

Data Quality Assurance

Kenneth H. Falter

Centers for Disease Control and Prevention,
Atlanta, Georgia

I. INTRODUCTION

The purpose of this chapter is to provide a systematic approach to the consideration and implementation of data quality assurance procedures into pharmaceutical research and development studies. To facilitate this, we will first place the scientific or research process into a systematic framework.

To use a "restaurant" metaphor, we will present a menu of data quality assurance "courses" and a "selection of items" within each course. The scientific or research process will be subdivided into 10 components (courses), each representing a general area of concern where data quality assurance may be implemented. Various techniques for assuring data quality will be offered as the menu selections within each course. In any given study or scientific endeavor, not all courses will be applicable for attention or for equal attention. In some studies, all data quality assurance can be consolidated into one or two steps and the quality of the "full meal" or scientific report assured. In other, more complex studies, data quality assurance issues may best be addressed at each of the 10 steps outlined in Section III. For most studies, the appropriate approach would probably lie somewhere between the above extremes. In the most unfortunate situation, the scientific process is nearly complete and no data quality assurance has been applied; then, all data quality assurance takes place at or near the end of the process.

This chapter will attempt to point out as many of the potential problem areas as possible and present potential solutions for these areas. It is not intended to imply that all of these potential problems must be explicitly addressed in all studies.

II. DEFINITION

We may operationally define data quality assurance as a process which assures that what is reported in the final scientific report accurately reflects the data as originally collected. This process may be implemented at several or all of the steps or it may be implemented as a one-shot approach at one step in the scientific process.

Data quality assurance is inherent to the scientific or research process. It must be present in order to assure that valid results are reported and that results are repeatable or reproducible in other experiments carried out under the same experimental conditions. In the pharmaceutical industry, data quality assurance takes on even more importance because of the uses made of the data. Data derived from pharmacology and animal safety trials are used as the basis for an application for an Investigational New Drug application (IND) filed with the Food and Drug Administration (FDA) in order to conduct initial trials of the drug in humans. Data collected in clinical trials are periodically submitted to the FDA by the pharmaceutical company which sponsors the clinical trials. Preclinical and clinical data, plus additional data on manufacturing and controls, are submitted to the FDA in a New Drug Application (NDA) and considered by them in order to approve a new drug for marketing. Data generated on the manufacturing process and in the quality control departments are further used to validate the purity, identity, and stability of the drugs which are being prescribed, marketed, and sold to patients. Thus, assurance of data quality is vital in the pharmaceutical industry because the data are used to support the development and marketing of new pharmaceuticals in the United States and worldwide.

III. THE SCIENTIFIC OR RESEARCH PROCESS

For purposes of this discussion, I define 10 steps which comprise the scientific or research process in a clinical trials context:

1. Planning the experiment
2. Data generation and acquisition
3. Data screening or forms checking
4. Data entry and key verification
5. Data editing
6. Data reduction
7. Statistical analysis

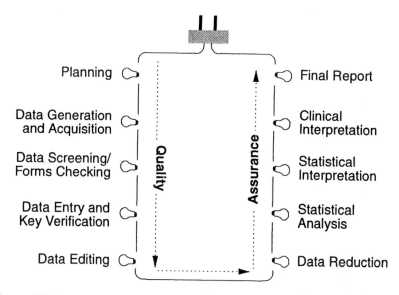

Figure 18.1 Data quality assurance: a series of activities.

8.   Statistical interpretation of results
9.   Clinical interpretation of results
10.   Final scientific report

The remainder of this chapter is devoted to examining quality assurance considerations for each of these 10 steps and presenting potential methods and procedures to apply. Some of the solutions have been used successfully by the author or his colleagues and some are proposed as approaches that may be worth looking into.

These 10 steps comprise a series of activities leading to the final study report. Continuing the restaurant metaphor, one bad course can ruin the entire meal. Alternatively, one can use the analogy of an electrical circuit with the 10 steps in series, like some old strings of Christmas lights (Figure 18.1). Then, a failure of data quality at any one step may lead to a lack of data quality in the final report.

Data quality assurance should be handled at each step, thus ensuring the integrity of the series circuit step by step. This is akin to assuring the quality of each bulb prior to adding it to the string. Or it can be delayed until the final step, when the final report is in preparation, at which time virtually every number can be proofread against the raw data in order to verify that the final report accurately and adequately reflects the original data. This is akin to plugging in the string of lights and, if they do not light, checking each bulb until the defect(s) is(are) found and fixed.

A compromise between the above two scenarios allows data quality assurance to be built into several *key* points in the system rather than into each and every one of the 10 points that have been identified. Data quality assurance at key points in this series can assure in a cumulative fashion the quality of all prior steps, and ultimately, the quality of the entire process. For instance, put 2 bulbs on the string and test the circuit. Then test again after 5 bulbs are added to the string, etc.

The amount of effort expended to ensure accurate and complete data should be commensurate with its importance and with the sensitivity of data analysis to errors in the data (Liepins, 1983). Thus, different data elements may demand different levels of quality assurance.

Two common techniques of data quality assurance are redundancy and sampling. Redundancy is a technique whereby methods are done in duplicate, or where slightly different questions or methodologies are used which should lead to the same answer. A difference in answers would indicate a problem to be resolved. A second technique, sampling, is used as an error-detection process as well. It is based on expected error rates and the allowable error rate of the outgoing material and uses statistical sampling inspection techniques to devise the scheme and the process for correcting errors. (I should mention that it is difficult to explain satisfactorily to upper management why there may still be errors in the data after all the data quality assurance that has gone on. They usually feel that the outgoing error rate should be zero.) Sampling also may involve randomly sampling the Case Report Forms (CRFs) to be checked.

Once errors are detected, a feedback, or backtracking process is utilized to trace the problem down through all prior steps so that it may be corrected at its source. A problem must be tracked down to its source so that any numbers, conclusions, etc., which follow from this original piece of data are corrected appropriately.

IV. PLANNING THE EXPERIMENT

The most important consideration in planning an experiment is to have a clear statement of objectives. With sophisticated devices, instruments, computers, and computer programs, it is possible to collect data on hundreds of variables and to process, summarize, and report these data fairly easily. However, good research is usually more targeted than this.

The objective of a study must clearly state a particular objective and which variables will be assessed to evaluate the objective. The objective should also indicate whether we are interested in estimating a parameter value, testing a hypothesis, or both. It is not sufficient to state the objective of a clinical trial as "to evaluate the safety and efficacy of compound X". A more reasonable objective may be: "to assess the comparative efficacy of drug X to drug Y in reducing blood pressure from baseline over a 12-week period". One might even raise questions about this objective and answers to these would indeed be spelled out in a paragraph which more fully elaborates on the objective.

Given a clear statement of objectives, it is then possible to lay out the methods, procedures and logistics of the following nine steps in the scientific or research process. The objectives will define which variables are to be measured and evaluated and these in turn will define the data which must be collected during the study. The details of carrying out a study are usually spelled out in a study protocol. Many people, including a statistician, will have input into this document. The protocol will usually describe the conduct of the study by the investigator as well as some of the data-handling and statistical analysis procedures which must be carried out. As such, it is a most critical document in that it contains vital information needed in the decision making regarding data quality assurance—how much, which techniques, and at what points in the process.

However, because of the many details necessary in data handling, data quality assurance, and data analysis, some companies have adopted the practice of having a separate protocol for the data-oriented aspects of the study. Whether the detailed steps are spelled out in the study protocol, in a data-handling protocol, or in separate planning documents, it is important to treat the study and all the ancillary steps which go on before, during, and after it, especially including the data quality assurance steps, as a unified process with detailed identification of the following:

1. The components of each step
2. Time estimates for each component
3. Resources needed—personnel, hardware, software
4. Interactions between personnel of different departments

The identification and analysis of possible sources of error should take place as part of the planning process (Bailar, 1983).

During this planning stage, the CRFs (or data collection forms or patient record forms) must be developed. It is the skill with which the questions are posed and the skill in laying out these questions on a CRF which often determine the success or failure of a study, no matter how good or how poor the study protocol may be. CRFs must address themselves to the objectives of the study and must collect sufficient data on the key variables so that the data-handling procedures can be applied expediently and the objectives of the study attained. Forms can be laid out quickly and efficiently using graphics/paint or desktop publishing software. The use of different fonts, shading, etc., at the first or second draft stage is now feasible and inexpensive. Thus, forms can be commented on early in the process regarding layout, ease of use, etc. Furthermore, the saving of forms, or modules, on computer disks simplifies production of CRFs for future studies of a similar nature, as the old form can be used "as is" or easily modified.

The forms must be complete in that all necessary questions are asked in order to meet the objectives. On the other hand, the forms must be concise in that extraneous information is not requested. Targeted research is not the place for a fishing expedition in which excess information is collected on a subject, patient, or animal at a given point in time.

The questions on the forms must be unambiguous. Murphy's Law surely operates here; if a question can possibly be misinterpreted and incorrectly answered, it will be. Many people subscribe to the KISS principle (Keep It Simple, Stupid). Also, forms must be easy to use and not become a bore or drag on the person filling them out. Efficiency and accuracy suffer in proportion to the difficulty of filling out the form. The form must flow smoothly in a logical sequence of collecting data following the manner in which the study is conducted. No one likes to flip back and forth between various pages of a CRF in order to record a series of observations. In addition, forms should not have much room for narrative. Most investigators will fill in all the narrative space you provide. The more space you provide, the more narrative you get and the more headaches you inherit in trying to interpret, categorize, tabulate, or computerize these responses. Narrative can be reduced by asking only the specific questions of interest and by providing a limited set of answers which can be checked off or filled in.

Forms must be amenable to easy entry into a computer. This can be accomplished by using forms which indicate columns of a computer record and codes to be entered. With the availability of data base packages and data entry packages that allow the set up of data entry screens that mimic the CRF, it is no longer necessary to indicate columns of a record. However, the length allowed for an answer which is to be filled in should be clearly visible on the CRF and on the entry screen. This can be done using underscores, one per character in the field, or by using boxes to be filled in. We usually try to avoid transcribing data from a CRF to a coding form prior to computer entry as this is a step which leads to errors. This will be discussed further in Section VI.

Since the planning step is probably the most crucial step in the research process, it should be tested out in a pilot study. In relatively small to moderately-sized studies, what may start out as a pilot study may become the main study (as long as no serious deficiencies develop). However, for long-term, large-scale Phase III-B or Phase IV studies, a pilot study is strongly recommended in order to determine if any key elements have been omitted or mishandled in the planning step. This is the easiest point at which to correct serious problems and, in fact, may be the only point at which such problems can be corrected.

V. DATA GENERATION AND ACQUISITION

Data quality assurance at the acquisition stage can only be handled well if one has a good understanding of the data generator, be it a machine, a human, or an interaction of the two.

Machines may generate either continuous or discrete data. Examples of continuous data generators are a Holter recording of a continuous electrocardiogram, whereby many hours of electrocardiographic information are collected on tape for future analysis, or a transducer implanted in an animal which is generating data continuously to either a tape or a computer. A sampling scheme is inherently built into each such machine as the data are read at small but discrete intervals (e.g., at .01 minute intervals). A discrete data generator might be an automated sphygmomanometer which takes a blood pressure reading at specific time intervals (e.g., every 15 minutes) and either visually displays or records the data on a graph, strip of paper, or magnetic tape.

Such machines should be carefully calibrated and tested under a variety of conditions similar to those which will take place during the study or experiment. Quality control checks should be run on these machines so that both their precision and accuracy are well-understood before the trial begins. Where practical and feasible, a sample of the readings should be generated by a second machine or by manual means in order to monitor the quality of the machine's output. In the case of Holter recordings, the machine readings are usually analyzed by a computer program in order to interpret the electrocardiogram. The accuracy and precision of this analysis is then an interactive function of both the algorithm used to analyze the recording and of the nature in which the Holter recording was generated. This can be checked by having the cardiographic tape analyzed at a second facility which might be using a different algorithm. In this way, the robustness of the algorithm is also tested.

The human as a data generator may be characterized by a patient or subject filling out a CRF or by a physician or investigator observing a subject, patient, or animal, and recording appropriate observations.

When a human subject is filling out a CRF, he is observing himself and the information which is recorded is subject to recall. The problem of recall has been handled in many sample survey settings. One way to help a person sharpen his recall is to ask a series of questions, only one of which is really of interest. For instance, signs and symptoms of an illness for the past month are of interest. The patient may be asked to record signs and symptoms not only for the past month, but also for the month prior to that. We might discard the information for the earlier month as that question was only asked to help the patient better focus on the prior month. This procedure should be spelled out in the study protocol. A good discussion of recall problems and of response effects to behavioral questions appears in Sudman (1983).

A nurse or doctor asking questions of a patient or subject and recording the answers leads to other interesting interactions. We are all aware of the nurse who reads through a list of 20 diseases asking if we or our family have ever had them. The list is usually read off at machine-gun speed with little chance for the patient to grasp the individual items being asked, let alone think about the answers. This may be all right for a routine annual physical exam, but it is certainly not all right when a clinical trial is being conducted and when the answers themselves may become very important for future analyses. This may be handled by training the observer who will be asking these questions. When a doctor is examining a patient and recording observations, we have the direct observation of the subject by the person who is recording the data and there is virtually no way to check this process out in any great detail except to instruct and train the investigator in what you are after.

In general, if the data generator is human, it is most important that this person understands the nature of the data, its importance, and its use in meeting the objectives of the trial. If the person generating the data is made to understand the importance of data quality and the fact that the identical data cannot be regenerated later, the quality of this type of data may be improved.

It may not be realistic to expect the data generator to have a copy of the CRF at the examining table or at the laboratory workbench. Indeed the original data may be recorded on worksheets or copies of the CRF. They would then be transposed later to the official CRF, usually by a staff member. In some cases, when the CRF is filled out at the patient's side, the handwriting may tend to be scribbled or unintelligible. In this situation, transposing is almost mandatory in order to assure that the data can be accurately read, interpreted, and processed. The ideal situation, however, is one in which the data are handled a minimum number of times and direct entry onto the CRF is a desirable aim where practical and feasible.

Many times we have used a typewritten copy of the transposed results in order to assure legibility. This provides a neat and understandable copy rather than a scribbled and sometimes unintelligible one. However, the typist may make errors in interpreting and typing the information onto the final form. Problems of accuracy must be recognized and dealt with. One way is to have both the scribbled, handwritten form and the neat, typed form delivered to the company for data quality assurance; one is then faced with reading the original, somewhat unintelligible form anyhow. There is no substitute for the initial recording of the data to be neat and intelligible as every step which transposes, copies, types, or interprets the data will inevitably lead to errors. Another possible solution is remote data entry. Then, instead of typing a legible CRF, the typist enters the data via a PC-based entry program. I will discuss this further in Section VIII.

A cardinal principle of data quality assurance is to detect and correct errors at the earliest possible moment in the process. Where clinical investigations are performed outside of the sponsor's facility, there is no substitute for on-site monitoring. Most companies utilize staff identified as clinical or medical research assistants or associates (CRAs or MRAs). Not only should these people be doing a careful monitoring job on site, but federal regulations on the obligations of sponsors and monitors of clinical trials require prescribed data quality assurance procedures. The CRAs must review the forms for legibility and completeness, and reconcile all problems on the spot with the investigator who recorded the information. If this information has been transposed from a work sheet or from a patient record onto the CRF, the CRA must be prepared to follow some type of statistical sampling scheme to compare the CRF information with patient record information in order to verify the accuracy of the transcription. If an error is found on a single form, those forms which were done immediately before or immediately after that one should similarly be proofread in order to try to detect a possible series of errors. Once the data are sent to the sponsoring firm and start through the various data processing steps, it becomes more difficult and costly, in terms of time, effort, and money to correct errors. Tracking them back to the doctor's office is time and energy consuming and we must also contend with the doctor's or nurse's recall of what went on and what was meant to be recorded when the patient was being examined.

The interaction of machine and human data generator would occur, for instance, when a nurse takes the patient's blood pressure and records it. Consistency is the keyword here. This is usually accomplished by the sponsor providing very clear and definite instructions on which sounds are to be utilized to determine the systolic and diastolic blood pressure. Similar problems occur when a doctor examines a patient's heart and lungs with a stethoscope. When a patient must receive an oral glucose tolerance test, it is important that the mechanics of the test and the exact sampling times be specified so that all such test results may be compared and analyzed as one homogeneous group.

Problems in data acquisition would ideally be identified during a pilot study and would be handled by referring back to the planning phase and then modifying or improving the planning, instructions, or CRFs as necessary in order to eliminate such problems. If a series of quality control tests must be performed on various machines which are generating data or if additional training is needed for physicians and nurses who are examining patients, these should be done before the actual study starts. In those instances in which a study is so small that a pilot study will not be done, very careful monitoring of the first several patients of each investigator can be done in order to quickly detect problems in data acquisition and to correct them before much time and money have been expended and before many patients have been subjected to the drug.

VI. DATA SCREENING AND FORMS CHECKING

Once data forms are received by the sponsor, they should be manually checked. Encoding can also be done at this time. Detecting errors at this point is important as this is the last step which precedes the computerization and analysis of the data. If remote data entry is being used, then this step should be done on-site prior to data entry.

Data screeners are a special breed. They should be intelligent, data-oriented people who have training and/or experience in three areas:

1. Data processing
2. The application area (i.e., medicine or physiology)
3. The ultimate use of the data

They should be "nitpickers" who will stick to a problem and track down a discrepancy or something which looks incorrect until it has been resolved and straightened out. It is very helpful if they have been involved in defining the output, data displays, and analyses of the data. They can screen the data not only for completeness, accuracy, and consistency, but also with an eye to whether it can be successfully processed to the desired end points.

Case record forms usually contain hundreds of variables and it is impossible to verify or screen each variable to the same degree of accuracy and care. Certainly, a fatigue factor would enter in and erode any such idealistic scheme which assumes that all variables are equally well screened. Thus, it is important to determine which variables are the most important and most critical in a study and to focus more attention on these. Sampling schemes should be set up to determine the frequency with which each variable will be verified (Naus, 1975). In fact, some variables may be better off not being examined at this point if they are easily amenable to computer editing.

Checklists are very handy for the data screeners as they may be handling data from many studies in the course of a day. Thus, a checklist for each study is very helpful to ensure that the proper procedures for that study are carried out. These checklists should also contain procedures defining what to do if data for a given patient seem incorrect or if a data set is incomplete. In some cases, it may pay to enter only correct data or correct CRFs into the computer and have the incorrect or incomplete data follow along later. In other instances, it may pay to enter all the data but to somehow flag those data that are known to be incorrect or that are questionable so that they may be corrected later on. In other cases, one may decide not to enter any data from a CRF which has key data items incorrect or missing. Such decisions must be made in the context of the particular study and data items involved, and with a keen appreciation of the data processing steps to follow. The problems of error correcting and the time necessary to track down and correct errors are critical in reaching this type of decision.

Problems which are found at this stage of the operation would be tracked back to the data generation and acquisition phase. Errors would be corrected as a result of this backtracking. All paperwork, both that remaining with the investigator and that which has been returned to the sponsor, must be corrected so that they remain consistent with one another. If many problems are being uncovered at this stage of the operation, planning may have to be reactivated as there may be a definite deficiency in the study protocol.

VII.   A MAJOR CHECKPOINT

An alternative to checking each step in the drug-development process is to audit the data at one or more key steps. An audit reviews all prior data quality assurance steps or, if none have been done, all prior operational steps. One appropriate time for an audit is after the data have been collected but before the data have been entered into the data base (Figure 18.2).

If there have been prior manual processing steps, then an audit at this time would provide reassurance that those steps were properly carried out and that data which were to be entered into the computer were essentially correct

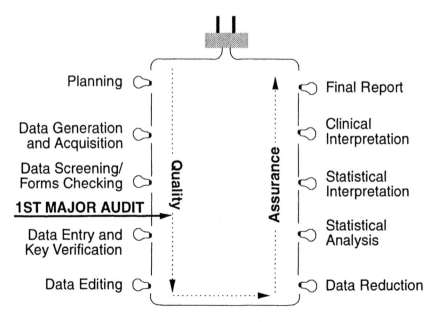

Figure 18.2 Data quality assurance with a major audit.

as well as meaningful. If previous data quality assurance steps have been instituted, then the audit might consist of a sampling of the data on the CRFs with appropriate checks for consistency with other variables. If no prior data quality assurance steps have been performed, then at this point, elements of all previously defined techniques would be brought to bear in order to provide some verification or comfort in the quality of the data which were generated and collected.

This is a logical place for such an audit, for if errors are found later in the process, not only must the computerized data files, or data base, be corrected, but so must the paperwork and the investigator's files as well as any intermediate or holding files involved in entering the data into the computer. Such extensive correcting is essential so that there is a consistent trail of data from the investigator's office to the computerized data base. We would like to avoid such detailed correction procedures by correcting data at the earliest possible point in the process.

An audit at this point (or at any point in the process) is expensive in both time and manpower if it is to be performed well. However, it is well worth the cost if it avoids major problems from showing up later. Such problems create long and intricate chains of backtracking that are even more costly. Thus, an audit should be built into the data processing schedule of events with appropriate time and manpower allocated for the task.

The auditors should have the same characteristics as the data screeners or verifiers. They should perform their tasks in exquisite detail, tracking down all obvious errors or, indeed, any suspicious data values. If this audit follows previous data quality assurance steps, then the auditors should be independent of the people who performed the earlier steps. They may work in the same department of the organization, but they should be different personnel whose main task is to perform such audits and to verify the quality of the work of the group. If this function is seen as an internal quality assurance step whose purpose is to avoid problems later on rather than as a policeman type of function, then the audit will have beneficial results in terms of feedback to the others in the group and result in improved performance in the long run.

VIII. DATA ENTRY AND KEY VERIFICATION

Data entry refers to entry of the data into a digital computer for purposes of data editing, data verification, data reduction, and statistical analysis. Data entry may be from "machine to machine" or from "human to machine"; the former will be referred to as machine data entry and the latter as manual data entry.

Machine data entry takes place when the data are generated by one machine (e.g., from a CHEM 24 analysis or a Holter recording) onto media (i.e., diskettes or magnetic tape) which are readable by a computer or are recorded directly onto an internal computer disk. Some form of redundancy is essential in order that lost or garbled data values may be recovered. This may be accomplished by having a printout from the original machine or by some other means. Auditing is important to make sure that complete values are not lost. This can usually be done by carrying an item count or by carrying some type of check total which may be the sum of a series of values or the sum of the bits used in the computer representation (Mayrend, 1978). At any rate, it is important to know that all values were transmitted or, if not, to identify and track down the missing values. The production of a hard copy by both the generating machine and by the receiving computer will allow for proofreading or sampling of the data to ensure total and correct transmission. Recovery of lost data may be accomplished by the redundancy means described above or, if the missing value is discovered quickly enough, by drawing another blood sample or remeasuring the subject or by similar measures.

We usually try to avoid transcribing data from a CRF to a coding form prior to computer entry as this is a step which leads to errors. This is no longer necessary as most data entry and data base software allow for the set-up of data entry screens which mimic the CRF. Data entry is facilitated by indicating exactly how many characters the field contains, by performing list and range checking at entry (point of entry checking), by using logic to determine which is to be the next question to be answered based on the answer to the previous question, and by placing the entry codes on the screen for ready reference by the data enterer.

Manual data entry is usually accomplished by direct entry into a computer via key-to-disk and is usually followed by a key verification operation if the data entry software provides such an option. If this option is lacking, the data can be independently re-entered and the resulting two data files compared line-by-line or variable by variable using an appropriate computer program. The second data entry, for either verification or for file comparison, must be performed by a different person than did the original entry.

In a study involving remote data entry with quality assurance done by the data coordinating center, double entry or verification may be done for a subset of variables considered of primary importance, or for sites not meeting quality standards (Bagniewska et al., 1986).

Optical scanning may represent a cross between machine and manual data entry. If handwritten forms are optically scanned, then the optical scanning and interpreting process should be subjected to quality assurance checking.

Once the data are in the computer, a hard copy of the file may be produced and a 100% proofreading of the hard copy against the original data records may take place. However, we have earlier pointed out the dangers and mistaken assumptions of a supposedly 100% proofreading scheme. Such a scheme may only be practical for a small file. For large data files, a sampling scheme should be utilized whereby the most important variables are sampled and proofread at a higher frequency than the less important variables. Uncovering of an error or a series of errors in such a sampling scheme would lead to a proofreading of a larger segment of the data. Character fields, where the enterer may have used judgement or discretion in what was entered, may be listed and proofread against the original forms by two people.

Perhaps one of the most important steps in the data entry operation is the creation of a high degree of morale and understanding among the data entry personnel. If they understand something about the application and the importance of the data, this understanding can lead to high morale and to a higher degree of accuracy than from an operation about which they have no knowledge. Errors found in the data entry step are usually corrected internally within this step.

IX. DATA EDITING

Data editing has traditionally been carried out after the data have been entered into computer files or into a data base. Usually the data are passed through various computer programs to perform such editing. However, modern data entry systems permit the data entry and data-editing steps to be combined, to a degree. Normally specifications are programmed to edit and/or scrutinize the data as they are entered and data which fail the editing test may or may not be allowed to be entered into the file at all.

Editing procedures should be fully and clearly documented. They should prevent modifying data in a manner that would not be taken into account in the statistical analysis (Marinez et al., 1984). There are at least four types of data editing which may take place:

1.   Range checking
2.   List checking
3.   Logical relationships
4.   Data display

Perhaps the most common type is range checking. In range checking, numerical variables are compared against reasonable ranges and data points which fall outside these ranges are rejected or flagged. For instance, an age should probably lie between 0 and 90 years. Values beyond 90 years may be accurate but should certainly be checked, as they are unusual. Laboratory values from blood chemistries, hematology, or CBC can be checked against reasonable ranges and values that fall above or below the ranges can be flagged as being high or low and further checked.

It is important when selecting or developing data entry/editing software to decide on which philosophy you want to implement regarding values falling outside the entry range. On the one hand, the entry software can allow the range to be exceeded by "forcing" the entry of an out-of-range value. The software should keep a log of all such forced values so that a report can be periodically generated and these values can be double checked. I personally prefer this approach to the one described below.

Alternatively, one may be stopped at an out-of-range value and not allowed to proceed until an in-range value is entered. If the value on the case report form is truly out of range, then we would have to leave the field blank in order to proceed with data entry. The out-of-range value would then have to be pursued separately and entered later. This step requires manual notation by the entry person of a value not entered and follow-up by others later. A report of all blank values can be generated for later checking and follow-up; however this would contain both legitimate blanks as well as blanks entered in place of out-of-range values.

The capability to designate different types of missing values is helpful in allowing one to generate reports for each type in order to keep track of follow-up which must be done. Types of missing values might include: a) missing because the field is blank, b) missing and efforts to identify the value and enter it will be pursued, c) missing and efforts to identify the value and enter it will not be pursued, and d) missing because the value was out of range and must be checked and corrected or confirmed.

The second type of edit check is list checking. This usually follows from the results of range checking. Once missing, out-of-range, or suspicious values are found, the next step is to identify and list the cases containing these values and the values of any related variables. Then the case report forms can be pulled for these cases and the problem resolved.

The third type of edit check is logical relationships. If one variable has a certain value, then an additional variable or variables may be expected to have a particular subset of values (including blanks). Such relationships can be checked by computer and if the expected relationship does not hold, then both variables can be printed out for further examination. Differential blood count values should be summed and printed out if the sum does not equal 100 or is not within a reasonable range (e.g., 98–102). The use of check digits on key fields such as ID number or specimen number can be considered a type of logical check (Hosking and Rochon, 1982).

Edit checks should always be performed to search for protocol violations. All inclusion and exclusion criteria should be checked. Visit date intervals and dosing schedules should be checked for compliance. Other context sensitive tests can be defined and are limited only by the protocol and the imagination of the data analyst.

A fourth type of editing is data display. Data may be displayed in specific arrays such that unusual values would stand out. For example, each variable may be displayed over time for each patient, animal, or experimental unit. Trends can then be quickly examined by eye and unusual values (outliers) quickly picked out. Another such listing might consist of a group of variables at one time point for each experimental unit. If several variables have a relationship to one another or measure different aspects of the same organ or body system, then these variables can be examined in the context of their relationship with one another and outliers and unusual values quickly spotted.

I should mention at this time that there are certain individuals who have a knack of looking at data displays and instantly spotting unusual or suspicious values. For these individuals, the unusual value literally "leaps off the page at them." Try to identify these individuals in your organization and use their talents accordingly.

The software or programming requirements for the above methods vary considerably. For range checking, list checking, and data displays, it is relatively simple and straightforward to produce general computer programs which are driven by a small set of specifications. For logical relationships and protocol violations, each application must be customized, and this requires either a new program for each application (which is very expensive) or a generalized data base package or management information system with a "macro-instruction" query language. Many such packages exist and are quite widely used throughout the industry.

Even though each type of clinical trial and each type of drug is different, there are many aspects of each trial that are identical or fairly common across the board. These aspects lend themselves to standardization of variable names, range checks, logical relationship checks, and data displays. These might include some or all of the following:

Admission information
Physical exam
Audiology, ECG, and ophthalmology screening
Hematology, CBC and blood chemistry
Vital signs
Adverse effect reporting
Death report
End of study form

All data corrections from this step onward must be captured in an audit trail. In addition to regulations requiring this, additional reasons for an audit trail are documentation of changes, management or control of changes, and ability to recover the data as of any point in time. The audit record should include, as a minimum, patient ID, form changed, visit number, field changed, old value, new value, why changed, by whom, and date changed.

Errors detected during the data-editing step usually force us to backtrack at least as far back as the data entry step, often to the data verification, and possibly even to the data acquisition step in order to correct such problems.

X. DATA REDUCTION

The data reduction step can be considered to be the last step in data input as well as the first step in the production of meaningful data output. Large volumes of data can be reduced to graphs or to simple, meaningful tables supplemented by appropriate summary statistics.

Examination of tables (such as described in Section IX) and their appropriate summary statistics is one way to identify unusual or outlying values. Looking at a series of standard deviations of measurements taken over time is a quick way to discover potential problems, as an abrupt change in standard deviation (either upward or downward) may indicate a data problem. The distribution of missing values for individual variables and for related groups of variables can provide important information about data quality (Stellman, 1989). During this step, one should ask the following three questions:

1. Do the results make sense?
2. Are the results as expected?
3. Are outliers suspected?

Both the statistician and the client should examine the output from this step in order to evaluate the data according to the above three questions. Basically, these questions ask us to "know our data". Any suspicious or aberrant results would be tracked down, usually to the data entry step and occasionally to earlier steps.

XI. ANOTHER MAJOR CHECKPOINT

The previous step completes all data input activities and is the last step before statistical analysis and interpretation of results. Hence, this might be the place for a second major audit to ensure the accuracy of the data entered into the computer (Figure 18.3). Characteristics of this audit would be similar to those described in Section VII and would basically encompass one or a combination of steps described in Sections VIII–X. A major audit at this point can be used in place of the three different sets of data quality assurance procedures utilized in data entry through data reduction or may be used in a "sampling" sense to supplement these previous steps.

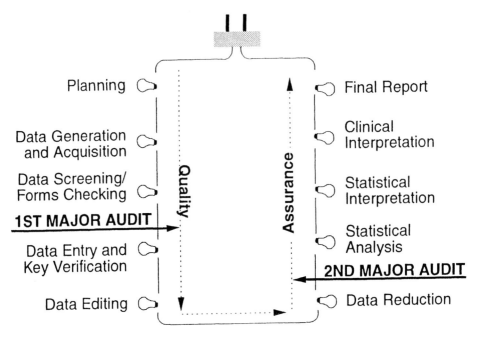

Figure 18.3 Data quality assurance with two major audits.

XII. STATISTICAL ANALYSIS

Data analysis should be carried out according to the methodology as spelled out in the planning step and in the study protocol. However, experiments frequently are not carried out exactly as specified in the protocol; therefore, the statistical methodology will usually have to be modified to accommodate protocol modifications and/or the loss of experimental units. Modifications to the protocol occur in a formal manner, involving input as well as reviews and approvals by appropriate staff and management. Statisticians and their management must be involved in the protocol modification and approval process as well as in the original protocol development and approval process.

A modification which is made in writing with appropriate review and sign-off by all involved parties, including the statistician, allows the statistician to prepare for the implications of such modifications. Modifications which the statistician finds out about at the data analysis step—often when the analysis either "blows up" or leads to ridiculous results—usually lead to a loss in time and productivity, just when the study is coming to a conclusion and when time is at a premium. The latter type of modification usually occurs because the experimenter did not realize that there was a statistical implication to the modifications. This can be avoided in several ways. One is to have a thoroughly informed statistician on the project team, who (hopefully) would be involved automatically in any protocol modification. Another is to have a process which requires statistical input and sign-off of any protocol modification.

Statistical analyses are only as good as the computer programs on which they are performed. It is imperative to have validated, tested, and documented computer programs or algorithms, regardless of whether the calculations are carried out on a personal computer, minicomputer, programmable calculator, calculator, or by hand.

Validity of analyses depend on satisfying the assumptions that are required by the analysis method. Unfortunately, analysis software does not usually automatically check assumptions about the distribution of the data. A statistician must be involved in this important aspect of analysis.

The results of analyses must be examined in order to see if they are reasonable and expected. This is another data quality assurance step as unusual results should lead to the same kind of checking and backtracking as have previous data quality assurance steps. Unusual results at this point will usually be tracked back to the data reduction step and, if necessary, to prior steps.

XIII. STATISTICAL INTERPRETATION OF RESULTS

An intelligent statistical interpretation of results requires an understanding of the application or project by the statistician. The statistician must ask whether the results make sense and whether they are consistent in the context of the application. Graphs are very helpful in interpreting results and evaluating trends. Interpretation should be couched in the language of the client, as the client is the ultimate consumer of the report and must be able to understand it.

Interpretation of results of analyses based on transformations of the data is a potential problem. Conclusions based on hypothesis testing of transformed data are easily carried back to the untransformed data. However, point estimates and confidence intervals are not easily "untransformed" and symmetry is usually lost. Therefore, these results must be carefully expressed in the original scale in order to facilitate interpretation by the client.

All tables and graphs which are manually produced, or have been edited electronically, must be carefully proofread. Nothing is as embarrassing as a graph with points which are incorrectly plotted. Wherever possible, direct computer output should be utilized and typing of numbers should be minimized. If the computer output cannot be annotated electronically as part of the program which produces the output, then an edit program can be used, or headings and labels can be placed directly onto the hard copy. Avoid typing hundreds of numbers as the proofreading task is almost impossible.

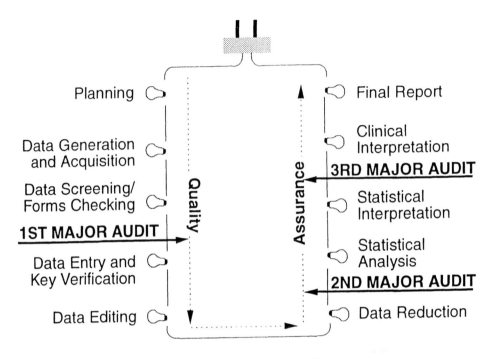

Figure 18.4 Data quality assurance with three major audits.

Errors which are detected due to unusual results will usually force us to backtrack to the data analysis step in order to uncover the problem. This might then force us to further backtrack through earlier steps, possibly back to the data collection step. This is a problem of great magnitude as the data collection may have taken place months earlier and it may now be impossible to track down problems at this point in time. That is the reason for major checkpoints at earlier steps in the process.

XIV. ANOTHER MAJOR CHECKPOINT

At this point, the results are about to leave the statistical department and be turned over to the client for use in the preparation of a final study report and, in many cases, for submission to a regulatory agency as a statistical appendix to the main report. An audit is called for at this point in order to be sure that the report or information which is released from the statistical department to the client is statistically sound and defensible (Figure 18.4).

The audit at this step would involve auditing statistical methodology, results, and interpretations. This audit should be carried out by statisticians in the department who were not involved with the report being audited. These statisticians should examine assumptions of the analysis and appropriateness of the methodology used. They should also examine the nature of data points which were excluded from the analysis and the reasons supporting such exclusions. They should examine the presentation of results for completeness, accuracy, and fairness.

In many cases, this step is included as a statistical department peer review and is not considered a major audit. No matter what it is called, it is most important that this step take place before further steps are taken with the statistical interpretation of results.

XV. THE CLIENT'S INTERPRETATION OF RESULTS

The client should review and interpret the results jointly with the statistician, asking (as the statistician did) whether the results make sense and whether they are consistent. The client should also look at the variables in meaningful groups in the context of the application and should utilize graphs and all other means to review and assimilate the results and determine whether they make sense.

At this point, if not previously done, statistical significance must be reconciled with biological or clinical significance. Discrepancies between the two usually point to deficiencies in the planning step in that either too many or too few experimental units (e.g., patients) were studied or the study had multiple end points and the sample size could not be accommodated for all of them. If too many patients were studied, a statistically significant result may be clinically unimportant. On the other hand, if too few patients were used, a result may appear to be clinically significant but may not be statistically significant. In this instance, the client may not claim significance but may describe an interesting finding or trend which needs further exploration in another study.

All unexpected results should be questioned by the client. Unexpected results may occur if the analysis is not appropriate or if there was some earlier misunderstanding as to what the data meant, what types of hypotheses were to be tested, or which parameters were to be estimated. Unexpected results may come about because of changes made during the experiment of which the statistician is unaware and which did not become obvious during the data analysis step. Such changes may render the analyses inappropriate. Re-analysis may require transformations of the data or may lead to different methods of analysis altogether. Misunderstandings can be clarified by restating the objectives of the experiment in clearer language. Informal changes must be formalized so that the statistical analysis is based on formal documents rather than verbal discussions of changes made during the experiment. As mentioned earlier, statisticians should be an integral part of the protocol amendment process.

All unexpected results must be resolved to the mutual satisfaction of the statistician and the client. Unresolved problems may lead back to an examination of the statistical interpretation or to a re-examination of the statistical analysis. In some cases, backtracking may carry to earlier steps in order to reconcile such problems.

XVI. FINAL SCIENTIFIC REPORT

The final scientific report must contain certain statistical elements. Ideally, these would be written by the statistician but in some cases they may be extracted from the statistical report and written by the client. In either case, a review and sign-off by the statistical department is an important means to validate or assure the integrity of the statistical aspects of the final report.

At a minimum, the final report should contain a valid description of the statistical methodology, including a statement describing the design of the experiment, the statistical analyses themselves, and the statistical interpretation of the results. Details of the analyses may be placed in an appendix to the report or in the company archives. However, summaries of the results should appear in the body of the report in all cases.

The report must be consistent with respect to client and statistical interpretation. The report should be carefully proofread by the statistician, by responsible members of the statistical department, and/or by a member of a Quality Assurance Unit, with respect to the following:

1. Any tables which are typed.
2. Any graphs which are produced manually or are electronic manipulations of the original electronically produced graph.
3. All statements containing numbers must be checked against the source tables or graphs.
4. All conclusions must be checked against the analyses which support them.
5. A sample of all raw data tables should be proofread against the original documents (e.g., CRFs).

Problems discovered at the last minute regarding missing data, additional data, or errors must be carefully backtracked and corrected. For these small elements, it may be necessary to go all the way back to the data-screening step, re-enter the data, re-edit it, rereduce it, reanalyze it along with all the rest of the data, re-interpret the results, both statistically and from the client's viewpoint, and recheck the final report. Errors or problems at this late stage are very expensive and can severely impact the issuance of the report. The adages that "haste makes waste", or, "there's no time to do it right, but there is always time to do it over" certainly apply at this stage of the game.

XVII. FINAL MAJOR CHECKPOINT

The final major checkpoint is an audit performed by the Compliance Unit or Quality Assurance Unit of the company (Figure 18.5). This audit would utilize aspects of all the elements described previously and would perform additional checks. They would see that all standard operating procedures were followed. They would sample and check data listings against CRFs, paying particular attention to corrections made to the forms and checking that these corrections were made to the data base and all subsequent analyses. They would check randomization treatment assignment in the data base against the randomization scheme for the study. They might check all adverse effects, deaths, and early terminations for completeness and accuracy of the data in the data base.

The findings of such an audit should be provided to all parties involved in the study. Each finding or problem should require from the group responsible

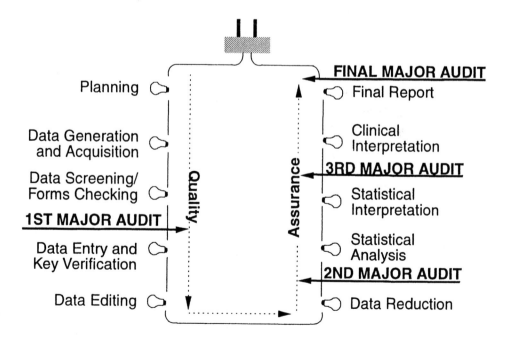

Figure 18.5 Data quality assurance for the entire study.

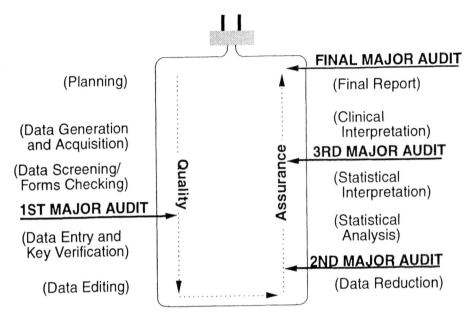

Figure 18.6 Data quality assurance for a small study.

a solution that is specific for the instance at hand, a solution that is general and is designed to prevent the problem in the future, the person responsible for the solutions, and a date by which the solutions will be implemented. They would then follow-up on all solutions for development and implementation.

XVIII.   A SMALL STUDY

For a small study, the amount of quality assurance and the number of checkpoints must be scaled to the size and scope of the study. I would recommend that quality assurance consist only of major audits at four selected points (Figure 18.6):

1.   Prior to data entry
2.   Prior to data analysis
3.   Prior to client interpretation and
4.   A proofreading of the final report combined with the final audit by the compliance unit

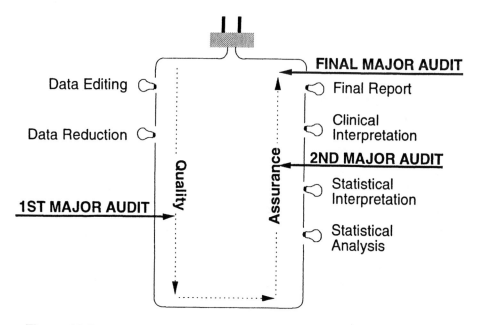

Figure 18.7 Data quality assurance after the study has been completed and the data entered.

XIX. AFTER THE STUDY IS COMPLETED

Occasionally, statisticians become involved only after the study has been completed and the data entered. In this case, the quality assurance steps we take in data editing and data reduction are especially vigorous, as they must find the accumulation of all errors which may have occurred in previous steps. The process would then look like Figure 18.7.

XX. SUMMARY

Careful documentation of the data quality assurance steps both as general procedures and in their application to a particular study constitutes a well-defined audit trail for that study. It is essential that such an audit trail be created and retained for future reference.

This chapter has tried to make the following points:

1. Quality must be consciously built into the scientific process. It does not happen by chance.
2. Identify and correct problems at the earliest practical step.

3. Problems in later steps may require successive backtracking through one or all previous steps.
4. Audits at major checkpoints can reduce the range of backtracking.
5. Do not compromise quality under any circumstance.

Since one research study usually leads to another, it is most important when a project is completed to hold a "postmortem" and feed forward to the next project. In this way, we can continually learn from our experiences and both refine our data quality assurance steps and improve the data quality of each successive project.

REFERENCES

Bagniewska, A., Black, D., Molvig, K., Fox, C., Ireland, C., Smith, J., and Hulley, S., (1986). Data quality in a distributed data processing system: The SHEP pilot study. *Controlled Clin. Trials*, 7:27-37.

Bailar, B., (1983). Error profiles: uses and abuses. In *Statistical Methods and the Improvement of Data Quality*, (Wright, T., Ed.), Academic Press, New York, pp. 117-130.

Hosking J. D. and Rochon, J., (1982). A comparison of techniques for detecting and preventing key-field errors. *Proc. of Stat. Comp. Sec. of the Amer. Stat. Assoc.*, American Statistical Association, Alexandria, Virginia. pp. 82-88.

Liepins, G., (1983). Can automatic data editing be justified? One person's opinion. In *Statistical Methods and the Improvement of Data Quality*, (Wright, T., Ed.), Academic Press, New York, pp. 205-213.

Marinez, Y. N., McMahan, C. A., Barnwell, G. M., and Wigodsky, H. S., (1984). Ensuring data quality in medical research through an integrated data management system. *Stat. in Med.*, 3:101-11.

Mayrend, G. R., (1978). Guide to EDP control. *J. Syst. Manag.*, 29:16-17.

Naus, J. I. (1975). *Data Quality Control and Editing*. Marcel Dekker, New York.

Stellman, S. D. (1989). The case of the missing eights. An object lesson in data quality assurance. *Amer. J. Epidemiol.*, 129:857-860.

Sudman, S. (1983). Response effects to behavior and attitude questions. In *Statistical Methods and the Improvement of Data Quality*, (Wright, T., Ed.), Academic Press, New York, pp. 85-115.

19

Managing CANDA Submissions in the 1990s

William S. Cash and Vita A. Cassese

*Pfizer Pharmaceuticals,
New York, New York*

I. INTRODUCTION

The Food and Drug Administration (FDA) reviews dozens of New Drug Applications (NDAs) every year. An average hard copy NDA consists of hundreds of volumes filled with thousands of pages of data, information, graphs, tables, and statistics. The increasing complexity of NDA submissions and the resultant complexity of their review is a concern of both the FDA and the pharmaceutical industry. With the advent of sophisticated computer technology, both pharmaceutical companies and the FDA have actively sought the use of computer technology to assist NDA reviews. Computer-Assisted New Drug Applications (CANDAs) are meant to improve the quality and the efficiency of the reviews, as well as to speed and simplify the FDA review process.

II. CANDA SUBMISSIONS TODAY

A. What Is a CANDA?

In its most basic form, a CANDA is the submission of NDA information in digitalized and/or image form. CANDAs consist of the same data or information available in the hard copy NDA submission. At present, CANDAs are meant to compliment the conventional hard-copy NDA submission.

The overall objectives of the Pharmaceutical industry and FDA for CANDA submissions is to increase the efficiency, quality, and speed of the NDA review process. Toward achieving these goals, CANDA submissions are designed to make it as easy as possible for the FDA reviewers to retrieve information and to do so without altering or interfering with the thought process used to review conventional hard-copy NDA submissions. Effective CANDAs, like conventional NDAs, provide the reviewer with the means to answer the questions generated by a review. The chief advantage of CANDAs is that reviewers can retrieve information and manipulate data included in the database without having to lose time leafing through printed tables and manually performing difficult calculations, or without the time-delaying process of petitioning the sponsor for an answer. Easy access to a computer database allows the reviewers to answer important questions relevant to the review for themselves as they arise. Contacting the sponsor with further questions is facilitated by electronic mail.

B. The Growing Trend of CANDA Submissions

The first company to submit a CANDA to the FDA was Abbott Laboratories, who worked with a third party vendor to develop an experimental electronic NDA in 1985. The initial results of the first tests were positive. The FDA reviewers were encouraged by the power and flexibility of the electronic data base, and by the ability to retrieve and tabulate data. However, the experiment pointed out the need for menu-driven CANDA submissions to allow the reviewers to answer basic questions and perform ad hoc queries of the database without having to learn a programming language. In addition, it was apparent that in order to ensure optimum use of the CANDA there should be a close interaction between the primary reviewer and the sponsor.

By mid-1987, the FDA had received 10 CANDA submissions for review by the FDA; by the summer of 1988, the number of CANDAs submitted for medical review reached 24, and by the end of 1991, that number increased to 65, and has continued to grow since. The growth of CANDA submissions seen in the late 1980s is predicted to continue well into the 1990s. By the beginning of the 21st century, it would not be surprising to see all of the NDA submissions accompanied by some form of CANDA submission.

While CANDAs involving clinical efficacy studies have been the most popular so far, the complete NDA submission can cover other topics ranging from chemistry and pharmacology to packaging and manufacturing. Any of these topics can be put in the form of a CANDA submission to the FDA.

C. Industry and FDA Support for CANDA Submissions

There is a clear commitment by both industry and the FDA to use and improve CANDAs. In 1988, at the first public conference to discuss the progress of

CANDAs, sponsored by the Pharmaceutical Manufacturers Association (PMA) and the FDA, the FDA strongly supported the use of the new technology and encouraged the pharmaceutical companies to actively pursue CANDA development to help shorten the review cycle of FDA submissions. This theme was stressed by former FDA Commissioner, Dr. Frank E. Young, in his keynote address when he stated, "We [the FDA] will work vigorously in developing this approach for not just the 21st century but the 1990s . . . resulting in complete and full automation of the new drug application and the clinical studies that lie behind it." In addition to emphasizing the time-saving advantages of CANDA submissions, the conference noted that economic considerations play an important role in promoting their use. Dr. Carl C. Peck, Director of the FDA's Center for Drug Evaluation and Research, emphasized this when he stated, "I think in the future CANDAs will be essential—we're going to be driven by technology and by economics. Developing drugs is becoming more and more expensive. One way of cutting down on expense would be faster review." Dr. David Kessler, Commissioner of the FDA, has stated, "The agency's goal is to have all [NDA] applications submitted electronically by 1995." This statement shows the overwhelming support that the FDA has towards CANDA technology and development in the 1990s.

D.   Computer-Assisted Submissions in Other Industries

The FDA is not the only government agency to seek computerized submissions from industry. The Securities and Exchange Commission (SEC), which fields millions of pages of reports and applications from financial institutions every year, is installing a paperless application system involving digital codes sent over telephone lines. Once stored in the SEC database, information can be retrieved and handled at will by certified SEC employees. The Internal Revenue Service and Social Security Administration are developing similar information gathering systems.

III.   MANAGING CANDA SUBMISSIONS: THREE PHASES

A.   Phase One: Planning a CANDA Submission

1.   Setting Realistic Goals and Objectives

It is important when considering and planning a CANDA submission that senior management within a company have clear expectations and objectives. The primary goals of CANDA submissions are to improve the quality and the efficiency of the review, and to reduce the time to approval. While the benefits are clear enough, CANDAs are expensive, time-consuming, and labor intensive. Management's expectations of a rapid drug approval following a CANDA submission must be tempered with a realistic assessment of how the FDA operates. Since the review process at the FDA goes through a series of steps, beginning with the primary reviewer and ending with secondary and tertiary reviewers, there are several opportunities for delays. Even if the CANDA cuts the time a primary reviewer spends with an NDA in half, that does not ensure that the drug will be approved twice as fast. If the priorities of the next round of reviewers do not allow for immediate review, the time saved by the CANDA is diminished. While poorly designed CANDAs have the potential for prolonging the submission process, there is little question that a well-designed CANDA can only help to expedite it.

Well-designed CANDAs also provide additional benefits to the sponsor beyond the primary goals of CANDA submissions. Using a CANDA database to monitor safety data while the data for the studies are actually being collected is particularly useful. Additionally, developing a CANDA helps to facilitate the later preparation of an NDA submission. CANDAs also reduce the amount of time needed by the sponsor to reply to questions from the FDA reviewers.

2. Premeeting with the FDA

Planning for a CANDA relies heavily on past experience and on interactions with the FDA. Given the range of computer expertise among the reviewers, it is always a good idea to meet with the reviewer to discuss which features should be included in a CANDA submission. By premeeting with the FDA, the sponsor's system designers and clinical support team can obtain a basic understanding of how the reviewer intends to use the system, and can assure the reviewer that the system can be tailored, within reason, to the individual needs.

Such a meeting with the FDA is best conducted in one of two ways. With a scaled-down prototype CANDA on hand, the sponsor can learn from immediate feedback which additional features the reviewer might like to see. For experienced sponsors who have already had a successful CANDA submission, a second approach would be to present a more complete CANDA system with the knowledge and confidence that their system provides the broad functionality desired by FDA reviewers. Even in this case, it should also be made clear that the functionality can be expanded during the course of the review process.

Premeeting with the FDA is an especially good idea for pharmaceutical companies who are submitting CANDAs for the first time. When this is the case, background knowledge of the content and design of previous CANDA submissions from other companies is essential. The meeting should be made with the proper FDA reviewing division. If, for example, the submission involves a cardiovascular drug, the primary reviewer in the Division of Cardio-Renal Drug Products would have to agree to a CANDA submission. Often there is an information systems specialist at the FDA who acts as an intermediary between the reviewer and the sponsor. The Director of Information Systems at the FDA can provide basic technical and policy consultation regarding CANDA submissions.

3. Time Requirements

Until a company completes its first CANDA, judging the time and manning requirements for assembling a CANDA and providing technical support for it once it is in the hands of the FDA is by necessity a rough approximation. Clearly, the better the sponsor's internal NDA database is managed, the easier it will be to move the data into a new CANDA database.

The time required to build a CANDA submission is based on its level of functionality, the complexity and size of the NDA, the sophistication of the NDA database, and the amount of data involved. Vendors and technical consultants can help speed the process of building of the database, although they would not be involved with data review. The time also depends on the number of technical and clinical support personnel the sponsor is willing to commit to the CANDA project.

A CANDA submitted to each of the three different divisions of the FDA—namely biopharmaceutical, medical, and statistical—will take more time to prepare. The more reviewers involved, and the amount of time each can devote to working with the sponsor on its CANDA, will also effect both the CANDA's preparation and its review time.

4. Manning Requirements

There are several types of personnel involved in building a CANDA. The core of the CANDA submissions group consists of systems and clinical research personnel. The overall system is designed by information systems experts, while the data are prepared and checked by data analysts, statisticians, and clinical personnel. Because of the significant commitment of time and resources involved, companies who are planning a long term commitment to CANDA submissions should consider forming a separate group in charge of CANDA submissions, the goal of which would be to build and support CANDA submissions prospectively.

5. Understanding the Needs of the FDA Reviewers

The builders of CANDA systems must take into account the wide range of computer know-how and expertise among the FDA reviewers, and provide flexible training and support to accommodate them. Some reviewers will use the system to process data and generate custom reports on subsets of patients, while others may use it to browse through data on case report forms or to retrieve summary text information within the NDA. Some reviewers may be interested in learning a program or query language in order to manipulate the database, while others may not. Because a system will be used by each reviewer in different ways, the perception of how helpful it is may vary widely.

6. Choosing the Proper System to Meet FDA Needs

There are three principle FDA reviewers of the clinical portion of a CANDA submission: the medical reviewer, the statistical reviewer, and the biopharmaceutical reviewer. Each category of review has different information needs and requirements. Because at present the biopharmaceutical reviewers rely primarily on their own system to manipulate NDA data, they currently ask only that the sponsor format NDA data in computer readable form. Medical and statistical reviewers, on the other hand, rely on the functionality provided by the sponsor's CANDA submission to facilitate their review.

7. Integrated Workstation

There are four types of functions which can be provided to a CANDA reviewer: browsing, performing statistical reviews, scanning or looking at an image, and manipulating text. When deciding on the functionality that should be provided in the software, a single system that provides for the four functions in a transparent user interface is an increasingly viable and worthwhile objective.

The four functions can be linked together in the integrated workstation. Integrated workstations provide the reviewer with several advantages. It is important, for example, not to have several different hardware stations to provide for those four functions in a reviewer's office, where space may be limited. A workstation accommodating four functions eliminates the need for the users to learn four different types of systems. Users also find it is useful to share data among the four functions. For instance, integrated workstations allow for data reviewed in the browse function to be passed easily into the statistical package for analysis.

While the FDA wishes to use its own computer hardware and software, their personnel will accept hardware on loan from industry to review a CANDA submission when necessary.

B. Phase Two: Building a CANDA Submission

1. System Requirements

CANDAs are operated by the FDA reviewers either at an independent workstation resident at the FDA or at terminals linked to a mainframe computer located at the pharmaceutical company or a vendor site. As in any contemporary information system, the software selection is of primary importance. Software selection should always reflect the needs of the FDA reviewers. For example, the software chosen for the biometrics division should have a strong statistical component in order to analyze data. The medical reviewers, on the other hand, are less interested in analyzing data, but instead are more focused on reviewing data, and generating descriptive statistics and graphical displays. Once the best software is chosen, choosing the proper hardware to drive the software becomes an important consideration. The first decision that needs to be made is whether to use a central mainframe computer, a multiuser minicomputer, or a stand-alone microcomputer to store the CANDA information.

Storing a CANDA on a company mainframe, or on a minicomputer that is dedicated to operating an inhouse CANDA, poses potential problems. Because remote computers are accessed by telephone lines, communication difficulties may exist. Data security may also pose a problem. When the sponsor uses the company's mainframe computer for CANDA submissions, steps have to be taken to ensure that only the pertinent CANDA data are made available to the reviewers, and that these data remain consistent and unchanged during the review process. Another security issue is to ensure that the transmission of the data is secure from unauthorized access.

While these problems can be eliminated if the CANDA is developed on a stand-alone personal computer, data storage and memory storage capacity of the personal computer are potential problems. Storage capacity should be taken into account when designing the CANDA to ensure that there is adequate response time and adequate storage capabilities.

Given the limitations of statistical software, a system has to be written that can provide fast enough responses and still be rich enough to allow for customized data reports and specific analyses based on ad hoc parameters. If reviewers are forced to wait protracted amounts of time for a response, they may choose to stop using the CANDA system entirely. For this reason, response time is critical.

2. Linking a CANDA to the Pharmaceutical Company Database

Generally speaking, information from the efficacy studies of a particular drug are assembled in a pharmaceutical company database (known as the NDA database) in computerized form. The NDA database is used by the company's own statisticians and reviewers to analyze and review the drug, and will eventually be used to prepare the data and tables in the CANDA submission.

If the pharmaceutical sponsor is providing the FDA with a stand-alone CANDA system, the data from the internal company database must be transferred to the CANDA database for the FDA. Unless underlying data structures of the company database management software and the CANDA database management software are the same, translating and transferring the data from one system to another can be time-consuming and will require rigorous quality control.

3. Data Preparation and Quality Control

Quality control of data transfer will continue to be a critical factor in building retrospective CANDAs. When CANDAs are assembled retrospectively, data are downloaded from the company database into a CANDA database and the software and data structures are often different. Quality control procedures must assure that the data have been accurately transferred and that analyses and results in the NDA submission can be reproduced from the CANDA database.

In the future, CANDAs will be assembled prospectively. When CANDAs are assembled prospectively, the data gathered in the company's internal NDA database are simultaneously gathered in the CANDA. By obviating the retrospective downloading process and therefore eliminating the need for a full-blown quality control program to check information transfer, prospective CANDAs can be assembled more quickly, efficiently, and inexpensively.

C. Phase Three: Implementing CANDA Submissions

1. Installation at FDA

Filing a CANDA is conditional on the FDA's willingness to accept the technical approach used by the sponsor. In order to accomplish this, the systems group at the FDA must be contacted to make clear which software and hardware is appropriate for the FDA's use.

The FDA's Center for Drug Evaluation and Research (CDER), which is committed to exploring the use of computer technology, has issued guidelines in an official statement with regard to procedures for implementing CANDA. In that statement, first printed in the Federal Register of Sept. 15, 1988, the CDER encourages drug sponsors to contact its Office of Management to identify and discuss potential problems involved with submitting a CANDA. As a follow-up, FDA and PMA have formed a number of CANDA working groups and from this effort a complete set of guidelines devoted to administrative and technical issues of CANDA submission has been completed.

Preliminary discussions with CDER officials are ordinarily conducted with the scientific personnel from the FDA division that will be reviewing the CANDA submission. If a third party is involved with building the CANDA submission, the sponsor should be prepared to identify the third party during this meeting. In addition to explaining the method used to store the data, the sponsor should be prepared to discuss the proposed hardware and software requirements of their submissions, as well as their plans for training the reviewer on their use. Currently, the CANDA submission does not preclude or affect in any way the required submissions of the conventional hard-copy NDA submission.

2. Training and Support of FDA Reviewers

All sponsors need to implement a training program to introduce their CANDA systems to the FDA. An effective training program will ensure that the CANDA is understood by the FDA and that it is indeed a useful tool in the review.

In general, the more menu driven a system is, the less training is needed. Most reviewers will expect formal training to be furnished by the sponsor and will request it if they are not familiar with the system. Training is most effective when spread out over a two to three month period in six or seven sessions of one hour each, although training sessions may be shorter once the reviewer is actively using the system.

Technical support may be given over the phone, by electronic mail, or in person. A formal procedure for fielding technical questions from a reviewer should be arrived at before the CANDA is released. Either the project manager or the Drug Regulatory Affairs division, should decide how calls for technical support will be handled.

3. Supporting FDA Reviewers

Pfizer has found that a team consisting of the CANDA project manager, the NDA project physician and statistician, a systems analyst, and a representative from the Drug Regulatory Affairs department is most effective in helping train and support an FDA reviewer. The CANDA project manager is responsible for overall training and support of the FDA reviewers in the use of a CANDA submission. The systems analyst provides the system support for the CANDA hardware and software. The physician and statistician are required to answer FDA questions on the CANDA data as it relates to the NDA submission.

IV. THE PFIZER PHARMACEUTICALS CANDA SYSTEM

A. Personal Computer Platform

Pfizer Pharmaceuticals' CANDA system is based on a personal computer. The PC platform was selected for several reasons. First, it has the capacity to easily integrate the optical, data review, analysis, and full text search functions in one workstation. The ability to handle data, text, and images is central to the workstation approach. An entire NDA submission encompassing many volumes of paper can be written and accessed on optical disks. Second, stand-alone PCs avoid queuing for access on a remote sponsor mainframe and response time is consistent. By circumventing the need for on-line data communications, the problems inherent with telecommunications are eliminated.

Other advantages of a PC-based system include the easy incorporation of familiar software. For computer novices, PCs are often easier to operate and less intimidating than main frames, and can be quickly mastered by beginners. Moreover, the trend for smaller and faster PCs can only mean that even larger CANDAs will be accommodated by more powerful PCs in the future. And finally, personal computer workstations are currently proving to be the most cost effective solution for CANDA submissions.

B. Hardware and Software

One personal computer chosen by Pfizer for its CANDA submission was the IBM PS/2 model 95 with 32 megabytes of random access memory and a 640 megabyte hard disk. An optical disk drive adds additional space for the optical scanning of the NDA. IBM's OS/2, version 2.0, is the operating System and the hard disk is partitioned for both file allocation table (FAT) and high performance file system (HPFS) processing. Presentation Manager is used as the graphical user interface. Most features of the CANDA system are accessed via the use of a mouse. The software packages include Oracle, a database management system with its own query language, and SAS for statistical and data analysis. Although a variety of utility software programs will be made available for PC applications including Lotus 123/G, Lotus Freelance, Word-Perfect, and Verity Topic, the reviewer can use whatever word processing package he or she feels comfortable with or is accustomed to using. Pfizer provides a Hewlett-Packard LaserJet III printer and telecommunications software for electronic mail. The communications software is helpful in downloading information to and from Pfizer to the FDA, and in transmitting messages and technical support.

V. FUNCTIONS OF THE PFIZER SYSTEM

The four main components of the Pfizer CANDA are the clinical review system, the laboratory review system, the biopharmaceutic review system, and the statistical review system. Each is supplemented with on-line help so that reviewers can navigate their way through the system without a manual nearby. The various function keys and pull-down menus are explained at the bottom or top of each screen. The help key also defines and describes what each data field means.

The fifth component, the Optical NDA Review, allows the reviewer access to images of the original case report, as well as images of the more than 40,000 pages in the NDA. The optical portion is easily accessed by placing the optical disk in the optical disk drive and identifying specific information for review. The sixth component provides the ability to search and retrieve the text of the summary sections which have been included with the submission. By including the optical portion of the CANDA with the four basic data components, the result is a truly integrated workstation that allows for data and image to be available on an easy-to-use system.

The overall structure of Pfizer's CANDA system is shown in Figure 19.1, and the main menu displayed when the reviewer first logs onto the system is displayed in Figure 19.2.

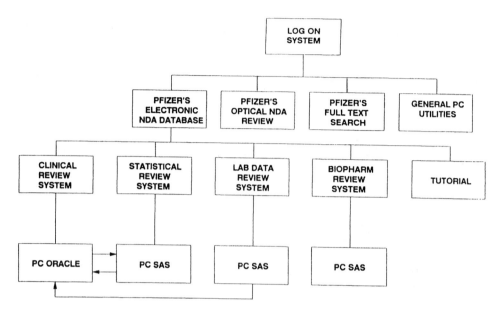

Figure 19.1 Overall structure of Pfizer's CANDA system.

A. The Clinical Review System

The clinical review system gives the reviewer access to all the clinical data in the NDA submission and the means to review and tabulate data, and to generate custom and ad hoc reports. The versatility of the system allows the user to inspect individual patient records, call up subsets of patient records, or retrieve the adverse reactions of a single patient then display the medication record of the same patient. Since all of these steps can be performed following simple commands from a menu on the screen, no programming skills are necessary.

The first step in reviewing case report form data is to select a clinical study from among the many drug efficacy studies documented in the CANDA. Next, a category of information is selected from a menu listing medical history, previous medication, adverse experiences, concurrent illness, plasma levels, and so on. If records from an individual patient or group of patients are specified, the screen displays the relevant information.

For example, if the reviewer requests the adverse experiences for a single patient, the system will show screens of all the case report form data that have been collected for the patient. Repeated use of a function key will display the adverse experiences in the order in which they occurred (Figure 19.3).

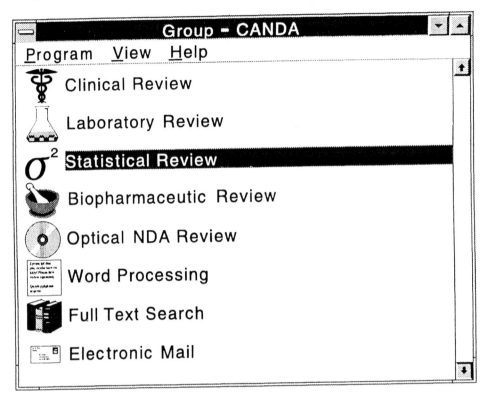

Figure 19.2 CANDA main menu.

The following example illustrates the kind of ad hoc report the FDA reviewer can generate from the system. The reviewer may wish to examine the adverse experiences in females with ages over 60 in a group of patients who received placebo in a clinical trial. The adverse experience screen is selected and the age, sex, and treatment group parameters are specified. Data may be sorted according to any of these parameters. Alternatively, the reviewer may request that all the adverse experiences occurring in females over 60 be sorted into placebo or active drug treatment groups. When the task is completed the screen will list the information and automatically display the totals of the number of patients and adverse experiences in each of the specified categories. The system also has a feature which allows the reviewer to display custom reports in graphic form. Once the custom report is generated it may be viewed, printed, or saved in a separate file for future use.

```
┌─────────────────────────────────────────────────────────────────────────┐
│ │                      ADVERSE EXPERIENCE                │                │
│ ├────────────────────────────────────────────────────┬──────────────────┤
│ │ Study: SAR-1  Study site: 84N0050  Patient No. 12   │ Drug  PLACEBO    │
│ │ Treat grp: PLACEBO   Sex F    Age 43      Race W     │ Dose 0  mg QAM   │
│ │ Status:  COM-A COMPLETED STUDY PER PROTOCOL         │ Tot/day     0 mg │
│ │ Evaluable safety:  Y      Evaluable efficacy:  Y    │ DSSD 1 of 7      │
│ ├────────────────────────────────────────────────────┼──────────────────┤
│ │                                                     │ Onset 10/08/88   │
│ │ Investigator's term:                                │ Stop 10/14/88    │
│ │                                                     │                  │
│ │     WHO included term SLEEPINESS                    │ Ongoing          │
│ │     WHO preferred term SOMNOLENCE                   │                  │
│ │     WHO body system PSYCHIATRIC DISORDERS           ├──────────────────┤
│ │           Cause:  (coded) 1 STUDY DRUG              │                  │
│ │           Cause:  specified                         │ Severity 1       │
│ │           Action taken:  1  NONE                    │ MILD             │
│ │           Outcome:  1   RECOVERED                   │                  │
│ │ Comment:                                            │                  │
```

Investigator's term:

WHO included term SLEEPINESS
WHO preferred term SOMNOLENCE
WHO body system PSYCHIATRIC DISORDERS

Cause: (coded) 1 STUDY DRUG
Cause: specified

Action taken: 1 NONE
Outcome: 1 RECOVERED

Comment:

F1 = Help   Ctrl - F3 = Main Menu   F4 = Select Recs   Dn/Up = Next/Prev Rec   F6 = Rpt/Graph
F2 = Values   F3 = Prev Menu   F5 = View Other Data for this Patient   F7 = Patient Set
└ HELPS ┘   └ EXITS ┘   └---- CRF REVIEW FUNCTIONS ---┘   └─OUTPUT─┘

Figure 19.3 Repeated use of a function key will display the adverse experiences in the order in which they occurred.

B. The Statistical Review System

The statistical review system is written in PC SAS and is completely menu driven (Figure 19.4). It is designed to facilitate the statistical review by providing the following features and options:

1. Easy access to both the original data on all patients and the data behind the tables presented in the statistical reports,
2. An easy method to identify and include data on patients excluded by the sponsor's analysis,
3. A readable source code for the models used to generate the analyses, and
4. The capability to transport data to various PC and mainframe software packages.

These principal features allow the reviewer to modify the status of a patient or to change the eligibility of patients excluded by the sponsor and then rerun customized analyses. With access to the SAS source codes and the original data, the reviewer can examine the statistical models and reproduce the analyses

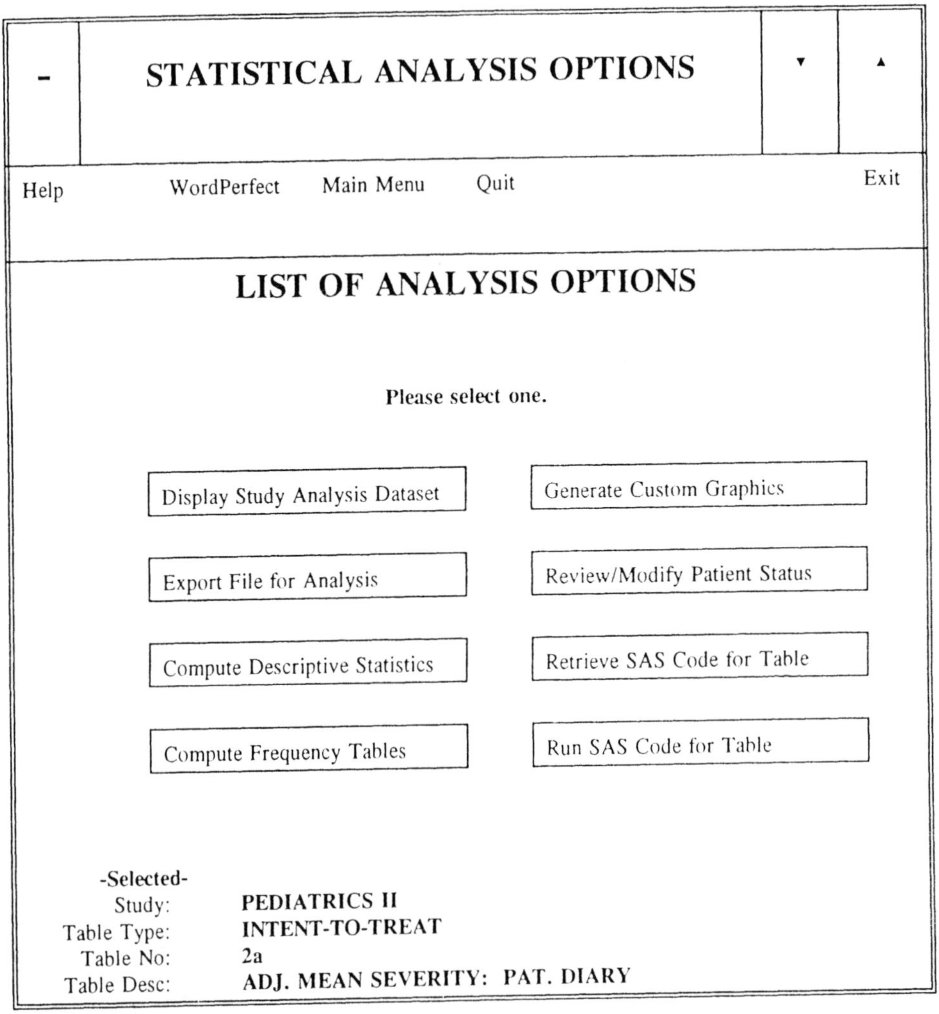

| - | STATISTICAL ANALYSIS OPTIONS | ▾ | ▲ |

Help   WordPerfect   Main Menu   Quit   Exit

LIST OF ANALYSIS OPTIONS

Please select one.

| Display Study Analysis Dataset | | Generate Custom Graphics |

| Export File for Analysis | | Review/Modify Patient Status |

| Compute Descriptive Statistics | | Retrieve SAS Code for Table |

| Compute Frequency Tables | | Run SAS Code for Table |

-Selected-
Study: **PEDIATRICS II**
Table Type: **INTENT-TO-TREAT**
Table No: 2a
Table Desc: **ADJ. MEAN SEVERITY: PAT. DIARY**

Figure 19.4 Statistical review system in PC SAS.

used in the preparation of the NDA at will. The reviewer also has the full capability of SAS for defining her/his own models or exporting the data into other statistical packages for further analysis.

C. The Laboratory Review System

The Pfizer laboratory review system, also developed in SAS and menu-driven, gives graphic displays of the laboratory data generated from clinical trials. It can generate a scatter plot of patients' pretreatment (baseline) versus post-treatment laboratory values (Figure 19.5). The reviewer can select and identify a group of patients from any specific region of the plot and generate descriptive statistics for the selected subset. The menu allows the reviewer to select and stratify patients according to such variables as percent change by age, sex, and weight, or by visit week.

The laboratory review system provides a simple link between clinical and laboratory data. This is particularly useful when a reviewer notices an abnormal laboratory value (e.g., high serum liver function) and wishes to examine all the clinical data for that particular patient. All identification numbers of the abnormal patients can be saved in a separate file which can be used in identifying the patients' case report forms when further review of the data is necessary.

D. The Biopharmaceutic Review System

The Biopharmaceutic Division of FDA has pushed for the adoption of the Field Interchange Specification (FIS) as a minimal standard for data submission to that division. This group has worked closely with the PMA CANDA task force to guide them towards a fully-functional system for data analysis and review. Pfizer has traditionally supplied this division with either ASCII or SAS datasets to facilitate their review.

The Biopharmaceutic Review System is designed primarily for the Biopharmaceutic reviewer to facilitate the pharmacokinetic review of Pfizer NDA submissions. This SAS-based facility supports the statistical and graphical analysis used in the NDA. Additionally, the biopharmaceutic review includes a Boolean Query facility, a Custom Graphic Generator, the ability to perform ad hoc descriptive statistics, a full-screen "cut and paste" of the study synopsis and statistical reports, and data exports to the FIS format.

E. Text Management and Optical Disk Storage

The summary sections of Pfizer's CANDAs, which describe the protocols and the design and results of clinical studies, have a full text search feature which relieves the reviewer of the burden of searching through volumes of hard copy reports and manuals. The reviewer can retrieve summary information from any

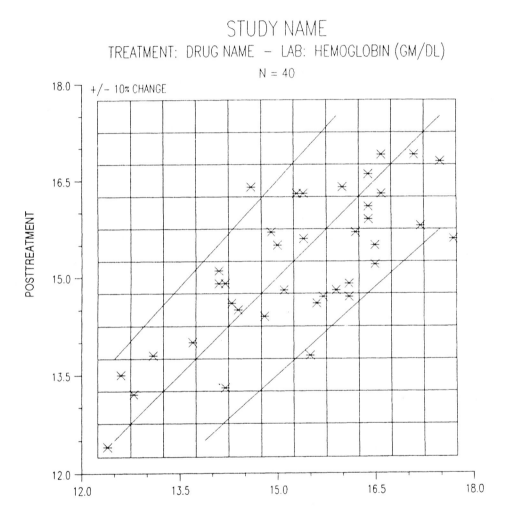

Figure 19.5 Scatter plot of patients' pretreatment (baseline) versus posttreatment laboratory values.

clinical study, handle it using the text management feature of the system, and assemble it in his own summary review document. The global word search feature also allows reviewers to select any word or subject title in a document, and find where it occurs throughout the whole CANDA.

Image storage is another feature which expedites the review process. Images of case report forms stored on an optical disk can be rapidly sorted and examined by a reviewer. Because they have the identical appearance as the original paper case report forms, reviewers can read physicians' notes and comments on unusual side effects which may have been overlooked during data analysis. Additionally the reviewer has access to the image of any page of the many volumes of the NDA submission.

F. Data Security

Because stand-alone systems, by definition, are not connected with other systems, security concerns are minimized. Passwords and other standard software safeguards limit access to authorized users and protect accidental erasure of information. The optical disk contains all the data and text of the full submission as well as all the case report forms. It can serve as a useful back-up against which reviewers may check the integrity of the database after they complete a session of complicated computer manipulations.

G. Electronic Mail

When FDA reviewers need to talk to Pfizer personnel concerning NDA submissions, an effective communications vehicle is electronic mail. The technology for electronic mail has been in use for several years and is an easily accessed feature of the Pfizer CANDA system. A menu option allows FDA reviewers to access electronic mail in order to instantly send messages or questions to Pfizer personnel. Replies can be generated by the sponsor and sent back to the FDA using standard word processing packages.

VI. FUTURE TRENDS IN CANDA SUBMISSIONS

A. Impact of New Technology

As hardware and software technology continue to evolve and good design principles become more widespread, the quality of CANDA systems available to sponsors will also improve. More sophisticated systems with increased speed, data handling, and processing capabilities will be more user-friendly and will require less time to master. The introduction of video imaging will simplify and help speed the review process. Voice-annotated tables and queries are also likely to become available by the end of the decade.

B. Standard Features and Standard Menus

In the past CANDA users have had to learn to operate several different systems. It is likely that the kind of standardization that recently occurred with word processing packages will occur with CANDAs in the 1990s. It is clear that features such as browsing through case report forms, image retrieval, and full text searches will be standardized. In the future, all CANDAs will have standardized systems for case report form review, statistical analysis, and imaging. Once this occurs, workstations offered by different sponsors will be very similar and easily mastered by all users. Because CANDAs will always be custom designed to some extent to accommodate the unique aspects of each drug development program, no two CANDA software packages will ever be identical in terms of menus and function keys. However, each will have a common functionality that will be easy to learn with the help of on-line instructions.

C. Prospective CANDAs

CANDA submissions have largely been built and submitted retrospectively. The clinical and laboratory data are accumulated into the sponsor's NDA and then painstakingly reassembled for the CANDA submission. In the 1990s, the data accumulation and the building of the CANDA will be a single integrated process. As the data are received, it will be added to the CANDA prospectively. When all the data are accumulated, the CANDA will for the most part be ready for FDA submission and review.

Assembling NDA submissions prospectively will help realize the full potential of CANDAs. Prospective CANDA data bases can be used to expedite the assembly of NDA submissions, and to prepare the sponsor for better communication and interaction with the FDA reviewers. Prospective CANDAs will quickly become less expensive than retrospective CANDAs, and less time and effort will be needed to ensure the integrity of the data submitted to the FDA.

D. Expansion Beyond Clinical Data

The use of CANDA submissions in the future will expand to areas beyond clinical data to include chemistry, pharmacology, preclinical data, and even packaging and manufacturing. With expanded submissions, the physical size of the hard-copy NDA submission could be significantly reduced in the future as the CANDA data base becomes an acceptable storage medium for data, text and images.

E. Full Potential of CANDAs

The potential of CANDA technology to enhance the review process is a reality today. As we look forward to the next decade, the technology is expected to expand significantly, thus providing an even greater certainty that CANDA systems will fulfill their potential. While the pharmaceutical industry can be encouraged by the current success of CANDAs, the challenge will be for industry and the FDA to work together over the coming years to maximize the potential of CANDA submissions. Achieving this goal will be to the benefit of all. The sponsors who are organized to produce well designed CANDAs today will be poised to quickly take advantage of this emerging technology, and will be able to build even stronger CANDA submissions in the future.

20

Quality Control

John R. Murphy and Charles B. Sampson

*Eli Lilly and Company,
Indianapolis, Indiana*

I. DRUG PRODUCT QUALITY

A. Introduction

Whenever human health in considered, the quality of health care is of foremost concern. The patient and the family need and expect high-quality health care. For example, one does not like to entertain the thought that a physician's diagnosis may be wrong, that a hospital chart is inaccurate, or that a treatment mix-up may have occurred. The drug obtained from the pharmacy is expected to cure the infection or relieve the pain or have its specified effect. People expect that the bottle has the specified number of tablets and that each tablet contains the specified quantity of the correct drug.

As patients in a sophisticated health care system, we may not be able to judge the quality of health care and consequently must trust that it is good. For example, it may be difficult to ascertain if the physician's diagnosis is correct or if each penicillin tablet is of the same potency. Contrast this to the situation where the door handle on your automobile falls off.

A free-market system could serve as an effective regulatory force on the manufacture of drug products if the patient or the pharmacist could easily determine product quality. However, this is not usually the case, and the pharmaceutical industry has a unique responsibility for the quality of its products.

Quality control in the pharmaceutical industry is mandated by law. Indeed, the organization of the quality control unit, much of its responsibilities, and the way it is to perform its functions are covered by federal regulations. Moreover, the regulations tend to outline a quality control function that emphasizes inspection and defect detection, and pharmaceutical quality control technology, needless to say, has historically been weighted toward this mode of thinking. Yet in spite of the regulatory emphasis on compliance and testing, the concept of total quality management is rapidly gaining a foothold in the industry. In today's pharmaceutical companies, people in all components of the organization are being empowered with increased responsibility and authority for the quality of the products and services delivered to internal and external customers, and it is not left solely to the quality control department to act as "policemen" to catch the defects. In today's environment, quality of products and services are of concern in all parts of the organization, from product design to process development, through distribution, sales, and marketing, rather than in the production component alone. Paradoxically, the regulations governing the industry have had both positive and negative influence in this transformation.

In this chapter, we will examine statistical and quality control issues that are more or less unique to the pharmaceutical industry. Concepts that fall into general statistical quality control are treated in the extensive quality control literature (see for example, Burr, 1976, 1979; Grant and Leavenworth, 1988; or Duncan, 1986). The reader is encouraged to look there for either more general or more detailed exposition of ideas touched upon here.

We will present a brief overview of the types of pharmaceutical dosage forms in use today and we will talk about some of the more important quality attributes. The role and influence of the Food and Drug Administration (FDA) and the *United States Pharmacopoeia* (USP) in assuring quality in drug products will be explored. Some unique problems of measuring quality attributes such as potency will be touched upon. A sampling of regulatory product specifications and testing requirements will be covered. The role of validation and the impact of expiration dating upon product release strategy will be discussed. Finally, some aspects of the application of Total Quality Control in a pharmaceutical setting will be outlined.

B. Drug Product Dosage Forms

The story begins when the pharmaceutical company has a compound which appears to be useful. One of the first questions is how to get this active compound into or onto the human body and to the site at which the action is desired. A surprising array of pharmaceutical dosage forms are available to the formulation scientist. There are aerosols, capsules, creams, elixirs, emulsions, extracts, fluid extracts, gels, implants, infusions, inhalations, injections, irrigations, lotions, lozenges, ointments, ophthalmic preparations, pastes, powders, solutions, spirits, suppositories, suspensions, syrups, tablets, tinctures, and transdermal systems. The 22nd edition of the *United States Pharmacopoeia* (1989) contains complete descriptions of all these pharmaceutical dosage forms. A few will be described here.

The *compressed tablet* is the most widely used dosage form in this country. Preparation of a compressed tablet is normally a large-scale production process beginning with the blending of active ingredients with one or more inactive ingredients (sometimes called excipients). These inactive agents are needed for coloring, disintegration, binding, flavoring, lubrication, and dilution. When the amount of active ingredient required for a dose is small, the active ingredient may be mixed with a diluent such as starch. Binders are added to improve the compression characteristics of powder mixtures or granulations. Disintegration substances are included to help the tablet break up after the patient has ingested it. Coloring agents allow for product identification and for aesthetic appeal. Lubricants aid the flow of material to the tablet press as well as prevent the press from binding up. Once a homogeneous mixture is obtained through adequate mixing, portions of the powder mixture are pressed to form the tablets.

Alternatively, the homogeneous mixture of active ingredients and excipients may be filled into small gelatin containers called *capsules*. Capsules may be either hard or soft and, of course, must be soluble. Hard gelatin capsules are manufactured by blending bone and pork skin gelatins having high gel strength. Soft gelatin capsules usually are filled with liquids or pastes.

Another class of widely used dosage forms are *parenterals*. Parenteral products are solutions which are administered other than through the alimentary canal; for example, they may be injected directly into the muscle or the veins. These types of dosage forms have their own special quality control problems. It is imperative that these solutions for injection are sterile, for even one nonsterile injection can have serious consequences.

Aerosols are designed to spray the medicine into the nose, mouth, or lungs, or onto the skin. They are packaged under pressure so that a precisely measured dose of the active ingredient is delivered when a valve is actuated. Components of an aerosol include the container, the propellant, the concentrate containing the active ingredient, the valve, and the actuator. An aerosol's performance depends on how each of these components performs individually and collectively.

A *syrup* is a colored and flavored aqueous solution of sugar in which a medicinal agent has been dissolved. An *elixir* is like syrup except that it also contains ethanol.

Popular topical dosage forms are creams, ointments, and pastes. A *cream* is a semisolid emulsion of either oil-in-water or water-in-oil type. *Pastes* and *ointments* are semisolid preparations usually containing medicinal substances and are applied externally. Some pastes are thick and stiff and do not usually flow at body temperature, and thus serve as protective coatings over the treated areas.

Transdermal delivery systems, which are applied to the intact skin like band-aids or patches, are a relatively new form of topical dosage form. The active drug diffuses from the drug reservoir and is absorbed through the skin into the circulatory system. The transdermal delivery system provides an example of a drug delivery feature called extended release. Other dosage forms with this feature are tablets and capsules which are specially formulated to release drug to the patient's system over an extended period of time. Each of these dosage forms has a unique quality control problem; the statistical aspects of assuring the quality of some of these dosage forms will be discussed.

C.   Chemical and Biological Considerations in Drug Product Quality

The features of a drug product that are the easiest to understand, measure, and control tend to fall under the classification of chemical or physical attributes. For example, we can measure the amount of active ingredient, the type and amount of excipient, the amount of impurity, the time it takes for a tablet or capsule to dissolve, the rate and duration of drug release from an extended release formulation, to name only a few. Moreover, we can determine the stability characteristics of a drug product throughout the process from manufacturing to patient use.

Consider, however, the quality characteristic of most interest to the patient: "Does the drug work?". Or more specifically, "Will these tablets or capsules or this injection cure the problem?" These issues are addressed through the conduct of large scale experiments called controlled clinical trials, where the efficacy and safety of the drug are carefully studied and documented in a large number of patients. Concurrently, the chemical and physical properties of the drug are determined and specified. Thus, if the manufacturer produces a uniform drug product that is equivalent to that proven to work in clinical trials, then the patient's expectation that the tablets or capsules or injection will work is reasonable.

All in all, this approach seems to work quite well; yet, the drug development process can yield less-than-satisfactory results, because the relationship between the physical/chemical properties of a drug substance and its safety and efficacy in the human body is very complex and only partially understood. Thus, it is possible for two drug products that appear to be chemically equivalent to fail to provide the same therapeutic effect. For example, this can occur when two different processes are used to synthesize the material or to formulate the drug product.

Some factors that affect therapeutic equivalence are beyond the control of the manufacturer and/or the physician, such as a patient's physiological constitution and personal habits. However, one step that can be taken is to ensure that a drug is bioavailable, meaning that the active ingredient is available in the body for transport to the site of therapeutic need. Bioavailability is studied by measuring the blood concentration of the compound over a period of time after the patient has taken the medicine by the prescribed route of administration. In this way, it is possible to determine that a particular formulation is capable of being absorbed into the patient's blood or lymphatic system where the drug can be transported to its intended site of action. (See also Chapter 12).

In summary, although it is not possible for the manufacturer to guarantee the patient that a specific tablet, capsule, or injection will cure the ill, the manufacturer can assure the patient that the dosage unit was produced by a controlled process that yields units satisfying the same chemical and physical specifications as dosage units shown experimentally to be bioavailable and therapeutically effective.

D. The Role of the USP and the FDA in Assuring Quality

1. The USP

For pharmaceutical manufacturers, marketed products are subject to recognized standards of identity, quality, strength, and purity set forth in various compendia around the world, such as the *United States Pharmacopoeia and National Formulary*, the *British Pharmacopoeia*, the *European Pharmacopoeia*, the *Japanese Pharmacopoeia*, and the *International Pharmacopoeia*. In the United States, the first USP was published in 1820 with the stated objective, ". . . to select from among substances which possess medicinal power, those, the utility of which is most fully established and best understood; and to form from them preparations and compositions, in which their powers may be exerted to the greatest advantage. . . ." The science of medicine and pharmacy, as well as the technology of chemical and pharmaceutical manufacturing, have changed dramatically in the intervening years, and so has the scope of the USP. From an initial publication consisting of a 272-page listing of 217 drugs worthy of recognition for guiding the pharmacist and physician in making simple mixtures and preparations, the USP has grown to a continuously revised document, for which the most recent revision (USP XXII – NF XVII, 1989) contains more than 2000 pages covering standards of identity, quality, strength, and purity for over 3000 drugs and substances.

In addition to the standards for drugs and pharmaceutical substances, which are called Official Monographs, the USP also contains sections called General Tests and Assays and General Information. These sections cover such subjects as analytical methods to be employed, how to design and analyze a biological assay, as well as general information on how to clean glassware, the laws and regulations governing drug manufacturing and distribution, stability considerations, validation of compendial methods, and sterility.

The standards published in the USP are recognized as official and binding by law in the United States under a set of regulations called the Food, Drug, and Cosmetic Act. These standards apply at any time in the life of a drug from production to consumption. Manufacturers are expected to develop and utilize release tests and specifications that assure that a drug will comply with compendial standards until its expiration date when stored as directed.

2. The FDA

The FDA is the federal agency which oversees the quality of drugs discovered, developed, and produced for use in the United States. Other countries around the world have similar agencies. It is through the FDA and similar agencies worldwide that the patient and physician have a say in the quality control of pharmaceutical products. Although the FDA has been given statutory mandate for the regulation and supervision of drug product quality, the role of the FDA in today's world can also be regarded as that of partnership with the pharmaceutical industry in providing cost effective, quality drugs. The regulations which guide the FDA and the pharmaceutical industry in the manufacturing process are the Current Good Manufacturing Practices (cGMPs), published in the Federal Register and in the USP. The FDA, through its field offices around the country, enforces the cGMP regulations through plant inspections, audits, and sample analyses.

The careful reader of the cGMPs will readily perceive that the regulations are quite general, provide minimal detail, and leave much to the discretion of the manufacturer. The regulations truly cannot guarantee drug product quality but they do provide a framework within which the FDA and industry work together to achieve that objective.

The FDA also publishes and updates guidelines to interpret, amplify, and clarify the intent of the cGMPs. One such guideline, entitled *Guidelines on General Principles of Process Validation* (FDA, 1987), makes the interpretation that process validation is a regulatory requirement and it provides insight into the Agency's thinking on the subject. What validation is and how it should be carried out have received a great deal of attention and emphasis over the last few years. Process validation and its relationship to pharmaceutical quality control will be discussed in a later section of this chapter.

Although the cGMPs do not specifically address each and every detail of pharmaceutical manufacturing, the FDA has statutory authority to enforce the regulations by means of product recalls, product seizure, plant closings, and civil and criminal penalties. A recent addition to the FDA review process is the strict enforcement of the authority to withhold approval to market a product until the agency is satisfied via a preapproval inspection that the process, equipment, and facilities have been validated and will conform to cGMP requirements.

E. The Measurement Process

1. Assay Methodology

When we think about the quality attributes of a drug product, first and foremost is the requirement that the dosage units have the right potency to be able to produce the desired therapeutic effect without inducing toxicity. The measurement of potency is accomplished via an indirect method called assay. The development and maintenance of assays to assure potency for the spectrum of drug products marketed today comprises a critical and substantial part of quality control in the pharmaceutical industry. Nearly every statistician in the industry is confronted at one time or another with assay issues.

The method of determining potency that relates most directly to what we desire to measure is bioassay, whereby the drug is administered to groups of animals such as rodents, rabbits, dogs, or cats. The strength of the substance is then calibrated by observing a specified indicative physiological response. Most bioassays are comparison assays, meaning that the potency of the unknown preparation is determined by comparison of the physiological response of animals given the unknown with the response of animals administered comparable amounts of a known preparation called the standard. The design and use of bioassay is covered in Chapter 3, and the statistical treatment of bioassay is a fascinating and well-developed subject.

Although there exist some drug products that can only be assayed biologically, for the most part bioassay has given way to assay methods that utilize other types of response systems, such as microorganisms, immunological responses of cultured animal cells, or more commonly, chemical properties of the drug itself. Even though testing the drug in a biological system intuitively seems to be closest to what we really need to measure, the inherent variability of the methodology, together with other disadvantages associated with the use of animals, has led to the practice of substituting other assay technology for bioassay as soon as practicable in the development cycle.

As sophisticated as assay methodology has become today, it is still an indirect measurement method that presents special problems when dealing with reliability, accuracy, and precision. It might appear, for example, from one viewpoint, that the assay can be considered a "black box" and that the quality control person really does not need to be concerned with exactly how this black box determines the amount of an ingredient or impurity in a dosage unit. From this angle, the assay is regarded more in terms of traditional quality control metrology, where the main concern is whether the black box performs in a repeatable and accurate manner. One might wonder, for example, whether different analysts can get comparable results on different instruments at different locations or points in time. On tne other hand, inside the black box, we will often find sophisticated chemical and/or biological processes being carried out, possibly in several steps. The way the components of these processes combine and contribute to assay variability must also be considered. The characterization and documentation of the performance of an assay, both inside and outside the black box, is a matter of interest and importance to the pharmaceutical industry and regulatory agencies. Assay characterization is the study and documentation of the contribution of the various sources of variability that affect the reliability, reproducibility, and precision of the method. Assay validation is demonstrating and documenting that the assay has the required reliability, reproducibility, and accuracy to measure what it is designed to measure.

Given the complexity of the measurement process called assay, it is not surprising that statistical methods have found wide acceptance and use in this field of application. Regression methodology, experimental design, variance components analysis, derivation of confidence, tolerance, or prediction limits, and interlaboratory study design and analysis are but a few of the common statistical tools in use today. Kohberger (1988) provides examples of some of the common statistical methods employed in assay characterization and validation.

2. Illustrative Example

Consider a chemical assay for the detection of an amount of active ingredient in a lyophilized vial. Analytical development has recommended the procedure to be used for assay of the final formulation prior to the release of the lot. The question is: Should quality control have faith in the results produced by this assay procedure? We shall sketch, in a generic manner, some of the procedures involved in performing a typical assay.

The analyst first accurately weighs the vial. Then the contents are dissolved and transferred to an appropriate volumetric flask. The empty dried vial is reweighed and its fill weight determined. An aliquot of the solution from the volumetric flask is transferred to a separatory funnel. The pH of the solution is appropriately adjusted with a buffer and the active drug substance extracted with multiple portions of an organic solvent. The solvent is passed over a drying agent, pooled, and concentrated to an appropriate volume to which an internal standard is added. The mixture is derivatized, if required, and the resulting compound measured by gas chromatography. Quantitation is accomplished using a standard of known purity treated in the same manner. A comparison is made of the active solution to the standard solution and an estimate of the potency is calculated.

It is apparent from this general description of the assay used for these lyophilized vials that analyst hand-to-eye coordination is an important factor in the extraction of the active ingredient. The method is thus vulnerable to an analyst-to-analyst variation. It is also quite possible that the gas chromatograph used for this assay will not be available at all times and that one or two additional instruments might have to be used for completing the assay. There are also a large number of reagents used in the assay. As the assay is used throughout the year, new bottles and new lots of reagents will be continually supplied to the personnel doing the assay. Consequently, there is the possibility that the reagents themselves could cause a variation in the performance of the chemical assay over a period of time. There may be variations from morning-to-afternoon, day-to-day, or week-to-week which are not easily explainable but are quite quantifiable. The question is: If a homogeneous mixture was sampled and assayed in a manner simulating the way in which the assay was going to be used, what variation would be expected due to just the assay alone?

The characterization of the assay described in this example lends itself easily to the traditional design of experiments using factorial arrangements. The factors in this experiment could be analyst, instruments, and time. The time factor could be broken into morning or afternoon, night shift or day shift, or week-to-week or day-to-day variation. Once the appropriate experiments have been designed and completed, the collected data can be subjected to an analysis of variance (ANOVA). The results of the ANOVA would indicate the significant main effects and interactions (if any). For some assays, components of variation can be estimated and "assay schemes" devised to minimize assay variation. For example, to obtain a 2% coefficient of variation, it may be necessary to assay each sample over several days using two or more analysts. Since there will frequently be a large number of batches created in many different drug dosage forms, all requiring the use of the assay, this assay scheme might be considered too cumbersome and too expensive to run. In this case, it may be more cost effective to direct efforts toward developing a more precise assay.

3. The Aberrant Assay Value

Dealing with unusual assay results is often a frustrating matter. There is little concern if the aberrant assay result can be positively traced to physical assay irregularities, (e.g., dropped test tube, power failure). However, if there is no apparent reason for the deviant result, one must decide if the decision-making procedures are best served by leaving the observation in the sample, removing it, or giving it low weight using some form of robust estimation methodology. One philosophical problem with discarding an assay result when there is no assignable reason for it to be in error is that the result may actually be correct although seemingly out of line. For example, if the activity of 96% of a batch of tablets were normally distributed about 200 mg with a standard deviation of 3 mg and if the activity of the other 4% were normally distributed around 220 mg with a standard deviation of 2 mg, then we would expect an occasional result which would appear to be unusually high. A sample of 100 tablets would typically yield a majority of assay results between 191 mg and 209 mg, with from 1 to 8 results ranging up to 214 mg or higher.

If the true potencies of tablets are uniform, with the heterogeneity caused by assay variation, it makes sense to cast out the aberrant assay values. However, if the population of tablet potency is dichotomous as described above and the assay is accurate, then to discard an assay value just because it was greater than 3 or 4 standard deviations from 200 mg would be to misrepresent the quality of the lot. Needless to say, an accurate characterization of the assay, as discussed earlier, would aid in determining whether an assay result should be discarded as aberrant.

There are a multitude of statistical tests for outliers, all based on underlying assumptions, the violation of which could render them useless. The USP (1989, p. 1503) has included some outlier methodology in its description of the design and analysis of biological assays. Beckman and Cook (1983) provide a good overview of the whole subject area of outliers.

The difficulty with all statistical outlier methods is that they judge whether a result is unusual relative to a model which predicts what results are reasonable to expect. As the example above demonstrates, what is entirely unreasonable according to one model might be quite realistic under a different model. The problem is that we are rarely provided with a blueprint of what the true underlying model is. Hence, although statistical methods might at first glance provide an objective method to judge whether a particular value is reasonable or not, we are always left with the nagging issue—"reasonable relative to what?"

Another criticism to the practice of rejecting outliers via statistical tests is that analysts can fall into the trap of not investigating special cause variation in assays. Thus, it is strongly recommended that no outlier test be performed without investigation into the potential causes for an unusual or aberrant result.

II. PHARMACEUTICAL QUALITY CONTROL

A. Introduction

The idea that quality must be designed and not tested into the product has real meaning when the concern is with quality assurance of a drug product. In the pharmaceutical industry, end product testing is usually expensive and most often destructive. In addition, end product testing is effective (and only marginally at that) in merely detecting nonconformances and it may come too late in the process to be of value in finding root causes and in preventing future occurrences. Lastly, the scrap and rework costs at the final product state are particularly high with a finished batch of a drug product. Nevertheless, sampling and acceptance procedures for the final dosage form still consume much attention and many resources in the pharmaceutical industry, because the critical nature of the products dictate rigorous gathering and analysis of quality data. Federal regulations also continue to require these activities. Even so, many in the industry now view the sampling and acceptance aspect of quality assurance more like a necessary and final step to a total quality system that relies primarily on careful product and process design and good process control. Thus, quality control for pharmaceutical manufacturing is a balanced activity that utilizes efficient final testing procedures to document and verify that the control measures have performed as designed.

In this section, we will discuss some aspects of traditional sampling and inspection and will examine some procedures published in the USP. Quality control issues associated with expiration dating and batch release procedures will be outlined briefly, and other quality control concepts unique to the pharmaceutical industry such as process validation and periodic product quality evaluation will be discussed. Finally the emergence of Total Quality Control in the pharmaceutical industry will be touched upon.

B. Acceptance Sampling

Suppose that production has manufactured 1,000,000 aspirin tablets. How should the decision be made that the lot may be released for distribution? The potency of an individual aspirin tablet can only be determined by an assay method which destroys the tablet. Consequently, even if an assay system is incredibly fast and inexpensive, one could not examine all the dosage units in the lot. It is reasonable to assay a representative sample of the 1,000,000 tablets and to combine the information from this sample with the process history to make a decision as to the quality of the lot. This is an example of a traditional acceptance sampling approach to quality control.

There are several important aspects in specifying an acceptance sampling plan. First, the *lot* or *batch* of dosage units must be specified. A lot or batch is defined as that part of a production run having uniform character and quality within specified limits. In the case of a drug product manufactured by a continuous process, it would be a specific amount produced in a manner that assures its having uniform character within specifications. Usually, the prime criterion for batch identification for a continuous production process is a unit of time which provides some guarantee that the batch has uniform character.

Acceptance criteria should be specified that are based on the product specifications and the sampling plan necessary for acceptance or rejection of a lot or batch. Product specifications might include the criteria for identifying an item as conforming or nonconforming. For example, a penicillin tablet might be declared nonconforming with respect to potency if the amount of active ingredient is not within plus or minus 15% of the amount claimed on the label. The *acceptable quality level* (AQL) and the *unacceptable quality level* (UQL) should be stated. The AQL is the highest percentage of nonconforming units that is acceptable as a process average. The UQL (also sometimes called the rejectable quality level, or RQL) is the percentage of nonconforming units for which a low probability of acceptance is desired. In some quality control references, the probability of acceptance at the AQL and UQL is specified to be 95% and 5%, respectively. Each important product quality characteristic should have an AQL or UQL or both.

Unless there is a 100% inspection with a perfect measurement system, any sampling plan can result in an incorrect decision. For example, a batch might be rejected even though the percent nonconforming is less than or equal to the AQL. The risk of rejecting an acceptable batch is appropriately called the producer's risk (or α error). The risk of accepting a batch with a percent nonconforming greater than or equal to the UQL is called the consumer's risk (or β error).

Let π denote the proportion of nonconforming units in a batch. Let $P(\pi)$ be the probability of passing a batch of quality π. The graph of $P(\pi)$ plotted against π is called the *operating characteristic curve* (OC curve). The definitions given, then, can be stated as $P(\text{AQL}) = 1 - \alpha$ and $P(\text{UQL}) = \beta$.

Acceptance sampling plans are classified by the type of dependent variable being measured. *Attribute sampling plans* are useful when the items being inspected are simply classified as conforming or nonconforming. Generally, few probability distribution assumptions are necessary for these plans. If the quality of an inspected item is measured by a continuous variable, such as weight or potency, then the associated inspection plans are called *variables sampling plans*.

A *fixed-sample-size* plan is a sampling plan in which a random sample of n items is chosen for inspection with the sample size n and the α error and/or β error being chosen and fixed before the random sample is drawn.

If the items are chosen one by one and the decision after inspection of each item includes the option to select another item for inspection, then we have a *sequential sampling plan*. Note that the sample size is a random variable while the α error and β error are fixed before sampling begins. A *Group sequential sampling plan* is a sequential sampling plan for which the decision to accept, reject, or continue is made after a group of items have been sampled and inspected.

A *double sampling plan* is one in which a sample of n_1 items is drawn, inspected, and then the decision is made to accept or reject the lot, or sample another n_2 items. The decision to accept or reject the batch after the second sample is based on the $n_1 + n_2$ items. Extensions of this thought process to multiple samples, each being of prechosen size, are called *multiple sampling plans.*

As was indicated earlier in this book, selection of the sample is extremely important. The method of selection will determine if proper inferences can be made from the sample to the total population. Most published sampling tables are prepared on the assumption that samples are drawn at random, meaning that any one of the remaining uninspected units of product has an equal chance to be sampled.

As one might expect, the selection of a simple random sample is very difficult in certain production environments. First of all, the production process may be continuous and there may be only one stage of manufacturing where it is economical to draw the sample. After the product is manufactured it may be placed in bottles, the bottles will be placed in cartons, and the cartons may be stacked on pallets. It may be prudent then to take a stratified random sample or a systematic random sample. To obtain a stratified random sample, the population is divided into subpopulations or strata, and a simple random sample is drawn from each stratum. Suppose the units of a population to be sampled are arranged in some order. For example, a systematic sample from tablets coming off one-by-one from a compressing machine might be required. If a sample of size n is required, the idea of a systematic sample is to take a unit at random from the first $k = N/n$ units and then take every k^{th} unit thereafter, where N is the total number of units in a population.

The most convenient type of sample or method of sampling is often not the correct one. If one is sampling from a bulk operation in which there are relatively large containers of powder and one would like to test to see if this powder is of the correct potency, then one must resist the temptation of taking the sample from just one part of the container—for example, the top. While statisticians are often adamant about the sampling protocol, oftentimes those in charge of obtaining samples may focus on sampling convenience and fail to see the purpose behind elaborate schemes to obtain a random sample. Examples of this are sampling the beginning, the middle, and the end of a production batch, sampling the four corners and the center of each tray, or selection from the top of a drum of powder without considering whether such sampling schemes are either adequate or necessary. Frequently, the times of sampling in production must be scheduled among other important duties of the production operators and this type of sampling can result in predictability and periodicity rather than the randomness that is required for a successful implementation of many of the published sampling plans. A requirement of any sampling scheme is that the sample be *representative of the batch* and some effort may be required to achieve this without imposing an undue burden on those who must take the samples.

C. Compendial Tests and Procedures

Many USP tests and procedures have the form and substance of statistical sampling and acceptance plans, but the USP cautions that they should not be regarded as such. In the Preface of USP XXII, p. xlv,

> Confusion of compendial standards with release tests and with statistical sampling plans occasionally occurs. Compendial standards define what an acceptable article is and give test procedures that demonstrate that the article is in compliance. . . . Some tests, such as those for Dissolution and Uniformity of dosage units, require multiple dosage units in conjunction with a decision scheme. These tests, albeit using a number of dosage units, are in fact the singlet determinations of those particular attributes of the specimen. These procedures should not be confused with statistical sampling plans.

Even though the USP discourages interpreting its procedures as statistical sampling and acceptance plans, many of them include decision rules for acceptability and therefore may be evaluated using much the same methodology as traditional sampling and acceptance plans.

Since the USP tests and procedures are the standard by which drug products will be evaluated after they are released for commercial distribution, they often are used as lot release criteria by pharmaceutical manufacturers. This practice, however, carries an economic risk, since there is no assurance that the product will pass future tests given only that it has met the USP criteria at release. Since the failure costs, including possible product recall and FDA censure, are quite high in the event that a product in the market is judged to be noncompliant, manufacturers often find it expedient to adopt "in-house" product release criteria that directly or indirectly provide a reasonable level of assurance that the drug product will meet the USP standards until its use or expiration date.

Even if the USP tests and procedures are not explicitly utilized as product release criteria, they may be used as a starting point from which companies develop release specifications. As a part of this exercise, it is often helpful to examine the features and operating characteristics of the USP tests and procedures. In the remainder of this section, we will examine two USP procedures that apply to many marketed drug products: Uniformity of dosage units and Dissolution.

1. Uniformity of Dosage Units

This procedure is described in the USP (pp. 1617–1619) and can be envisioned as a two-stage combined attribute/variables sampling and acceptance plan with classified nonconformances. There are actually two different procedures differing only in the acceptance criteria, one for tablets, etc., and another for capsules, etc. To carry out the test, it is necessary to obtain a sample of 30 dosage units and to determine the amount of active ingredient per dosage unit expressed as a percent of the amount claimed on the label. Most often this can be carried out by assaying the entire dosage unit directly, but it may involve a calculation derived from weighing and assaying each unit.

The procedure itself is not simple to understand, and the manner in which it is described in the USP adds to the difficulty. While the USP does not define or use terms like ruberic mean or *USP reference value*, we define them below to explain the procedure.

Definitions. $\overline{X}$ is the arithmetic average of the assay results. The *RSD* is the relative standard deviation (arithmetic standard deviation, divided by $\overline{x}$) expressed as a percent. The ruberic mean is the average of the monograph tolerances.

USP reference value. If the monograph tolerances are symmetric about label claim, then the USP reference value is equal to label claim. If the monograph tolerances are not symmetric about label claim, then the USP reference value is equal to:

   a.   *label claim* whenever $\overline{X}$ < (label claim)
   b.   *X-bar* whenever (label claim) < $\overline{X}$ < (ruberic mean)
   c.   *the ruberic mean* whenever $\overline{X}$ > (ruberic mean)

The following is a description of the procedure for tablets.

Stage 1

   Assay 10 tablets and calculate the RSD and USP reference value from the assay results.

   If all assay results are within 85% to 115% of the USP reference value and the RSD is not more than 6.0%, the uniformity is acceptable.

   If one or more assay results are outside of 75% to 125% of the USP reference value or more than one assay result is outside of 85% to 115% of the USP reference value, the uniformity is not acceptable.

   If neither of the above criteria are met, a second stage test is performed.

Stage 2

   Assay 20 additional tablets and calculate the RSD and USP reference value from all 30 assay results combined.

   If all 30 individual assay results are within 75% to 125% of the USP reference value and at most one assay result is outside 85% to 115% of the USP reference value and the RSD is not more than 7.8%, the uniformity is acceptable. Otherwise the uniformity is not acceptable.

An alternate description of the procedure is possible if we additionally define:
When an assay result falls outside 75% to 125% of the USP reference value it is classified as a *major nonconformance*.
When an assay value falls outside 85% to 115%, but not outside 75% to 125% of the USP reference value, it is classified as a *minor nonconformance*

With these conventions, the decision criteria for tablets may also be described as follows:

Stage 1
If there are no major and no minor nonconformances out of 10 assay results and the RSD is not more than 6.0%, the uniformity is acceptable.
If there is one or more major nonconformances or there is more than one minor nonconformance in 10 assay results, the uniformity is not acceptable.

If neither criterion is satisfied, a second stage test is performed.

Stage 2
If there are no major nonconformances and no more than one minor nonconformance in the 30 assay results and the RSD is not more than 7.8%, the uniformity is acceptable.
If there are one or more major nonconformances, or there is more than one minor nonconformance, or the RSD is greater than 7.8%, the uniformity is not acceptable.

A tabular summary of the decision rules is presented in Table 20.1.
The USP Dose Uniformity Test provides implicit control of average potency as well as explicit control of potency variation, so it might be natural to think in terms of both spread and location when evaluating the performance of the procedure. Thus, one approach to examining the operating characteristics is to quantify the probability that the procedure determines acceptable uniformity across a spectrum of hypothetical populations indexed by a grid of true means and relative standard deviations. To derive the operating characteristic curves, Monte Carlo simulations are necessary, because analytical approaches are intractable, even with the assumption of underlying normal distributions.
Figure 20.1 illustrates the operating characteristics for the USP procedure applied to tablets for a case where the regulatory tolerances are symmetric about label claim. Figure 20.2 shows operating characteristics for an example where the regulatory tolerances are asymmetric.
When evaluating a particular strategy for lot release, it may be useful to compare the operating characteristics of the candidate plan for sampling and

Table 20.1 USP Tablet Dose Uniformity

Stage 1	Major nonconformances			
(N=10)	0			> 0
	Minor nonconformances			
	0	1	> 1	
RSD ≤ 6.0%	Pass	Stage 2	Fail	Fail
RSD > 6.0%	Stage 2	Stage 2	Fail	Fail
Stage 2	Major nonconformances			
(N = 10 + 20)	0			> 0
	Minor nonconformances			
	0	1	> 1	
RSD ≤ 7.8%	Pass	Pass	Fail	Fail
RSD > 7.8%	Fail	Fail	Fail	Fail

acceptance with that of the USP procedure. Such an approach would not be unlike a supplier evaluating an outgoing materials plan against the incoming materials plan of his customer in order to determine the likelihood that his material would be found acceptable by the customer's plan.

Since the USP Dose Uniformity Test involves assay measurements, it is appealing to consider traditional sampling and acceptance by variables plans for lot release strategies. One difficulty associated with this approach is that the USP procedure does not involve specifications in the usual quality control sense. For example, 85% to 115% are not specifications, but are only part of the USP acceptance criteria, and therefore, there is no such thing as percent nonconforming for the USP procedure. One possible strategy might be to compare the operating characteristics of one or more standard one-stage and/or two-stage variables plans with that of the USP procedure on a common basis such as that illustrated in Figures 20.1 and 20.2.

Sampson and Breunig (1971) and Sampson et al. (1970) examined how the continuous nature of assay measurements could be exploited to improve the information gained from content uniformity data. They further explored the feasi-

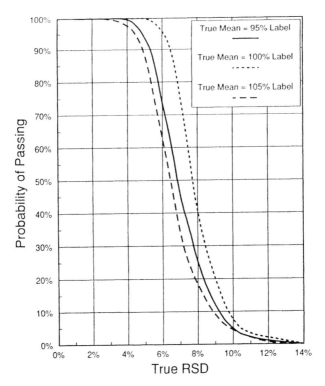

Figure 20.1 USP dose uniformity OC curves (monograph tolerance = 90–110% of label.)

bility of devising more efficient sampling and acceptance strategies than the USP procedure, by utilizing traditional quality control methodology and concepts.

2. Dissolution

The USP Dissolution Test (USP 1989, pp. 1578–1579) is regarded as a primary quality control tool for evaluating bioavailability. In the Preface of the USP (p. xliii),

> Experience has demonstrated that where a medically significant dif-
> ference in bioavailability has been found among supposedly identical
> articles, a dissolution test has been found efficacious in discriminat-
> ing among these articles. . . . such a discriminating test is satisfactory
> because the dissolution standard can exclude definitively any unac-
> ceptable article. Therefore, no compendial requirements for animal
> or human tests of bioavailability are necessary.

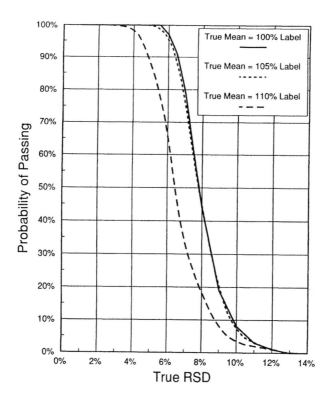

Figure 20.2 USP dose uniformity OC curves (monograph tolerance = 90–110% of label.)

To carry out the procedure, a number of dosage units are tested in one of two types of apparatus described in detail in the USP, which are intended to provide uniform temperature and agitation. Basically, a dosage unit is placed in a container of specified volume, dimensions, and geometry, containing a product-specific dissolution medium. The medium is stirred or agitated for a specified period of time, after which, a sample of the liquid is taken and assayed for the active ingredient. Typically, six dosage units are tested at one time. Acceptable dissolution results can be obtained at any one of three separate stages as outlined in Table 20.2.

The Q-value in the table is product-specific, but typical values are 75% or 80% of label claim. Instructions are, "Continue testing through the three stages unless the results conform at either S_1 or S_2." As Pheatt (1980) notes, the test entails implicit failure criteria at all stages. For example, if at stage S_1, the assay for one of the six units falls below Q–25%, or if the assays for two

Table 20.2   USP Dissolution Requirements.

Stage	Number tested	Acceptance criteria
S_1	6	Each unit is not less that $Q + 5\%$
S_2	6	Average of 12 units ($S_1 + S_2$) is equal to or greater than Q and no unit is less than $Q - 15\%$
S_3	12	Average of 24 units ($S_1 + S_2 + S_3$) is equal to or greater than Q, not more than 2 units are less than $Q - 15\%$, and no unit is less than $Q - 25\%$

or more units fall below, Q–15%, then there is no point in continuing the test, since failure at stage S_3 is already assured. As a practical matter, if units are tested in groups of six, then there is actually a stage (e.g., $S_{2.5}$) between stages S_2 and S_3 for which testing would be terminated if the failure criteria of stage S_3 have already been exceeded.

The USP Dissolution Test may be regarded as a three-stage combined variables/attribute sampling and acceptance plan. An earlier version of the test, involving only two stages, was interpretable in terms of attributes. For the

Table 20.3   Probability of Passing the Test

Deviation of the mean	Σ units[1]				
	$\Sigma = 1\%$	$\Sigma = 2\%$	$\Sigma = 5\%$	$\Sigma = 10\%$	$\Sigma = 15\%$
−1.0	0.0	0.0	0.0	0.0	0.0
−0.8	0.4	0.3	0.3	0.2	0.1
−0.6	2.6	2.0	2.0	1.3	0.2
−0.4	12.3	10.7	9.3	5.6	1.5
−0.2	37.1	34.1	30.0	21.0	6.5
0.0	71.0	65.6	63.3	51.1	18.4
0.2	92.8	90.2	88.9	79.2	40.2
0.4	99.1	98.5	98.4	93.4	60.2
0.6	99.9	99.9	99.9	98.1	77.3
0.8	100.0	100.0	100.0	99.5	87.3
1.0	100.0	100.0	100.0	99.6	94.2
1.2	100.0	100.0	100.0	99.7	97.5

[1] The true population mean is this value times the true Σ away from Q. For example, for $\Sigma = 10\%$, the true mean takes the values $Q - 10\%$, $Q - 8\%$, ..., $Q + 12\%$.

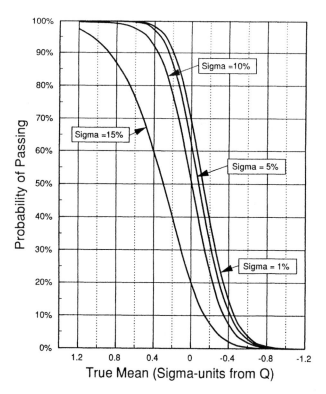

Figure 20.3 USP dissolution test operating characteristic curves.

present test, the operating characteristics can be studied through Monte Carlo simulation, since analytical approaches have proven intractable. For the simulations, as Pheatt (1980) shows, it is reasonable to assume an underlying normal distribution of dissolution results. For the range applicable in practice, the operating characteristics do not depend on the particular value of Q, and the operating characteristics may therefore be regarded only as a function of the variability and the population location relative to Q. Table 20.3 displays the results of our Monte Carlo simulation. In the table, Σ is expressed in units of percent of label claim.

These results are presented graphically in Figure 20.3, where the abscissa is plotted in reverse order so that the operating characteristic curve is displayed consistent with common practice. In general, it appears that when the true mean is one-sigma above Q, the units will almost surely pass the USP Dissolution Test, and when the true mean is one-sigma below Q, the units will almost surely fail the test. Any dissolution test being considered for lot release by a prudent manufacturer should have similar features.

D.   Stability Issues in Quality Control

Drug products must be expiration dated because, in the marketplace, they may undergo changes whereby some of the active ingredient may be converted to a form that is pharmacologically less active, inactive, or even toxic. We say that the drug "loses potency". In order to provide assurance to the consumer that the drug will maintain its strength and purity throughout its shelf life, manufacturers gather, analyze, and interpret data on the stability of their products. Stability studies involve storing samples of the product in a controlled environment, periodically removing some of the stored containers, and assaying the drug substance for potency and other relevant characteristics.

Some interesting and controversial statistical issues arise when considering the inferences that can or should be drawn from stability data but it is not the intent of this section to present a detailed discussion of stability data analysis. Part of the problem is that stability studies are needed for more than one purpose and the optimal design and analysis in one case may be inappropriate or inadequate in the other. A more complete discussion of statistical methods for stability data analysis is given in the next chapter.

In research and development for example, the primary interest is in gathering data to decide what product formulations are appropriate, which of the packages will adequately maintain potency and preserve product integrity, what are the underlying kinetic mechanisms of degradation, what are the chemical products of the degradation process, and how many months or years of shelf life should be assigned. Since much of this information will determine the eventual formulation and packaging of the product, it will typically be part of the documentation submitted to the FDA in a NDA and for that reason this type of stability data gathering and analysis is often referred to as "NDA Stability".

For quality control, on the other hand, the issues of product formula, package type, shelf life, etc., have been more or less settled, and the objective is to determine appropriate release specifications to give reasonable assurance that the drug product will remain within regulatory tolerances throughout its shelf life. The task is made difficult by assay variability and prediction uncertainty. The amount of potency loss that a particular batch of a drug product will experience must be predicted from the stability data for other previously manufactured batches of the same product.

In order to understand some of the issues involved, consider the following scenario. For a certain product, the USP Monograph specifies that the potency is to be within 90% to 110% of label claim throughout its shelf life. A batch of this product has been recently manufactured. A random sample of 10 units has been assayed for potency, and the average of the 10 results is 94.8%, with a standard deviation of 1.6%. The stability data base contains data from 15 previous batches, and examination of this data shows that ten of the batches lost an estimated 2% to 3% over the shelf life, two batches lost 0%, one batch lost 4%, and two batches have been within stability requirements for only about half of the shelf life, one of which will lose 5% if it continues at the present rate, and the other of which will "gain" 1% if the trend continues. The average loss for all 15 batches is 2.4%. Let us assume the batch in question will not be released unless it can be reasonably assured that the true average potency is and will remain above 90% throughout expiry. What is our recommendation?

Our recommendation will depend on whether we believe that all batches of this product degrade at effectively the same rate or that different batches degrade at truly different rates. In the first instance, the release strategy needs to allow for the uncertainty in the assay and uniformity of the lot, whereas in the second case, the additional issue of variability in the degradation rate must be considered.

Suppose there is good reason to believe that the variation observed in the losses for the 15 batches is random variation attributable to assay variation and nonhomogeneity, and that a one-sided 95% confidence limit on the average loss is 3.5%. Then we could reason that any batch, including the one in question, starting out with a true average above 93.5% of label claim is acceptable. We could further reason that since a lower one-sided 95% confidence limit on the true mean for this batch is $94.8 - (1.833)(1.6)/(3.162)$ or 93.9% (mean minus $t_{(.95, 9df)}$ times standard deviation divided by square root of n), that the batch is acceptable since we are reasonably confident that the batch mean is now above 93.5% and will remain above 90% of label claim throughout its shelf life.

Suppose, on the other hand, that the difference among the losses for the 15 batches are too great to be attributable only to assay and homogeneity. What strategy is now reasonable? We might try estimating a "worst case" slope or loss. For example, the worst observed loss might be used, which, for the present case, is 5%. Using the same reasoning as above, we would release no batch with a true batch average below 95% of label claim and this approach would therefore lead us to recommend not releasing the batch, since the 95% LCL is 93.9% of label. Using some form of confidence bound on the worst observed slope would be even more conservative and would lead to the same reject recommendation.

To highlight some additional issues, suppose we want to use a single potency release limit for releasing all batches. Suppose, further, that we have the historical release assay results for the last 25 batches of this product that the within-batch standard deviations are "in control", and that the pooled within-batch standard deviation is 2.2% (with 225 degrees of freedom). Then, for the uniform loss rate scenario with a fixed sample size of 10, the release limit would be 90% + 3.5% + (1.65)(2.2)/(3.162) = 94.6% of label claim. Any batch, for which the average assay result of 10 randomly selected units is above 94.6%, would be acceptable. What if we get a batch for which the within-batch standard deviation is 4%? Are we still comfortable with 94.6% as a release limit? To allow for that eventuality, we might recommend that if there is reason to believe that the variation is greater than expected for a given batch, then the release limit be adjusted upward to compensate. On the other hand, suppose we get a batch for which the average of 10 potency assays is 94.5% of label claim, but the observed within-batch standard deviation is 1.0%. Is this batch truly unacceptable? Calculating a lower confidence limit for this batch as a stand-alone exercise, we get 94.5 − (1.833)(1.0)/(3.162) = 93.9%, which is above 93.5%. In this case, the batch seems to meet the general requirements, even though its average is below an acceptance limit based on the historical pooled within-batch standard deviation. What seems clear is that measurement variation is an important factor in lot release and that some form of monitoring and/or adjustment is warranted.

When a product seems to have differential losses from one batch to the next, it is possible to hypothesize a stochastic process in which the batch slopes are random observations from an underlying distribution. Some of the implications of the "random slopes" model have been examined by Murphy and Weisman (1990), and it appears that reasonable lot release strategies can be derived from this approach.

As a general rule, the type of stability data most useful for quality control purposes is different than that needed for "NDA Stability". For quality control, more batches in the stability data base and fewer observations per batch are preferable. In quality control, it is important to get a good representative sampling of the total potency loss during the approved shelf life across as many batches as possible and there is little concern regarding the shape of the degradation curve. The variability of the loss across batches is of major concern because it impacts the batch release strategy directly.

E. Process Validation

Validation as a formal activity is closely identified with the pharmaceutical industry, although the basic concepts are universally applied in many industries. As defined by the FDA (1987),

> (Validation is) . . . establishing documented evidence which provides a high degree of assurance that a specific process will consistently produce a product meeting its predetermined specifications and quality attributes.

At an intellectual level, validation simply means performing good development. At the operational level, terms such as "establishing documented evidence" and "high degree of assurance" tend to have elements of subjectivity leading to differing interpretations. Consequently, validation has been and continues to be a source of some frustration for both the FDA and the pharmaceutical industry. In theory, a company having effective processes for quality assurance, cGMP compliance, and product/process development, should be readily able to meet validation requirements. In practice, there has been sharp disagreement about what constitutes adequate documentation, among other things.

In this section, our intention is neither to discuss the controversial history of validation nor to provide a detailed description of process validation itself. Instead, we will state and expand on three main concepts and mention some of the statistical tools and methods that may apply.

1. Reduced Reliance Upon Endpoint Testing

Consider the case of heat sterilization, where a unit is either sterile or it is not. A simple binomial calculation shows, for example, that in order to be 95% confident that the true proportion of nonsterile units in the batch is less than one in a thousand, we must observe 0 sterility failures from a sample of nearly 3000 units. Now, testing 3000 or more units out of every batch is not really feasible, but more to the point, a nonsterility level as high as .001 is clearly unacceptable, and 95% confidence is highly inadequate. In short, an adequate level of assurance of sterility simply cannot be achieved without destructively testing every unit of the batch. Instead, since it has been demonstrated that the reduction of viable organisms is a predictable mathematical function of time and temperature, sterility is assured when the target unit is subjected to the necessary temperature for a sufficient time. For heat sterilization, therefore, validation means documenting that each unit of each batch consistently gets the necessary time and temperature, and for this case, validation means that the time-temperature profile of the sterilization cycle is relied upon rather than sterility testing of individual units.

This same concept can be applied to "nonsterile" processes, where documenting that critical process parameters are kept within tolerances can provide much of the assurance that the finished units meet established specifications. For example, consider the case of an oral antibiotic for children, where a dry mixture of active ingredient and excipients is filled into a bottle to which the pharmacist later adds water to create a suspension of pleasantly flavored liquid. Here, it is critically important that the bottle contain the right amount of antibiotic because the entire therapy consists of multiple spoonfuls from the same bottle. If end product testing were the primary basis for gaining assurance of the correct potency of the bottles in a batch, then in order to achieve acceptable producer and consumer risks, a relatively large representative sample of the valuable finished units must be obtained and destructively measured with a time-consuming and expensive assay. On the other hand, suppose there is (documented) assurance that: a) the correct amount of active ingredient and excipient were mixed together in a homogeneous blend, b) the mixture was carefully handled so as to prevent segregation or settling, and c) the weight of material filled into the bottles was charted and controlled within tight tolerances throughout the filling process. Intuitively, in the latter case, a manufacturer can utilize a smaller sample for lot release testing. The statistician would explain that a smaller sample size leading to larger producer risk with the same or lower consumer risk is acceptable because the chance of rejecting a batch is small when the process is unlikely to produce nonconforming units. In effect, it is practically a foregone conclusion that the batch produced by this controlled and validated process consists of bottles having the correct amount of antibiotic and the final sampling and lot release strategy serves merely to confirm that fact.

2. The Role of Development in Process Validation

Although the concept of process validation is intuitive and relatively simple, applying it in a full-scale production environment presents some nontrivial logistical difficulties. For example, from one viewpoint, it is perfectly reasonable to expect that the process be demonstrated to produce acceptable product when critical process variables are anywhere in their established tolerance ranges, even at the extremes. Yet, a production manager may be understandably reluctant to force the process parameters to their extremes simultaneously, because he intuitively knows that there has never been and will likely never be any batch where every control parameter has been allowed to wander to one of its extremes. Moreover, there is justifiable discomfort in conducting bold experimentation (deliberately upsetting the system) for a process that seems to be working well, and even more so for the first several batches of a new process.

For reasons stemming from a narrow regulatory interpretation that product produced by an inadequately validated process violates cGMPs, pharmaceutical manufacturers have found it prudent to restrict formal validation activities in production scale equipment to a minimum number of batches (typically three). The statistician will immediately recognize that it is hopeless to expect to study and quantify the effects and interactions of the important variables in only three experimental runs.

Thus, the meat of process validation must be performed before a process reaches full-scale production. The relationship of important product quality characteristics and critical process parameters should be determined early in the development cycle. This activity is closely related to what is called quality function deployment in the current literature. Statistical screening and response surface experiments can be run in the laboratory and pilot plant to identify and characterize the effect of the critical process variables. At this scale, the process can be pushed to the edge of failure (and beyond) without severe penalty, permitting realistic tolerances to be determined. Thus, although current interpretations of regulatory guidelines stipulate that the validation of a process includes the execution and approval of a formal protocol utilizing a minimum of three batches from actual production, this final step is only a formality when it follows a well-documented development program.

3. Process Validation as a Continuing Activity

If process validation actually begins long before the execution of a formal three-production-batch protocol, it similarly continues afterwards because a well-controlled process provides the necessary data to monitor and improve its performance. In fact, in pharmaceutical manufacturing, validation batches are often indistinguishable from ordinary batches except that the validation batches have been so designated and may have some additional testing performed. For some processes, data obtained in the course of normal production may be adequate to meet the requirements of validation and the evaluation of such existing data with a formal pre-defined protocol is termed retrospective validation. Whether or not the practice is called validation or is performed and documented under a formal protocol, the data that accrues as a result of process monitoring, evaluated as an aggregate against the process history, forms the basis of a continuing record of process adequacy. However, in order to make process monitoring a tool for continuing validation and process improvement, it is necessary to collect and utilize the right kind of measurements. The statistician with expertise in gathering, analyzing, and interpreting data can be of significant help in making cost-effective use of process and product monitoring data.

The cGMP Regulations actually stipulate that periodic evaluation be done. In Subpart J, 211.180, (e),

> Written records required by this part shall be maintained so that data therein can be used for evaluating, at least annually, the quality stan- dards of each drug product to determine the need for changes in drug product specifications or manufacturing or control procedures. Written procedures shall be established and followed for such evaluations ...

Although the regulations tend to emphasize the clerical and inspection aspect, the fact that periodic review is required by law provides a starting point from which a continuing validation program can be structured. The advantage this affords over the one-at-a-time batch-by-batch review (which is also required by regulation) is that the process history may be examined from a more global perspective, where general trends and features are more apparent.

The final chapter on process validation has yet to be written. Much was accomplished in the decade of the 1980s yet there is still more to be done before process validation is consistently understood, interpreted, and put into practice by both the FDA and the pharmaceutical industry as a cost-effective quality improvement tool providing value-added benefit to the patient.

F. Total Quality Control

1. Introduction

The 1992 publication catalog from the American Society for Quality Control lists 57 book titles under the subject heading called Total Quality Management. Almost daily, somewhere in the world, a seminar on Total Quality Control (TQC) in one form or another is being presented. The Total Quality Management processes of corporations such as Motorola, IBM, and Xerox have won them the prestigious Malcolm Baldridge National Quality Award. A statistician, Dr. W. Edwards Deming, and a quality management expert, Dr. Joseph Juran, relative unknowns outside their professions only a few years ago, are now internationally sought-after consultants. These and other signs point to nothing less than a quality revolution effecting a transformation in fundamental business practices of companies in the United States and Europe over the next few years. Some companies in the pharmaceutical industry have started or are starting the process of integrating Total Quality into their business practices.

It is quite beyond the scope of this chapter to attempt to present a complete discourse on the subject. The reader interested in learning more about Total Quality is invited to begin by reading Deming (1982, 1986), Juran (1964, 1988), Crosby (1979, 1984), Feigenbaum (1991), and Ishikawa (1985) and going from there.

Although TQC goes by different names such as *Total Quality Management, Total Quality, Quality Improvement, World Class Manufacturing,* and *Companywide Quality Control*, many or most of the following basic elements are common:

Quality improvement
— is good business strategy (profitable)
— focuses on the customer's needs and expectations
— can and must be managed
— must be part of the corporate culture
— is a corporatewide team effort
— is continual (Japanese call this *Kaizen*)
— results from energized, knowledgeable, empowered employees

— results from elimination of wasted time and materials
— results when processes are in control
— results when variation is understood and reduced

In the remainder of this section, we shall discuss TQC in the context of statistics in the pharmaceutical industry. We shall first examine the influence of regulations on the implementation of TQC, and we will conclude with a look at how statistics and statisticians fit into the picture.

2. TQC Under cGMP Constraints and FDA Regulations

There can be no doubt that the FDA through its interpretation and enforcement of regulations including the cGMPs has had a positive impact on the quality of drugs marketed in the United States today. On the other hand, there are those who believe that the current interpretation and application of regulations by the FDA and similar governmental agencies worldwide tend to hinder the growth of quality improvement in the pharmaceutical industry.

First, the regulations are behind the times with respect to the function of a quality control unit in today's modern business. In the cGMPs, for example, the only group given statutory responsibility for quality is the quality control organization and much of that division's work is described in terms of inspecting, double-checking, verifying, auditing, etc. The regulations tend to promulgate the idea of Quality Control as policemen to catch the defects by vigilant surveillance and elaborate final product testing. The concepts of cooperation, teamwork, defect prevention, and quality improvement are basically not to be found in the regulations.

In some cases, the cGMPs are far too detailed and tend to institutionalize dated technology. Examples include the prescribed method for label reconciliation and the requirements in several places for manual checking and verification, both of which could be replaced by more effective and less costly computer controls. In these and other instances, the regulations focus more on who is responsible if something goes wrong rather than on preventing the problem.

In the 1980s, the FDA recognized that assurance of quality could not be attained by inspection and end-product testing alone and they began actively promoting the idea of process validation, which we defined previously and discussed briefly. At a basic intellectual level, there is little disagreement about the wisdom of understanding and characterizing processes better in order to more effectively control and improve them. In addition, there is no question that documenting, preserving, and transmitting such knowledge and expertise is essential. It is even possible to include process validation under the umbrella of *designing and building quality in*, whereby the essential product attributes are identified and tied to the operating ranges of critical process parameters. Unfortunately, the potential of process validation as a quality improvement tool has not been realized to the extent possible, due in part to the regulatory climate in which validation has been promoted. There has been an emphasis on documentation, paperwork, and compliance. In some cases, there have been misguided interpretations as to what validation is. For example, at one time, there was regulatory insistence that computer systems validation included software source code inspection, which is nothing more than an example of the "inspect it to make sure it's right" mentality. Thus, in an ironic twist, process validation, which in its purest form is a concept entirely consistent with TQC, is at present no small source of frustration to both the FDA and to the industry it regulates.

Lastly, the present regulatory climate is not conducive to process improvement because process changes are discouraged. "Significant" process changes must be approved by the FDA before they can be implemented. Approval is obtained by means of a formal supplement to the NDA, which can be a slow and inefficient process. Imagine how quality improvement would have fared in the Japanese automobile or electronics industries if each significant process or product improvement had required government approval before being implemented.

In spite of these and other negative influences, TQC will be integrated into the pharmaceutical industry, because the goals of the industry, the FDA, and society are basically the same: availability of safe, efficacious, and *cost-effective* medicines and medical devices. In today's environment, one sees an increasing emphasis on cost effectiveness, and TQC is viewed more and more as a vehicle to effect improvement in that product attribute.

3.   The Role of Statistics and Statisticians in TQC

Deming has emphasized the need for industry to utilize statistical methods and statistical expertise. He is sharply critical of American management for failing to understand the nature and interpretation of variation. Deming has even advised companies to hire competent statistical expertise and deplores the faulty application of improperly taught statistical methods.

Joiner (1985) issues a direct and controversial challenge to statisticians working in industry that they must prepare themselves to assume a leadership role in the "transformation of North American Industry". Joiner, a disciple of Deming, asserts that since good decisions result from intelligent collection and analysis of data, statisticians are uniquely qualified to assist in this activity. Joiner further observes that statisticians can help management separate people-induced variation from process variation, and can thus contribute toward solving problems by objectively finding root causes. Others (Hahn and Godfrey, 1985), commenting on Joiner's article, suggest that although statisticians have much to offer, their role is more likely to be one of promoting statistical thinking in their organizations rather than leading the transformation.

Snee (1990) expands on the concept of statistical thinking as a major contribution to total quality. Snee defines statistical thinking as "thought processes, which recognize that variation is all around us and present in everything we do; all work is a series of interconnected processes; and identifying, characterizing, quantifying, controlling, and reducing variation provide opportunities for improvement." Snee further advocates that statistical thinking can be utilized in all levels of the organization, not only at the operational level. He emphasizes that managers need to understand the difference between special-cause variation (problem to be fixed) and common-cause variation (defect in the system).

In spite of Deming's stature and influence, it is unlikely that we will see many pharmaceutical companies elevate statisticians to the level of power that he suggests. Similarly, most statisticians will not find themselves leading the transformation as suggested by Joiner. However, quality control statisticians in the pharmaceutical industry will continue to experience a transformation of their roles in their organizations. The common theme from these writers and other experts is that statistical thinking is at the foundation of TQC. Thus, over the next few years we should expect to see quality control statisticians functioning more and more as purveyors of statistical thinking and less and less as generators and interpreters of sampling and acceptance strategies. We are already beginning to hear pharmaceutical company executives talk openly about the need to make data-based decisions and how essential it is to understand and reduce variation. It seems we are talking about an expanded management vocabulary that includes, in addition to money, the currency of variation. Statistical tools for problem solving, statistical process control, and statistical design of experiments for process analysis and improvement, are being utilized more and more as means to achieve these ends. It is an exciting time for statisticians in the industry, presenting both an opportunity and a challenge.

ACKNOWLEDGMENTS

The authors wish to thank Dr. Wendell Smith and Dr. Doris Weisman for valuable comments and suggestions.

REFERENCES

American Society for Quality Control, 1992 Publications Catalog, ASQC Quality Press, Milwaukee, Wisconsin.

Beckman, R. J., and Cook, R. D. (1983), Outliers..........s, *Technometrics*, 25:119-149.

Burr, I. W. (1976), *Statistical Quality Control Methods*, New York, Marcel Dekker.

Burr, I. W. (1979), *Elementary Statistical Quality Control*, New York, Marcel Dekker.

Crosby, P. B (1979), *Quality is Free*, New American Library, New York.

Crosby, P. B (1984), *Quality Without Tears*, New American Library, New York, NY.

Current Good Manufacturing Practices, Code of Federal Regulations, Title 21 (21 CFR 210, 211, & 229), United States Government Printing Office, Washington, DC.

Deming, W. E. (1982), *Quality, Productivity, and Competitive Position*, MIT Press, Cambridge, MA.

Deming, W. E. (1986), *Out of The Crisis*, MIT Press, Cambridge, MA.

Duncan, A. J. (1986), *Quality Control and Industrial Statistics*, 5th ed., Richard D. Irwin, Inc., Homewood, IL.

FDA (1987), *Guidelines on general principles of process validation*, Food and Drug Administration, Center for Drugs and Biologics, Office of Compliance, Rockville, MD.

Federal Food, Drug, and Cosmetic Act, Code of Federal Regulations, Title 21 (21 CFR 1, 201, 301, 303, 501-528, 703-707), United States Government Printing Office, Washington, DC.

Feigenbaum, A. V. (1991). *Total Quality Control*, 3rd ed. revised, McGraw-Hill, New York.

Godfrey, A. B. (1985). Comment on: The key role of statisticians in the transformation of North American Industry, by B. L. Joiner. *The Amer. Statist.*, **39**:231-232.

Grant, E. L. and Leavenworth, R. L. (1988), *Statistical Quality Control*, 6th ed., McGraw-Hill, New York.

Hahn, G. J. (1985). Comment on: The key role of statisticians in the transformation of North American Industry, by B. L. Joiner. *The Amer. Statist.*, **39**:229-231.

Ishikawa, K. (1985), *What is Total Quality Control? The Japanese Way*, Prentice-Hall, Englewood Cliffs, NJ.

Joiner, B. L. (1985), The key role of statisticians in the transformation of North American industry, *The Amer. Statist.*, **39**:224-234.

Juran, J. M. (1964), *Managerial Breakthrough*, McGraw-Hill, New York.

Juran, J. M. (Ed.) (1988), *Quality Control Handbook*, 4th ed., McGraw-Hill, New York.

Kohberger, R. C. (1988), Manufacturing and quality control. In *Biopharmaceutical Statistics for Drug Development*, (K.E. Peace, Ed.), Marcel Dekker, New York.

Murphy, J. R. and Weisman, D. A. (1990), Using random slopes for estimating shelf life, *1990 Proc. Biopharm. Sec. Amer. Stat. Assoc.*, American Statistical Association, Alexandria, Virginia, pp. 196-203.

Pheatt, C. B. (1980), Evaluation of U.S. Pharmacopeia sampling plans for dissolution, *J. Quality Technol.*, **12**:158-164.

Sampson, C. B. and Breunig, H. L. (1971), Some statistical aspects of pharmaceutical content uniformity, *J. Quality Technol.*, **3**:170-178.

Sampson, C. B., Breunig, H. L., Comer, J. P., and Broadlick, D. E. (1970), Characterization of the Content Uniformity Plan, *J. Pharm. Sci.*, **59**:1653-1655.

Snee, R. D. (1990), Statistical thinking and its contribution to total quality, *The Amer. Statist.*, **44**:116-120.

United States Pharmacopeia (1989). 22nd rev. ed. Mack Publishing Co., Easton, PA.

SUGGESTED ADDITIONAL READING

Box, G. E. P. (1989), Quality improvement: an expanding domain for the application of scientific method, *Phil. Trans. R. Soc. Lond. A*, **327**:139-152.

Chapman, K. G. (1983), A suggested validation lexicon, *Pharm. Tech.*, **7**:51-57.

Chapman, K. G. (1984), The PAR approach to process validation, *Pharm. Tech.*, **7**:51-57.

FDA (1987), *Guidelines for Submitting Documentation for the Stability of Human Drugs and Biologics*, Food and Drug Administration, Center for Drugs and Biologics, Office of Research and Review, Rockville, MD.

Imai, M. (1986), *Kaizen*, Random House, New York, NY.

Johnson, N. L. and Leone, F. C. (1964). *Statistics and Experimental Design*, Vol 1, Wiley, New York.

Lieberman, G. J. and Resnikoff, G. J. (1955). Sampling plans for inspection by variables. *J. Amer. Stat. Assoc.*, **50**:457-516

Olsen, T.N.T, and Lee, I. (1966), Application of statistical methodology in quality control functions of the pharmaceutical industry, *J. Pharm. Sci.*, **55**:1-14.

Schilling, E. G. (1982). *Acceptance Sampling in Quality Control*, Marcel Dekker, New York.

Snee, R. D. (1986), In pursuit of total quality, *Quality Progress*, **Aug**:25-31.

Tribus, M. (1989), The statistician's stake in the managerial revolution, *Phil. Trans. R. Soc. Lond. A*, **327**:9-20.

Willig, S. H., Tuckerman, M. M., and Hitchings, W. S. (1975). *Good Manufacturing Practices for Pharmaceuticals*. Marcel Dekker, New York.

21A

Stability of Drugs

Room Temperature Tests

Karl K. Lin, Tsae-Yun D. Lin, and Roswitha E. Kelly

United States Food and Drug Administration,
Rockville, Maryland

I. INTRODUCTION

The law requires that an expiration dating period be indicated on the immediate container label for every human drug and biologic on the market. The expiration dating period is defined as the time interval that a drug or biological product is expected to remain within the approved specifications after manufacture. The intention of this regulation is to devise and implement systems and procedures that provide a high probability that each package or dose of a pharmaceutical product will have the same characteristics and properties within reasonable acceptable limits to ensure both clinical safety and efficacy of the formulation.

A comprehensive stability testing plan is an essential and pertinent extension of the quality assurance program. The expiration date is a direct application and interpretation of the knowledge gained from stability testing. Stability of a drug has been defined as the ability of a particular formulation, in a specific container, to remain within its physical, chemical, therapeutic, and toxicological specifications. Assurances that the product in its container will be suitably stable for an anticipated shelf life must come from an accumulation of data on the packaged drug. These stability data involve selected parameters which, taken together, form the stability profile.

In 1987 the FDA issued the *Guideline for Submitting Documentation for the Stability of Human Drugs and Biologics*. The Guideline provides recommendations for the design, statistical analyses, and interpretations of results of stability studies to establish appropriate expiration dating period(s) and product storage requirements. A stability study is usually conducted during the Investigational New Drug (IND) and the New Drug Application (NDA) stages of a new drug development, and during the Product License Application (PLA) stage of a biological product. The study continues after the drug is approved and/or the license of the biological product is granted.

There are three steps in the stability analysis. The first step is to collect the assay results at several time intervals for the samples stored under appropriate conditions. The second step is to choose the appropriate model to describe the relationship between assay results and time. The third step is to establish an expiration dating period based on the information from all batches assayed.

Since different batches of a drug or biological product may have different degradation patterns, the FDA requires that at least three batches (and preferably more) should be tested to allow for an estimate of batch-to-batch variability and to test the hypothesis that a single expiration dating period is justifiable for all batches. The specification that at least three batches be tested is a minimal requirement representing a compromise between statistical and practical considerations. In order to infer expiration dates to all future batches, the tested batches should represent a simple random sample taken from the entire existing population of batches. In practice, however, the tested batches are often the first few batches produced and sometimes they may be even research or pilot-scale batches. If research or pilot batches are used, they should have the same characteristics as the production-scale batches and should be validated by results from production batches as soon as these data become available.

In order to determine with reasonable assurance the nature of the degradation curve, the sample times should be chosen so that any degradation can be adequately characterized. The recommended testing schedule is every three months during the first year, every six months during the second year, and yearly thereafter. For drug products predicted to degrade rapidly, more frequent sampling is necessary. The Guideline (FDA, 1987) also encourages testing of a larger number of replicates at the later sampling times, because this will move the average sampling time toward the desired expiration dating period.

In general, the degradation curve of a product characteristic can be adequately represented by a linear, quadratic, or cubic function in the arithmetic or logarithmic scale. In the simplest case, the degradation of a drug product is explained by a linear relationship between the drug product characteristic of interest (e.g., strength) and time. If the drug characteristic is expected to decrease (or increase) with time, the shelf life is estimated as the time period at which the lower (upper) 95% one-sided confidence bound for the mean degradation curve intersects the acceptable lower (upper) specification limit. The confidence bound is obtained by using the ordinary least squares method. For some drug products, the concentration might decrease initially and then increase. In this case, the degradation pattern would not be linear and more complicated statistical methods would be needed.

If batch-to-batch variability is small, it would be advantageous to combine the data from the individual batches into one overall estimate, which usually results in a longer allowable expiration dating period. Combining the data should be supported by preliminary testing of batch similarity. If it is inappropriate to combine data from several batches, the overall expiration dating period may depend on the minimum time which a batch may be expected to remain within acceptable limits.

In many stability studies submitted to FDA, the storage times actually observed are shorter than the requested expiration dates. Precautions should be taken in extrapolating the stability of drug products beyond the actual data period. For the extrapolation to be valid beyond the observed range, the assumed degradation relationship must continue to apply throughout the estimated expiration dating period.

Estimating drug shelf-life from accelerated stability testing has been widely proposed and discussed in the literature. The FDA Guideline states that:

> ... in the absence of sufficient data at the proposed storage condition, stress testing will be accepted for the grant of a tentative expiration dating period, provided adequate information concerning stability of the drug substance has been submitted. The recommended stress testing conditions are: 40°C, or as appropriate for a particular drug product and 75 percent relative humidity. Samples should be analyzed initially and at 1, 2, and 3 months. The parameters described under Section III.B. should be considered when collecting stability data for various drug products. Available long-term stability data should be included and reported as outlined in Section VII of the Guideline. If the results are satisfactory, a tentative expiration dating period of up to 24 months may be granted.

The FDA Guideline also indicates that when changes in the drug product's formulation, in the supplier of a drug substance, or in the container-closure for a marketed drug product are proposed, accelerated data demonstrating comparability with the previously approved drug product plus the standard commitment to long-term stability studies may suffice. For important changes in products known to be relatively unstable, 6 months of data at the normal recommended storage temperature as well as the data from accelerated conditions may also be required. The above situation also applies when there is a change to a new manufacturing facility for the identical drug product using similar equipment.

However, limitations of data sources and uncertainty regarding the assumptions underlying the procedure limit the application of this approach to a few cases. In the current Guideline, long-term studies under ambient conditions and regression techniques are generally used in determining an expiration dating period for drugs and biologics. Accelerated stability testing is used only in Abbreviated New Drug Applications (ANDAs) and Supplements to New Drug Applications. Chapter 21B (Stability of Drugs: Accelerated Storage Tests, by Davies and Hudson) has a detailed discussion on this topic.

We shall focus on the statistical analyses mentioned in the Guideline. The rest of this material is organized as follows. In Section II, we present the statistical analyses applied to determine the expiration dating period based on the information from all batches. The models are described and then the statistical tests for pooling stability batch data are shown. Construction of the confidence bounds for the mean degradation curves and estimation of the shelf-life of the product are presented in Sections II.C and II.D. In Section II.E, the consequences of incorrectly pooling stability data are studied. Numerical examples are given in Section II.F. Finally, in Section III, estimating drug shelf-life with a random effects model is briefly discussed.

II. DETERMINING THE EXPIRATION DATING PERIOD BASED ON THE INFORMATION FROM ALL BATCHES

A. The Models

In this section, we assume that the shape of the degradation curve is negative, (i.e., the drug characteristic decreases as time increases). Other cases can be treated similarly. Currently, the FDA and many pharmaceutical companies use the ordinary least squares method to analyze long-term stability study data under room temperature conditions. The simple linear regression model for this approach is

$$Y_{ij} = \alpha_i + \beta_i X_{ij} + \varepsilon_{ij}$$

for $i = 1, 2, .. I$ batches and $j = 1, 2, .. J_i$ time points where

Y_{ij} = the product characteristic at the jth time point (in months or weeks) of the ith batch of a drug.

X_{ij} = time (in months or weeks) of the jth measurement of the ith batch.

ε_{ij} = errors in measurements of the ith batch which are assumed to be distributed as $N(0, \sigma_i^2)$.

α_i and β_i are the regression parameters to be estimated and are assumed to be fixed.

Y_{ij} most often represent the contents of the active ingredient expressed in percent label claim or a transformation thereof. If a product has more than one active ingredient, the analysis is performed on each ingredient separately. For some products, pH, level of impurities, etc., are very important measures of stability and are similarly analyzed, usually in their original units or an appropriate transformation which may improve linearity.

In practice, however, there may be a slight or moderate variation among the degradation patterns of different batches and the individual degradation lines may have different intercepts and/or slopes. The collected stability data can be fitted with one of the following models:

1. Model 1: common slope and common intercept (Figure 21A.1, a),
   $Y_{ij} = \alpha + \beta X_{ij} + \varepsilon_{ij}$
2. Model 2: common slope but different intercepts (Figure 21A.1, b),
   $Y_{ij} = \alpha_i + \beta X_{ij} + \varepsilon_{ij}$
3. Model 3: common intercept but different slopes (Figure 21A.1, c),
   $Y_{ij} = \alpha + \beta_i X_{ij} + \varepsilon_{ij}$
4. Model 4: different intercepts and different slopes (Figure 21A.1, d),
   $Y_{ij} = \alpha_i + \beta_i X_{ij} + \varepsilon_{ij}$

B. Model Selection: Tests for Pooling Stability Data

Batch similarity of the degradation curves is assessed by fitting linear regression models to the data of the individual batches and applying statistical tests for equality of slopes and/or zero-time intercepts to these models. If the degradation curves are similar, it is desirable to pool the data in order to obtain a more precise estimate of the expiration dating period.

The tests for similarity of degradation curves include the following steps:

1. Fit the I individual regression equations

$$Y_{ij} = a_i + b_i X_{ij} \qquad \text{for each i and } j = 1, 2, ..., J_i$$

where

$$b_i = \frac{\sum (X_{ij} - \overline{X}_{i.})(Y_{ij} - \overline{Y}_{i.})}{\sum (X_{ii} - \overline{X}_i)^2}$$

$$a_i = \overline{Y}_. - b_i \overline{X}_{i.}$$

and compute the sum of squares for variation SS_{i1} about each line. Pooling the SS_{i1}'s, SS_1 can be formed as follows:

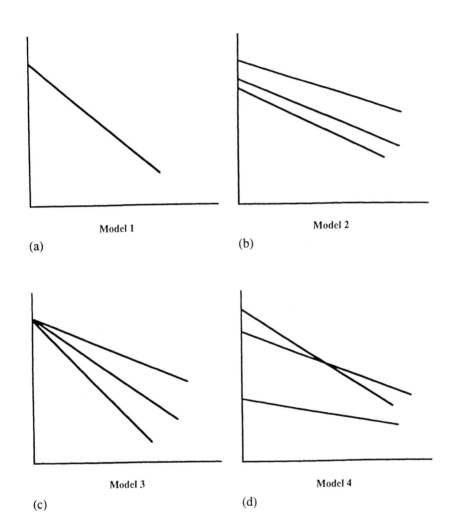

Figure 21A.1 Linear degradation models.

$$SS_1 = \frac{\sum [(J_i - 2)SS_{i1}]}{\sum (J_i - 2)}$$

2. Fit the I individual regression equations with different intercepts and a common slope, that is,

    $$Y_{ij} = \bar{a}_i + \bar{b}X_{ij} \qquad \text{for each i and } j = 1, 2, ..., J_i$$

    where

    $$\bar{b} = \frac{\sum\sum(X_{ij} - \bar{X}_{i.})(Y_{ij} - \bar{Y}_{i.})}{\sum\sum(X_{ij} - \bar{X}_{i.})^2}$$

    $$\bar{a}_i = Y_{i.} - \bar{b}X_{i.}$$

3. Fit the regression equation for the batch means $(\bar{X}_{i.}, \bar{Y}_{i.})$, $i = 1, 2, ..., I$. The equation is expressed as follows:

    $$\bar{Y}_{i.} = \hat{a} + \hat{b}\bar{X}_{i.} \qquad \text{for } i = 1, 2, ..., I$$

    where

    $$\hat{b} = \frac{\sum(\bar{X}_{i.} - \bar{X}_{..})(\bar{Y}_{i.} - \bar{Y}_{..})}{\sum(\bar{X}_{i.} - \bar{X}_{..})^2}$$

    $$\hat{a} = \bar{Y}_{..} - \hat{b}\bar{X}_{..}$$

4. Fit a single regression line assuming that all batches have identical degradation curves. The fitted regression can be expressed as follows:

    $$Y_{ij} = a + bX_{ij} \qquad \text{for } i = 1, 2, ..., I \text{ and } j = 1, 2, ..., J_i$$

    where

    $$b = \frac{\sum\sum(X_{ij} - \bar{X}_{..})(Y_{ij} - \bar{Y}_{..})}{\sum\sum(X_{ij} - \bar{X}_{..})^2}$$

    $$a = \bar{Y}_{..} - b\bar{X}_{..}$$

Table 21A.1 Analysis of Variance Table

Source of variation	Sums of squares	Degrees of freedom	Mean squares
Between slope of regression of means and common slope	$SS_{slopemn}$	1	$MS_{slopemn}$
About regression of group means	SS_{regmn}	$I - 2$	MS_{regmn}
Between individual slopes	$SS_{slopein}$	$I - 1$	$MS_{slopein}$
About individual regression equations	SS_{resin}	$\Sigma J_i - 2I$	MS_{resin}
About regression equation using all observations	SS_{res}	$\Sigma J_i - 2$	MS_{res}

5. Partition the total sum of squares, SS_{tot}, into sum of squares due to regression, SS_{reg}, and residual sum of squares, SS_{res}, using the data of all batches as follows:

$$\underset{SS_{tot}}{\sum\sum(Y_{ij} - \overline{Y}_{..})^2} = \underset{SS_{reg}}{\sum\sum(\hat{Y}_{ij} - \overline{Y}_{..})^2} + \underset{SS_{res}}{\sum\sum(Y_{ij} - \hat{Y}_{ij})^2}$$

6. Partition SS_{res} further into the following four components:

$$\sum\sum(Y_{ij} - a - bX_{ij})^2 = \sum\sum[Y_{ij} - (a_i + b_iX_{ij})]^2 +$$

$$\sum(b_i - \overline{b})^2 \sum(X_{ij} - \overline{X}_{i.})^2 + \sum J_i[\overline{Y}_{i.} - (\hat{a} + \hat{b}\overline{X}_{i.})]^2 +$$

$$\frac{(\overline{b} - b)^2}{\left[\sum J_i(\overline{X}_{i.} - \overline{X}_{..})^2\right]^{-1}\left[\sum\sum(X_{ij} - \overline{X}_{i.})^2\right]^{-1}}$$

The above components are expressed by the following notations:

$$SS_{res} = SS_{resin} + SS_{slopein} + SS_{regmn} + SS_{slopemn}$$

7. Construct test statistics and perform statistical tests by using the above sums of squares. The following analysis of variance table lists their degrees of freedom and mean squares (Table 21A.1).

There are two statistical tests for pooling data of individual batches. The rationale for the use of the level of significance of 0.25 for these tests is discussed by Bancroft (1964).

a. Test for Equality of Slopes

The statistic $MS_{slopein}/MS_{resin}$ which is distributed as $F(I - 1, \Sigma J_i - 2I)$ is for testing of the null hypothesis that the individual regression lines have a common slope (or that a model with separate intercepts and separate slopes fits the data better than a model with separate intercepts and common slope). To accept the null hypothesis, the test should have a p-value of 0.25 or greater.

b. Test for Equality of Intercepts Given Parallel Lines

The statistic $(MS_{slopemn} + MS_{regmn})/MS_{resin}$ which is also distributed as $F(I - 1, \Sigma J_i - 2I)$ is for testing of the null hypothesis that individual parallel lines also have a common intercept (or that a model with separate intercepts and the estimated common slope fits the data better than a model with common intercept and common slope). To accept the null hypothesis, the test should have a p-value of 0.25 or greater.

At the end of the above two hypothesis tests, one of the following models is selected for modeling the degradation pattern of a drug product:

i.   Separate intercepts and separate slopes (Model 3 or Model 4) if the null hypothesis of equal slopes is rejected,
ii. Separate intercepts and common slope (Model 2) if the null hypothesis of equal slopes is not rejected but the null hypothesis of equal intercepts given parallel lines is rejected,
iii. Common intercept and common slope (Model 1) if both null hypotheses of equal slopes and of equal intercepts are not rejected.

C. Construction of Confidence Bounds for the Mean Degradation Curves

Based on drug product characteristics and if Model 1 is used, the following three types of confidence bounds can be constructed:

1.   For product characteristics expected to decrease with time, the following 95% one-sided lower confidence bound CB_L for the mean degradation curve would be used:

$$\hat{Y}_{ij} - t(f,0.95) \ \sqrt{\hat{V}(\hat{Y}_{ij})}$$

where $\hat{Y}_{ij}$ is the predicted value of Y_{ij}.
$t(f,0.95) = 95^{th}$ percentage point of the t distribution with f degrees of freedom.

$$\hat{V}(\hat{Y}_{ij}) = \frac{1}{n} + \left[\frac{(X_{ij} - \overline{X}_{..})^2}{\sum\sum(X_{ij} - \overline{X}_{..})^2} \right] s^2$$

is the estimated variance of $\hat{Y}_{ij}$.

$\overline{X}_{..}$ is the mean of X_{ij} summing over i and j.
s^2 = mean square error about the regression model under Model 1.

2. For drug product characteristics expected to increase with time in the ingredient (e.g., there may be an upper limit on the amount of certain degradation products), the 95% one-sided upper confidence bound CB_U for the mean degradation curve would be used.

$$\hat{Y}_{ij} - t(f,0.95) \quad \sqrt{\hat{V}(\hat{Y}_{ij})}$$

3. For drug product characteristics which could increase or decrease with time in the ingredient, the two-sided 95% confidence bounds CB_L and CB_U would be used.

$$\hat{Y}_{ij} \pm t(f,0.95) \quad \sqrt{\hat{V}(\hat{Y}_{ij})}$$

where t(f, 0.95) = 95th percentage point with f degrees of freedom.

If Model 2, Model 3, or Model 4 is selected, the above confidence bounds can be constructed similarly for each batch with necessary changes of degrees of freedom and statistics.

D. Determination of Expiration Dating Periods

If the degradation pattern was modeled by Model 1, then expiration dating periods of a drug product are determined by comparing the 95% confidence bounds constructed in Section C with the established upper and/or lower specifications SP_U and SP_L of the pertinent product characteristics of the batch of the drug product. In case (1) of Section C, the expiration dating period is determined by the intersection of the 95% lower confidence bound and the lower specification of the product characteristic (i.e., $CB_L = SP_L$; Figure 21A.2). If CB_L intersects SP_L twice, the longer period will be the estimated period. In case (2), the expiration dating period is determined by the intersection of the 95% upper confidence bound and the upper specification limit of the product characteristic (i.e., $CB_U = SP_U$). Finally, in case (3), the expiration dating period is determined by the minimum of the intersections of the 95% upper confidence

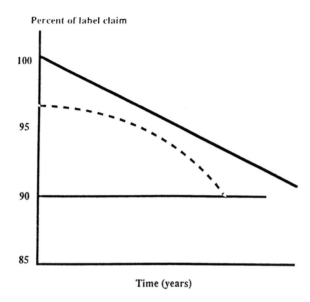

Figure 21A.2 95% one-sided lower confidence interval for the mean degradation curve.

bound and the upper specification limit and of the 95% lower confidence bound and the lower specification limit.

    If a drug product contains more than one active ingredient or several product characteristics that were studied (e.g., pH, impurities), then the expiration dating period of the drug product is estimated by the minimum of the estimated expiration dating periods of the active ingredients or other product characteristics. Under Model 2, Model 3, or Model 4, the final expiration dating period of the drug product is estimated by the minimum of the expiration dating periods of the individual batches.

E. Consequences of Pooling Stability Data When They Cannot Be Pooled

Assuming that 3 batches were sampled, and that the three degradation curves have identical slopes but different intercepts, then the true model will be Model 2 which can be written as follows:

$$Y_{ij} - \alpha + V_i + \delta W_i + \beta X_{ij} + \varepsilon_{ij} \qquad \text{for } i = 1, 2, 3, \text{ and } j = 1, 2, ..., J_i.$$

where

$(V_i, W_i) = (1, 0)$ when $i = 1$
$\qquad\quad = (0, 1)$ when $i = 2$
$\qquad\quad = (0, 0)$ when $i = 3$

If Model 1 (common slopes and common intercepts)

$$Y_{ij} = \alpha + \beta X_{ij} + \varepsilon_{ij} \qquad \text{for } i = 1, 2, 3; \ j = 1, 2, ..., J_i \text{ for each } j$$

is used to estimate the expiration dating period of the drug product, then the estimated period will be biased.

The design matrices of the above two models are as follows:

$$
X_{M2} =
\begin{bmatrix}
1 & 1 & 0 & x_{11} \\
\cdot & \cdot & \cdot & \cdot \\
\cdot & \cdot & \cdot & \cdot \\
\cdot & \cdot & \cdot & \cdot \\
1 & 1 & 0 & x_{1J_1} \\
1 & 0 & 1 & x_{21} \\
\cdot & \cdot & \cdot & \cdot \\
\cdot & \cdot & \cdot & \cdot \\
\cdot & \cdot & \cdot & \cdot \\
1 & 0 & 1 & x_{2J_2} \\
1 & 0 & 0 & x_{31} \\
\cdot & \cdot & \cdot & \cdot \\
\cdot & \cdot & \cdot & \cdot \\
\cdot & \cdot & \cdot & \cdot \\
1 & 0 & 0 & x_{3J_3}
\end{bmatrix}
\qquad
X_{M1} =
\begin{bmatrix}
1 & x_{11} \\
\cdot & \cdot \\
\cdot & \cdot \\
\cdot & \cdot \\
1 & x_{1J_1} \\
1 & x_{21} \\
\cdot & \cdot \\
\cdot & \cdot \\
\cdot & \cdot \\
1 & x_{2J_2} \\
1 & x_{31} \\
\cdot & \cdot \\
\cdot & \cdot \\
\cdot & \cdot \\
1 & x_{3J_3}
\end{bmatrix}
$$

If the true model for the degradation pattern of the three batches is Model 2, which can be rewritten as

$$Y_{ij} = (\alpha + \beta X_{ij}) + (V_i + \delta W_i) + \varepsilon_{ij}$$

and Model 1 is used, the least square estimators of the regression coefficients are

$$\hat{\beta}_A = (X_A'X_A)^{-1} X_A'Y$$

and

$$E(\hat{\beta}_A) = \beta_A + [(X_A'X_A)^{-1} X_A'X_B] \quad \beta_B \neq \beta_A$$

where Y is the observation vector.

$$\beta_A = [\alpha, \beta]', \quad X_A = [C_1, C_4]$$

$$\beta_B = [\gamma, \delta]', \quad X_B = [C_2, C_3]$$

and

$$C_m = \text{the } m^{th} \text{ column of } X_{M2}.$$

It is seen that $X_A\beta_A$ is a biased estimate of the true regression curve. The extent of the bias in the estimate depends on the stability data as well as the true model of the degradation pattern.

F. Numerical Examples

Most relationships between a product characteristic and its change over time can be characterized by a linear model in the original or in a log scale. In this section we present the usual partitioning of the total variation due to regression that FDA statisticians look at and the decision rules employed to see which model best fits the data. FDA programs are currently written in SAS and may be obtained by sponsors.

Examples of Models 1, 2, and 4 (Model 3 being treated the same as Model 4) from reviews of approved products are now presented to give the reader an awareness of how FDA applies the methods described above. The total variance is partitioned into previously defined sums of squares for which appropriate mean square ratios form F-statistics which test the hypotheses shown in Table 21A.2.

The decision rules with respect to pooling regression lines from individual batches obey the following conventions:

If the p-value of test C < .25, Model 3 or 4 is applicable
If the p-value of test C ≥ .25, then test for B
If the p-value of test B < .25, Model 2 is applicable
If the p-value of test B ≥ .25, Model 1 is applicable

Table 21A.2 Hypothesis for Model Selection

Source	Alternative hypothesis	Null hypothesis
A	Separate intercepts, separate slopes	Common intercept, common slope
B	Separate intercept, common slope	Common intercept, common slope
C	Separate intercept, separate slopes	Separate intercept, common slope
D	Residual	
E	Full model	

Model 1 represents the case when a common regression line fits the data of all batches. Below are given the potency data expressed in percent label claim (LC) for three batches of an approved drug product, the resulting analysis of variance, and the calculations of the confidence bands on which the determination of the expiration dating period depends. It can be seen that for this particular product the requested two year expiration dating period is easily supported by the data (Table 21A.3; Figure 21A.3).

Model 2 represents the situation where the data from all batches support a common degradation slope but not a common intercept. In this example, the sponsor also requested a two year expiration dating period but did not have data at all suggested time points (Table 21A.4; Figure 21A.4). Preliminary tests indicate a model with separate intercepts but a common slope ("Analysis of Covariance Model").

Model 3 is the case where the data from the individual batches support neither a common intercept nor a common degradation slope. For this example we present two graphs which demonstrate the diverse degradation patterns of the batches studied (Table 21A.5; Figure 21A.5).

III. APPLYING A RANDOM EFFECTS MODEL FOR DETERMINING THE EXPIRATION DATING PERIOD

As mentioned in the previous section ordinary linear regression techniques are generally used in determining an expiration dating period for drug and biological products. The current procedure for determining the shelf-life assumes the slope and intercept are fixed effects. This limits the validity of the inference. The Guideline states that "if there is evidence of severe violation of the assumptions in the data, an alternate approach may be necessary to accomplish the objective of determining an allowable expiration dating period with a high degree of confidence that the period does not overestimate the true time during which the drug product remains within specifications." In an effort to correct this problem, it was proposed (Stegeman et al., 1987; Brandt et al., 1989; and Chow et al., 1989,

Table 21A.3 Data: Mean Concentration (in percent LC)

Time (months)	Batch number		
	1	2	3
0	100	100	100
1	100	99	99
3	101	100	101
6	102	102	101
12	99	101	100
18	97	100	96
24	99	98	99

Statistical analysis:

Source	SS	DF	MS	F	P
A	2.287	4	.572	.283	.884
B	1.143	2	.571	.283	.758
C	1.144	2	.572	.283	.758
D	30.294	15	2.020		
E	208815.705	6	34802.618		

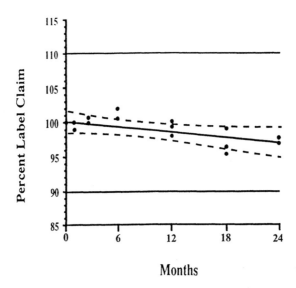

Figure 21A.3 95% two-sided confidence interval for the mean degradation curve (Numerical example of model 1).

Table 21A.4 Data: Mean Concentration (in percent LC)

Time (weeks)	Batch number		
	1	2	3
0	105.4	101.8	105.1
4	-	-	104.5
8	-	-	103.6
13	-	-	104.9
26	-	-	106.3
39	102.8	99.9	-
52	103.1	101.0	104.5
78	102.8	99.2	103.7
104	102.2	99.1	-

Statistical analysis:

Source	SS	DF	MS	F	P
A	42.786	4	10.696	16.281	.0001
B	41.680	2	20.840	31.721	.0000
C	1.106	2	.553	.842	.4570
D	7.227	11	.657		
E	180194.623	6	30032.437		

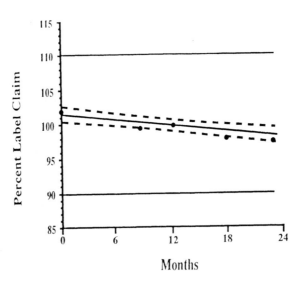

Figure 21A.4 95% two-sided confidence interval for the mean degradation curve (Numerical example of model 2).

Table 21A.5 Data: Mean Concentration (in percent LC)

Time (months)	Batch number				
	1	2	3	4	5
0	95.7	95.7	99.2	101.5	100.5
3	97.8	96.4	98.6	100.5	98.1
6	95.5	94.4	96.5	97.4	95.6
9	99.8	98.7	-	96.9	-
12	95.5	93.2	92.5	91.7	92.5

Statistical analysis:

Source	SS	DF	MS	F	P
A	57.732	8	7.216	2.503	.068
B	9.929	4	2.482	.861	.513
C	47.803	4	11.951	4.145	.022
D	37.485	13	2.883		
E	215215.455	10	21521.546		

1990) to apply a random effects model to determine the expiration dating period for drugs. In order to infer expiration dates to all future batches, the tested batches must represent a random sample taken from the entire existing population of batches. Since there is inherent batch to batch variability, it follows that any batch effects in the model for drug stability must be considered as random.

A random effects, simple linear regression model for this approach (Model R1) is

$$Y_{ij} = \alpha_i + \beta_i X_{ij} + \varepsilon_{ij}$$

where α_i and β_i are random effects and ε_{ij} is the random error term.
The assumptions associated with this model are as follows:

1. $\alpha_i \sim N(a, \sigma_a^2)$
2. $\beta_i \sim N(b, \sigma_b^2)$
3. $\varepsilon_{ij} \sim N(0, \sigma^2)$
4. All random effects are mutually independent.
5. The sampling pattern is the same for all batches.

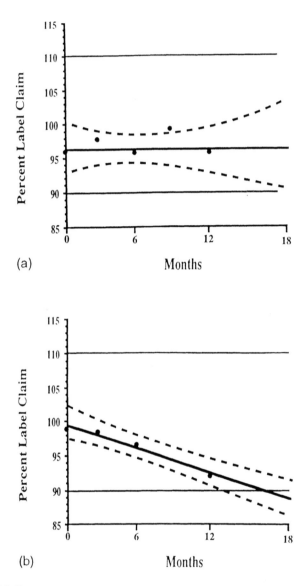

(a)

(b)

Figure 21A.5 (a) 95% two-sided confidence interval for the mean degradation curve (Numerical example of model 3). (b) 95% two-sided confidence interval for the true degradation curve (Numerical example of model 3).

Note that the random effects model has two components of variation in addition to that contained in the fixed effects model. Therefore, the random effects model results in a wider confidence interval and a more conservative estimate of the shelf-life. Several methods have been proposed to estimate shelf-life based on a random effects model. The first method is performed by obtaining the maximum likelihood estimates for each of the parameters and then construct one-sided confidence intervals. The method can be described mathematically in the paragraphs that follow.

The assumptions associated with the random effects model are utilized in the derivation of the mean, variance, and covariance of the Y_{ij}'s. The assumptions involving the distributions of the batch effects imply

$$E(Y_{ij}) = E(\alpha_i + \beta_i X_{ij} + \varepsilon_{ij})$$

$$= a + b X_{ij}$$

$$\text{Var}\,(Y_{ij}) = \text{Var}\,(\alpha_i + \beta_i X_{ij} + \varepsilon_{ij})$$

$$= \sigma_a^2 + \sigma_b^2 X_{ij}^2 + \sigma^2$$

Assume the mutual independence of the ε_{ij}, then

$$\text{Cov}\,(Y_{ij}, Y_{ij'}) = \text{Cov}\,(\alpha_i + \beta_i X_{ij} + \varepsilon_{ij},\ \alpha_i + \beta_i X_{ij'} + \varepsilon_{ij'})$$

$$= \text{Cov}\,(\alpha_i, \alpha_i) + \text{Cov}\,(\beta_i X_{ij}, \beta_i X_{ij'})$$

$$= \sigma_a^2 + \sigma_b^2 X_{ij} X_{ij'}$$

Thus, the covariance matrix of the observation vector of the ith batch is

$$\text{Cov}\,(\underline{Y}_i) = \sigma^2 I_i + \sigma_a^2 J_i + \sigma_b^2 \underline{X}_i \underline{X}_i'$$

where

$$
\underline{Y}_i = \begin{bmatrix} Y_{i1} \\ Y_{i2} \\ . \\ . \\ . \\ Y_{in_i} \end{bmatrix} \quad \text{and} \quad \underline{X}_i = \begin{bmatrix} X_{i1} \\ X_{i2} \\ . \\ . \\ . \\ X_{in_i} \end{bmatrix}
$$

I_i is an $n_i \times n_i$ identity, J_i is an $n_i \times n_i$ matrix of ones.

The vector $\underline{Y}_i$ contains the values for the percent of effective drug, while the vector $\underline{X}_i$ contains the sampling times.

As previously shown $E(\underline{Y}_i) = a1_i + b\underline{X}_i = \underline{\tau}_i$. Assuming the data follow a multivariate normal distribution, the likelihood function for the data in the ith batch is

$$
L(\underline{Y}_i) = \left[\frac{1}{|2\pi\Sigma_i|^{(n_i/2)}} \right] e^{(-1/2)(\underline{Y}_i - \underline{\tau}_i)' \Sigma_i^{-1}(\underline{Y}_i - \underline{\tau}_i)}
$$

Since the batches are mutually independent,

$$
L(\underline{Y}_1, \underline{Y}_2, ..., \underline{Y}_n) = L(\underline{Y}_1)L(\underline{Y}_2) \cdots L(\underline{Y}_n)
$$

Also, since the Σ_i's are equal, the log likelihood for the entire set of data is

$$
\ln L = -\sum_{i=1}^{k} \frac{n_i}{2} \ln |2\pi| - \sum_{i=1}^{k} \frac{n_i}{2} \ln |\Sigma_i| - \frac{1}{2} \sum_{i=1}^{k} (\underline{Y}_i - \underline{\tau}_i) \Sigma_i^{-1} (\underline{Y}_i - \underline{\tau}_i)
$$

Next, the log-likelihood will be differentiated to obtain the likelihood equations which can be set to zero and solved for maximum likelihood estimates for each of the parameters. The iterative approach needs to be used to obtain the maximum likelihood estimates. Once convergence is obtained the parameter estimates will then be used to construct an appropriate tolerance interval. For example, the constructed interval could be a one-sided 95% tolerance interval with 80% of the population having at least a potency of 90%.

If there is batch-to-batch variation, then the slopes and the intercepts of the regression lines differ among batches. Although one can estimate the shelf-life based on fixed effects model for each batch, valid inferences can not be made to future batches. If $\sigma_a^2 = 0$ and $\sigma_b^2 = 0$, then Model R1 reduces to Model 1 in Section II. If $\sigma_a^2 > 0$ but $\sigma_b^2 = 0$ (different intercepts but common slope), the method of estimation of shelf-life based on the ordinary least squares method is still valid since

$$Y_{ij} = a + b\,X_{ij} + \varepsilon_{ij}{}^*$$

with $\varepsilon_{ij}{}^* = \alpha_i - a + e_{ij}$ being independent N $(0, \sigma_a^2 + \sigma^2)$. However, if $\sigma_b^2 > 0$, then the ordinary least squares method is inappropriate. Thus, it is crucial to test the hypothesis that $\sigma_b^2 = 0$ as the first step in stability analysis. Chow and Shaw (1989) have proposed three test procedures for batch-to-batch variation in the stability analysis of a drug product and presented applications to data from new drug application and marketing production stability studies. The test procedures were clearly described in Chow and Shaw's paper and the interested reader can get the detailed discussion from their paper. Based on a simulation study of the test power, Chow and Shaw gave the following guidance for uses of the test procedures under various I (number of batches) and J (number of sampling points):

1.   When I is large or moderate (for example, marketing stability analysis), test procedure 1 is recommended.
2.   When I is small (for example, NDA stability analysis) but J is larger than 5, test procedure 3 is recommended.
3.   When both I and J are small (J = 4 or J = 5), test procedure 2 is recommended.
4.   Test procedure 2 provides a quick examination of batch-to-batch variation and is robust against non-normality. Its restriction on J, however, limits its utility.

Based on the weighted least squares method, Chow and Shaw (1990) proposed two methods for assessing shelf-lives for the marketing stability study. In the first method, the average shelf-life was estimated. Compared with the method used in the Guideline, this method is somewhat less conservative. The second method is to use a tolerance correction. First the lower specification limit η is increased to a higher level η^*, which is a function of η and some bounds (or estimated bounds) of the CV of batches. The estimated lower bound of shelf-lives is then equal to the estimated average shelf-lives which is obtained by treating η^* as the lower specification limit. Chow and Shaw stated that this method is more conservative than the average approach but is much less conservative than the method used in the Guideline.

As mentioned previously, due to statistical and practical considerations, there are only a few batches (usually three to six) tested for a few years (usually one to two) in a stability study. Applying the random effects model to determine the expiration dating periods based on a few data points can not be expected to be accurate. In addition, the Guideline indicates that the production process is expected to remain within tight specification limits, and any changes in the process or in the material (e.g., change in facility location, change in supplier of new material, etc.) requires the expiration dating period to be reevaluated. Should the new data result in a shorter estimate of shelf-life, this new estimate would apply to the product from then on. Unless there are more batches and sampling time points available, the current procedure recommended by the FDA is still considered as appropriate, although it appears to be conservative.

IV.   PROBLEMS ENCOUNTERED IN THE STATISTICAL REVIEW OF STABILITY STUDIES

The FDA has issued guidelines on the design, analysis, and interpretation of stability studies. When sponsors fail to comply with these guidelines, it results in delays in approving the drug or biological product. It is hoped that by pointing out the most commonly encountered deficiencies, the sponsors' attention will be drawn to these issues and will be remedied in future submissions of stability studies.

A.   Lack of Statistical Analysis by Manufacturer

At a disturbingly high rate sponsors submit assay values in their stability section of NDA submissions to the FDA but fail to perform a statistical analysis on the data. In such cases the FDA statistician must analyze the data him/herself instead of only verifying the appropriateness and correctness of the sponsor's approach. With FDA's limited resources this additional review effort results in unnecessary delays and workload backups. The Industry is, therefore, requested to submit complete and appropriate statistical analyses and corresponding graphs for all pertinent product characteristics studied under their stability protocol. As mentioned above, the FDA's computer programs are available upon request.

B.   Statistical Procedure

Sponsors sometimes use statistical procedures other than those described in the FDA Guideline and neglect to provide FDA with the references on the methods used. This, of course, leads to unnecessary delays in the review process. If batch-to-batch variation is small (i.e., if degradation patterns of individual batches are similar), it would be advantageous to combine the data of all sampled batches in the estimation of the expiration dating period. However, the similarity of degradation patterns (i.e., equality of slopes and intercepts) has to be first tested by statistical procedures. In order to obtain a longer expiration dating period, sponsors often combine data of individual batches in their statistical analyses without testing for batch similarity first. Assuming Model 1 (common slope and intercept) when actually Model 2 (common slope but separate intercepts) is true, results in a biased estimator of the slope.

The FDA Guideline suggests that the tests for batch similarity are performed at a level of significance of $p = 0.25$. In some submissions the sponsor argues that this level is too restrictive. One needs to consider, however, that these preliminary tests on establishing the correct model are performed on the same data set as the final analysis. Therefore, a high p-value protects against false positive results (i.e. against incorrectly pooling data which are likely to come from different models).

C. Batch Sampling

One recommendation in the FDA Guideline that is particularly difficult to fulfill is the requirement to use a random sample from the population of production batches to study stability. The reason for employing randomly selected batches is that the estimated expiration dating period based on the sampled batches is applicable to all future production batches. In practice, however, just the first several production size lots are put on stability. If these lots are truly representative of future batches, they may be acceptable in lieu of randomly selected lots for practical and safety considerations. If a truly random sample of batches were to be selected, the determination of an expiration dating period could be delayed for years. In addition, any safety problems that may develop during the scale-up from pilot to production batches could remain undiscovered for an intolerably long time. The compromise is, therefore, to take truly random samples from each lot that is put on stability.

To allow for some estimate of batch-to-batch variability and to test the hypothesis that a single expiration dating period is justifiable for all batches, FDA requires that at least three batches, and preferably more, are included in the stability study. Many studies submitted by sponsors do not meet this requirement. From a statistical point of view, more than three batches would improve the estimate of batch-to-batch variability. From a practical point of view the sponsor is rarely inclined to expend more than the necessary minimum of resources. It is clear, however, that batch-to-batch variability cannot be assessed at all if only one batch is studied. Studying two batches results in a statistically poor estimate of batch-to-batch variability. Therefore, the minimum requirement of three batches is a compromise of statistical and practical considerations.

D. Sampling Times and Number of Assays

The current FDA Guideline recommends that stability testing be done at three month intervals the first year, at six month intervals the second year, and yearly thereafter. This schedule ensures that possible safety problems, such as unexpected loss of potency, are detected early. The failure to meet this recommendation may result in insufficient data points to yield reasonably precise estimates of expiration dating periods.

Associated with this time schedule of assaying a product is the number of data points per batch being recorded. Generally, sponsors report only the mean assay value per time point per batch. Presenting the individual assay values will increase the statistical precision of the estimates of the parameters of the degradation curve. The sponsor can also improve the length of the expiration dating period by performing more assays at the end of the expiration dating period and, therefore, moving the grand assay mean closer to the end of the study period.

E. Extrapolation Beyond the Observed Data

In the estimation of an expiration dating period of a drug product, the observed data are used to fit a regression model and to construct a 95% lower, upper, or two-sided confidence interval around the mean degradation line. The estimate of the expiration dating period is the point of the earliest intersection of either confidence interval with either the upper or lower product specification limit. This estimate of an expiration dating period is simply the forecasting of a particular time point when the mean drug characteristic is likely to be still within the prescribed range. If a sponsor submits hard data for only a few months and requests a two or three year expiration dating period, he makes the implicit assumption that the degradation pattern seen in the early months will continue throughout the estimated shelf life. This assumption can only be verified by the collection of data over the total range of the requested expiration dating period and is not satisfied by the fact that the extrapolation provided for a long expiration dating period.

F. Improper Pooling of Data of Differing Container-Closures

Due to the possibility of interaction between drug and container-closures and the introduction of leachables into the drug formulation during storage, it is suggested in the Guideline that stability data should be developed for each type of immediate container-closure proposed for marketing. FDA reviewers often find that sponsors pool data from different types of container-closures without demonstrating similar degradation curves from each container-closure type.

When a manufacturer markets many different dosage forms, strengths, and container-closure types of the same product, it may appear too burdensome to study three batches for each dosage form/strength/container combination. Some manufacturers have started to apply the "matrix" concept which identifies a subsample of combinations to be studied. Although such innovations in design are encouraged, certain assumptions of homogeneity of degradation patterns across strengths, batches, and container types need to be shown to hold before such an approach becomes feasible. Early discussion with FDA reviewers is especially encouraged before the sponsor embarks on such a design.

ACKNOWLEDGMENT

This work has been produced by the authors, Karl K. Lin, Tsae-Yun Daphne Lin, and Roswitha E. Kelly, in the capacity of federal government employees, as part of their official duty, and is in the public domain, and is not subject to copyright. The views expressed in this chapter are those of the authors and are not necessarily those of the Food and Drug Administration.

REFERENCES

Bancroft, T. A. (1964), Analysis and inference for incompletely specified models involving the use of preliminary test(s) of significance, *Biometrics*, **20**:427-442.

Brandt, A., Collings, B. J., and Carter, M. W. (1989), The development of estimators for parameters in a random effects model for shelf-Life. Presented at the Annual American Statistical Association.

Food and Drug Administration, (1987), *Guideline for Submitting Documentation for the Stability of Human Drugs and Biologics*, U.S. Department of Health and Human Services, Rockville, MD.

SUGGESTED ADDITIONAL READING

Brownlee, K. A. (1965), *Statistical Theory and Methodology in Science and Engineering*, John Wiley & Sons, New York.

Carstensen, J. T., and Nelson, E. (1976), Terminology regarding labeled and contained amounts in dosage forms, *J. Pharmaceut. Sci.*, **65**(2):311-312.

Chow, S. C. and Shao, J. (1989). Test for batch-to-batch variation in stability analysis. *Statistics in Med.*, **8**:883-890.

Chow, S. C. and Shao, J. (1990). Estimating drug shelf-life with random batches. *Biometrics*, **47**(3):1071-1079.

Davies, O. L., and Budgett, D. A. (1980), Accelerated storage tests on pharmaceutical product: Effect of error structure of assay and errors in recorded temperature, *J. Pharm. Pharmacol.*, **32**:155-159.

Draper, N. R. and Smith, H. (1968), *Applied Regression Analysis*, John Wiley & Sons, New York.

FDA Center for Veterinary Medicine, (1987), *Applied Regression Analysis*, John Wiley & Sons, New York.

King, S. P., Kung, M., and Fung, H. (1984), Statistical predication of drug stability based on nonlinear parameter estimation, *J. Pharmaceut. Sci,*, **73(5)**:657-662.

Ruberg, S. J. and Stegeman, J. W. (1987), Pooling data for stability studies: testing the equality of batches, *Biometrics*, **47(3)**:1059-1069.

Yang, W. (1981), Statistical treatment of stability data, *Drug Devel. and Indust. Pharmacy*, **7(1)**:63-77.

21B

Stability of Drugs

Accelerated Storage Tests

Owen L. Davies

*University College of Wales,
Aberystwyth, Wales*

Harry E. Hudson

*Imperial Chemical Industries Ltd.,
Macclesfield, Cheshire, England*

I. INTRODUCTION

The stability of a drug may be defined as that property which enables it to maintain its physical, chemical, and biological properties when subjected to a variety of challenges (e.g., heat, light, and moisture). In the context of this chapter we are concerned with the loss in strength or potency of the drug, and hence stability may be mathematically defined as a differential quotient between concentration and time:

$$\text{Stability} = \frac{dy(t)}{dt}$$

445

where y(t) is the concentration at time t. Consideration of the above leads to definition of shelf life, which is the period in which the product remains acceptable for use. With respect to potency, a loss not exceeding 10% may be acceptable.

Stability studies on a new drug per se and the formulations containing it are an integral part of its development program, and data from these studies must be available in order to effect registration with the appropriate authorities. In the absence of data, potential hazards face the patient:

1.   Due to decomposition, insufficient active agent may be present to have the desired therapeutic effect.
2.   The decomposition products may be toxic and hence detrimental to the patient.
3.   The decomposition products may have a synergistic effect with the action of the drug and produce undesirable side effects.

A general description of stability studies, the physical and chemical parameters which may require monitoring and their relationship to a pharmaceutical formulation, are adequately presented by Willig et al. (1975) and are not referred to again here. The aim of this chapter is to study the three variables: drug potency, time and temperature with respect to a statistical appraisal of these parameters, and the influence the analytical procedure may have on the results.

Drug regulatory authorities do not normally specify in detail what stability studies should be done or what data are required. The usual procedure is for them to suggest general lines which tend to have a wide interpretation. Over the years, stability programs have been designed on the basis of experience, good science, and, where possible, informal discussions with the authorities concerned. This enables the research pharmacist to use his initiative, to be innovative in experimental design but nevertheless appreciate the limitations of his techniques. It is with the last two factors that statistics plays a major role.

Stability programs are designed to study the drug per se and the formulated product. Initially accelerated storage conditions are employed to predict the rate of potency fall under ambient conditions. Essentially the test involves exposing the samples under study to three or four elevated temperatures chosen from the range 50–100°C and fitting the data usually to either the zero-order or first-order rate equations. The zero-order equation is

$$y(t) = y_0 - Kt \tag{21B.1}$$

where y_0 and y(t) are the potencies initially and after time t, and K is the rate constant; y(t) plotted against t is linear with a slope of $-K$. The first-order equation is

$$\log_e y(t) = \log_e Y_0 - Kt$$

(21B.2)

that is, $\log_e y(t)$ plotted against t is linear with slope $-K$.

In order to relate decomposition to temperature, the rate constants are substituted in the Arrhenius equation:

$$K = Ae^{-E/RT}$$

(21B.3)

where A is the frequency factor, E the activation energy, T the temperature in degrees absolute, and R the gas constant. In logarithmic* form, Equation 21B.3 becomes

$$\log_e K = \log_e A - \frac{E}{RT}$$

(21B.4)

that is, log K plotted against $1/T$ is linear with slope equal to $-E/R$.

The first-order model occurs much more frequently in practice than the zero-order model.

From Equations 21B.3 or 21B.4 and 21B.1 and 21B.2 it is possible to predict the time it will take for the drug to decompose x% at temperature T, that is, define the shelf life of the drug or formulation. Use of this model enables shelf life predictions to be made at a very early stage of drug development, which is advantageous when submitting data to regulatory authorities and when considering marketing potential.

A computer program (Clark and Hudson, 1968), now revised, is employed to provide a full statistical analysis of the data, including a test of the validity of the Arrhenius equation, and gives predictions for the shelf life and 95% confidence limits at ambient temperatures. This analysis is most important since small errors in conducting the tests may be magnified, due to the long extrapolation involved with respect to temperature. Also, reactions may occur at elevated temperatures which do not occur at lower temperatures, in which case the Arrhenius law would not be applicable.

Other procedures for predicting the shelf life of a drug are to be found in the literature (e.g., Toothill, 1961; Rogers, 1963; Lordi and Scott, 1965; Kirkwood, 1977). These procedures are modifications of the Arrhenius equation. In some cases where the data are insufficient to test it, the equation is assumed in order to discriminate quickly between the stability of a number of systems—a

---

* All logarithms in this chapter are to the base e.

typical example is the construction of a stability chart which has been used with advantage in the early development of a formulation.

Another model which is sometimes used is represented by the Eyring equation (Kirkwood, 1977), which may be expressed in the form

$$\log K(T) = \log \left(\frac{k}{h}\right) + \frac{S}{R} - \frac{H}{RT} + \log T$$

(21B.5)

where H is the heat of activation, S the entropy of activation, and k and h are Boltzmann's constant and Planck's constant, respectively. This compares with

$$\log K(T) = \log A - \frac{E}{RT}$$

(21B.6)

for the Arrhenius model. The essential difference between the two models is the presence of the term log T, which indicates the dependence of the frequency factor, or its equivalent, on temperature. This effect, however, is quite small and can usually be ignored in accelerated storage tests. The models based on the Eyring equation do not introduce any new features into the statistical considerations in this chapter, which concentrates on the Arrhenius models. However, the methods in this chapter may be applied with little modification to the Eyring models.

Irrespective of the procedure adopted, errors from many sources can occur (e.g., sample preparation, storage, and examination) which may significantly reduce the accuracy of the prediction (Clark and Hudson, 1968).

Registration authorities require the results of accelerated storage tests to be substantiated by storage under ambient conditions. The procedures we have found to be acceptable by registrations authorities on an international basis (Hudson, 1966) is to store the drug in suitable containers under the following conditions: exposed to daylight at room temperature and in the dark at 4°C, room temperature, 37°C, and 50°C. Also, the hygroscopicity profile of the drug is determined by exposing the drug to controlled humidities at room temperature and at 37°C (to simulate tropical storage). Samples are monitored after 1, 2, 3, 6, 9, 12, 18, 24, 36, 48, 60, and 72 months, assuming the results of these tests or the accelerated studies do not dictate that more frequent examinations would be prudent.

Assuming storage conditions are adequately defined and controlled, the analytical examination is probably the most important part of a stability program. The analytical methods which may be used are manifold but essentially must be specific for the drug or alternatively for its decomposition products. From experience, peace of mind may be achieved by monitoring the active agent after separation from the decomposition products, usually by some chromatographic procedure or by a radiochemical technique which is dependent on the introduction of a ^{14}C isotope into the drug prior to study (Hudson, 1966; Hudson and Jones, 1972a, 1972b; Kitson et al., 1976).

II. FITTING MODELS TO DATA: SOME PRELIMINARY CONSIDERATIONS

A drug, when stored in an accelerated storage test, is contained in hermetically sealed containers called ampoules. There are two main situations which arise in the assay methods in these tests:

1. Each ampoule stored at an elevated temperature, when assayed, is examined side by side with an ampoule stored for the same time at a low temperature where no appreciable deterioration can occur. The potency of the former is then expressed as a percentage of the latter.
2. Each ampoule stored at an elevated temperature and also the drug prior to storage are both assayed directly or side by side against an independent standard. The initial test, or the mean if replicated, is made equal to 100 and the remaining potencies scaled accordingly.

These will be referred to as "situation 1" and "situation 2", respectively.

An example of situation 1 is the storage test on Levamisole in solution at pH 4.0 (Dickenson et al., 1971). The potency determinations after storage at various times and temperatures are given in Table 21B.1.

The data are represented graphically in Figures 21B.1(a) and 21B.1(b). Figure 21B.1(a) plots the actual potencies at various times of storage at temperatures 101, 90, 80, and 71.5°C, and Figure 21B.1(b) plots the log of the potencies. It is required to decide which of the two models—zero-order or first-order—is the appropriate one, and then to estimate the parameters of the model.

The essential difference statistically between the two situations 1 and 2 is that in 2 the initial potency, although scaled to equal 100, is an observation subject to error, whereas in 1 the initial potency is by definition exactly 100.

In recent years, 1 has become the more common situation. There are variations in this situation where, for example, ampoules taken at the same time from two or more temperatures are compared side by side with one or more ampoules used as a standard. The potencies of the samples assayed together are then correlated and have to be taken into account in the log likelihood expression when fitting the model (Davies, 1980). These variations will not be dealt with specifically in this chapter.

It is apparent on inspection of the graphs (Figures 21B.1(a) and 21B.1(b)) that the model is first-order. In Figure 21B.1(b), the plots for each temperature

Table 21B.1 Results of Accelerated Storage Test on Levamisole Stored Under Oxygen at pH 4.0

Temperature							
101°C		90°C		80°C		71.5°C	
Time	Potency	Time	Potency	Time	Potency	Time	Potency
2	67.0	4	76.5	8	84.8, 85.9	12	91.3
4	45.1, 45.3	6	68.0	12	79.0	18	83.3
6	30.0	8	59.1, 58.3	16	70.5	24	78.5
7	24.5, 26.4	10	51.0	21	61.7, 62.7	30	74.2
8	19.7	12	45.0	27	52.8, 52.3	34	73.3, 73.4
9	16.0	16	33.9	30	49.6		
		20	25.8, 26.2	34	41.5		
		24	20.0, 20.4				
		26	17.1				

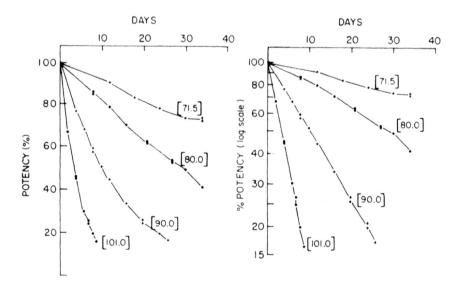

Figure 21B.1 Plots of observed potencies of Table 21B.1 on linear and log scales against times of sampling from ovens kept at elevated temperatures. (Numbers in brackets represent temperatures of ovens in °C.) (a) on the left, (b) on the right.

are seemingly linear, whereas for Figure 21B.1(a), there is a distinct curvature for potencies below 50%. Departures from linearity in the plots of actual potencies do not become apparent until the potencies have dropped below 60 or even 50%. Unless the potencies have dropped below 50% for one or more temperatures, it may be difficult to discriminate between zero-order and first-order. For the more stable formulations it may take too long for the potencies to drop below 50% even when stored at 100°C. but for such formulations the shelf life would clearly be satisfactory, whichever model is adopted.

An important consideration in fitting a model to data is the error structure of the observations. The observations are potency determinations, and it is clear from the various procedures used to determine potency that the error distribution is likely to be log-normal, that is, the logarithm of the potency is normally distributed.

III. ANALYTICAL METHODS AND THEIR ERROR STRUCTURE

In any stability study it is essential that the analytical method be meaningful, that is it will distinguish the drug from its decomposition products, and either one or the other will be monitored. For this chapter we have confined our attention to monitoring the drug. Two approaches are open: a) direct measurement of the drug and b) measurement of the drug after separation from decomposition products.

Analytical methods coming under (b) are usually quite sophisticated and probably more prone to effects from change in solvents, operator variation, and day-to-day manipulative techniques. It is therefore important that such samples be assayed alongside a standard so variations simultaneously occur during the assay with this and the sample.

Irrespective of the approach used, the standard deviation of the assay is normally of the order of 1 to 2%.

It is reasonable to postulate three sources of error in the determination of potencies (Davies and Budgett, 1980):

1. Those which give rise to a standard deviation proportional to the potency.
2. Those which give rise to a standard deviation proportional to the amount of deterioration.
3. Those which give rise to a standard deviation independent of the potency.

Denoting the variances of the three types of error by σ_0^2, σ_1^2, and σ_2^2, respectively, then the combined variance of potency determination is

$$V(y) = \sigma_0^2 y^2 + \sigma_1^2(y_0 - y)^2 + \sigma_2^2 y_0^2$$

where y is the potency at time t and y_0 the initial potency. Here y is log-normal so we need the variance of log y. This is

$$V(\log y) = \sigma_0^2 + \sigma_1^2 \left(\frac{y_0}{y} - 1 \right)^2 + \sigma_2^2 \left(\frac{y_0}{y} \right)^2 \tag{21B.7}$$

In a well-conducted test σ_1 should be small, and it can be considered as being absorbed in σ_0 and σ_2. The formula then reduces to

$$\sigma_0^2 + \sigma_2^2 \left(\frac{y_0}{y} \right)^2 \tag{21B.8}$$

In situation 1, where both the sample and the standard ampoules are analyzed side by side, we find

Variance of y_0, the measurement on the standard $= (\sigma_0^2 + \sigma_2^2)y_0^2$

Variance of y, the measurement on the sample $= \sigma_0^2 y^2 + \sigma_2^2 y_0^2$

Whence

$$\text{Variance of log potency} = (2\sigma_0^2 + \sigma_2^2) + \sigma_2^2 \left(\frac{y_0}{y} \right)$$

This is of the same form as Equation 21B.8, and it shows that when Equation 21B.8 is applied to situation 1, $\sigma_2^2/\sigma_0^2 \le 1$. Now σ_1^2 is assumed to be absorbed in σ_0^2 and σ_2^2, and if appreciable, would increase the ratio σ_2^2/σ_0^2, which could then exceed unity. The ratio could exceed unity when the standard is replicated and could equal the degree of replication.

Table 21B.2 shows how the standard error of log potency increases in Equation 21B.8 as y decreases for various relative values of σ_0^2 and σ_2^2 including the extreme value $\sigma_0^2 = 0$. The figures have been scaled so that the standard error of log y_0 is .01, that is, 1% of potency.

Table 21B.2 shows that the standard error increases rapidly for potencies below 30%, and the rate of increase is greater for larger ratios of σ_2^2/σ_0^2. The last column gives the limit values when $\sigma_0^2 = 0$.

It is an observed fact that the coefficient of variation of potency determinations tends to increase for decreasing potency, and this has to be taken into account. The above method is a reasoned way of expressing this increase. The method is flexible, and by assigning appropriate ratios of values to σ_0, σ_1, and

Table 21B.2 Illustration of the Dependence of the Standard Error on Potency (Standard Error Expressed as Percentage of Potency)

Percentage potency	σ_2^2/σ_0^2				
	2	1	0.5	0.25	$\sigma_0^2 = 0$
100	1.0	1.0	1.0	1.0	1.0
80	1.2	1.1	1.1	1.1	1.3
50	1.7	1.6	1.4	1.3	2.0
30	2.8	2.5	2.1	1.7	3.3
20	4.1	3.6	2.9	2.4	5.0
10	8.2	7.1	6.1	4.6	10.0

σ_2 we may cover the range from a constant coefficient of variation to a constant standard deviation for the whole range of potencies.

There may be other methods of explaining and expressing this relationship between coefficient of variation and potency in specific situations, but it is essential that this relationship be taken into account in the fitting procedures which follow.

The preferred method of fitting a model to data is by "maximum likelihood." The other method commonly used is "least squares," but when the observations are normally distributed, the two methods are equivalent.

The general method adopted, therefore, is to fit the model to the logarithms of the potencies by least squares. Both models—zero- and first-order—are nonlinear in two of the parameters: the energy of activation and the frequency factor in the Arrhenius equation. This brings in the question of fitting nonlinear models, and much has been published in this field. For completeness, we give in the Appendix to this chapter a discussion of the main features of linear and nonlinear fitting. For further details the reader is referred to standard textbooks (e.g., Sprent 1969).

Very briefly, nonlinear fitting is an iterative procedure, being repeated applications of linear methods on the assumption that the likelihood surface is locally linear. This requires starting values for the parameters obtained by a quick approximate method, and the success of the process often depends on how good an approximation the starting values are. An important step in the procedure is to transform the likelihood surface by a simple process called "reparameterization," which is usually achieved by moving the origin of the variables, in this case, the potency and the temperature. The effect of the transformation is to reduce the correlation between the parameters and improve the convergence of the iterative procedure. (The reader is advised to read the Appendix before the next section.)

IV. APPLICATION TO ACCELERATED STORAGE MODELS

A. First-Order Model

The rate equation representing first-order deterioration is

$$\frac{dy}{dt} = -Ky \tag{21B.9}$$

where $K = ae^{-b/T}$. This results in the model

$$\log y = \log y_0 - ae^{-b/T} t \tag{21B.10}$$

In situation 1, y_0 is 100 and in situation 2, y_0 is a parameter. Denote $\log y$ by Y and apply the reparameterization for the Arrhenius expression given by Equation 21B.A10 in the Appendix. This results in the expression

$$Y = Y_0 - e^{A - b(1/T - \lambda)} t \tag{21B.11}$$

where $A = \log a$. In situation 2 we may eliminate the parameter Y_0 by taking deviations from the mean. This results in the expression

$$Y = \overline{Y} - (e^{A - b(1/T_i - \lambda)} t - M) \tag{21B.12}$$

where

$$\overline{Y} = \frac{\sum \omega_i Y_i}{\sum \omega_i}$$

$$M = \frac{\sum \omega_i e^{A - b(1/T_i - \lambda)} t_i}{\sum \omega_i}$$

$$\omega_i = \frac{1}{V(Y_i)}$$

The summation is taken over all observations. This completes the reparameterization for situation 2. For situation 1, the reparameterized model is simply

$$Y = \log 100 - e^{A - b(1/T - \lambda)} t \tag{21B.13}$$

We may represent the two models respectively as follows:

$$F(A,b,T,t) = Y - \log 100 = -e^{A - b(1/T - \lambda)} t \tag{21B.14}$$

$$F(A,b,T,t) = Y - \overline{Y} = -e^{A - b(1/T - \lambda)} t + M \tag{21B.15}$$

In order to start the Gauss estimation procedure we need initial estimates for A and b. These can usually be derived quite simply by graphical means. A numerical method is given later, and this has the advantage that the main part of the method is needed for the purpose of testing the Arrhenius assumption.

Denote the initial values by $A(1)$ and $b(1)$ and expand Equations 21B.14 and 21B.15 about $A(1)$ and $b(1)$ to first-order. This results in

$$F(A,b,T,t) = F(1) + \frac{\partial F(1)}{\partial A} \partial A + \frac{\partial F(1)}{\partial b} \partial b \tag{21B.16}$$

where $F(1)$ denotes substitution of $A(1)$, $b(1)$ after differentiation.

Let Y_i denote the response for the ith observation and T_i, t_i the temperature and time; then $\underline{D}_i = Y_i - [C + F(A(1), b(1), T_i, t_i)]$ where $C = \log 100$ for situation 1 and $\overline{Y}$ for situation 2.

Using matrix notation and for brevity, F_i to denote $F(A(1), b(1), T_i, t_i)$, for both situations, the adjustments δA and δb are given by

$$
\begin{bmatrix} \delta A \\ \\ \delta b \end{bmatrix} =
\begin{bmatrix} \sum \omega_i \left(\frac{\partial F_i(1)}{\partial A} \right)^2, & \sum \omega_i \frac{\partial F_i(1)}{\partial A} \frac{\partial F_i(1)}{\partial b} \\ \\ \sum \omega_i \frac{\partial F_i(1)}{\partial A} \frac{\partial F_i(1)}{\partial b}, & \sum \omega_i \left(\frac{\partial F_i(1)}{\partial b} \right)^2 \end{bmatrix}^{-1}
\begin{bmatrix} \sum \omega_i D_i \left(\frac{\partial F_i(1)}{\partial A} \right) \\ \\ \sum \omega_i D_i \left(\frac{\partial F_i(1)}{\partial b} \right) \end{bmatrix}
\tag{21B.17}
$$

The inverse of the matrix $X'X$ is simply

$$\frac{1}{\Delta}\begin{bmatrix} \sum \omega_i \left[\dfrac{\partial F_i(1)}{\partial b}\right]^2, & -\sum \omega_i \dfrac{\partial F_i(1)}{\partial A}\dfrac{\partial F_i(1)}{\partial b} \\[3mm] -\sum \omega_i \dfrac{\partial F_i(1)}{\partial A}\dfrac{\partial F_i(1)}{\partial b}, & \sum \omega_i \left[\dfrac{\partial F_i(1)}{\partial A}\right]^2 \end{bmatrix} \qquad (21B.18)$$

where Δ is the determinant of the matrix.

The expressions for δA, δb readily follow and the values of the parameters to use for the next iteration are

$$A(2) = A(1) + \delta A$$

$$b(2) = b(1) + \delta b \qquad\qquad (21B.19)$$

The residual sum of squares is calculated for $A(1)$, $b(1)$ and $A(2)$, $b(2)$ from which the reduction in the residual sum of squares brought about by the improved estimates $A(2)$, $b(2)$ is apparent. The procedure is repeated until the reductions in the residual sum of squares and/or $\delta A/A$ and $\delta b/b$ are very small, for example, 10^{-4}.

The variance-covariance matrix of A and b is simply the matrix (Eq. 21B.18) in which the final estimates of A and b have been substituted. The error variance is incorporated in the elements of this matrix because the weights used are the inverses of the variances of the observed potencies.

The residual sum of squares divided by its degrees of freedom gives an internal estimate of the error variance. The degrees of freedom are the number of observations minus the number of parameters estimated, being 2 for situation 1 and 3 for situation 2. For the same reason as above, the expected value of the internal error variance is unity. The significance can be tested by regarding the sum of squares as a χ^2 with the above degrees of freedom. If significant, and there is no reason to doubt the model, Equation 21B.18 has to be multiplied by the internal error variance in order to give the variances and covariance of A and b.

B. Estimation of Shelf Life

The shelf life of a drug is the time taken for the drug to lose 10% of its potency when stored under ambient conditions. Usually a conservative estimate is taken which is the lower 95% confidence limit. Then the conclusion is made that the drug is unlikely to lose 10% or more of its potency by the end of this period.

More generally, let T_0 be the temperature specified for storage and $p\%$ be the permitted maximum deterioration. The estimated storage time for both situations 1 and 2 is derived by solving the equation

$$e^{A - b(1/T_0 - \lambda)} t = \log\left(\frac{100}{100 - p}\right) = \alpha(p)$$

(21B.20)

We see that the initial potency does not appear explicitly. The storage time is

$$t = \alpha(p)e^{-A + b(1/T_0 - \lambda)}$$

$$\log t = \log \alpha(p) - A + b\left(\frac{1}{T_0} - \lambda\right)$$

(21B.21)

$$V(\log t) = V(A) + \left(\frac{1}{T_0} - \lambda\right)^2 V(b) - 2\left(\frac{1}{T_0} - \lambda\right) \text{cov}(A,b)$$

These enable the lower confidence limit for log t to be derived, and the antilogs give the estimated shelf life and its lower confidence limit.

The variance of log t is taken rather than the variance of t because it has been shown from extensive simulation (Booth, 1977) that log t is more nearly normally distributed than t. For similar reasons we earlier made the transformation $A = \log a$. Also, as shown later, A is only slightly nonlinear.

C. Initial Values for the Parameters in a First-Order Model

Approximate values for the parameters A and b to be used as starting values in the Gauss procedure are derived from the log of the Arrhenius relationship:

$$\log K = A - b\left(\frac{1}{T} - \lambda\right)$$

(21B.22)

Values of K are calculated for each temperature, and using these values, A and b are calculated by linear least squares, weighting each observed log K by the inverse of its variance. The method of weighting is shown later, but sufficiently approximate values can usually be obtained without the refinement of weighting. To estimate the various K values, we note that for a first-order model, the log potency is linear for each temperature and the lines emanate from the same initial point Y_0. Denoting the slope of the line for temperature T_j by K_j, the set of lines may be represented

$$Y = Y_0 - K_j t, \quad j = 1, 2, ..., r$$

(21B.23)

This set of lines may also be represented by the single equation

$$Y = Y_0 - \sum_j K_j t_j(i)$$

(21B.24)

where the t values are zero except for the corresponding K, for example, when $j = 1$, all t values for $j > 1$ will be zero, etc.

For situation 2, the K values and y_0 are linear and we may apply standard linear regression methods to estimate them, giving, in matrix notation, the following:

$$\begin{bmatrix} Y_0 \\ K_1 \\ K_2 \\ \cdot \\ \cdot \\ \cdot \\ K_r \end{bmatrix} = \begin{bmatrix} \sum \omega, \sum \omega t_1, \sum \omega t_2, ..., \sum \omega t_r \\ \sum \omega t_1, \sum \omega t_1^2, \quad 0, ..., \quad 0 \\ \sum \omega t_2, \quad 0, \sum \omega t_2^2, ..., \quad 0 \\ \cdot \\ \cdot \\ \cdot \\ \sum \omega t_r, \quad 0 \quad 0, ..., \sum \omega t_r^2 \end{bmatrix}^{-1} \begin{bmatrix} \sum \omega Y \\ \sum \omega Y_1 t_1 \\ \sum \omega Y_2 t_2 \\ \cdot \\ \cdot \\ \cdot \\ \sum \omega Y_r t_r \end{bmatrix}$$

(21B.25)

and the residual sum of squares is

$$RSS = \sum \omega Y^2 - Y_0 \sum \omega Y - K_1 \sum \omega Y_1 t_1 - ... - K_r \sum \omega Y_r t_r$$

(21B.26)

$\sum\omega$, $\sum\omega Y$, and $\sum\omega Y^2$ are taken over *all* the observations; $\sum\omega t_j$, $\sum\omega Y_j t_j$, and $\sum\omega t_j^2$ apply only to the observations for the jth temperature. The K values of Equations 21B.25 and 21B.26 are of opposite sign to the K values of Equations 21B.24; Equations 21B.25 and 21B.26 apply to situation 2 where Y_0 is a parameter.

An alternative method with some advantages is to eliminate Y_0 by taking deviations from the means:

$$\text{mean } Y = \overline{Y} = \frac{\sum \omega Y}{\sum \omega}$$

(21B.27)

$$\text{mean } t_j = \frac{\sum \omega t_j}{\sum \omega}$$

The summations have the same meaning as above. The K values are given by

$$K = [X'X]^{-1}X'Y \tag{21B.28}$$

where the leading diagonal elements of the matrix $X'X$ are

$$\sum \omega t_j^2 - \frac{\left(\sum \omega t_j\right)^2}{\sum \omega}, \quad j = 1, 2, 3, 4$$

and the off diagonal elements (j,i) or (i,j) are given by

$$\frac{-\left(\sum \omega t_i\right)\left(\sum \omega t_j\right)}{\sum \omega}, \quad i \neq j$$

The jth element of the vector $X'Y$ is

$$\sum \omega Y_j t_j - \frac{\left(\sum \omega Y_j\right)\left(\sum \omega t_j\right)}{\sum \omega}$$

The summations have the same meaning as previously.
The residual sum of squares is

$$RSS_1 = \sum \omega(Y - \overline{Y})^2 - K'X'Y \tag{21B.29}$$

All calculations need to be carried out on a computer for which there are program packages available.

For situation 1, Y_0 is a constant and not a parameter. The K values are then independent and are given by

$$K_j = \frac{\sum \omega Y_j t_j}{\sum \omega t_j^2} \tag{21B.30}$$

where the summation is taken over all the observations of the jth temperature.

The final step of this procedure for obtaining approximate values for A and b is the straightforward one of fitting the linear model

$$\log K_j = A - b\left(\frac{1}{T_j} - \lambda\right) \tag{21B.31}$$

When not weighted. this gives

$$b = \frac{\sum [\log K_j - \text{mean}(\log K_j)][1/T_j - \text{mean}(1/T_j)]}{\sum [1/T_j - \text{mean}(1/T_j)]^2}$$

(21B.32)

and

$$A = \text{mean}(\log K_j) + b\left[\text{mean}\left(\frac{1}{T_j}\right) - \lambda\right]$$

where, in this case, mean $(1/T_j)$ is an unweighted mean.

Some improvement results from weighting the log K values by the inverse of their variances. We find

$$\text{var}(\log K) = \frac{\text{var}(K_j)}{K_j^2}$$

and the weight to be used for the numerator and denominator of Equation 21B.32 is

$$W_j = \frac{K_j^2}{\text{var}(K_j)}$$

The variances of K_j are given by the diagonal elements of the inverse matrix of Equation 21B.25. For the case where y_0 is a parameter, the K values are correlated and so the above least-squares estimates are approximate.

For situation 1 where y_0 is a constant, the K values are independent, and

$$\text{var}(K_j) = \frac{1}{\sum \omega t_j^2}$$

giving

$$W_j = K_j^2 \sum \omega t_j^2$$

This suggests an alternative and probably improved method of allocating a value to λ. This can be chosen to make A and b independent or nearly so in Equation 21B.31, by using the weights W_j in mean $(1/T)$. Thus

$$\lambda = \frac{\sum W_j(1/T_j)}{\sum W_j} \tag{21B.33}$$

A further possibility is to choose λ to make the estimates of A and b in the nonlinear fitting of the model linearly independent. This is achieved by choosing λ to make the off-diagonal elements of Equation 21B.18 zero, that is

$$\sum \omega_i \frac{\partial F_i}{\partial A} \frac{\partial F_i}{\partial b} = 0 \tag{21B.34}$$

In situation 1, $F_i = -e^{A - b(1/T_i - \lambda)} t_i$, and

$$\frac{\partial F_i}{\partial A} = F_i$$

$$\frac{\partial F_i}{\partial b} = -\left(\frac{1}{T_i} - \lambda\right) F_i$$

Equation 21B.34 then becomes

$$\sum \omega_i (F_i)^2 \left(\frac{1}{T_i} - \lambda\right) = 0 \tag{21B.35}$$

whence

$$\lambda = \frac{\sum \omega_i F_i^2 (1/T_i)}{\sum \omega_i F_i^2} \tag{21B.36}$$

Now $F_i = Y_i - Y_0$, where $Y_0 = \log 100$. Equation 21B.36 then becomes

$$\lambda = \frac{\sum \omega_i (Y_i - Y_0)^2 (1/T_i)}{\sum \omega_i (Y_i - Y_0)^2} \tag{21B.37}$$

The Y_i values should be "expected" values. Initially, we substitute the observed values for $(Y_i - Y_0)$ and this will usually be good enough but closer values can be obtained from the calculated values of the K's. These considerations may also be applied quite straightforwardly to situation 2.

There is an alternative method of fitting the first-order model which does not require the above reparameterization (Sid Ahmed, 1975; Kirkwood, 1977). The likelihood function in both situations 1 and 2 is a function of A and b. A may be eliminated from the two maximum likelihood equations, or the normal equations in least squares terminology, to give a function of b only. A search is made around the initial value of b given by Equation 21B.32. A one-dimensional Gauss procedure should suffice, but repeated halving of a suitable interval around the initial value of b will always result in the maximum likelihood estimate of b to any given accuracy. This method might be preferred in some situations. It would still be necessary to evaluate the elements of the matrix (Eq. 21B.18) in order to obtain the variances and covariance of the estimates of the parameters.

Kirkwood (1977) considers the situation which could arise in some biological products, where appreciable deterioration may occur in the material stored at a low temperature and used as a standard in situation 1. For first-order deterioration, and denoting the temperature at which the standard is stored by T_0, then

$$Y_i = (e^{A - b/T_i} - e^{A - b/T_0})\, t_i + \varepsilon_i$$

It is probable that the reparameterization for the Arrhenius expression (Equation 21B.11) will enable the Gauss minimization procedure to be used. This has not been tested, and until this is done, the method recommended by Kirkwood and outlined above should be used.

D. Test of the Arrhenius Model

A simple test exists for the Arrhenius assumption. In fitting the set of straight lines to estimate the separate K values, the Arrhenius law is not assumed. The only assumption made is that the deterioration is first-order for each temperature. The sum of squares of the residuals for this model is RSS_1 of Equation 21B.29. After fitting the first-order model with the Arrhenius assumption, we obtained the sum of squares of the residuals. Denote this by RSS2.

We may now construct the ANOVA table (Table 21B.3). Now M_1 may be compared with M_2 for significance by the F test. If significant, the Arrhenius law is in doubt. A plot of log K against 1/T should always be undertaken—the nature of the departure, if any, from the Arrhenius law will then become apparent. Apparent departures, from the Arrhenius law could arise when there are two competing reactions with different activation energies and may be observed when one of the reactions occurs both homogeneously and heterogeneously. The homogeneous reaction usually has a high energy of activation and occurs at the higher temperatures; the heterogeneous reaction predominates

Table 21B.3 Analysis of Variance[a]

Source of variation	Sum of squares	Degrees of freedom[a]	Mean square
Departures from Arrhenius	$RSS_2 - RSS_1$	$r - 2$	M_1
About separate lines	RSS_1	$N - r - n + 2$	M_2
About first-order model	RSS_2	$N - n$	

[a] number of parameters estimated = n.

at the lower temperatures. Under these circumstances the Arrhenius plot assumes a curve, which may sometimes be resolved into two parts each of which is virtually linear. Alternatively, departures may arise due to incorrect assignment of the order of reaction or inaccurate measurement of the storage temperature, which should be correct to ± 0.2°C or a standard deviation of 0.1°C.

We may extend the analysis further, because if the weights used truly represent the inverse of the analytical error variance, then the expected values of the mean squares, all assumptions being satisfied, will be unity. If, therefore, M_2 is significantly greater than unity by the F-test with the appropriate degrees of freedom then either

1. The log potency versus time is not linear, or
2. The analytical testing error variances have been underestimated.

E. Zero-Order Model

When the data appear to satisfy a zero-order model and it is required to fit this model, the method of fitting differs somewhat from that used for a first-order model.

The zero-order model is

$$y = y_0 - ae^{-b/T} t$$

where y is the actual potency at time t and temperature T in degrees absolute, y_0 the initial potency, and a, b the Arrhenius constants. Since the potency determinations are log-normal, the model should preferably be fitted in its logarithmic form, that is

$$Y = \log y = \log (Y_0 - ae^{-b/T} t)$$

$$(21B.38)$$

All parameters are now nonlinear. The model can be reparameterized as follows:

$$Y = \log y_0 + \log \left(1 - \frac{a}{y_0 e^{-b/T}} t \right)$$

$$(21B.39)$$

Substituting Y_0 for $\log y_0$, and further reparameterizing the Arrhenius part of the expression, we get

$$Y = Y_0 + \log (1 - e^{A - b(1/T - \lambda)} t)$$

$$(21B.40)$$

where $A + b\lambda = \log (a/y_0)$ and

$$\lambda = \frac{\sum \omega i (1/Ti)}{\sum \omega i}$$

There are now only two nonlinear parameters. The resemblance between this model and the form used for the first-order model may be seen from the fact that for small deteriorations

$$Y = Y_0 - e^{A - b(1/T - \lambda)} t$$

As in the first-order model and for similar reasons, the zero-order model is reparameterized further for situation 2 by taking deviations from the mean. This eliminates Y_0 and the model takes the form

$$Y = \overline{Y} + [\log (1 - e^{A - b(1/T - \lambda)} t) - \text{mean}]$$

$$(21B.41)$$

where

$$\overline{Y} = \frac{\sum \omega_i Y_i}{\sum \omega_i}$$

$$\text{mean} = \frac{\sum \omega_i \log (1 - e^{A - b(1/T_i - \lambda)} t_i)}{\sum \omega_i}$$

This reparameterization is not required for situation 1.

The Gauss method described in the Appendix or any suitable minimization procedure may now be applied to estimate A and b, and also to derive the variances and covariance of A and b, and the sum of squares of the residuals.

F. Estimation of Shelf Life for Zero-Order Model

Using the following form for the zero-order model,

$$Y = Y_0 + \log (1 - e^{A - b(1/T - \lambda)} t) \tag{21B.42}$$

the shelf life for temperature T_0 and p% drop in activity is the solution of

$$\log (1 - e^{A - b(1/T_0 - \lambda)} t) = \log \frac{y}{y_0} = \log \left(\frac{100 - p}{100} \right) \tag{21B.43}$$

That is,

$$e^{A - b(1/T - \lambda)} t = \frac{p}{100}$$

$$t = \frac{p}{100} e^{-A + b(1/T_0 - \lambda)}$$

$$\log t = \log \left(\frac{p}{100} \right) - A + b \left(\frac{1}{T_0} - \lambda \right) \tag{21B.44}$$

and

$$V(\log t) = V(A) + \left(\frac{1}{T_0} - \lambda \right)^2 V(b) - 2 \left(\frac{1}{T_0} - \lambda \right) \text{cov} (A,b) \tag{21B.45}$$

The confidence limits for log t and, therefore, for t, the shelf life, can then be readily determined. Here again, Y_0 does not appear explicitly, but of course it is implicit in the estimate of A, and the shelf life will depend upon the initial potency.

G. Test for the Arrhenius Assumption for Zero-Order Model

In order to test the Arrhenius assumption it is necessary to the model

$$y = y_0 - K_j t, \quad j = 1, 2, ..., r \tag{21B.46}$$

where K_j is the rate constant for the jth temperature.

In terms of log potency, the model is

$$Y = Y_0 + \log \left(1 - \frac{K_j}{y_0} t \right), \quad j = 1, 2, \ldots, r \tag{21B.47}$$

For simplicity, we replace (K_j/y_0) by K_j in subsequent expressions. Just as for the first-order model and under the same assumptions, the set of Equations 21B.47 may be replaced by the single equation

$$Y = Y_0 + \sum \log (1 - K_j t_j) \simeq Y(1) - \sum \left[\frac{t_j}{1 - K_j(1)t_j} \right] \delta K_j \tag{21B.48}$$

where $K_j(1)$ are the starting values of K_j and $Y(1)$ the corresponding expected value of Y. All values of K_j must, of course, be less than $1/t_j$.

Estimates of K_j may now be obtained using the Gauss procedure. Starting values for K_j may be determined readily by graphical means by the approximate method of Section IV.H. For situation 1, where Y_0 is a constant, the K values are independent but still nonlinear. The minimization procedure also gives the sum of squares of the residuals which can be used as in the first-order model, to test the Arrhenius assumptions. Usually the estimation of the K values in Equation 21B.48 precedes the estimation of the Arrhenius constants A and b and they are used to derive starting values for the latter.

H.   Approximate Method for Zero-Order Model

The approximate method disregards the fact that y is log-normal and assumes it to be approximately normal. The assumption of normality for actual potency is reasonable for potencies down to 20% of initial potency and probably will not lead to appreciable error for potencies below this provided they do not form a high proportion of the observations taken, as, of course, they are very unlikely to do.

With this assumption we may fit the zero-order model

$$y = y_0 - e^{A - b(1/T - \lambda)} t \tag{21B.49}$$

directly by least squares using the same procedure as for the first-order model. The appropriate weighting factors must, of course, be used, these being now the inverses of the variances of the actual potencies.

The same procedure could also be used to estimate the K values for the various temperatures in order to obtain starting values for A and b to be used in the minimization process and also to derive the ANOVA table to test the Arrhenius assumption. The estimates of the K values derived in this way, when divided by y_0, can be used as starting values in the theoretically more precise method of Section IV.E.

I. Numerical Example

The results of the application of the above methods to the data of Table 21B.1 are given in the Appendix.

Figure 21B.2 gives plots of the likelihood surface for A and b, or more accurately, the surface of the residual sums of squares which is the negative of the former. Figure 21B.2(a) relates to the Levamisole data of Table 21B.1, and Figure 21B.2(b) relates to data on a drug which did not satisfy the Arrhenius law, and for which the model was a poor fit. In both cases the contours are closed and approximate to ellipses. There is a ridge in both cases, but this is approximately linear and definitely not stationary or even nearly so. The parameters A and b are clearly only slightly nonlinear. Such surfaces would not give rise to any difficulty when applying the Gauss minimization procedure, provided the initial values are reasonable, as they would be when derived by the method of Sections IV.C et seq.

It is seen that some correlation still remains between the two parameters, but it is not large enough to cause any difficulty. It is probable that the correlation would be further reduced and even eliminated by using the alternative

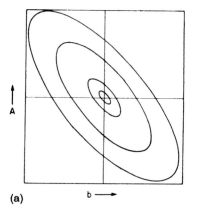
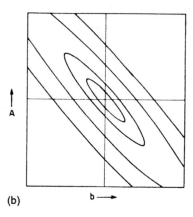

(a)   b ⟶   (b)   b ⟶

Figure 21B.2 Likelihood surface: least square estimates are the origin in each graph, range is ± 3 times the standard deviation. (a) Data which satisfy the Arrhenius law (b) Data which do not satisfy the Arrhenius law.

forms for λ given by Equations 21B.33 and 21B.37, but this refinement is not really needed.

The fact that the likelihood contours approximate to ellipses also indicates that the estimates of A and b are nearly normal. This means also that log K is nearly normal. Since for situation 1 the K values are also independent, a least-squares estimation of A and b of Equation 21B.31, using the appropriate weights, should give satisfactory estimates of these parameters. This has now been examined on several sets of data with the conclusion that the approximate method gives results for situation 1 almost identical with the theoretically more accurate method except when the Arrhenius law was clearly not applicable.

The appropriate weight for a given K is

$$W = \hat{K}^2 \sum \omega_i t_i^2$$

where i refers to the ith observation for the corresponding temperature. Since

$$\log \frac{100}{\hat{y}_i} = \hat{K}t_i$$

it follows that

$$W = \sum \omega_i \left[\log \frac{100}{\hat{y}_i} \right]^2$$

where $\hat{y}_i$ is the expected percent potency for the ith observation.

This is an interesting result because it shows that high potencies supply little information on K. Since the ω_i values decrease for decreasing potency in the manner indicated in Table 21B.2, there is an optimum potency which depends on the relative values of σ_0, σ_1, and σ_2. For $\sigma_1 = 0$ and $\sigma_2 = \sigma_0$, the optimum point is around 32.5% potency. For $\sigma_0 = 0$ and $\sigma_1 = 0$, the optimum is around 37.5%. It can be shown empirically that the optimum is fairly flat and potencies within the range 55% down to 20% are acceptable (Davies and Budgett, 1980). This has a bearing on the design of accelerated storage tests considered in Section VI.

V. WHICH MODEL?

A decision has to be taken as to whether the deterioration is approximate to zero-order or first-order. Usually when the deterioration is monitored through two half-lives, that is, covering deterioration of 75% or more, for two or more of the temperatures, the order of the reaction is apparent from visual examination of the graphs of (1) potency against time and (2) log potency against time. For deteriorations of less than 50%, discrimination may not be apparent and an objective method is needed to carry out the choice. A fairly obvious method of doing this is to compare the residual sums of squares after fitting the two models separately. The model giving the smaller residual sum of squares is the more likely.

The residual sum of squares may also be used to decide the degree of discrimination between the models supplied by the observations. For if the expected log potencies by the two models are denoted by $\hat{Y}_i(0)$, $\hat{Y}_i(1)$ respectively, for observations Y_i, then the maximum likelihoods of the respective models, given the observations, are proportional to

$$e^{-(1/2)\Sigma\omega_i\,[Y_i - \hat{Y}_i(0)]^2}; \quad e^{-(1/2)\Sigma\omega_i\,[Y_i - \hat{Y}_i(1)]^2} \tag{21B.50}$$

This assumes that the weights ω_i are inverses of the true variances—in other words, the error variances are known.

The positive values of the difference between the log likelihoods is

$$\frac{1}{2}\left|\sum \omega_i[Y_i - \hat{Y}_i(0)]^2 - \sum \omega_i[Y_i - \hat{Y}_i(1)]^2\right| \tag{21B.51}$$

It is generally accepted that if this is greater than 3, there is clear discrimination between the models.

Suppose prior information is available on the two models and this is capable of being expressed as probabilities, that is

Prior probability of zero–order model = p

Prior probability of first–order model = 1 – p

If the maximum likelihoods of the two models given the observations are L(0) and L(1), then we can give the posterior probabilities of the models. These are:

$$\left\{ \begin{array}{l} \dfrac{pL(0)}{pL(0) + (1 - p)L(1)} \text{ for zero–order model} \\[3mm] \dfrac{(1 - p)L(1)}{pL(0) + (1 - p)L(1)} \text{ for first–order model} \end{array} \right\} \tag{21B.52}$$

Discrimination is usually indicated when there is a substantial difference between these two probabilities (e.g., ≥ 0.9, or even ≥ 0.95, for the favored model).

If no prior information is available, both models are assumed equally likely before test, in which case p above would be equal to $\frac{1}{2}$, but of course we do know that a first-order model is much more likely.

This is a somewhat simplified and approximate procedure. More precisely, one should take into consideration the prior distribution for the model and for all parameters of the model, including the variance, particularly when the latter is not known precisely. A full discussion of the more rigorous procedures is given by Kanemasu (1972, 1973).

VI. DESIGN

The design of an experiment is simply the coordinates, that is, the chosen values of the independent variables on which the observations are taken. In accelerated storage tests, these are the oven temperatures and the times at which the samples are assayed.

The *efficiency* of a design is related to the amount of information supplied on the measure or measures which the experiment is designed to supply. In accelerated storage, the main purpose is to estimate the shelf life. The efficiency could then be expressed as the amount of information per unit of cost, or more generally, the cost of obtaining the required amount of information. For example, the experiment supplies an estimate of the shelf life together with its variance. The inverse of the variance measures the amount of information supplied on the shelf life, and this needs to be sufficient for the purpose of arriving at an appropriate decision.

The costs could be fairly complicated and could vary in composition and amounts with different companies. The main ingredients are usually:

1. The cost of carrying out the assays
2. The cost of the material for test and making up the ampoules
3. The cost of storing at elevated temperatures
4. A "penalty" cost for the duration of the experiment, because an early result is more valuable than a delayed result.

Careful consideration has to be given to what costs should be associated with any given experiment. Normally, overheads should not be included if the tests are carried out by the company developing the product, and in such cases marginal costs should usually be used, in other words, the difference in costs incurred by the company in carrying out and in not carrying out the test. For a company which supplies a testing service to others, different considerations may apply. It would not be practicable to go into detail on this matter here, except to say that a measure of subjectiveness has often to be used in estimating and apportioning costs. This applies particularly to item 4, above.

In a completely linear model, the amount of information supplied by a design on any linear function of the estimated parameters depends only upon the design and the error variance of the observations. This one makes it possible to determine an optimum design for the experiment whatever the estimates of the parameters turn out to be. This however, does not apply to nonlinear models in general and to those used in stability testing in particular. The variance of the shelf life is a function, not only of the design, but also of the estimated values of the parameters A and b. The problem of deriving optimum or near-optimum designs is a difficult one, and an impossible one if it is required to fix the design before the start of the experiment.

There are, however, some general principles derived from linear models which apply to nonlinear models, for example:

1. Since extrapolation is made with respect to temperature, as wide a range including as low a temperature as practicable should be used.
2. The amount of information increases with the duration of the experiment.
3. Costs increase with the duration of the experiment.

Since the weight associated with an observation decreases for low potencies, principles 1 and 2 tend to be conflicting requirements. Principles 2 and 3 are certainly conflicting requirements. It can be readily shown empirically that for any given requirement of accuracy, there is an optimum for the duration of the tests.

Intelligent application of the above principles should enable reasonable designs to be proposed; in any case, it is readily possible to assess the efficiency of any given design, and therefore to compare the efficiencies of different designs. This can form the basis of a sensible search.

It is necessary to bear in mind that the assumptions have to be tested, for example: Does the Arrhenius Law apply? Is the deterioration zero-order or first-order? And so forth. The Arrhenius law necessitates the use of at least three different oven temperatures suitably spaced.

It is known that compounds for test vary over a wide range in storage properties at elevated temperatures and that a good design for one product could be poor for another. The only real answer must be a sequential approach to the design.

Here again there are certain general principles which can be used as a guide. For instance, one needs to get information on parameters A and b which can be acted upon as early as possible in order to decide on the subsequent observations. Several reasonable initial strategies can be proposed, but further research work is needed on this question.

Predevelopment studies often give a crude indication of the drug's stability profile, and this, linked with experience, usually provides sufficient information to suggest a suitable range of temperatures to use in the accelerated storage test. This is the main hurdle in designing the test. Normally, four temperatures at intervals of 10°C are sufficient. Four ovens are usually needed because of possible errors in temperature recordings. Simulations have shown that six ampoules in each of four ovens are more than adequate for almost all situations.

When deterioration can be assumed first-order, which is almost always the case, and also, when testing situation 1 applies, it is clear that one should aim at getting as many observations as practicable in the near-optimum range up to the limit of the optimum duration of the test. It may not be possible to achieve this for the first few samples. For testing situation 2, a proportion of the observations have to be taken at zero time, the remainder being taken as far as practicable in the near-optimum range.

If it is required to test whether or not the deterioration is linear, observations would have to be taken at intermediate times.

APPENDIX

A1.   Some General Considerations of Linear Models

A linear model in one dependent variable has the form

$$\hat{y} = b_1X_1 + b_2X_2 + \ldots + b_rX_r \tag{21B.A1}$$

where $\hat{y}$ is the dependent variable (e.g., potency determinations), and the X's are the independent variables or functions of independent variables (e.g., as in a polynomial expression). For any given observation, y and all the X's take on numerical values independent of the parameters.

Let y_i denote the observed response for the ith observation; then

$$y_i = \sum_{j=1}^{r} b_jX_j(i) + \varepsilon_i, \quad i = 1, 2, \ldots, n \tag{21B.A2}$$

where $X_j(i)$ is the value of the jth independent variable for the ith observation, and ε_i is the experimental error. It is generally assumed that the ε values are independent, normally distributed with zero mean.

It is more convenient to use a matrix notation in the study of linear models. Equation 21B.A2 then becomes

$$Y = B'X + E \qquad (21B.A3)$$

where Y is the vector of the responses, B the vector of the parameters, X the $n \times r$ matrix of the independent variables $[X_j(i)]$, and E the vector of the experimental errors or residuals as it is often called. Since X specifies the conditions of the observations, it is referred to as the design matrix.

The theory of least squares results in the following expressions:

Estimates of the parameters: $B = (X'X)^{-1}X'Y \qquad (21B.A4)$

Variance and covariance of the estimates of the parameters: $(X'X)^{-1}\sigma^2$

where σ^2 is the experimental error variance. The $(X'X)$ matrix is symmetrical and positive-definite (assuming nonsingular). The sum of squares of the residual is

$$Y'Y - B'(X'X) \qquad (21B.A5)$$

When σ^2 is not known, an estimate of it is given by dividing the residual sum of squares by the degrees of freedom given by $(n - r)$. The sum of squares of the residuals is, of course, the value of the "least squares." Another expression for the sum of squares of the residuals is

$$RSS = (Y - \hat{Y})'(Y - \hat{Y}) \qquad (21B.A6)$$

where, as before, Y is the vector of the observations and $\hat{Y}$ is the expected value given by the model. Thus $(Y - \hat{Y})$ is the vector of the residuals.

In Equations 21B.A4, 21B.A5, and 21B.A6, the assumption is made that all observations have the same error variance. When, as in stability testing, this is not the case, it is necessary to weight the elements of X'X and X'Y and Y'Y by the inverse of the variance of the corresponding observation. The same result is obtained by multiplying the elements of X, Y, and $\hat{Y}$ by the square root of the inverse of the variance—that is, the inverse of the standard deviation—of the corresponding observation.

There is one further quantity which more or less completes a statement of the theory of least squares, and that is the determinant:

$$|X'X|$$

The inverse of this determinant is proportional to the volume of the ellipsoid which represents the confidence region of the parameters. In this sense, the determinant is proportional to the amount of information on the parameters supplied by the design. This quantity, divided by the cost of obtaining the observations, is then a relative general measure of the efficacy of the design when compared with other designs. This criterion is often called D optimality in the literature.

A2. Application to Nonlinear Models

A nonlinear model may be represented by

$$Y = F(B', X') + E \tag{21B.A7}$$

where B is the vector of the parameters, X the vector of the independent or derived variables, B' and X' are the transpose of B and X, Y the vector of the observations, and E the vector of the experimental errors or residuals.

The least-squares estimate of the parameters B are given by those values which minimize the weighted sum of squares of the residuals

$$\sum_i \omega_i [Y_i - F(B', X'_i)]^2 \tag{21B.A8}$$

where

$$Y_i = i^{th} \text{ response}$$
$$F(B', X'_i) = \text{expected value given by the model}$$
$$\omega_i = \text{inverse of the variance of the } i^{th} \text{ observation}$$

As shown in Section A1, the error variance may not be the same for all observations, and account has to be taken of this in the expression for the sum of squares of the residuals.

The minimization of Equation 21B.A8 cannot be done analytically but requires an iterative procedure. There are several methods and computer programs in existence which will carry out this minimization. They all require initial estimates of the parameters. These are usually rough estimates derived by graphical or simple numerical means or may even be subjective estimates. The minimization process then derives a set of closer estimates of the parameters by making certain assumptions about the nature of the model (e.g., that the model is locally linear in the parameters). The process is repeated with the new estimates to get still closer estimates, and this continues until stable values are obtained.

An uncritical use of these methods can give rise to misleading results, and consideration should be given to changing the form of the model, that is, reparameterizing the model, to avoid this eventuality. A good example which has a direct application to the models used in accelerated storage is given by

Box (1960). This concerns the estimates of the parameters a and b in the following form for the Arrhenius equation:

$$K = ae^{-b/T} \tag{21B.A9}$$

where b replaces (E/R). Now T, the oven temperature, usually ranges from 323–373°A, that is a range of 50°C, which is small compared with the mean temperature. Under these conditions, a very high correlation will arise between the estimates of a and b, and this could give rise to difficulties in the minimization procedure. A simple way of reducing this correlation to acceptable levels is to move the origin of (1/T) from 0 to (mean 1/T). The Arrhenius expression then becomes

$$a'e^{-b(1/T-\lambda)} \tag{21B.A10}$$

where

$$\lambda = \frac{\sum \omega_i(1/Ti)}{\sum \omega_i}$$

that is, λ is the weighted mean of (1/T). This formula for l has given satisfactory results in all applications encountered by the authors. Alternative and probably better formulas involving different weighting systems are given by Equations 21B.33 and 21B.37. The symbol λ is used throughout this chapter to represent the weighted mean of the reciprocal of the absolute temperatures.

A further improvement arises if a' is expressed as its logarithm. The Arrhenius expression then becomes

$$e^{A-b(1/T-\lambda)} \tag{21B.A11}$$

where $A = \log a'$. The reason for this improvement is that estimates of A are more nearly normally distributed than estimates of a'. The form of Equation 21B.A11 for the Arrhenius expression has been used throughout this chapter.

A further reparameterization which is usually worthwhile when one of the parameters is linear is to express the observed responses and their expected values as deviations from their means. This is certainly the case for both zero- and first-order models for situation 2, but of course is not needed for situation 1. A minimization program may then be safely applied to estimate the parameters and their variances and covariances.

A3.   Gauss Minimization Procedure

One of the simplest yet effective methods of minimizing the residual sum of squares is the method due to Gauss, commonly referred to as the Gauss-Newton method. Briefly, the method is as follows:

Let $F(B', X')$ denote the reparameterized model, where B' is the set of parameters $(b_1, b_2, ..., b_r)$, and X' the set of independent variables $(X_1, X_2, ..., X_r)$. For the ith observation, the observed response is Y_i and the expected response is $F(B', X'(i))$.

In the Gauss method, the function F is expanded to first-order around a set of values $b_1(1), b_2(1), ..., b_r(1)$ for the parameters, giving

$$Y = F_1 + \sum \frac{\partial F_1}{\partial b_j} \delta b_j$$

(21B.A12)

on the assumption that the function is locally linear in the parameters. Here F_1 represents the function in which the numerical values of parameters have been substituted. Similarly, $\partial F_1/\partial b_j$ is the value of the differential after substitution. Now F_1 and each $\partial F_1/\partial b_j$ are then functions only of the independent variables. For any given observation for which the independent variables take on the appropriate values, the functions F_1 and each $\partial F_1/\partial b_j$ then become numerical values. Thus Y is a linear model in the parameters δb_j which can therefore be estimated by the application of the standard regression methods outlined in Section A1. Appropriate weighting factors must be used when all observations do not have the same variance.

Starting from given values $b_j(1)$ for b_j, the above procedure has estimated δb_j, and if the model were truly locally linear, the least-squares estimates of the parameters would be

$$b_j = b_j(1) + \delta b_j, \quad i = 1, 2, ..., r$$

(21B.A13)

As it is, b_j will in general be closer to the least-squares estimates than $b_j(1)$. The procedure can now be repeated with the new values of b_j to obtain still closer approximations. Repetition is continued until stable values for the estimates of the parameters are obtained. Usually the parameter values are considered stable when either the adjustments δb_j or the reduction in the sum of squares of the residuals are acceptably small.

The next step is to derive the variance-covariance matrix of the parameters. Here again we fall back on the method used for linear models. We assume that the model is locally linear in the parameters around the least-squares estimates. Substituting the estimates obtained for the parameters b_j, say, in the final iteration, we get in general

$$Y - F = \sum \frac{\partial F}{\partial \beta_j} \delta \beta_j$$

(21B.A14)

and for the ith observation

$$Y_i - F(i) = \sum \frac{\partial F(i)}{\partial \beta_j} \delta \beta_j$$

(21B.A15)

The design matrix is the $n \times r$ matrix with the general element $\partial F(i)/\partial \beta_j$; here $i = 1, 2, ..., n$ represent the rows of the matrix, and $j = 1, 2, ..., r$ the columns. The $X'X$ matrix will then have the general element

$$\sum_{i=1}^{n} \frac{\partial F(i)}{\partial \beta_j} \frac{\partial F(i)}{\partial \beta_k}$$

The matrix has, of course, the dimension $r \times r$ where r is the number of parameters.

This matrix has to be inverted and multiplied by σ^2 to give the variances and covariances of the estimates of the parameters. That is

$$\text{Variance–covariance matrix} = (X'X)^{-1}\sigma^2$$

(21B.A16)

The residual sum of squares is $\Sigma[Y_i - F(i)]^2$, which, if required, may be divided by $n - r$ to give an estimate of σ^2.

Excellent discussions of nonlinear least squares and of the Gauss method in particular are given in Box (1957, 1960) and Box and Kanemasu (1972). These publications point out that if the initial estimates of the parameters are far removed from the least-squares estimates, the Gauss procedure may converge very slowly and may even diverge. The authors recommend in such cases that the vector defined by the b values be explored to find the point along the vector which reduces the residual sum of squares by the greatest amount. Halving and, if need be, doubling the adjustments δb and choosing the one which gives the least residual sum of squares, provided this is less than the residual sum of squares found after the previous iteration, would deal with difficult situations. In order to ensure a reduction in the residual sum of squares, it is possible in extreme cases that repeated halving of the Gauss adjustments might be required.

Difficulties of the above sort have not been encountered by the authors, who have applied the methods recommended in this chapter to a wide range of stability tests, even when the model is a poor fit. Convergence is usually obtained in three to six iterations. In case of doubt, one can fall back on the method used by Kirkwood (1977).

Analysis of Data of Table 21B.1

Sum of squares of log potency about first-order model = 64.08
Sum of squares of log potency about zero-order model = 2920.46
Decomposition is first-order.
Log frequency factor = 33.86   Standard deviation = 0.30
Activation energy/R = 1318.   Standard deviation = 109.9
Covariance = 32.87

Prior information on experimental error:

$\sigma_0 = 0.010$, $\sigma_1 = 0.000$, $\sigma_2 = 0.007$

Analysis of Variance

Source	Sum of squares	df	Mean square
About model	56.41	33	1.71
About regressions	47.64	31	1.54
Departure from Arrhenius	8.77	2	4.38
Error			1.00

Departure from Arrhenius assumption is not significant.

There is evidence that the internal estimate of error exceeds the prior estimates. The standard errors and confidence limits are therefore based on the internal estimate of error.

Individual Values of K

Temperature°C	K	log K
101.0	0.2041	−1.589
90.0	0.0690	−2.673
80.0	0.0246	−3.706
71.5	0.0104	−4.562

Shelf Life: Expected Time to 10% Loss in
Potency and Lower 95% Confidence Limit:

Temperature °C	Time (days)	Confidence limit (days)
37.0	792	717
23.0	6022	5365

ACKNOWLEDGMENT

The authors wish to thank Mr. S. H. Ellis for valuable comments and suggestions.

REFERENCES

Booth, A. (1977) Private communication.

Box, G. E. P. (1957). Use of statistical methods in the elucidation of basic mechanisms. *Bull. Inst. Int. Stat.*, **36**(3):215.

Box, G. E. P. (1960). *Fitting Empirical Data.* MCR Technical Summary Report no. 151.

Box, G. E. P. and Kanemasu, H. (1972). *Topics in Model Building, Part II, On Nonlinear Least Squares.* Tech. Rpt. no. 321, Dept. Stat., Univ. Wisconsin, Madison.

Clark, C. J. and Hudson, H. E. (1968). Acclerated storage tests: Practical considerations in their conduct and analysis. *Manuf. Chem. Aeros. News*, **39**:25-27.

Davies, O.L. (1980). Note on regression with correlated responses. *Biometrics*, **36**:551-552.

Davies, O. L. and Budgett, D. A. (1980). Accelerted storage tests on pharmaceutical products: Effect of error structure of assay and errors in recorded temperature. *J. Pharm. Pharmacol.*, **32**:155-159.

Dickenson, N. A., Hudson, H. E., and Taylor, P. J. (1971). Levamisole: Its stability in aqueous solutions at elevated temperatures. *Analyst*, **96**:248-253.

Hudson, H. E. (1966). A scheme for assessing the stability of a new drug prior to formulation. *Manuf. Chem. Aeros. News*, **37**: 40-41.

Hudson, H. E., and Jones, M. F. (1972a). The use of radioisotopically labelled compounds in drug substance and formulated product stability studies: (a) Studies with the drug substance. *Analyst*, **97**:723-725.

Hudson, H. E. and Jones, M. F. (1972b). The use of radioisotopically labelled compounds in drug substance and formulated product stability studies: (b) Application to pharmaceutical formulations. *Analyst*, **97**:726-727.

Kanemasu, H. (1972). *Topics in Model Building, Part 3: Posterior Probabilities of Candidate Models in Model Discrimination.* Tech. Rpt. no. 322. Dept. Stat., Univ. Wisconsin, Madison.

Kanemasu, H. (1973). *Topics in Model Building, Part 4: Some Problems in Model Discrimination.* Tech. Rpt. no. 323. Dept. Stat., Univ. Wisconsin, Madison.

Kirkwood, T. B. L. (1977). Predicting the stability of biological standards and products. *Biometrics,* **33**:736-742.

Kitson, G. E., Hudson, H. E., and Dickenson, N. A. (1976). Assay methods for use in stability studies of 2-amino-6-methyl-5-oxo-4-n-propyl-4,5-dihydro-[1,2,4] triazolo [1,5-a] pyrimidine. *Analyst,* **101**:463-468.

Lordi, N. G., and Scott, N. W. (1965). Stability charts: Design and application to accelerated stability testing of pharmaceuticals. *J. Pharm. Sci.,* **54**:531-537.

Rogers, A. R. (1963). An accelerated storage test with programmed temperature rise. *J. Pharm. Pharmacol.,* **15(Suppl)**: 101T-105T.

Sid Ahmed, M. E. R. (1975). M.Sc. thesis. University of Wales, Aberystwyth.

Sprent, P. (1969). *Models in Regression and Related Topics.* Methuen's Monographs on Applied Probability and Statistics. Methuen, London.

Toothill, J. P. R. (1961) A slope ratio design for accelerated storage tests. *J. Pharm. Pharmacol.,* **13(Suppl)**:75T-86T.

Willig, S. H., Tuckerman, M. M., and Hitchings, W. S. (Eds.) (1975). Stability. In *Good Manufacturing Practices for Pharmaceuticals: A Plan for Total Quality Control.* Marcel Dekker, New York, Chap. 13.

22

When and How to Do Multiple Comparisons

Charles W. Dunnett and Charles H. Goldsmith

*McMaster University,
Hamilton, Ontario, Canada*

I. INTRODUCTION

In many drug experiments in the pharmaceutical industry, at least two drugs or at least two levels of one drug are considered. As a consequence of these experiments, questions related to picking out drugs that are different from the others or the determination of what dose level is different from the others are often generated. However, it is rare that some overall test of a null hypothesis provides the researcher with specific enough details to answer the questions of interest. Because overall tests tend to average out real effects with negligible effects, they may fail to detect important features. Even if an overall test is significant, further analyses may be necessary to determine which specific differences among the treatments are clinically important. These further analyses generally constitute making many (multiple) comparisons among the treatments in order to detect those effects of prime interest to the researcher.

II. DESCRIPTION AND TAXONOMY OF MULTIPLE-COMPARISON PROCEDURES

A. Terms

Suppose we have a drug trial with k distinct treatments (different dosage levels, for example) that are randomly allocated to distinct experimental units (rats,

481

patients, baboons, etc). Suppose y_{ij} is the response of the jth experimental unit to the ith treatment. If the responses are quantitative (interval or ratio scale), it is common to summarize the responses of the n_i experimental units to the ith treatment by their *arithmetic mean*, $\bar{y}_i$.

A **linear combination** of treatment means is defined as

$$L_1 = c_1\bar{y}_1 + c_2\bar{y}_2 + \ldots + c_k\bar{y}_k \quad \left(= \sum_{i=1}^{k} c_i\bar{y}_i \right) \tag{22.1}$$

where the c_i are given constants. If $c_1 + c_2 + \ldots + c_k$ equals zero

$$\left[\sum_{i=1}^{k} c_i = 0 \right]$$

this linear combination is known as a *contrast*. A contrast is said to be *pairwise* if exactly 2 of the coefficients c_i are nonzero (one being the negative of the other in value and the other $k - 2$ being zero).

If there is a second contrast

$$L_2 = d_1\bar{y}_1 + \ldots + d_k\bar{y}_k \quad \left(= \sum_{i=1}^{k} d_i\bar{y}_i \right) \tag{22.2}$$

where $d_1 + d_2 + \ldots + d_k$ equals zero

$$\left[\sum_{i=1}^{k} d_i = 0 \right]$$

the two contrasts L_1 and L_2 are said to be orthogonal if they are uncorrelated, that is, provided

$$\frac{c_1 d_1}{n_1} + \frac{c_2 d_2}{n_2} + \ldots + \frac{c_k d_k}{n_k} = 0 \quad \left[\sum_{i=1}^{k} \frac{c_i d_i}{n_i} = 0 \right]$$

A set of $k - 1$ (or fewer) contrasts is said to be an *orthogonal set of contrasts* provided all possible pairs of contrasts in the set are orthogonal.

A contrast is said to be an *a priori* contrast if its coefficients were determined before the results of the experiment were analyzed, whereas a contrast is an *a posteriori* contrast if the coefficients were only formulated after the results of the experiment had been looked at (implying that the coefficients were chosen after seeing patterns in the results).

A treatment is said to be a *control* if it is the standard of comparison in the experiment. A control may be a *placebo* if it contains no pharmacologically active agent or it may be some active treatment that has become the *standard* comparison in that clinical, experimental, research, or therapeutic area.

Numbers assigned to objects in the measurement process can have various properties which determine the kinds of things that can be done with the measurements in their analysis. The four measurement scales are nominal, ordinal, interval, and ratio. A measurement is said to be *nominal* if the measurements serve only to distinguish the objects being measured into mutually exclusive and exhaustive groups. The most common examples of nominal scales in pharmaceutical trials are the presence or absence of various outcomes such as side effects, death, stroke, tumor, etc., where presence is indicated by 1 and absence by 0. If the measurements taken on the objects can be ordered in some sensible way, then the measurements are said to possess the ordering property in addition to distinguishability and so the scale of measurement is said to be *ordinal*. Severity scales and quality-of-outcome judgments are common ordinal scales used in drug trials. Measurements made on nominal or ordinal scales are commonly called *qualitative* measurements because they tend to reflect qualitative aspects of the things being measured. Measurements made on an *interval* scale have in addition to the distinguishability and ordering properties, the property that intervals of the same length chosen from various locations on the scale have equal meaning. Measurements such as time at a location and temperature on a Fahrenheit or Celsius scale in a drug study are interval measurements. If in addition to these three previously mentioned properties (distinguishability, ordering, equality of intervals), the measurements are such that the ratio of the two observations gives the sense of their relative magnitude, the scale of measurements is said to be *ratio*. This latter property also generally means that the zero point on the scale corresponds to the absence of the item or count being measured. Heights, weights, volumes, concentrations, and time intervals are common examples of ratio scales in pharmaceutical studies.

If the responses y_{ij} are at least ordinal (ordinal, interval, or ratio), they can be ranked from smallest to largest and these *ranks* are denoted by r_{ij}. For each of $i = 1, 2, ..., k$, the *rank sums* are

$$R_{i\cdot} = r_{i1} + r_{i2} + ... + r_{in_i} \quad \left(= \sum_{j=1}^{n_i} r_{ij} \right)$$

and *mean ranks* are

$$\bar{R}_i = \frac{R_{i\cdot}}{n_i}$$

The *overall mean rank* is $\bar{R} = (N + 1)/2$, where

$$N = \sum_{i=1}^{k} n_i$$

Multiple comparisons with ordinal scales generally use contrasts of rank means instead of the response means or contrasts of rank sums if the sample sizes are equal.

If the observations are made on a nominal scale (with 1 denoting the presence and 0 the absence of the trait being studied) then the mean response $\bar{y}i$ is the proportion of the sample possessing the trait. With these nominal scales, multiple comparisons are based on contrasts of proportions. Although it is clear that multiple comparisons can be used with all kinds of measurements, the type of procedure and its properties and the critical value table used will vary somewhat.

As a test of the significance of a contrast, the computed value of the contrast is compared with that of a certain *allowance*: if the absolute value of the contrast exceeds its allowance, it is significant at a chosen level of probability (statistical significance). Confidence limits may also be obtained by adding the allowance to and subtracting it from the value of the contrast. The differences between the various multiple comparison procedures lie in the way in which the respective allowances are determined.

Before we indicate how the allowances are calculated it is essential to understand the concept of *error rates*. By *error* one means the error of rejecting a null hypothesis when it is true (the so-called Type I error).

Tukey (1953) distinguished among three error rates. For our purposes, however, we need to be concerned with only two of them, the comparison error rate and the family (or, more correctly, "familywise") error rate. A family is a set of comparisons, usually all the comparisons made in a given experiment. It can also be a subset of these or, in principle, it may even be all comparisons in a series of experiments. The determination of what constitutes a family is difficult and is a controversial issue among statisticians. For further elaboration of this point, the reader is referred to Miller (1981, pp. 31–35).

The following definitions are based on a long series of such comparisons made over many experiments, when the null hypothesis for each comparison is in reality true:

$$\text{Comparison error rate} = \frac{\text{Number of comparisons leading to rejecting of null hypothesis}}{\text{Total number of comparisons}}$$

$$\text{Family (familywise) error rate} = \frac{\text{Number of families in which one or more null hypotheses are rejected}}{\text{Total number of families}}$$

Consider first a standard t test for comparing two treatment means. This consists of determining the difference between the two means and dividing by its standard error. The null hypothesis that the two groups have the same mean is rejected if the absolute value of the result exceeds the appropriate critical value tabulated for the Student's t distribution at the probability level α. Thus, the allowance for the difference between the two means is the critical value of Student's t multiplied by the standard error of the difference between the two means. Such a test controls the comparison error rate at the level α.

Now consider the effect of performing multiple t tests in the same experiment, each one at the probability level α. Provided that the contrasts are selected in advance (i.e, they are not determined on the basis of the results of the experiment), it will still be true that among many such tests in a large number of experiments a proportion a of those for which the null hypothesis is actually true will be falsely labelled significant. This outcome is not altered by the fact that the tests performed within the same experiment are correlated to some extent. Thus, for multiple t tests carried out at a fixed level of significance, α, the comparison error rate is controlled at the level α.

The *family error* rate takes into account the entire set of comparisons, the family being the unit. It measures the relative frequency with which families containing one or more comparisons falsely labelled significant occur among all families. This is appropriate in situations in which conclusions are based on the whole set of comparisons in the family. These are situations in which the existence of a single error within a family might jeopardize the conclusions. For example, suppose an experiment is performed to compare several new treatments with a standard with the object of selecting one of them for future use. The family of comparisons of interest might consist of each new treatment versus the standard. The experimenter would be led seriously astray if the comparison in error happened to be the one involving the selected treatment, since it could result in the replacement of the standard by a treatment that was not really superior to it.

When the family error rate is controlled, the critical allowance depends upon the number of treatment groups but not upon the actual number of comparisons within the family. Thus, without affecting the nominal value of the error rate, some of the methods can be extended to a wider class of contrasts so that additional comparisons can be added after observing how the data have turned out.

In a family of multiple comparisons, one can make correct or incorrect decisions. For a family F, the error rate is P(F), the probability of making an incorrect decision and hence $1 - P(F)$ is the probability of making a correct decision, when the null hypothesis is true. Suppose there are s individual statements in the family which may be correct or incorrect. If $P(S_j)$ is the probability that the jth statement is incorrect, then the Bonferroni inequality

$$1 - P(F) \geq 1 - P(S_1) - P(S_2) - \dots - P(S_s) = \left[1 - \sum_{j=1}^{s} P(S_j) \right] \tag{22.3}$$

can be used to place a lower bound on the probability of making a correct decision for the family. For example if $s = 5$

$$P(S_j) = \alpha_j = \alpha = 0.05, \ 1 - P(F) \geq 1 - 5\alpha = 0.75$$

Notice that this inequality does not rely on the knowledge of joint relationships among the decisions. Games (1977) gives a table based on Sidák's (1967) inequality which produces a better lower bound. Sidák's inequality is

$$1 - P(F) \geq [1 - P(S_1)] [1 - P(S_2)] \dots [1 - P(S_s)] = \prod_{j=1}^{s} [1 - P(S_j)] \tag{22.4}$$

and with the same values used in the evaluation of the Bonferroni inequality, the Sidák bound is

$$1 - P(F) \geq (1 - \alpha)^5 = (0.95)^5 = 0.77$$

In drug studies, some or all of k treatments are sometimes related to each other in a quantitative way. If certain treatments are related as if they were measured on either an interval or ratio scale of measurement, the functional relationship among these treatments can be exploited to determine functional relationships among the responses to these treatments. Quite often this occurs when the experimenter is interested in a dose-response relationship and may use functions such as orthogonal polynomials to determine the degree of the polynomial that best fits the dose-response curve.

B. Descriptions of Multiple-Comparison Procedures

In what follows, the $\bar{y}_1$, $\bar{y}_2$, ..., $\bar{y}_k$ refer to the observed treatment means and s^2 to an estimate of variance (from an analysis of variance table) that is based on, say, f degrees of freedom (df). Also, s is the positive square root of s^2 and is known as the standard deviation. These means and standard deviation summarize the experimental results of interest for making treatment comparisons, regardless of what experimental design was used to generate the data. It will be assumed that the variances within each treatment group are homogeneous. Where this is not true, the formulas quoted will require modification, as described in some of the references given.

1. The Least-Significant-Difference and Multiple t Test Procedures

The least-significant-difference (LSD) procedure and the multiple t test are based on the well-known Student's t test. For any contrast, say L_1 the allowance is

$$\text{LSD} = t_{\alpha,f} \, s_c = t_{\alpha,f} \, s \, \sqrt{c_1^2/n_1 + ... + c_k^2/n_k} \tag{22.5}$$

where s_c is the standard error of the contrast, and $t_{\alpha,f}$ is the appropriate two-tailed upper α point of Student's t distribution with f df. If the intention is to compare two treatments in a pairwise comparison, such as treatment 1 and treatment 2 as the difference between their means, such as

$$\bar{y}_1 - \bar{y}_2 \quad (c_1 = +1, c_2 = -1; c_j = 0, j = 3, 4, ..., k)$$

the allowance becomes

$$\text{LSD} = t_{\alpha,f} \, s_d = t_{\alpha,f} \, s \sqrt{n_1^{-1} + n_2^{-1}} \tag{22.6}$$

where s_d is the standard error of the difference between the two means. Use of the LSD procedure for *a priori* comparisons with quantitative data will lead to the occurrence of results falsely labelled significant in a proportion α of the comparisons; that is, the procedure controls the comparison error rate at the level α. If the absolute value of the computed contrast exceeds the allowance, the contrast is said to be statistically significant and an appropriate interpretation can be made.

2. S Method

The S method, otherwise known as Scheffé's (1953) fully-significant-difference (FSD) method controls the familywise error rate at any desired level α. This is achieved by replacing $t_{\alpha,f}$ in the formula for the LSD allowance by

$$\sqrt{(m-1)F_{\alpha,\,m-1,f}}$$

where $F_{\alpha,m-1,f}$ is the upper α point of the F distribution with $m-1$ numerator df and f denominator df.

$$[\text{Note: when } m = 2, \sqrt{(m-1)F_{\alpha,m-1,f}} = \sqrt{F_{\alpha,1,f}} = t_{\alpha,f}]$$

Here m is the number of means involved in the family of comparisons ($m \leq k$), while the allowance for any contrast, say, L_1, is given by

$$FSD = s_c\sqrt{(M-1)F_{\alpha,m-1,f}} = s\sqrt{(m-1)F_{\alpha,m-1,f}(c_1^2/n_1 + \ldots + c_m^2/n_m)}$$

$$(22.7)$$

Since the S method controls the familywise error rate at the a level for all possible contrasts that can be formed from the treatment means it is not necessary to specify *a priori* the particular contrasts of interest. *A posteriori* contrasts suggested by the data can be added and tested without affecting the nominal value of the error rate.

3. T Method

Like the S method, the T method, otherwise known as Tukey's (1953) wholly-significant-difference (WSD) method, controls the family error rate. It is particularly well suited when the main interest is in testing all the differences between $m \leq k$ treatment means, when they are taken in pairs, because the allowances for these contrasts are considerably smaller than those for the S method. The allowance for any contrast, say, L_1, is given by

$$WSD = q_{\alpha,m,f}\,s\,\frac{\displaystyle\sum_{i=1}^{m}\sum_{j=1}^{m} c_i^+ c_j^- \sqrt{n_i^{-1} + n_j^{-1}}}{\displaystyle\frac{1}{2}\sum_{i=1}^{m} |c_i|}$$

$$(22.8)$$

where $|c|$ represents the absolute value of the constant c, $c_i^+ = \max (c_i, 0)$, $c_j^- = \min (c_j, 0)$ and $q_{\alpha,m,f}$ is the upper α point of the distribution of the Studentized range for m means and f df. For a difference such as $\bar{y}_1 - \bar{y}_2$, the allowance simplifies to

$$\text{WSD} = q_{\alpha,m,f} \, s \, \sqrt{\tfrac{1}{2}(n_1^{-1} + n_2^{-1})} \tag{22.9}$$

4. Orthogonal Contrasts

Another method, also due to Tukey (1953), controls the family error rate when the family consists of a set of orthogonal contrasts (TOC: Tukey orthogonal contrasts). To obtain the required allowance, $t_{\alpha,f}$ in Equation 22.5 for the LSD is replaced with $t'_{\alpha,p,f}$, the upper α point of the distribution of the Studentized maximum modulus, where p is the number of contrasts in the orthogonal set, with f degrees of freedom. The Studentized maximum modulus is the maximum of p independent normal variates with mean zero, divided by an estimate of the standard error. This distribution coincides with Student's t for a single comparison (p = 1) but its percentage points exceed those of Student's t for two or more orthogonal contrasts in the set. With this method, the allowance for any contrast, L_1, in a set of p mutually orthogonal contrasts is

$$\text{TOC} = t'_{\alpha,p,f} \, s_c = t'_{\alpha,pf} \, s \, \sqrt{c_1^2/n_1 + \ldots + c_k^2/n_k} \tag{22.10}$$

This concept can be extended to test other contrasts that can be expressed as linear combinations of the p contrasts in the orthogonal set without modifying the error rate.

Orthogonal contrasts arise in analysis of variance when a treatment sum of squares is partitioned into single-degree-of-freedom components. In a clinical trial with four distinct treatments, for example, the treatment sum of squares could be partitioned into three mutually orthogonal contrasts of interest. One way to obtain a single overall test is to add together these three orthogonal sums of squares and apply an F test with 3 and f df. Another way is to test the p = 3 contrasts separately by using this multiple-comparison procedure based on the Studentized maximum modulus distribution. The latter tests will have the same Type I error probability as the F test, but will have greater power to detect a failure in validity that has affected one or more of the specified orthogonal contrasts.

5. Comparisons with a Control or Standard

Many times in clinical trials and dose-response studies, one of the groups has a special status and the comparisons of interest are those between each of the other treatment means and the specified mean. Examples of such comparisons are between several new treatments and a control or standard treatment, and between a new drug and several reference standard drugs. Let $\bar{y}_1$ be the mean for the specified group; then the contrasts of primary interest are $\bar{y}_2 - \bar{y}_1$, $\bar{y}_3 - \bar{y}_1$, ..., $\bar{y}_k - \bar{y}_1$. For control of the family error rate, the upper α point of a multivariate Student's t distribution is required in place of $t_{\alpha,f}$ in the LSD allowance (Eq. 22.5). The allowance for this procedure (MCC after Multiple Comparisons with a Control) is

$$MCC = t''_{\alpha,k-1,f} s_d = t''_{\alpha,k-1,f} s \sqrt{n_i^{-1} + n_1^{-1}} \tag{22.11}$$

where $t''_{\alpha,k-1,f}$ is the upper α point of a $(k-1)$-variate Student's t distribution with correlations $\rho_{ij} = \lambda_i \lambda_j$, where $\lambda_i = 1/\sqrt{1 + n_1/n_i}$, and f degrees of freedom. The allowance can be extended to test the control against any weighted mean of the treatment means.

6. Stepwise (Step-Down and Step-Up) Tests

In testing a set of k contrasts, such as a set of orthogonal contrasts or a set of contrasts representing the differences between several treatments and a specified treatment, the power for detecting true differences can be increased by performing the tests in "stepwise" fashion. Denote by t_i the statistic for testing the ith contrast, ranked in ascending order so that $|t_1| \leq ... \leq |t_k|$, where $|t_i|$ denotes the absolute value (the value with any − sign deleted) of t_i. (For 1-sided tests, the actual t's instead of their absolute values should be ordered.) Denote by c_1, ..., c_k a set of critical values to be used in the tests, c_i being the critical value to be used with t_i. In stepwise testing, the tests are carried out sequentially. Step-down tests start with the largest one, t_k, and continue downwards towards t1, until a nonsignificant result is obtained. At this point testing stops (and all remaining tests are automatically nonsignificant). In other words, the ith contrast is declared significant provided that $|t_i| \geq c_i$, $|t_{i+1}| \geq c_{i+1}$, ..., and $|t_k| \geq c_k$. The reason for the increase in power is that the c's become smaller as i decreases, making it easier to find significance. In fact, the smallest one, c_1, is the same as used in the LSD test.

Step-up tests, on the other hand, start with t_1 and continue upwards towards t_k. As long as a nonsignificant result occurs, testing continues to the next t_i. As soon as a significant result is obtained, the testing stops (and all remaining tests are automatically significant). In other words, the ith contrast is declared significant if at least one of $|t_1|$, $|t_2|$, ..., $|t_i|$ exceeds its critical value.

The critical constants needed for the tests depend upon the particular contrasts being tested. The set of c's required for step-up tests will ordinarily be slightly larger than the c's required for step-down tests, except for c_1 which is the same for both. Whether a step-down or a step-up procedure is better in terms of power depends on how many of the k contrasts are expected to correspond to real effects. If only one or a few are in this category, step-down is better but if all or most of them are, step-up is better [see Dunnett and Tamhane (1992a), who consider the case of testing a set of equally-correlated contrasts]. In Section III.B, we describe a practical example.

7. Multiple-Range Tests

The multiple-range tests—known as the Student-Newman-Keuls (SNK) after Newman (1939) and Keuls (1952) and Duncan (DCN) procedures—may be considered as a modification of the T method although they were developed independently. They are stepwise tests and are used when pairwise comparisons of the treatment means are of interest.

We illustrate these procedures for the case where the means are based on equal sample sizes. The k treatment means are ranked from the lowest to the highest:

$$\bar{y}_{[1]} \le \bar{y}_{[2]} \le, ..., \le \bar{y}_{[k]}$$

To test the difference between any two means, say

$$\bar{y}_{[i]} - \bar{y}_{[i']} \quad (i > i')$$

the WSD allowance is $q_{\alpha,k,f} s_y$, no matter which pair is being compared. The SNK procedure, on the other hand, uses the allowance

$$SNK = q_{\alpha,k',f} s_y \tag{22.12}$$

where k' is the number of means lying between and including the two being compared (i.e., $k' = |i - i'| + 1$, where i and i' represent the rank order of the two means being compared). Thus the SNK allowance is smaller than the WSD allowance except for comparisons of two extreme means, in which case they are equal. Here s_y is the standard deviation divided by the square root of n. Once a pair of ordered means $(\bar{y}_{[i']}, \bar{y}_{[i]})$ is tested and found to be nonsignificant, no other pair of treatments whose means fall between $\bar{y}_{[i']}$ and $\bar{y}_{[i]}$ can be declared to be significant.

In Duncan's multiple-range test, the allowance is

$$DCN = q_{\alpha',k',f} s_y \tag{22.13}$$

where k', as in the SNK allowance (Eq. 22.12), is the number of means between and including the two being tested, while α' is equal to $1 - (1 - \alpha)^{k'-1}$. This choice of the percentage point of the Studentized range statistic is based on the concept of *error rate per degree of freedom*, introduced by Duncan (1955). The DCN allowances (Eq. 22.13) are smaller than the WSD (Eq. 22.8) and the SNK (Eq. 22.12) allowances but larger than those of the LSD (Eq. 22.5) method.

Duncan has also developed another method for comparing all treatments in pairs, called an *adaptive k-ratio t test*, in which the size of the allowance is a function of the value of the F-statistic for testing treatment homogeneity (Duncan and Brant, 1983). However, multiple range tests continue to be popular among practitioners, particularly in the agricultural areas, although there has been considerable criticism of their indiscriminant use (Bryan-Jones and Finney, 1983).

8. Confidence Intervals

It has been assumed so far that interest is in significance testing. The estimate of the value of each contrast is simply the observed value of the contrast; confidence limits are easily obtained by adding and subtracting the value of the appropriate allowance. There is the same choice between methods of determining the allowance in confidence interval estimation as in significance testing; the confidence coefficient, $1 - \alpha$, may apply to each confidence interval separately or jointly to all the confidence intervals in a family. Thus corresponding to the comparison, question, and experiment error rates of significance testing procedures, we have comparison, question, and experiment confidence coefficients for interval estimation. There is no confidence interval procedure corresponding to either the SNK or the DCN multiple-range test.

9. Nonparametric Procedures

The multiple-comparison procedures discussed so far have dealt exclusively with quantitative data. However, many times in clinical trials or dose-response animal studies, the data that need to be analyzed are only qualitative.

Using the ranks of observations, r_{ij}, and the mean ranks R_i, the experiment-wise error rate is controlled and the allowance for a one-way design is approximately (KWA: after Kruskal and Wallis)

$$KWA = q_{\alpha,k,\infty} \sqrt{\frac{k(kn+1)}{12}}$$

$$(22.14)$$

where $q_{\alpha,k,\infty}$ is the upper α percentile point of the range of k independent standard normal random variables. Here two treatments are said to be significantly different if the absolute value of the difference of their mean ranks is larger than the allowance KWA. The procedure does not apply to general contrasts.

If one wishes to compare the $k - 1$ treatments with a control when the data are ranks, controlling the experiment-wise error rate yields an allowance that is approximately (KWC: after Kruskal-Wallis control)

$$KWC = |m|_{\alpha,k-1,\frac{1}{2}} \sqrt{\frac{k(kn + 1)}{6}}$$

(22.15)

where $|m|_{\alpha,k-1,\frac{1}{2}}$ is the upper α point of the maximum absolute value of $k - 1$ standard normal random variables with correlation $\frac{1}{2}$. Both of these are based on the Kruskal-Wallis (1952) test.

If the ranks came from a two-way design where the k treatments are ranked within each of n blocks, the experiment-wise control of the error rate for all pairwise comparisons yields an approximate allowance of

$$FRI = q_{\alpha,k,\infty} \sqrt{\frac{k(k + 1)}{12}}$$

(22.16)

where $q_{\alpha,k,\infty}$ is again, the upper α point of the range of k independent standard normal random variables (FRI: Friedman).

When treatment 1 is the control or standard and the paired comparisons of interest all involve treatment 1, the ranks derived from a two-way design provide an experiment-wise control of the error rate with the approximate allowance

$$FRC = |m|_{\alpha,k-1,\frac{1}{2}} \sqrt{\frac{k(k + 1)}{6n}}$$

(22.17)

where $|m|_{\alpha,k-1,\frac{1}{2}}$ is the upper α point of the maximum absolute value of $k - 1$ standard normal random variables with the correlation $\frac{1}{2}$ (FRC: Friedman control). Both of these allowances are based on the Friedman (1937) test statistic. When ties are present in the ranks, these procedures need to be modified (Hollander and Wolfe, 1973).

10. Multiple Comparisons Between Dose Levels and Zero Dose

In the special multiple-comparisons case where the object is to establish the minimum dose at which some undesirable response is first observed when compared to a zero dose, the procedure proposed by Williams (1971, 1972) and Shirley (1977) is appropriate. These procedures assume that the dose-response curve is monotonic even though the observed data may not be monotonic.

If the first treatment 1 is the control, and the other $k - 1$ treatments are a series of increasing doses of the same therapeutic agent, then for the Williams test, the response to the ith dose is determined by

$$\hat{M}_i = \max_{2 \leq u \leq i} \; \min_{i \leq v \leq k} \; \sum_{j=u}^{v} \frac{\bar{y}_j}{v - u + 1} \tag{22.18}$$

for $i = 2, 3, ..., k$. This estimate guarantees that $\hat{M}_2 \leq \hat{M}_3 \leq ... \leq \hat{M}_{k-1} \leq \hat{M}_k$. Once these estimates are obtained, the differences $\hat{M}_i - \bar{y}_1$, $i = k, k - 1, ..., 3, 2$ are successively compared to the allowances (WIL: Williams)

$$\text{WIL} = \bar{t}_{\alpha,k-1,f} \, s_y \, \sqrt{2} \tag{22.19}$$

where $\bar{t}_{a,k-1,f}$ is the upper α percentage point of the averaged Student's t distribution for $k - 1$ dosage levels (apart from the control) and f residual degrees of freedom. Where the usual normality assumption is not justified in the data, Shirley (1977) outlined a similar procedure based on the means of ranks of the responses rather than the responses themselves.

11. Multiple Comparisons of Proportions (0–1 Data)

If the k experimental groups contain binary responses so that the means $\bar{y}i$ are proportions of experimental units in that group with the characteristics of interest, Knoke (1976) showed that multiple comparisons based on the F distribution such as Scheffé's procedure (Eq. 22.7) are suitable provided $n \geq 10$ and the true proportions lie between 0.2 and 0.8.

When a two-way design has been used to obtain the proportions of interest, Bhapkar and Somes (1976) showed how the procedures of weighted least squares and suitably chosen χ^2 values can be used to construct allowances based on the Scheffé-type multiple comparison.

12. Other Methods

We have described the most commonly used multiple-comparison tests but there are several others that have been proposed in the literature as well. A method has been described by Gabriel (1964) which tests groups of means on the basis of the values of their sums of squares, as in an analysis of variance, rather than dealing with contrasts. Krishnaiah (1979) describes what he calls a *finite intersection* method for testing a set of contrasts chosen a priori; the critical region for these contrasts is the intersection of the usual critical regions for testing the individual contrasts with its size equated to $1 - \alpha$. If the set of contrasts coincides with that of one of the standard methods, such as all differences from a control or all pairs of treatment differences, his method coincides with the standard methods. If it is not possible to determine the critical region exactly, he uses Bonferroni or Sidák bounds as an approximation. Hochberg and Rodriguez (1977) have devised a class of "intermediate" procedures for testing contrasts which are, in a sense, midway between the Tukey T method and the Scheffé S method. They are based on enlarging the set of contrasts of primary interest to the investigator from the set of all pairwise comparisons, on which the T method is based, to include additional contrasts of a specific type, such as all differences between single treatment means and means of two treatments or all differences between means of pairs of treatments. The method produces shorter allowances than either the T method or the S method for contrasts of the particular type specified. Several authors, such as Bradu and Gabriel (1974) and Johnson (1976), have advocated multiple-comparison procedures for testing contrasts of the interaction effects in two-way experimental designs but the examples used to illustrate the methods raise questions concerning their actual utility in practice.

13.   Allocation of Observations Among the Treatment Groups

For most of the methods, the contrasts of interest are symmetric in the treatment groups and there is no reason for having certain means estimated more precisely than others by allocating unequal numbers of observations to the treatment groups. An exception occurs when the investigator has special interest in one of the treatment means, such as the mean for control or standard. Intuitively, one should allocate a higher proportion of the observations to the control or standard if the investigator is interested mainly in the treatment versus control contrasts. The problem of optimum allocation in this situation has been investigated by Bechhofer (1969) and Bechhofer and Nocturne (1972). They showed that when the variances are equal in the various treatment groups the optimum allocation approaches asymptotically the familiar "square root" rule, namely, the ratio of the number of observations assigned to the control to the number assigned to any treatment should be $\sqrt{k}$, where k is the number of noncontrol treatments. This is optimum as the size of the experiment approaches infinity; the optimum can be quite different than this when the sample sizes are small and error rates larger than the customary 1% or 5% are adopted.

C.   Summary Taxonomy

The purpose of Table 22.1 is to summarize all the techniques discussed in Section II.B of this chapter by considering the allowance short-form name, the error rate, whether it is pairwise- or general-contrast-oriented, and the key distribution that is needed to compute the allowance.

Table 22.1 Summary Taxonomy of Multiple Comparison Procedures

Allowance	Error rate	Type of contrast	Design	Parametric/ Nonparametric	Distribution
LSD	Comparison	General	Any	Parametric	t
FSD	Family	General	Any	Parametric	F
WSD	Family	General	Any	Parametric	Studentized range
TOC	Family	Orthogonal	Any	Parametric	Studentized maximum modulus
MCC	Family	Control	Any	Parametric	Multivariate t
SNK	Family	Pairwise	Any	Parametric	Studentized range
⌠DCN	df	Pairwise	Any	Parametric	Studentized range
⌡KWC	Experiment	Pairwise	Pairwise	Nonparametric	Normalized range
KWC	Experiment	Control (pairwise)	One-way	Nonparametric	Correlated multivariate normal
FRI	Experiment	Pairwise	Two-way	Nonparametric	Normalized range
FRC	Experiment	Control (pairwise)	Two-way	Nonparametric	Correlated multivariate normal
WIL	Experiment	Control	Any	Parametric	Averaged t

D. Common Sources of Tables

When it comes to applying multiple comparison procedures one needs to refer to tabulations of percentage points of the needed distributions. For the Student's t and F distributions, these can be found in many elementary statistics books, so none will be recommended here. Table 22.2 outlines the common sources of tables.

III. MULTIPLE-COMPARISON TESTS IN PRACTICE

Recall that our experiment has been performed to study the effects of k different treatments. The experimenter will be interested in making inferences about the "true" values of the treatment effects, which may take the form of either tests of hypotheses concerning certain contrasts or the estimation of confidence limits on their population values. There may be several contrasts which are of interest to the experimenter, but this does not necessarily mean that one of the multiple-comparisons procedures must be used to make the inferences. The choice between using a multiple-comparison test or an ordinary t test (or some equivalent method) depends on whether it is more appropriate to control the family error rate or the comparison error rate. If the user cannot decide which error

Table 22.2 Multiple Comparison Critical Table Sources

Distribution	Symbol	f	k	100%	Reference[a]
Studentized range	$q_{\alpha,k}$	v	r	5,1	Miller(1981):234
Studentized maximum modulus	$t'_{\alpha,p}$	v	k	5	Miller (1981):239
Multivariate t	$t''_{\alpha,k-1}$	v	p	20,10,5,1	Bechhofer & Dunnett (1988)
Studentized range	$q_{\alpha',k}$	v	k	5,1	Miller (1981):243 $[\alpha' = 1-(1-\alpha)^{k-1}]$
Normalized range	$q_{\alpha,k,\infty}$	∞	k	5,1+	Hollander & Wolfe (1973):330
Correlated multivariate normal	$|m|_{\alpha,k-1,\frac{1}{2}}$	∞	ℓ	5,1	Hollander & Wolfe (1973):365
Averaged t	$\bar{t}_{\alpha,k-1}$	v	k	5,1	Williams (1971)

[a] Number following the colon is a page number in the reference if a book.

rate is more appropriate, some statisticians would advise the user to use a multiple-comparison procedure on the grounds that it is the conservative course of action to take, since setting the family error rate at some value a ensures that the comparison error rate will be less than α.

However, it should be remembered that these error rates refer to Type I error (rejecting null hypotheses which are true). But there are other types of error to be concerned about as well; namely, failing to reject null hypotheses which are false (Type II errors) and the use of a multiple-comparison procedure in place of a standard test such as a t test will result in an increase in the probability of making a Type II error. In other words, the power of the test decreases.

A. Is a Multiple-Comparison Procedure Needed?

Here, we will consider some typical examples arising in pharmaceutical research to illustrate some of the reasons for using (or not using) multiple-comparison procedures. In general, the use of an appropriate multiple-comparison test to make inferences concerning treatment contrasts is indicated in the following situations:

1. To make an inference concerning a particular contrast which has been selected on the basis of how the data have turned out.

2. To make an inference which requires the simultaneous examination of several treatment contrasts.
3. In "data dredging," namely, assembling the data in various ways in the hope that some interesting differences will emerge.

On the other hand, multiple-comparison procedures are usually not appropriate when the particular contrasts to be tested are selected in advance and are reported individually rather than as a group. In such situations, the comparison error rate is usually of primary concern and the standard tests of significance can be used, rather than a multiple-comparison test.

For simplicity, we will assume in the examples considered in the rest of this section that the responses observed in an experiment give rise to quantitative data, that the latter satisfy the standard assumptions of model I (treatments fixed) analysis of variance, and that there is equal replication of each of the k treatments. Thus (as in Section II.B), the data from the experiment can be summarized in the form of a set of values: $\bar{y}_1, \bar{y}_2, ..., \bar{y}_k$; s^2 where $\bar{y}_i$ is the mean response for the ith treatment ($i = 1, 2,..., k$) and s^2 is a variance estimate based on f degrees of freedom. We represent by s_d the standard error of any treatment difference $\bar{y}_i - \bar{y}_j$. For a one-way design with n observations on each treatment, for example, we would have $f = k(n - 1)$ and $s_d = s\sqrt{2/n}$.

1. Testing a Selected Contrast

When a set of experimental data is examined, it often happens that some particular feature of the configuration of treatment means may catch one's eye and the question arises as to whether or not it is significant. Consider, for example, the data shown in Table I of Dunnett (1964), which represent measurements of the percentage fat content of breast muscle in cockerels on 4 different treatments: 1, 2, 3, and 4. The following mean values were obtained:

$$\bar{y}_1 = 2.493, \bar{y}_2 = 2.398, \bar{y}_3 = 2.240, \bar{y}_4 = 2.494$$

The birds which received treatment 1 were untreated controls while treatments 2, 3, and 4 were particular drugs. Each treatment mean shown is based on 20 independent values. The error mean square was $s^2 = 0.1086$ with $f = 64$ degrees of freedom, calculated from the analysis of variance, from which the standard error of the difference between any two treatments is $s_d = \sqrt{0.1086 \times 2/20} = 0.104$. The analysis of variance F test for testing the between-treatments mean square was significant, although it is not necessary to do a preliminary F test before proceeding with multiple comparisons (as some texts recommend). Performing a preliminary F test may miss important single effects that get diluted (averaged out) with other effects.

On examining these results, the experimenter was somewhat surprised at the low value for group 3 and asked the question "Is it significantly different from the control?" Thus, he is asking whether the contrast $\bar{y}_1 - \bar{y}_3 = 0.253$ is significantly different from zero.

To test the significance of a contrast, its observed value is compared with an allowance which is calculated by one of the methods described in Section II. Using the LSD, we would have LSD $= t_{\alpha,f} s_d$, where $t_{\alpha,f}$ is the upper α point of Student's t distribution with $f = 64$ degrees of freedom. From tables of Student's t percentage points, we find that $t_{.05,64} = 2.00$ and the LSD allowance becomes 0.208, so the contrast $\bar{y}_1 - \bar{y}_3 = 0.253$ would be judged significant. However, the LSD is inappropriate as an allowance in this case, because it does not take into account the fact that $\bar{y}_1 - \bar{y}_3$ is a selected contrast: it was chosen specifically because the value of $\bar{y}_3$ was observed to be low.

To decide what is the appropriate allowance to use in this case, it is necessary to determine the family of contrasts from which this one was selected. In the present case, it if can be assumed that only contrasts in which there are differences from the control would be considered for statistical testing, then the family consists of $\bar{y}_1 - \bar{y}_2$, $\bar{y}_1 - \bar{y}_3$, and $\bar{y}_1 - \bar{y}_4$. The appropriate allowance to the selected contrast would be MCC $= t''_{\alpha,k,f} s_d$; taking $t''_{.05,4,64} = 2.41$ from Table II in Dunnett (1964), we obtain MCC $= 0.251$. Thus, the contrast of interest is still significant but barely so.

On the other hand, if other contrasts might also have caught the experimenter's eye and the configuration of observed treatment means turned out differently, then the family would have to be enlarged and a different multiple-comparison test would have to be used. For instance, if any difference between two means was to be included then the T method of Tukey should be used and the allowance would become WSD $= q_{\alpha,k,f} s_y = (3.74)(0.074) = 0.277$ where $q_{.05,4,64} = 3.74$ is obtained from tables of the Studentized range.

The need to test selected contrasts arises frequently in pharmaceutical research. Consider, for example, the following situation. Suppose a new drug is under development and many chemicals of related structure to a known active compound can be synthesized. It may happen that several potential candidates become available and a choice has to be made to decide which one should be carried through to the clinical trial stage.

Let k be the number of candidates, where $k \geq 2$, and suppose an experiment is to be performed to measure a particular response, with the one producing the highest mean response to be selected. Note that it is not a question of determining one which is significantly better than the others but simply of picking the one which produces the highest observed mean. There are interesting statistical problems involved in the design of the experiment so that the best candidate will have sufficiently high probability of producing a higher mean than any of the others. We will not consider these here, but the interested reader is referred to Gibbons et al. (1977)

In such an experiment, a control or placebo treatment is often included for the purpose of estimating the "no-effect" response level. Let $\bar{y}_0$ be the no-effect treatment mean and the observed treatment means be $\bar{y}_1, \bar{y}_2, ..., \bar{y}_k$. Then the experimenter chooses $\bar{y}_{max} = \max \{\bar{y}_1, \bar{y}_2, ..., \bar{y}_k\}$ and the drug corresponding to $\bar{y}_{max}$ is the one chosen to undergo further development in preparation for clinical trials. Having chosen it, however, the company management might wish to be assured that $\bar{y}_{max}$ is significantly different from $\bar{y}_0$. What p-value can be associated with $\bar{y}_{max} - \bar{y}_0$? Clearly, this is a selected contrast and the family consists of all $\bar{y}_i - \bar{y}_0$, $i = 1, 2, ..., k$. Thus the appropriate p-value is determined from the MCC test previously discussed. Calculations of $t'' = (\bar{y}_{max} - \bar{y}_0)/s_d$ reference to the k-variate t distribution, with the degrees of freedom associated with s_d, tabulated in Bechhofer and Dunnett (1988) would give the required value of p. Note that if a different number n_1 of observations had been obtained on the control then the number n on each treatment, the $s_d = s \sqrt{n^{-1} + n_1^{-1}}$ and the tables would be entered with correlation coefficient $\rho = 1/(1+n_1/n)$.

2. Comparisons Between a New Drug and Active and Placebo Controls

This example illustrates that there can be situations where the set of contrasts of interest in an experiment forms more than one family. Suppose that a pharmaceutical company has developed a new drug which it wishes to market. To obtain permission from the regulatory authorities to do this, detailed information about the new drug must be submitted, including evidence that it possesses the claimed activity and that it is either equally effective or superior to existing drugs already on the market. Such evidence is obtained from clinical trials in which the efficacy of the new drug is compared with one or more reference drugs and a placebo. The purpose of the placebo, in addition to providing a baseline for measuring the efficacy of the new drug, is to verify that the trial is able to distinguish between the placebo and the known active drugs (referred to as the sensitivity of the trial). Of course, for ethical reasons it may not always be possible to include a placebo.

One way to design the efficacy trials would be to set up a number of trials, in each of which there are two treatment groups, one receiving the new drug and the other one of the reference drugs, and perhaps a third group receiving the placebo. However, a more efficient experimental design is to include all the treatments of interest together in each clinical trial. Suppose there are k (≥ 1) reference drugs. Then each trial provides a set of k + 2 observed treatment means: $\bar{y}_1, \bar{y}_2, ..., \bar{y}_k$, the means for the k reference drugs, $\bar{y}_T$ the mean for the test drug and $\bar{y}_0$ the mean for the placebo. Comparisons can be made either between the observed means within each trial or between appropriate mean values over the entire set of trials. In either case, multiple comparisons are involved and the statistician is faced with the problem of deciding which is the proper test to use.

It is necessary to consider first which treatment comparisons are of interest to the pharmaceutical company and to the regulatory agency. Although any contrast among the $k + 2$ treatment means is potentially of some interest, it must be remembered that the main purpose of the trial is to: 1) demonstrate that the new drug is active relative to the placebo, 2) verify the sensitivity of the trial by comparing the known actives with the placebo, and 3) compare the efficacy of the new drug with that of the reference drugs. Other comparisons, such as comparisons among the known actives, are superfluous to the main purpose of the trial. Thus, the comparisons of primary interest are limited to the particular set of $2k + 1$ contrasts and form three families: 1) $\bar{y}_T - \bar{y}_0$ representing the difference between the test drug and placebo, 2) $\bar{y}_1 - \bar{y}_0$, ..., $\bar{y}_k - \bar{y}_0$ representing the differences between the reference drugs and placebo, and 3) $\bar{y}_T - \bar{y}_1$, ..., $\bar{y}_T - \bar{y}_k$ representing the differences between the test drug and the reference drugs.

Inferences concerning this set of contrasts can be made either by performing significance tests or by calculating confidence limits. For significance testing, appropriate null hypotheses have to be specified. Since the pharmaceutical firm must establish that the new drug is active, the first step is to compare it with the placebo by testing the null hypothesis that the difference of the true mean of the new drug and the true mean of the placebo is zero; the alternative hypothesis is that the true difference is nonzero (or perhaps positive). The statistic for testing the null hypothesis is the contrast $\bar{y}_T - \bar{y}_0$, and since this contrast is chosen because it is pertinent for this particular null hypothesis and not because it is selected on the basis of the observed data as in the previous example (even though it may indeed be the largest contrast in the set), the appropriate allowance for determining its significance is the LSD (Eq. 22.5). Use of the LSD test at a particular significance level α will guarantee that the probability of the pharmaceutical company finding a significant difference between its new drug and the placebo, when the new drug is really not different from the placebo, does not exceed α. This error rate is not affected by the fact that there are other treatments in the experiment.

Now, unless this particular treatment contrast $\bar{y}_T - \bar{y}_0$ is sufficiently significant (i.e., has a sufficiently small p-value associated with it) to establish that the response to the new drug is indeed different from the placebo response, the testing of the remaining contrasts involving $\bar{y}_T$ is rather academic. Thus the testing of the remaining contrasts is conditional on a significant difference having been obtained for $\bar{y}_T - \bar{y}_0$.

The next set of contrasts to be tested pertains to the sensitivity of the trial, namely, $\bar{y}_1 - \bar{y}_0$, ..., $\bar{y}_k - \bar{y}_0$. Since these are known actives, we expect all of them to test significant. If any of them fail to show significance then we have failed to establish the sensitivity of the trial, at least with respect to the particular reference drugs that do not differ significantly from the placebo. Berger (1982) had the insight to realize that the type of test where all null hypotheses are to be rejected, to establish that a particular state of affairs pertains, requires a different type of multiple comparison test. The null hypothesis to be tested is that at least one of the differences between the true mean of a reference standard and the true mean of the placebo is zero, versus the alternative hypothesis that all the differences are positive (for a 1-sided test), or non-zero (for a 2-sided test). As Berger showed, the correct test for each is an ordinary comparisonwise test at level α (e.g., the LSD). This was considered in detail by Laska and Meisner (1989) and was the basis of their "min" test. Thus, each of the contrasts $\bar{y}_i - \bar{y}_0$ is tested for significance by the LSD and sensitivity is established if all the $\bar{y}_i - \bar{y}_0$ exceed their allowance.

Dunnett and Tamhane (1992b) have proposed the use of a step-up test (see Section II.B.6), with the critical value c_1 identical with the LSD so that the step-up test also establishes sensitivity when the min test does. The advantage of the step-up test is that if one or more of the $\bar{y}_i - \bar{y}_0$ fail to exceed the LSD allowance, it might still be possible to establish "partial sensitivity" with respect to some of the reference drugs. For example, if $\bar{y}_1 - \bar{y}_0$ does not reach significance with respect to its allowance based on c_1 but $\bar{y}_2 - \bar{y}_0$ is significant with respect to c_2, then we can conclude that, although the first reference standard fails to satisfy the sensitivity criterion, all the remaining standards do (we assume we have ordered the reference drugs according to their t-values as described for stepwise testing).

While ordinarily it is not necessary for the pharmaceutical company to establish a significant difference between their new drug and any of the reference drugs to justify it being allowed to go on the market, it would clearly be to their advantage to be able to do so if they can. The null hypothesis to be tested is that the new drug is not superior to any of the reference drugs; the alternative is that the new drug is superior to at least one of the reference drugs. Rejection of the null hypothesis would indicate that the new drug is superior to at least one of the reference drugs and thus strengthen the case of the pharmaceutical company to have it allowed on the market.

To test the null hypothesis with a specified upper bound on the probability of rejecting the null hypothesis when it is true, an extension of the step-down procedure described for equal sample sizes in Miller (1981, p. 86) can be used; see Dunnett and Tamhane (1991). The t-statistics for the set of contrasts $\bar{y}_T - \bar{y}_1, ..., \bar{y}_T - \bar{y}_k$ are ordered (as described in Section II.B.6) and compared with critical values $c_1, c_2, ..., c_k$. These critical values are the same as those used in the MCC allowance (Eq. 22.11). The largest t statistic t_k is thus compared with the MCC allowance for k comparisons; if it is significant, the next largest t_{k-1} is compared with the MCC allowance for $k-1$ comparisons, and so on. This step-down procedure is analogous to the stepdown of the Studentized range in the SNK procedure (Eq. 22.15).

Thus, the three families of contrasts for comparing first the test drug with the placebo, then the known active drugs with the placebo, and finally the test drug with the known active drugs can each be tested separately using an appropriate α-level test for each family. Since each family is tested conditionally on significance having been found for any previous families, the use of level α for each family ensures that the overall error rate is also less than α. So there is no need to consider the entire set of $2k + 1$ contrasts as a single family, which would cause a decrease in the power. This is discussed in more detail in Dunnett and Tamhane (1992b).

For example, suppose $s = 2$ and the following treatment means are obtained: $\bar{y}_0 = 6.9$, $\bar{y}_T = 10.2$; $\bar{y}_1 = 7.6$, $\bar{y}_2 = 9.0$, with $s_d = 1.22$ and $f = 20$ degrees of freedom. Then the new drug would be declared significantly different from the placebo since $(\bar{y}_T - \bar{y}_0)/s_d = 2.70$, which exceeds the $\alpha = 0.01$ point (two-tail) of Student's t with 20 df ($t_{.01,20} = 2.086$). Next, it would be declared superior to drug 1 since $\bar{y}_T - \bar{y}_1 = 2.6$ exceeds the allowance MCC $= t''_{.05,2,20}\, s_d = (2.03)(1.22) = 2.5$, but not to drug 2 since $\bar{y}_T - \bar{y}_0 = 1.2$ does not exceed the allowance MCC $= t''_{.05,1,20}\, s_d = (1.72)(1.22) = 2.1$ (using one-tail values for these tests, due to the fact that the alternative hypothesis is one-sided).

3. Combination Drugs

A problem that has been of special concern to pharmaceutical companies and to the regulatory authorities is the use of combination drugs. There are many possible reasons for combining more than one drug together in the same tablet or capsule. Sometimes a second drug is added to counteract a possible side effect of the main drug; sometimes the range of action can be increased by including more than one drug, such as in controlling an infection where it is not known which drug would be the best one to use against the particular microorganism causing the infection. Whatever may be the reason for combining two or more drugs together, the regulatory agency would require evidence from the sponsor of the product that the combination is more useful medically than any subcombination of the ingredients.

The whole question of demonstrating the efficacy of a combination drug product is very complex and we will consider only one aspect of it here to illustrate the use of simultaneous inference. Suppose there is just one response to the drug to be considered, such as a fall in systolic blood pressure. If a combination drug is proposed to bring about this response it would be necessary to establish that the combination is significantly better than each of the others. This would involve simultaneous inferences concerning the differences between the mean for the combination drug (e.g., drug 1) and the means for each of the subcombinations (e.g., drugs 2, 3, ..., k) assumed to be $k - 1$ in number. We would need to test the null hypothesis that the true mean of drug 1 was less than or equal to the true mean of drug i, i = 2, 3, ..., k, for at least one subcombination versus the alternative that the true mean of the combination drug was greater than all of the true means of the subcombinations. Note the similarity between the null hypothesis being tested here and the one for testing for sensitivity in the previous example. Here, rejection of the null hypothesis would require each of the differences $\bar{y}_1 - y_i$, i = 2, 3, ..., k, to fall in the rejection region rather than merely the largest one. The critical value needed to test these differences, chosen to achieve a test of size a for all true differences consistent with the null hypothesis, is based on the α-point of univariate (not multivariate) Student's t. For further discussion see Hochberg and Tamhane (1987, p. 13) and Patel (1991).

4. Data Dredging

We hesitate to include an example of data dredging since pharmaceutical experimenters are not supposed to stoop to that sort of thing! It is true that, in most experiments, the main treatment comparisons of interest are determined in advance of doing the experiment. The purpose of doing the experiment is usually to estimate certain prespecified treatment effects and not to look for new leads to follow up. Nevertheless, having tested the main treatment effects of interest, it can be tempting to look at the data in other ways just in case "something might show up." This may be particularly the case in medical experiments involving human patients since a large amount of subsidiary information is usually available about the patients and so it may be worthwhile to examine other effects. In a trial to compare two or more treatments, for example, in addition to comparing the mean responses of the treatment groups, the experimenter may wish to make additional comparisons between subgroups of patients. For instance, the experimenter may wish to determine whether male patients responded more than females, whether the age of the patient was a factor, and so on. This process can generate quite a large number of comparisons to be made in addition to the primary comparisons which were the main reasons for doing the experiment.

The best way to treat such comparisons is in an ad hoc manner, without trying to attach any p-value to them or doing any formal test of significance. Any effect that looks interesting (by virtue of it being large compared with its standard error) should be looked upon as something to be studied further in another trial. However, if the experimenter must calculate a p-value, it would be necessary to take account of the actual number of comparisons that have been attempted and use the Bonferroni method to arrive at a p-value, which would be an upper bound on the actual value of p. Shafer and Olkin (1983) have considered this problem and obtained a mathematical justification for the use of the Bonferroni method in this way.

For example, suppose a clinical trial is done to compare two drugs, denoted by A and B, with n = 100 subjects assigned to each of the treatments. The main comparison in the trial would be the mean response on drug A, $\bar{y}_A$ versus the mean response on drug B, $\bar{y}_B$ and the standard error of $\bar{y}_A - \bar{y}_B$ could be determined. Now suppose the subjects are split according to their sex and measures of the treatment differences obtained separately for the males and females, giving $\bar{y}_{A,m} - \bar{y}_{B,m}$ for the males and $\bar{y}_{A,f} - \bar{y}_{B,f}$ for the females. Similar splits could be obtained on the basis of age, disease severity, and a number of other factors, each one providing a measure of the treatment difference and a standard error depending on the number of subjects. Selecting a particular treatment difference because it is large compared to its standard error would have to be assessed by referring the value of its $(\bar{y}_A - \bar{y}_B/s_d$ to Student's t distribution with the appropriate degrees of freedom and multiplying the p obtained from the t distribution by the total number of such comparisons made, according to the Bonferroni method. Thus, if the 0.005 point of Student's t were reached and 30 such tests had been made, an upper bound on the p-value to be associated with the comparison would be (0.005) (30) = 0.15.

5. Drug Screening

Drug screening is another example where multiple comparisons are made within the same experiment since in routine drug screening there are usually several new drugs being tested simultaneously. Let k be the number of the test drugs in the trial, and $\bar{y}_i$ the response for the ith drug (i = 1, 2, ..., k). Often there is a control group, or a group treated with a standard drug, producing a response y_0 and the decision to accept the drug, to reject the drug, or to repeat the test is based on the magnitude of the difference $\bar{y}_i - \bar{y}_0$ (i = 1, 2, ..., k), or on its cumulative mean value if the drug has been tested in previous stages.

Although one could use a multiple-comparison procedure for reaching a decision on each observed difference $\bar{y}_i - \bar{y}_0$, there is no real justification for doing so. For a particular drug screening test, its important property is the proportion of drugs which are correctly classified as active or inactive in the long run. Thus the comparisonwise error rate is of concern, not the familywise or experimentwise error rate. The reason for including several drugs in the same experiment is not to do multiple comparisons but simply because it is more economical than testing them one at a time.

B. Analysis of a Randomized Trial

The following fictitious randomized trial was constructed to illustrate a variety of different multiple-comparison procedures with the same set of data.

This trial was mounted to test out a new nonsteroidal antiinflammatory drug (NSAID) called Arthritol. Animal studies had shown that the relative

Table 22.3 Results of Arthritol Randomized Trial

		Placebo	Aspirin (20 mg)	Arthritol dose (mg)				
				10	15	20	25	
Patient number $i =$		1	2	3	4	5	6	Total
$j =$	1	1.0^a	1.3	2.1	1.9	2.5	6.5	
	2	−0.6	2.7	1.1	1.0	2.0	3.0	
	3	0.7	2.1	2.4	0.9	4.0	2.4	
	4	1.4	0.7	0.1	1.7	4.0	3.5	
	5	1.0	3.6	0.1	1.9	1.9	4.3	
	6	1.8	1.9	−0.1	1.5	3.3	3.3	
	7	0.2	3.9	−0.3	2.2	4.3	2.3	
	8	1.7	−0.8	0.8	3.7	2.1	2.7	
	9	0.4	2.2	−0.6	0.1	2.8	2.6	
	10	1.0	1.9	0.6	−0.1	2.4	4.7	
	11	0.2	2.8	0.3	0.6	2.7	4.7	
	n	11	11	11	11	11	11	66
$\sum_{j=1}^{11} y_{ij}$		8.8	22.3	6.5	15.4	32.0	40.0	125.0
$\sum_{j=1}^{11} y_{ij}^2$		12.18	62.59	12.95	33.08	100.54	162.16	383.50
y_i		0.80	2.03	0.59	1.40	2.91	3.64	1.89
s_i		0.72	1.32	0.95	1.07	0.86	1.29	1.50

[a] Observations generated by adding a N(0,1) random variable to the effects 0.5, 2.0, 0.6, 1.3, 3.0, and 4.1, respectively. Random N(0,1) values obtained from the Rand Corporation (1966) tables.

Table 22.4 ANOVA of Arthritol Trial

Source	df	SS	MS	F	P
Mean	1	236.742			
Treatments (cfm)[1]	5	79.4521	15.89	14.17	0.001
Residual	60	67.3054	1.12		
Total	66	383.5			
Mean	1	236.742			
Total (cfm)	65	146.75			

[1] cfm = corrected for the mean.

potency of Arthritol was estimated to be 0.90 when compared with plain aspirin. Consequently, a one-way randomized trial was designed with a placebo, a standard of 20 mg aspirin, and four doses of Arthritol: 10, 15, 20, and 25 mg. Sample size calculations dictated that n = 11 patients were randomly allocated to each of the six different treatment groups.

The outcome measure chosen was the pooled index developed by Smythe et al. (1977), which was measured by an independent assessor (IA) in these Rheumatoid Arthritis (RA) patients before and after a 2-week treatment period. The IA was unaware of the treatment group to which any patient belonged.

The results of this trial are displayed in Table 22.3. Preliminary analyses showed there was no difference in the measures of variation [Hartley, 1950: $F_{max} = 3.36$, critical F_{max} ($\alpha = 0.05.$, k = 6, v = 10) = 6.92]. For purposes of estimating the variance to be used in the multiple-comparison procedures, a one-way analysis of variance was calculated and is displayed in Table 22.4.

Although this ANOVA showed that the treatments were quite different (p < 0.001), this was anticipated. Before the study was conducted, however, the following questions were of interest:

1. Is any dose of Arthritol different from placebo?
2. What is the lowest dose of Arthritol that is better than placebo?
3. Is any dose of Arthritol different from aspirin at 20 mg?
4. What is the nature of the dose-response of Arthritol? Is it linear, or does it need a higher polynomial?

It was also agreed that once a specific multiple-comparison procedure was chosen to answer a question, the testing would be done at $\alpha = 0.05$.

To answer question 1, the MCC allowance (Eq. 22.11) is employed where $s = \sqrt{1.121} = 1.059$, $f = 60$, $k = 4 + 1 = 5$, the 4 doses of Arthritol plus the placebo. Then $t''_{.05,4,60} = 2.51$.

$$\text{MCC} = \frac{(2.51)(1.059)(\sqrt{2})}{\sqrt{11}} = 1.133$$

The paired comparisons of interest are

$$\bar{y}_3 - \bar{y}_1 = -0.21,\ \bar{y}_4 - \bar{y}_1 = 0.60,\ \bar{y}_5 - \bar{y}_1 = 2.11,\ \bar{y}_6 - \bar{y}_1 = 2.84$$

Doses 20 and 25 mg of Arthritol are significantly different from placebo but doses 10 and 15 are not significantly different from placebo.

To answer question 3, the same allowance MCC could be used. This time the paired comparisons of interest would be

$$\bar{y}_3 - \bar{y}_2 = -1.44,\ \bar{y}_4 - \bar{y}_2 = -0.63,\ \bar{y}_5 - \bar{y}_2 = 0.88,\ \bar{y}_6 - \bar{y}_2 = 1.61$$

Consequently, 10 mg of Arthritol is significantly lower than the 20 mg of aspirin and the 25 mg of Arthritol is significantly higher than the aspirin dose. Also, both the 15 and 20 mg doses of Arthritol are not significantly different from the aspirin dose.

Question 2 can be dealt with by using the WIL (Eq. 22.19) allowance. Here $\alpha = 0.05$, $k - 1 = 4$, $f = 60$ so $\bar{y}_{.05,4,60} = 1.78$. Now

$$\text{WIL} = \frac{(1.78)(1.059)(\sqrt{2})}{\sqrt{11}} = 0.804$$

Then

$$\hat{M}_3 = \frac{\bar{y}_1 + \bar{y}_3}{2} = 0.695,\ \hat{M}_4 = \bar{y}_4 = 1.40,\ \hat{M}_5 = \bar{y}_5 = 2.91,\ \hat{M}_6 = \bar{y}_6 = 3.64$$

Hence

$$\hat{M}_6 - \bar{y}_1 = 2.84,\ \hat{M}_5 - \bar{y}_1 = 2.11,\ \hat{M}_4 - \bar{y}_1 = 0.60,\ \hat{M}_3 - \bar{y}_1 = -0.105$$

By comparing these differences to the allowance WIL $= 0.804$, a dose of 20 mg of Arthritol or more gives a significantly higher response than the placebo.

To answer question 4, a set of orthogonal contrasts is applied to the four treatment means $\bar{y}_3$, $\bar{y}_4$, $\bar{y}_5$, and $\bar{y}_6$. Since the linear, quadratic, and cubic

Table 22.5 Orthogonal Polynomials of Arthritol Doses

| Contrast name | Arthritol (mg) | | | | $\sqrt{c_3^2+c_4^2+c_5^2+c_6^2}$ | TOC | Contrast value |
| | $\bar{y}_3 =$ 0.59 | $\bar{y}_4 =$ 1.40 | $\bar{y}_5 =$ 2.91 | $\bar{y}_6 =$ 3.64 | | | |
	c_3	c_4	c_5	c_6			
Linear	−3	−1	1	3	$\sqrt{20} = 4.47$	3.513	10.66
Quadratic	1	−1	−1	1	$\sqrt{4} = 2$	1.571	−0.08
Cubic	−1	3	−3	1	$\sqrt{20} = 4.47$	3.513	−1.48

polynomial terms can be fitted to the four means, $p = 3$, $\alpha = 0.05$, $f = 60$, and the critical value used is

$$t'_{.05,3,50} = 2.46 \text{ for TOC} = \frac{(2.46)(1.059)}{\sqrt{11}} \sqrt{c_1^2 + \ldots + c_6^2}$$

In all these contrasts, $c_1 = c_2 = 0$. Table 22.5 displays the actual computations.

Since only the linear contrast exceeds its critical value given by TOC, the evidence indicates that response of Arthritol can be described as linear in the dose range given. Strictly speaking, this conclusion should cause the experimenter to revise the answers to questions 1 and 3! If the response function is linear down to zero, any dose is different from the placebo and bioassay methods would be used to estimate the dose that gives the same response as aspirin. This could also be used to estimate the dose that is equivalent to placebo.

REFERENCES

Bechhofer, R. E. (1969). Optimal allocation of observations when comparing several treatments with a control: Multivariate analysis, II. *Proceedings 2nd International Symposium on Multivariate Analysis*, (Krishnaiah, P. R., Ed.), Academic Press, New York, pp. 463-473.

Bechhofer, R. E. and Dunnett, C. W. (1988). Tables of percentage points of multivariate Student t distributions. *Selected Tables in Mathematical Statistics*, **11**:1-371.

Bechhofer, R. E. and Nocturne, D. J-M. (1972). Optimal allocation of observations when comparing several treatments with a control, II: 2-2-sided comparisons. *Technometrics*, **14**:423-436.

Berger, R. L. (1982). Multiparameter hypothesis testing and acceptance sampling. *Technometrics*, **24**:295-300.

Bhapkar, V. P. and Somes, G. W. (1976). Multiple comparisons of matched proportions. *Comm Stat: Theory and Methods, Ser. A*, **5(1)**:17-25.

Bradu, D. and Gabriel, K. R. (1974). Simultaneous statistical inference on interactions in two-way analysis of variance. *J. Amer. Statist. Assoc.*, **69**:428-436.

Bryan-Jones, J. and Finney, D. J. (1988). On an error in "Instructions to Authors". *Hort. Sci.*, **18**:279-282.

Duncan, D. B. (1955). Multiple range and multiple F-tests. *Biometrics*, **11**:1-42.

Duncan, D. B. and Brant, L. J. (1983). Adaptive t tests for multiple comparisons. *Biometrics*, **39**:790-282.

Dunnett, C. W. (1964). New tables for multiple comparisons with a control. *Biometrics*, **20**:482-491.

Dunnett, C. W. and Tamhane, A. C. (1991). Step-down multiple tests for comparing treatments with a control in unbalanced one-way layouts. *Statistics in Medicine*, **10**:939-947.

Dunnett, C. W. and Tamhane, A. C. (1992a). A step-up multiple test procedure. *J. Amer. Statist. Assoc.*, **87**:162-170.

Dunnett, C. W. and Tamhane, A. C. (1992b). Comparisons between a new drug and active and placebo controls in an efficacy clinical trial. *Statistics in Medicine*, **11**:1057-1063.

Friedman, M. (1937). The use of ranks to avoid the assumption of normality implicit in the analysis of variance. *J. Amer. Statist. Assoc.*, **32**:675-701.

Gabriel, K. R. (1964). A procedure for testing the homogeneity of all sets of means in analysis of variance. *Biometrics*, **20**:459-477.

Games, P. A. (1977). An improved t table for simultaneous control on g contrasts. *J. Amer. Statist. Assoc.*, **72**:531-534.

Gibbons, J. D., Olkin, I., and Sobel, M. (1977). *Selection and Ordering Populations: A New Statistical Methodology*, J Wiley & Sons, New York.

Hartley, H. O. (1950). The maximum F-ratio as a short-cut test for heterogeneity of variances. *Biometrics*, **37**:308-312.

Hochberg, Y. and Tamhane, A. C. (1987). *Multiple Comparison Procedures*, John Wiley & Sons, New York.

Hochberg, Y. and Rodriguez, G. (1977). Intermediate simultaneous inference procedures. *J. Amer. Statist. Assoc.*, **72**:220-225.

Hollander, M. and Wolfe, D. A. (1973). *Nonparametric Statistical Methods*, J Wiley & Sons, New York.

Johnson, E. E. (1976). Some new multiple comparison procedures for the two-way AOV model with interaction. *Biometrics*, **32**:929-934.

Keuls, M. (1952). The use of the Studentized range in connection with an analysis of variance. *Euphytica*, **1**:112-122.

Knoke, J. D. (1976). Multiple comparisons with dichotomous data. *J. Amer. Statist. Assoc.*, **71**:849-853.

Krishnaiah, P. R. (1979). Some developments on simultaneous test procedures. In: *Developments in Statistics, Vol. 2*, (Krishnaiah, P. R., Ed.), North Holland Publishing Co., Amsterdam, Chap. 4, pp. 157-201.

Kruskal, W. H. and Wallis, W. A. (1953). Eratta to "Use of Ranks in one-criterion variance analysis." *J. Amer. Statist. Assoc.*, **48**:907-911.

Laska, E. M. and Meisner, M. J. (1989). Testing whether an identified treatment is best. *Biometrics*, **45**:1139-1151.

Miller, R. G. Jr. (1981). *Simultaneous Statistical Inference. 2nd ed.*, Springer-Verlag, New York.

Newman, D. (1939). The distribution of the range in samples from a normal population, expressed in terms of an independent estimate of standard deviation. *Biometrika*, **31**:20-30.

Patel, H. I. (1991). Comparison of treatments in a combination therapy trial. *J. Biopharmaceutical Statist.*, **1**:171-183.

Rand Corporation (1966). *A Million Random Digits with 100,000 Normal Deviates*, Collier-MacMillan, New York.

Scheffé, H. (1953). A method for judging all contrasts in the analysis of variance. *Biometrika*, **40**:87-104.

Shafer, G. and Olkin, I. (1983). Adjusting p-values to account for selection over dichotomies. *J. Amer. Statist. Assoc.*, **78**:674-678.

Shirley, E. (1977). A non-parametric equivalent of Williams' test for contrasting increasing dose levels of a treatment. *Biometrics*, **33**:386-389.

Sidák, Z. (1967). Rectangular confidence regions for the means of multivariate normal distributions. *J. Amer. Statist. Assoc.*, **62**:626-633.

Smythe, H. A., Helewa, A., and Goldsmith, C. H. (1977). "Independent assessor" and "pooled index" as techniques for measuring treatment effects in rheumatoid arthritis. *J. Rheumatol.*, **3(2)**:144-152. (Reprinted in *Physiotherapy Can.*, **30**:175-180, 1978.)

Tukey, J. W. (1953). *The problem of multiple comparisons.* Unpublished manuscript. Department of Mathematics, Princeton University.

Williams, D. A. (1971). A test for differences between treatment means when several dose levels are compared with a zero dose control. *Biometrics*, **27**:103-117.

Williams, D. A. (1972). The comparison of several dose levels with a zero dose control. *Biometrics*, **28**:519-531.

23

Adjustment of P-Values for Multiplicities of Intercorrelating Symptoms

Satya D. Dubey

United States Food and Drug Administration,
Rockville, Maryland

I. THE MOTIVATION FOR THE METHOD

In the U.S. Food and Drug Administration (FDA), the reviewing clinical statisticians frequently encounter the problem of statistically evaluating the effectiveness of a new drug in the presence of multiple endpoints or symptoms. The number of endpoints, or symptoms, varies from one drug class to another. For example, in the case of antidepressant drugs a sponsor of a new drug application may choose to use a Symptom Evaluation Form consisting of 24 symptoms (Figure 23.1) and a Patient Self-Rating Symptom Scale, consisting of 35 symptoms (Figure 23.2).

The most common clinical conditions for which antidepressant drugs are evaluated are the various depressive syndromes. Patients with a depressive syndrome usually manifest depressed mood plus a number of associated symptoms. These are:

Depressed mood characterized by any of the following:
    sad, low, blue, despondent, hopeless, gloomy
Anhedonia—inability to experience pleasure
Poor appetite or weight loss
Sleep difficulty (insomnia or hypersomnia)
Loss of energy; fatigue; lethargy
Agitation

Depression	Loss of interest
Anxiety	Crying spells
Loss of appetite	Weight loss
Agitation	Somatic concern
Psychomotor retardation	Somatization
Fatigue	Constipation
Irritability	Dizziness
Phobic obsessive-compulsive	Palpitations
Self-depreciation	Headaches
Decreased libido	
Guilt feelings	
Insomnia	
Hopelessness	
Ruminations	
Hostility	

Figure 23.1 Symptom evaluation form.

Retardation

Decrease in libido

Loss of interest in work and usual activities

Feelings of self-reproach or guilt

Diminished ability to think or concentrate such as slowed thinking or mixed-up
     thoughts.

Thoughts of death and/or suicide attempts

Feelings of helplessness and hopelessness

Anxiety or tension

Bodily complaints

These symptoms are generally correlated (FDA, 1977).

In the case of rheumatoid arthritis, efficacy of a drug is assessed on the basis of many endpoints. Sixty-eight different joints are assessed under the rubric "number of painful or tender joints." Sixty-six joints are assessed under the number of swollen joints. In addition, duration of morning stiffness, grip strength, time required to walk 50 feet, and erythrocyte sedimentation rate are also assessed. For studies of six months or longer, the American Rheumatism Association (ARA) functional capacity, ARA anatomical stage, x-ray at beginning and end of the trial, and serum samples from the beginning and end of trial may also be assessed (FDA, 1977).

Sweating	Loss of sexual interest
Trouble breathing	Feeling easily annoyed
Scared	Poor appetite
Difficulty speaking	Feeling critical of others
Low energy	Difficulty making decisions
Pains in heart/chest	Difficulty sleeping
Trouble remembering	Feeling hopeless
Hot/cold spells	Feeling blue
Blaming self	Feeling lonely
Lump in throat	Temper outbursts
Feeling fearful	Headaches
Numbing/tingling	Heart pounding
Avoid certain things	Trouble concentrating
Do things slowly	Mind going blank
Heavy feeling	Thoughts of ending one's life
Fainting/dizziness	
Crying easily	
Nervousness	
Feelings easily hurt	
Constipation	

Figure 23.2 Patient self-rating symptom scale.

In the PARIS II study, a randomized multicenter clinical trial in heart disease, endpoints considered were total death, coronary death (death due to recent or acute cardiac event) and coronary incidence (coronary death or definite nonfatal myocardial infarction). Each of the three endpoints was considered at one year and at the end of the study. Thus six correlated endpoints were considered for the study (Karrison, 1985).

Generally speaking, in the case of chronic diseases, a battery of inter-correlated clinical endpoints and many laboratory test results are evaluated for determining the efficacy of a new drug (Eggar, 1985). Thus the field of clinical trials abounds with multiple correlated endpoints or symptoms.

The probability of obtaining a nominally significant result in the absence of any real treatment effect is enhanced if many endpoints are considered. Bonferroni's inequality has been frequently used for adjusting the P-values in the presence of multiple correlated endpoints. But this is an ultraconservative method. The degree of conservativeness that is associated with this method is a function of the number of endpoints (k), the degree of intercorrelations among these endpoints as well as the significance levels associated with the k endpoints or symptoms (γ_i, i = 1, 2, ..., k). If k is not too large, say k ≤ 5 and γ_i's are small, say $\gamma_i \leq 0.1$, then Bonferroni's inequality may not yield a bad result.

An extensive analysis of this problem in a broader context involving Bonferroni and related inequalities was reported by Dubey (Dubey, 1972). By applying the results of that paper one can show that the Bonferroni inequality yields ultraconservative results in many practical situations. This undesirable property of the Bonferroni inequality is widely recognized among statistical researchers.

Let me now briefly outline a theoretical framework of the basic problem which has been considered by previous researchers. For the sake of simplicity, let γ be the significance level of each of the k variables (endpoints or symptoms) instead of choosing γ_i (i = 1, 2, ..., k) to be the significance level of the ith variable. Let us further assume that the k variables are stochastically independent, hence uncorrelated. Under the null hypothesis of equal treatment effects, the probability of Type I error, based on k stochastically independent variables, is given by

$$1 - (1 - \gamma)^k \tag{23.1}$$

which is the probability of observing at least one statistically significant result at the γ significance level. Let α be the overall probability of Type I error for the k-variable experiment. Then

$$\alpha = 1 - (1 - \gamma)^k \tag{23.2}$$

could be regarded as an "adjusted" P-value for the multiple uncorrelated k variables. If the variables are fully correlated then we have k=1, which makes Equation 23.2 as $\alpha = \gamma$. That is, the overall significance level of the experiment (α) is exactly equal to the significance level of a single variable (γ).

For small γ and small k, expression $(1 - \gamma)^k$ can be satisfactorily approximated by $1 - \gamma k$, which reduces Equation 23.2 to

$$\alpha \simeq 1 - (1 - \gamma k) = \gamma k$$

In case of k correlated variables, the Bonferroni inequality provides

$$\alpha \leq \gamma k \tag{23.3}$$

Thus the adjustment in P-value, based on the Bonferroni inequality, is equivalent to setting

$$\alpha = \gamma k \tag{23.4}$$

From Equation 23.4, we obtain

$$\gamma = \frac{\alpha}{k} \tag{23.5}$$

It is a very widely used adjustment especially when k is small. For large k's and widely used α values, Equation 23.5 gives very unrealistic adjustments.

Among the many researchers who have struggled with this problem over several decades, Mantel reports a suggestion by Tukey (Mantel, 1980) which states that Tukey's intention is that we should behave as though we were making $k^{1/2}$ independent tests instead of k tests where k is the number of intercorrelating variables. Armitage suggested (Armitage, 1985) to use a value k', somewhere between 1 and k and remarked that it would be possible to investigate how k' varies with different patterns of correlation. He further stated that for a large k it would be a difficult task not only because of the computing involved but because of the unlimited choice of correlation structure. He concluded by remarking that perhaps there are some quite different approaches to his situation which are worth exploring. In this paper an approach has been taken which provides a reasonable answer to this problem in a more manageable manner.

II. THE METHOD

Let $r_{i,j}$ be the (i,j) element of a correlation matrix (i, j = 1, 2, ..., k) of k intercorrelating variables. Let $r_{i.}$ be an average correlation, based on the correlation coefficients between the ith variable (symptom) and the remaining (k − 1) variables (i = 1, 2, ..., k). The k average correlations may be an arithmetic mean or a median, suitably chosen on the basis of relevant expected correlation structure.

Now consider

$$\alpha = 1 - (1 - \gamma)^m \tag{23.6}$$

where $m = k^{m - r_{i.}}$ and i = 1, 2, ..., k. For fully correlated variables, define $r_{i.} = 1$ for all i = 1, 2, ..., k, which gives m = 1 and hence $\alpha = \gamma$. For fully uncorrelated variables, define $r_{i.} = 0$ for all i = 1, 2, ..., k, which gives m = k and hence $\alpha = 1 - (1 - \gamma)^K$

These two conditions provide correct boundary results. For partially correlated variables, define

$$r_{i.} = \sum_{i \neq j} \frac{r_{i,j}}{k-1} \tag{23.7}$$

which will ensure $0 < r_{i.} < 1$. From Equation 23.6 we obtain

$$\gamma = 1 - (1 - \alpha)^{1/m} \tag{23.8}$$

where γ, α, and m are as defined earlier.

Equations 23.6 and 23.7, which involve relevant correlation coefficients, are similar to the one given in the paper by Armitage (1985), where he uses x $(0 < x < 1)$ as a dummy variable.

Remarks

1. If the correlation coefficients are highly homogeneous, which may be true when symptoms are selected in such a manner that they are expected to be either highly correlated or lowly correlated when we will have $r_{i.}$'s very close to each other, and hence we may choose to replace $r_{i.}$ by $r_{..}$, where

    $$r_{..} = \sum_{i=1}^{k} \frac{r_{i.}}{k}$$

    This decision rule will simplify the operational procedure associated with the proposed method without any significant loss in efficiency.
2. If, on the other hand, the correlation coefficients are highly heterogeneous, which may be true when symptoms are selected in such a manner that they are expected to be different from each other then $r_{i.}$ as defined by Equation 23.7 should be used. This decision rule will reflect the actual situation accurately.
3. Under the formulations, as given by Equations 23.6 and 23.8, Tukey's suggestion amounts to using $r_{..} = 0.5$ or taking $r_{1.} = r_{2.} = ... = r_{k.} = 0.5$.

III. SOME ANALYTICAL PROPERTIES OF THE METHOD

Using Equation 23.6, the following properties are mathematically established; proofs are omitted for the sake of brevity.

1. Overall significance level (α) is a decreasing function of an average correlation measure $r_{i.}$ for a given number of symptoms (k) and a fixed γ (significance level of a symptom).

2. Overall significance level (α) is an increasing function of k (number of symptoms) for a fixed $r_{i.}$ and a fixed γ (significance level of a symptom).

3. Overall significance level (α) is a decreasing function of $r_{i.}$ and an increasing function of k for a fixed γ.

4. Overall significance level (α) is an increasing function of a single symptom significance level (γ).

Now using Equation 23.8, the following properties are mathematically established. Again, proofs are omitted.

5. Single symptom significance level (γ) is an increasing function of an average correlation measure $r_{i.}$ for a given number of symptoms (k) and a fixed α (overall significance level).

6. Single symptom significance level (γ) is a decreasing function of k (number of symptoms) for a fixed $r_{i.}$ and a fixed α (overall significance level).

7. Single symptom significance level (γ) is an increasing function of $r_{i.}$, an average correlation measure, and decreasing function of k (number of symptoms) for a fixed α (overall significance level).

8. Single symptom significance level (γ) is an increasing function of overall significance level (α).

IV. COMPUTATIONAL RESULTS

Tables 23.1 and 23.2 (in Appendix to this chapter) are computed using Equation 23.6. They provide overall significance levels (α) when testing correlated symptoms at significance levels ($\gamma = .01, .05$) for $r_{i.} = 0, .05, .10, .25, .50, .75, .90, .95, 1.00$ and k = 1, 2, ..., 40.

Tables 23.3, 23.4, and 23.5 (in Appendix to this chapter) are computed using Equation 23.8. They provide choices of significance levels for testing individual symptoms when overall significance levels ($\alpha = .01, .05, .10$) are specified for $r_{i.} = 0, .05, .10, .25, .50, .75, .90, .95, 1.00$ and k = 1, 2, ..., 40.

V. SOME OBSERVATIONS FROM THE COMPUTATIONAL RESULTS

Tables 23.1–23.5 provide a wealth of useful information regarding the behavior of the overall significance level (α) and individual symptom significance level (γ) as a function of the number of symptoms (k) and the average correlation measurer $r_{i.}$. A few selected results are stated here.

1. For $1 \le k \le 40$ and $0.90 < r_{i.} < 1.0$, α remains .01 when γ is selected as 0.01 (Robustness Property).

2. For $1 \le k \le 40$, and $r_{i.} = 0.75$, α ranges from 0.01 to .025 when γ is selected as 0.01.

3. For $1 \le k \le 40$ and $r_{i.} = 0.50$, α ranges from 0.01 to 0.06 when γ is selected as 0.01.

4. For $1 \le k \le 40$ and $r_{i.} = 0.25$, α ranges from 0.01 to 0.15 when γ is selected as 0.01.

5. For $1 \le k \le 40$ and $0 \le r_{i.} \le 0.1$, α ranges from 0.01 to 0.33 when γ is selected as 0.01.

A. General Observation (Based on Tables 23.1 and 23.2)

For high average correlation measures $r_{i.}$ and small γ, α fluctuates much less than it fluctuates for low average correlation measures and high γ (Robustness Property).

6. For $1 \le k \le 40$ and $0.2 \le r_{i.} \le 1.0$, choose γ in the range of 0.01 to .001 when α is selected as 0.01 (Robust Result).

7. For $1 \le k \le 40$ and $0.75 \le r_{i.} \le 1.0$ (high average correlation measures), choose γ in the range of 0.01 to 0.004 when α is selected as 0.01.

8. For $1 \le k \le 40$ and $.25 \le r_{i.} < 0.75$ (medium average correlation measures), choose γ in the range of 0.01 to 0.0006 when α is selected as 0.01.

9. For $1 \le k \le 40$ and $0 \le r_{i.} < 0.25$ (low average correlation measures), choose γ in the range of 0.01 to 0.0003 when α is selected as 0.01.

10. For $1 \le k \le 40$ and $0 \le r_{i.} \le 1$, γ varies from .05 to .0013 when α is chosen as 0.05.

B. General Observation (Based on Tables 23.3–23.5)

For a specified α, highly correlated symptoms provide much less fluctuation in y compared to the case when these symptoms are lowly correlated.

VI. APPLICATION OF THE METHOD

In this section, an application of the present method will be illustrated by means of an example.

   Example. On the basis of calculated correlation coefficients between 10 symptoms, the following average correlation measures are obtained.

i	$r_{i.}$
1	.82
2	.85
3	.87
4	.89
5	.90
6	.95
7	.97
8	.93
9	.98
10	.86

Since these average correlation measures vary between 0.82 and 0.98, it seems reasonable to obtain $r_{..}$.

$$r_{..} = \frac{1}{10} \sum_{i=1}^{10} r_{i.} = 0.90$$

Now for $i = 10$ and $r_{..} = 0.90$, we obtain for $\alpha = .05$ from Table 23.2, $\gamma = .04$, that is, choose the significance level of an individual symptom as .04. On the other hand, Bonferroni's inequality yields $\gamma = \alpha/k = .05/10 = .005$, a very conservative result. Again, for $i = 10$ and $r_{..} = 0.90$ we obtain for $\gamma = .01$, and from Table 23.1, $\alpha = .01$. That is, the overall significance level α does not materially change from γ. However, Bonferroni's inequality yields $\alpha = \gamma k = (.01)(10) = .001$, which would be an unnecessary adjustment in P-value.

Several other applications of the present method involving the adjustment for the multiplicity of statistical tests in toxicity studies as well as when actual correlation measures between symptoms are unavailable, but ranking of symptoms is available are beyond the scope of this chapter.

VII. CONCLUSION

A new method has been presented which permits applied statisticians to choose a significance level (γ) for k correlated symptoms in order to ensure a specified overall significance level (α) of the entire study. Conversely, the new method also enables applied statisticians to determine the overall significance level (α) of the entire study for a chosen significance level (γ) for k individual correlated symptoms. Five tables have been presented for applying the new method. These tables cover a variety of usual cases. Several properties of the method have been mentioned along with some robust results. It has been shown that the new method provides more reasonable results than the widely used Bonferroni's inequality under similar circumstances. The method utilizes average correlation measures associated with k correlated items in a manageable manner.

ACKNOWLEDGMENT

The author acknowledges the valuable assistance of his colleague Ms. Judy Chen of the Statistical Evaluation and Research Branch in the computation of tables. The views expressed in this chapter are those of the author and not necessarily those of the U.S. Food and Drug Administration. This work was produced by the author in his capacity as a Federal employee and, as such, is not subject to copyright.

REFERENCES

Armitage, P., (1985), Two areas of controversy in the design and analysis of clinical trials, *Recent Topics on Clinical Trials Evaluation*, **13(Suppl. III)**:29-40.

Dubey, S. D., (1972), *Statistical Contributions to Reliability Engineering*, ARL 72-0120, Aerospace Research Laboratories, Wright-Patterson Air Force Base, Ohio, Contract F 33615-71-C-1174.

Egger, M. J., (1985) *Correlated Measures of Disease Outcome in Clinical Trials: Effects on Type I Errors*, Abstract p-65, Sixth Annual Meeting of the Society for Clinical Trials (May 12-15, 1985), New Orleans, Louisiana

FDA, (1977), *Guidelines for the Clinical Evaluation of Antidepressant Drugs*, FDA 77-3042; GPO 017-012-00247-1, U.S Food and Drug Administration, Rockville, MD.

FDA, (1977), *Guidelines for the Clinical Evaluation of Anti-Inflammatory Drugs (Adults and Children)*, FDA 78-3054; GPO 017-012-00258-7, U.S Food and Drug Administration, Rockville, MD.

Karrison, T., (1985), *The Use of Randomized Tests for Multiple Endpoints in Clinical Trials*, Abstract 79, Sixth Annual Meeting of the Society for Clinical Trials (May 12-15, 1985), New Orleans, Louisiana.

Mantel, N., (1980), Assessing laboratory evidence for neoplastic activity, *Biometrics*, **36(3)**:381-399.

VIII. APPENDIX

See Tables 23.1–23.5.

Table 23.1 Overall Significance Level α When k Correlated Symptoms Are Tested at Significance Level $\gamma = 0.01$

| | \multicolumn{9}{c}{$r_{i.}$ = Average correlation measure between the ith symptom and the remaining (k − 1) symptoms.} | | | | | | | | |
k	0.00	0.05	0.10	0.25	0.50	0.75	0.90	0.95	1.00
1	0.0100	0.0100	0.0100	0.0100	0.0100	0.0100	0.0100	0.0100	0.0100
2	0.0199	0.0192	0.0186	0.0168	0.0141	0.0119	0.0107	0.0104	0.0100
3	0.0297	0.0281	0.0267	0.0226	0.0173	0.0131	0.0112	0.0106	0.0100
4	0.0394	0.0368	0.0344	0.0280	0.0199	0.0141	0.0115	0.0107	0.0100
5	0.0490	0.0453	0.0419	0.0330	0.0222	0.0149	0.0117	0.0108	0.0100
6	0.0585	0.0536	0.0492	0.0378	0.0243	0.0156	0.0120	0.0109	0.0100
7	0.0679	0.0618	0.0563	0.0423	0.0262	0.0162	0.0121	0.0110	0.0100
8	0.0773	0.0699	0.0632	0.0467	0.0280	0.0168	0.0123	0.0111	0.0100
9	0.0865	0.0778	0.0700	0.0509	0.0297	0.0173	0.0124	0.0112	0.0100
10	0.0956	0.0857	0.0767	0.0549	0.0313	0.0177	0.0126	0.0112	0.0100
11	0.1047	0.0934	0.0833	0.0589	0.0328	0.0181	0.0127	0.0113	0.0100
12	0.1136	0.1010	0.0898	0.0627	0.0342	0.0185	0.0128	0.0113	0.0100
13	0.1225	0.1086	0.0962	0.0665	0.0356	0.0189	0.0129	0.0114	0.0100
14	0.1313	0.1160	0.1024	0.0702	0.0369	0.0193	0.0130	0.0114	0.0100
15	0.1399	0.1234	0.1086	0.0737	0.0382	0.0196	0.0131	0.0114	0.0100
16	0.1485	0.1306	0.1147	0.0773	0.0394	0.0199	0.0132	0.0115	0.0100
17	0.1571	0.1378	0.1208	0.0807	0.0406	0.0202	0.0133	0.0115	0.0100
18	0.1655	0.1449	0.1267	0.0841	0.0417	0.0205	0.0133	0.0115	0.0100
19	0.1738	0.1519	0.1326	0.0874	0.0429	0.0208	0.0134	0.0116	0.0100
20	0.1821	0.1589	0.1384	0.0907	0.0440	0.0210	0.0135	0.0116	0.0100
21	0.1903	0.1658	0.1442	0.0939	0.0450	0.0213	0.0135	0.0116	0.0100
22	0.1984	0.1726	0.1498	0.0971	0.0460	0.0215	0.0136	0.0117	0.0100
23	0.2064	0.1793	0.1554	0.1002	0.0471	0.0218	0.0137	0.0117	0.0100
24	0.2143	0.1860	0.1610	0.1032	0.0480	0.0220	0.0137	0.0117	0.0100
25	0.2222	0.1926	0.1665	0.1063	0.0490	0.0222	0.0138	0.0117	0.0100
26	0.2300	0.1991	0.1719	0.1093	0.0500	0.0224	0.0138	0.0118	0.0100
27	0.2377	0.2056	0.1773	0.1122	0.0509	0.0226	0.0139	0.0118	0.0100
28	0.2453	0.2120	0.1826	0.1151	0.0518	0.0229	0.0139	0.0118	0.0100
29	0.2528	0.2183	0.1879	0.1180	0.0527	0.0231	0.0140	0.0118	0.0100
30	0.2603	0.2246	0.1931	0.1209	0.0536	0.0232	0.0140	0.0118	0.0100
31	0.2677	0.2308	0.1983	0.1237	0.0544	0.0234	0.0141	0.0119	0.0100
32	0.2750	0.2370	0.2034	0.1265	0.0553	0.0236	0.0141	0.0119	0.0100
33	0.2823	0.2431	0.2085	0.1292	0.0561	0.0238	0.0142	0.0119	0.0100
34	0.2894	0.2491	0.2135	0.1320	0.0569	0.0240	0.0142	0.0119	0.0100
35	0.2966	0.2551	0.2185	0.1346	0.0577	0.0241	0.0142	0.0119	0.0100
36	0.3036	0.2610	0.2234	0.1373	0.0585	0.0243	0.0143	0.0120	0.0100
37	0.3106	0.2669	0.2283	0.1400	0.0593	0.0245	0.0143	0.0120	0.0100
38	0.3174	0.2727	0.2331	0.1426	0.0601	0.0246	0.0144	0.0120	0.0100
39	0.3243	0.2785	0.2379	0.1452	0.0608	0.0248	0.0144	0.0120	0.0100
40	0.3310	0.2842	0.2427	0.1477	0.0616	0.0250	0.0144	0.0120	0.0100

Table 23.2 Overall Significance Level α When k Correlated Symptoms Are Tested at Significance Level $\gamma = 0.05$

$r_{i.}$ = Average correlation measure between the ith symptom and the remaining $(k-1)$ symptoms.

k	0.00	0.05	0.10	0.25	0.50	0.75	0.90	0.95	1.00
1	0.0500	0.0500	0.0500	0.0500	0.0500	0.0500	0.0500	0.0500	0.0500
2	0.0975	0.0943	0.0913	0.0826	0.0700	0.0592	0.0535	0.0517	0.0500
3	0.1426	0.1335	0.1288	0.1103	0.0850	0.0653	0.0556	0.0527	0.0500
4	0.1855	0.1742	0.1636	0.1350	0.0975	0.0700	0.0572	0.0535	0.0500
5	0.2262	0.2107	0.1961	0.1576	0.1084	0.0738	0.0585	0.0541	0.0500
6	0.2649	0.2453	0.2268	0.1785	0.1181	0.0771	0.0595	0.0546	0.0500
7	0.3017	0.2780	0.2559	0.1981	0.1269	0.0800	0.0604	0.0550	0.0500
8	0.3366	0.3091	0.2834	0.2165	0.1350	0.0826	0.0612	0.0553	0.0500
9	0.3698	0.3387	0.3097	0.2340	0.1426	0.0850	0.0619	0.0556	0.0500
10	0.4013	0.3669	0.3346	0.2506	0.1497	0.0872	0.0625	0.0559	0.0500
11	0.4312	0.3938	0.3585	0.2664	0.1564	0.0892	0.0631	0.0562	0.0500
12	0.4596	0.4193	0.3813	0.2816	0.1628	0.0911	0.0636	0.0564	0.0500
13	0.4867	0.4438	0.4031	0.2961	0.1688	0.0928	0.0641	0.0566	0.0500
14	0.5123	0.4671	0.4239	0.3101	0.1746	0.0945	0.0646	0.0568	0.0500
15	0.5367	0.4893	0.4439	0.3236	0.1802	0.0960	0.0650	0.0570	0.0500
16	0.5599	0.5105	0.4631	0.3366	0.1855	0.0975	0.0654	0.0572	0.0500
17	0.5819	0.5308	0.4815	0.3491	0.1906	0.0989	0.0658	0.0574	0.0500
18	0.6028	0.5502	0.4992	0.3613	0.1956	0.1003	0.0662	0.0575	0.0500
19	0.6226	0.5688	0.5162	0.3730	0.2004	0.1016	0.0665	0.0577	0.0500
20	0.6415	0.5865	0.5325	0.3844	0.2050	0.1028	0.0669	0.0578	0.0500
21	0.6594	0.6035	0.5482	0.3954	0.2095	0.1040	0.0672	0.0580	0.0500
22	0.6765	0.6197	0.5633	0.4061	0.2138	0.1051	0.0675	0.0581	0.0500
23	0.6926	0.6353	0.5778	0.4165	0.2181	0.1062	0.0678	0.0582	0.0500
24	0.7080	0.6501	0.5918	0.4266	0.2222	0.1073	0.0681	0.0584	0.0500
25	0.7226	0.6644	0.6052	0.4364	0.2262	0.1084	0.0683	0.0585	0.0500
26	0.7365	0.6780	0.6182	0.4460	0.2301	0.1094	0.0686	0.0586	0.0500
27	0.7497	0.6910	0.6307	0.4553	0.2340	0.1103	0.0688	0.0587	0.0500
28	0.7622	0.7035	0.6427	0.4644	0.2377	0.1113	0.0691	0.0588	0.0500
29	0.7741	0.7155	0.6543	0.4732	0.2414	0.1122	0.0693	0.0589	0.0500
30	0.7854	0.7270	0.6655	0.4819	0.2449	0.1131	0.0695	0.0590	0.0500
31	0.7961	0.7380	0.6763	0.4903	0.2484	0.1140	0.0698	0.0591	0.0500
32	0.8063	0.7485	0.6867	0.4985	0.2519	0.1148	0.0700	0.0592	0.0500
33	0.8160	0.7586	0.6968	0.5065	0.2552	0.1157	0.0702	0.0593	0.0500
34	0.8252	0.7682	0.7065	0.5143	0.2585	0.1165	0.0704	0.0593	0.0500
35	0.8339	0.7775	0.7158	0.5220	0.2617	0.1173	0.0706	0.0594	0.0500
36	0.8422	0.7864	0.7248	0.5295	0.2649	0.1181	0.0708	0.0595	0.0500
37	0.8501	0.7949	0.7336	0.5368	0.2680	0.1188	0.0710	0.0596	0.0500
38	0.8576	0.8031	0.7420	0.5439	0.2711	0.1196	0.0711	0.0597	0.0500
39	0.8647	0.8109	0.7501	0.5509	0.2741	0.1203	0.0713	0.0597	0.0500
40	0.8715	0.8184	0.7580	0.5577	0.2770	0.1210	0.0715	0.0598	0.0500

Table 23.3 Choice of Significance Level γ for Testing k Correlated Symptoms When Overall Significance Level α = 0.01

$r_{i.}$ = Average correlation measure between the ith symptom and the remaining (k − 1) symptoms.

k	0.00	0.05	0.10	0.25	0.50	0.75	0.90	0.95	1.00
1	0.0100	0.0100	0.0100	0.0100	0.0100	0.0100	0.0100	0.0100	0.0100
2	0.0050	0.0052	0.0054	0.0060	0.0071	0.0084	0.0093	0.0097	0.0100
3	0.0033	0.0035	0.0037	0.0044	0.0058	0.0076	0.0090	0.0095	0.0100
4	0.0025	0.0027	0.0029	0.0035	0.0050	0.0071	0.0087	0.0093	0.0100
5	0.0020	0.0022	0.0024	0.0030	0.0045	0.0067	0.0085	0.0092	0.0100
6	0.0017	0.0018	0.0020	0.0026	0.0041	0.0064	0.0084	0.0091	0.0100
7	0.0014	0.0016	0.0017	0.0023	0.0038	0.0062	0.0082	0.0091	0.0100
8	0.0013	0.0014	0.0015	0.0021	0.0035	0.0060	0.0081	0.0090	0.0100
9	0.0011	0.0012	0.0014	0.0019	0.0033	0.0058	0.0080	0.0090	0.0100
10	0.0010	0.0011	0.0013	0.0018	0.0032	0.0056	0.0080	0.0089	0.0100
11	0.0009	0.0010	0.0012	0.0017	0.0030	0.0055	0.0079	0.0089	0.0100
12	0.0008	0.0009	0.0011	0.0016	0.0029	0.0054	0.0078	0.0088	0.0100
13	0.0008	0.0009	0.0010	0.0015	0.0028	0.0053	0.0077	0.0088	0.0100
14	0.0007	0.0008	0.0009	0.0014	0.0027	0.0052	0.0077	0.0088	0.0100
15	0.0007	0.0008	0.0009	0.0013	0.0026	0.0051	0.0076	0.0087	0.0100
16	0.0006	0.0007	0.0008	0.0013	0.0025	0.0050	0.0076	0.0087	0.0100
17	0.0006	0.0007	0.0008	0.0012	0.0024	0.0049	0.0075	0.0087	0.0100
18	0.0006	0.0006	0.0007	0.0011	0.0024	0.0049	0.0075	0.0087	0.0100
19	0.0005	0.0006	0.0007	0.0011	0.0023	0.0048	0.0075	0.0086	0.0100
20	0.0005	0.0006	0.0007	0.0011	0.0022	0.0047	0.0074	0.0086	0.0100
21	0.0005	0.0006	0.0006	0.0010	0.0022	0.0047	0.0074	0.0086	0.0100
22	0.0005	0.0005	0.0006	0.0010	0.0021	0.0046	0.0074	0.0086	0.0100
23	0.0004	0.0005	0.0006	0.0010	0.0021	0.0046	0.0073	0.0086	0.0100
24	0.0004	0.0005	0.0006	0.0009	0.0020	0.0045	0.0073	0.0085	0.0100
25	0.0004	0.0005	0.0006	0.0009	0.0020	0.0045	0.0073	0.0085	0.0100
26	0.0004	0.0005	0.0005	0.0009	0.0020	0.0044	0.0072	0.0085	0.0100
27	0.0004	0.0004	0.0005	0.0008	0.0019	0.0044	0.0072	0.0085	0.0100
28	0.0004	0.0004	0.0005	0.0008	0.0019	0.0044	0.0072	0.0085	0.0100
29	0.0003	0.0004	0.0005	0.0008	0.0019	0.0043	0.0072	0.0085	0.0100
30	0.0003	0.0004	0.0005	0.0008	0.0018	0.0043	0.0071	0.0084	0.0100
31	0.0003	0.0004	0.0005	0.0008	0.0018	0.0043	0.0071	0.0084	0.0100
32	0.0003	0.0004	0.0004	0.0007	0.0018	0.0042	0.0071	0.0084	0.0100
33	0.0003	0.0004	0.0004	0.0007	0.0017	0.0042	0.0071	0.0084	0.0100
34	0.0003	0.0004	0.0004	0.0007	0.0017	0.0042	0.0070	0.0084	0.0100
35	0.0003	0.0003	0.0004	0.0007	0.0017	0.0041	0.0070	0.0084	0.0100
36	0.0003	0.0003	0.0004	0.0007	0.0017	0.0041	0.0070	0.0084	0.0100
37	0.0003	0.0003	0.0004	0.0007	0.0017	0.0041	0.0070	0.0084	0.0100
38	0.0003	0.0003	0.0004	0.0007	0.0016	0.0040	0.0070	0.0083	0.0100
39	0.0003	0.0003	0.0004	0.0006	0.0016	0.0040	0.0069	0.0083	0.0100
40	0.0003	0.0003	0.0004	0.0006	0.0016	0.0040	0.0069	0.0083	0.0100

Table 23.4 Choice of Significance Level γ for Testing k Correlated Symptoms When Overall Significance Level $\alpha = 0.05$

$r_{i.}$ = Average correlation measure between the ith symptom and the remaining $(k - 1)$ symptoms.

k	0.00	0.05	0.10	0.25	0.50	0.75	0.90	0.95	1.00
1	0.0500	0.0500	0.0500	0.0500	0.0500	0.0500	0.0500	0.0500	0.0500
2	0.0253	0.0262	0.0271	0.0300	0.0356	0.0422	0.0467	0.0483	0.0500
3	0.0170	0.0179	0.0189	0.0223	0.0292	0.0382	0.0449	0.0474	0.0500
4	0.0127	0.0136	0.0146	0.0180	0.0253	0.0356	0.0437	0.0467	0.0500
5	0.0102	0.0111	0.0120	0.0152	0.0227	0.0337	0.0427	0.0462	0.0500
6	0.0085	0.0093	0.0102	0.0133	0.0207	0.0322	0.0420	0.0458	0.0500
7	0.0073	0.0080	0.0089	0.0118	0.0192	0.0310	0.0413	0.0455	0.0500
8	0.0064	0.0071	0.0079	0.0107	0.0180	0.0300	0.0408	0.0452	0.0500
9	0.0057	0.0063	0.0071	0.0098	0.0170	0.0292	0.0403	0.0449	0.0500
10	0.0051	0.0057	0.0064	0.0091	0.0161	0.0284	0.0399	0.0447	0.0500
11	0.0047	0.0052	0.0059	0.0085	0.0153	0.0278	0.0396	0.0445	0.0500
12	0.0043	0.0048	0.0055	0.0079	0.0147	0.0272	0.0392	0.0443	0.0500
13	0.0039	0.0045	0.0051	0.0075	0.0141	0.0267	0.0389	0.0441	0.0500
14	0.0037	0.0042	0.0048	0.0071	0.0136	0.0262	0.0386	0.0440	0.0500
15	0.0034	0.0039	0.0045	0.0067	0.0132	0.0257	0.0384	0.0438	0.0500
16	0.0032	0.0037	0.0042	0.0064	0.0127	0.0253	0.0381	0.0437	0.0500
17	0.0030	0.0035	0.0040	0.0061	0.0124	0.0249	0.0379	0.0435	0.0500
18	0.0028	0.0033	0.0038	0.0059	0.0120	0.0246	0.0377	0.0434	0.0500
19	0.0027	0.0031	0.0036	0.0056	0.0117	0.0243	0.0375	0.0433	0.0500
20	0.0026	0.0030	0.0035	0.0054	0.0114	0.0240	0.0373	0.0432	0.0500
21	0.0024	0.0028	0.0033	0.0052	0.0111	0.0237	0.0371	0.0431	0.0500
22	0.0023	0.0027	0.0032	0.0050	0.0109	0.0234	0.0370	0.0430	0.0500
23	0.0022	0.0026	0.0030	0.0049	0.0106	0.0232	0.0368	0.0429	0.0500
24	0.0021	0.0025	0.0029	0.0047	0.0104	0.0229	0.0366	0.0428	0.0500
25	0.0020	0.0024	0.0028	0.0046	0.0102	0.0227	0.0365	0.0427	0.0500
26	0.0020	0.0023	0.0027	0.0044	0.0100	0.0225	0.0364	0.0426	0.0500
27	0.0019	0.0022	0.0026	0.0043	0.0098	0.0223	0.0362	0.0426	0.0500
28	0.0018	0.0022	0.0026	0.0042	0.0096	0.0221	0.0361	0.0425	0.0500
29	0.0018	0.0021	0.0025	0.0041	0.0095	0.0219	0.0360	0.0424	0.0500
30	0.0017	0.0020	0.0024	0.0040	0.0093	0.0217	0.0358	0.0423	0.0500
31	0.0017	0.0020	0.0023	0.0039	0.0092	0.0215	0.0357	0.0423	0.0500
32	0.0016	0.0019	0.0023	0.0038	0.0090	0.0213	0.0356	0.0422	0.0500
33	0.0016	0.0018	0.0022	0.0037	0.0089	0.0212	0.0355	0.0422	0.0500
34	0.0015	0.0018	0.0021	0.0036	0.0088	0.0210	0.0354	0.0421	0.0500
35	0.0015	0.0017	0.0021	0.0036	0.0086	0.0209	0.0353	0.0420	0.0500
36	0.0014	0.0017	0.0020	0.0035	0.0085	0.0207	0.0352	0.0420	0.0500
37	0.0014	0.0017	0.0020	0.0034	0.0084	0.0206	0.0351	0.0419	0.0500
38	0.0013	0.0016	0.0019	0.0033	0.0083	0.0204	0.0350	0.0419	0.0500
39	0.0013	0.0016	0.0019	0.0033	0.0082	0.0203	0.0349	0.0418	0.0500
40	0.0013	0.0015	0.0019	0.0032	0.0081	0.0202	0.0348	0.0418	0.0500

Table 23.5 Choice of Significance Level γ for Testing k Correlated Symptoms When Overall Significance Level $\alpha = 0.1$

$r_{i.}$ = Average correlation measure between the ith symptom and the remaining $(k - 1)$ symptoms.

k	0.00	0.05	0.10	0.25	0.50	0.75	0.90	0.95	1.00
1	0.1000	0.1000	0.1000	0.1000	0.1000	0.1000	0.1000	0.1000	0.1000
2	0.0513	0.0531	0.0549	0.0607	0.0718	0.0848	0.0936	0.0968	0.1000
3	0.0345	0.0364	0.0384	0.0452	0.0590	0.0769	0.0901	0.0949	0.1000
4	0.0260	0.0278	0.0298	0.0366	0.0513	0.0718	0.0876	0.0936	0.1000
5	0.0209	0.0226	0.0244	0.0310	0.0460	0.0680	0.0858	0.0926	0.1000
6	0.0174	0.0190	0.0208	0.0271	0.0421	0.0651	0.0843	0.0918	0.1000
7	0.0149	0.0165	0.0181	0.0242	0.0390	0.0627	0.0831	0.0912	0.1000
8	0.0131	0.0145	0.0161	0.0219	0.0366	0.0607	0.0820	0.0906	0.1000
9	0.0116	0.0130	0.0145	0.0201	0.0345	0.0590	0.0811	0.0901	0.1000
10	0.0105	0.0118	0.0132	0.0186	0.0328	0.0575	0.0803	0.0896	0.1000
11	0.0095	0.0107	0.0121	0.0173	0.0313	0.0562	0.0796	0.0892	0.1000
12	0.0087	0.0099	0.0112	0.0162	0.0300	0.0550	0.0789	0.0889	0.1000
13	0.0081	0.0092	0.0104	0.0153	0.0288	0.0540	0.0783	0.0885	0.1000
14	0.0075	0.0086	0.0098	0.0145	0.0278	0.0530	0.0777	0.0882	0.1000
15	0.0070	0.0080	0.0092	0.0137	0.0268	0.0521	0.0772	0.0879	0.1000
16	0.0066	0.0075	0.0087	0.0131	0.0260	0.0513	0.0767	0.0876	0.1000
17	0.0062	0.0071	0.0082	0.0125	0.0252	0.0506	0.0763	0.0874	0.1000
18	0.0058	0.0067	0.0078	0.0120	0.0245	0.0499	0.0759	0.0871	0.1000
19	0.0055	0.0064	0.0074	0.0115	0.0239	0.0492	0.0755	0.0869	0.1000
20	0.0053	0.0061	0.0071	0.0111	0.0233	0.0486	0.0751	0.0867	0.1000
21	0.0050	0.0058	0.0068	0.0207	0.0227	0.0480	0.0748	0.0865	0.1000
22	0.0048	0.0056	0.0065	0.0103	0.0222	0.0475	0.0744	0.0863	0.1000
23	0.0046	0.0053	0.0062	0.0100	0.0217	0.0470	0.0741	0.0861	0.1000
24	0.0044	0.0051	0.0060	0.0097	0.0213	0.0465	0.0738	0.0860	0.1000
25	0.0042	0.0049	0.0058	0.0094	0.0209	0.0460	0.0735	0.0858	0.1000
26	0.0040	0.0048	0.0056	0.0091	0.0205	0.0456	0.0732	0.0856	0.1000
27	0.0039	0.0046	0.0054	0.0089	0.0201	0.0452	0.0730	0.0855	0.1000
28	0.0038	0.0044	0.0052	0.0086	0.0197	0.0448	0.0727	0.0853	0.1000
29	0.0036	0.0043	0.0051	0.0084	0.0194	0.0444	0.0725	0.0852	0.1000
30	0.0035	0.0042	0.0049	0.0082	0.0191	0.0440	0.0722	0.0850	0.1000
31	0.0034	0.0040	0.0048	0.0080	0.0187	0.0437	0.0720	0.0849	0.1000
32	0.0033	0.0039	0.0046	0.0078	0.0185	0.0433	0.0718	0.0848	0.1000
33	0.0032	0.0038	0.0045	0.0076	0.0182	0.0430	0.0716	0.0847	0.1000
34	0.0031	0.0037	0.0044	0.0075	0.0179	0.0427	0.0714	0.0845	0.1000
35	0.0030	0.0036	0.0043	0.0073	0.0177	0.0424	0.0712	0.0844	0.1000
36	0.0029	0.0035	0.0042	0.0071	0.0174	0.0421	0.0710	0.0843	0.1000
37	0.0028	0.0034	0.0041	0.0070	0.0172	0.0418	0.0708	0.0842	0.1000
38	0.0028	0.0033	0.0040	0.0069	0.0169	0.0415	0.0706	0.0841	0.1000
39	0.0027	0.0032	0.0039	0.0067	0.0167	0.0413	0.0704	0.0840	0.1000
40	0.0026	0.0032	0.0038	0.0066	0.0165	0.0410	0.0703	0.0839	0.1000

24

Screening Compounds for Clinically Active Drugs

Charles E. Redman[*]

*Lilly Research Laboratories,
Indianapolis, Indiana*

Charles W. Dunnett

*McMaster University,
Hamilton, Ontario, Canada*

I. INTRODUCTION

The pharmaceutical industry screens for active drugs as an ongoing process. Screening of chemical compounds can be defined as a procedure that sifts through a large number of compounds in hope of finding a rare nugget (a potentially interesting compound). Compounds are studied in rapid and routine experiments (in vivo or in vitro) which often fall short of what is considered good experimentation.

Throughout the chapter *potentially active*, *active*, and *interesting* will be considered synonyms. These words apply to compounds at the earliest stage of drug development and they indicate a unique response in a screening model. Compounds can fail to become drugs for literally hundreds of reasons as drug development proceeds. Searching for active compounds also employs directed research using structure-activity models and relationships and using molecular biologic insights into the structure of important molecules.

---

*Current affiliation: HCA Medical Research, Nashville, Tennessee

II. RISK ASSESSMENT

The statistical aspects for finding interesting compounds are similar to screening for disease in asymptomatic people or to developing an improved strain of grain or animal. Each pharmaceutical company has an inventory of compounds and the capability of making a number of new compounds. For a particular area of interest (e.g., antitumor screening) there is a potential of N compounds of which an unknown small proportion P is truly active. The ideal screen will pass all PN active compounds to a definitive antitumor test to determine levels of activity and reject the remaining $(1 - P)N$ compounds as not active for that particular antitumor activity.

The terminology and probabilistic modeling introduced is based on the following assumptions:

1.   A compound is either potentially active or not active.
2.   There is a finite number of compounds, N, that can be tested over a fixed period of time.
3.   There is a fixed, but unknown, proportion P which are active (P may change from compound series to series, or for each screen).

Since some ideas may be new to the reader, simple 2×2 tables will be used to introduce terminology and notation, as illustrated by Table 24.1. If an ideal screening model existed, then there would be little uncertainty in the experimental results and we would have b (number of false positives) and c (number of false negatives) both zero. However, screening models are subject to experimental variation and we are not likely to find this to be the case.

The marginal values for the truth columns are PN and $(1 - P)N$, but the values within the cells are not obvious and neither are the row marginals. To understand how these values arise, let us consider the situation in Table 24.1. It is apparent that

$$PN = a + c$$

$$(1 - P)N = b + d$$

$$P = \frac{a + c}{a + b + c + d}$$

and

$$N = a + b + c + d$$

The proportions $s = a/(a + c)$ and $f = d/(b + d)$ are of interest and are estimates of sensitivity and specificity, respectively. Sensitivity (s) is the con-

Table 24.1 Realistic Screen with Standard Terminology

Screening result	Truth		Total
	Active	Not active	
Active	True positive a	False positive b	a + b
Not active	False negative c	True negative d	c + d
Total	a + c PN	b + d (1 − P)N	a + b + c + d N

Note: Lowercase letters (a, b, c, d) represent numbers of compounds that fall into each cell.

ditional probability that the screen will classify the compound as interesting, given that it is active. Similarly, specificity (f) is the conditional probability that the screen will declare an agent not interesting, given that it is inactive. Looking back on Table 24.1, note that

$$\text{Sensitivity} = s = \frac{\text{True positives}}{(\text{True positives} + \text{False negatives})} = \frac{a}{a+c}$$

$$\text{Specificity} = f = \frac{\text{True negatives}}{(\text{True negatives} + \text{False positives})} = \frac{d}{d+b}$$

Desirable properties for a realistic screening model will be to have false negatives and false positives near zero, meaning that both sensitivity and specificity are near 1. Morrison (1985) presents a discussion on these ideas as they apply to asymptomatic disease and early clinical diagnostic screening techniques, and Lau (1991) explores multiple screens on the same subjects.

Let us see how the concepts of sensitivity and specificity help fill the cells of the 2 × 2 table. Thus, we are able to construct conceptual values for all four cells given estimates of sensitivity and specificity. For example:

$$a = \frac{a}{a+c} \frac{a+c}{a+b+c+d} \frac{(a+b+c+d)}{1} = s \times P \times N$$

The estimates of the proportions $v = a/(a + b)$ and $g = d/(c + d)$ are also of interest and are called positive and negative accuracy, or positive and negative

Table 24.2 Conventional Risk Concepts

Hypothesis statistical test	Truth (null hypothesis)	
	False	True
Reject	Power $1 - \beta$	Type I error α risk
Accept	Type II error β	$1 - \alpha$

predictive value, respectively. Ideally, we should have both of these conditional probabilities equal to unity.

The value of positive accuracy v is considered critical by screening investigators because screening models are often judged by the proportion of compounds which are declared interesting and continue to show activity in definitive testing. The amount of effort, time, facilities, and money devoted toward definitive testing is often 10 to 20 times greater than the initial screening expenditures on a per compound basis. Therefore, positive accuracy is an important consideration.

These concepts should not be entirely novel to readers familiar with hypothesis testing. The concepts relate to conventional α and β risks as shown in Table 24.2. Rejecting a false null hypothesis is similar to passing an active compound as interesting, since the null hypothesis in screening is that a compound has no activity (i.e., it is the same as the placebo control). These ideas are discussed in standard statistics texts in sections on p-values and statistical errors.

III. OPERATING CHARACTERISTIC CURVE

In a more realistic situation, there are not just two kinds of compounds, active and not active, but an entire range of activities from none to high activity. So the true activity of a compound is an unknown parameter that can be designated by φ. A particular value of φ, say φ_0, denotes zero or a low level of activity while another value, φ_1, denotes a level of activity that is interesting, meaning that it is sufficiently high to warrant further study. Compounds with an activity level φ between φ_0 and φ_1 are not of high enough activity to be interesting and the experimenter is indifferent as to how the screening result turns out. Compounds with activity φ_1 correspond to the active column in the 2×2 table and compounds with activity φ_0 correspond to the not active column. But there are other values of φ as well, and of interest is the probability that a compound having a particular activity level φ will be classed as active by the screening test. This is shown by the *Operating Characteristic Curve* (or OC Curve) which is simply a plot of the probability of being accepted against φ. The concept is identical with that used in acceptance sampling in statistical quality control. At the value φ_1, we want the probability of accepting the compound to be high, say $1 - \beta$, whereas at the value φ_0, we want the probability of accepting the compound to be small, say α. Here, β denotes the risk of missing an interesting compound, while α denotes the risk of accepting a poor compound for further study. These correspond to the producer's and consumer's risks in acceptance sampling.

The OC Curve is a useful thing to be aware of for any screening procedure and any good statistician can work it out from existing data. Often it is possible to choose a good screening procedure by comparing several possibilities in terms of their OC curves and their relative costs.

IV. SCREENING PROCEDURE

To illustrate these concepts, an article by Dunnett (1972) presents an antitumor model that will be utilized throughout this chapter. Dunnett assumes that P = .004 and N = 10,000; therefore, PN = 40.

In this antitumor screen, cancer cells are implanted in mice; three mice are treated with a compound for 10 days and then the tumors are removed and weighed. For comparison, a similar group of six untreated mice (controls) is experimentally handled in the same manner but no compound is administered. The first example illustrates a criterion point of 2 standard errors (70 mg) between mean tumor weights of the treated and control mice. The decision value for passing interesting compounds is called the criterion point. The assumption is made that each of the 40 active compounds is capable of reducing the tumor weight by 70 mg. Based on the selection criterion of 70 mg and the assumed screening parameters, Table 24.3 can be visualized, where

$$\text{Sensitivity} = s = \frac{20}{40} = .500$$

$$\text{Specificity} = f = \frac{9733}{9960} = .977$$

$$\text{Positive accuracy} = v = \frac{20}{247} = .081$$

$$\text{Negative accuracy} = g = \frac{9733}{9753} = .998$$

Table 24.3 Illustrative Antitumor Screen (70 mg)

	Truth		
Screening result	Active	Not Active	Total
Active	True positive 20	False positive 227	247
Not active	False negative 20	True negative 9,733	9,753
Total	40	9,960	10,000

Thus, from the viewpoint of our four measures of an ideal screen, specificity and negative accuracy are high while sensitivity and especially positive accuracy leave opportunity for improvement.

A. Optimization of Selection Criterion (Illustrative Example)

Dunnett's article discusses a plan to optimize on the decision difference. Based on definitive testing, which is assumed equivalent to 10 times the research effort for the screen (called work units), Dunnett proposes the ratio of the 20 active compounds passed to 12,470 work units to delineate the true positives: 10,000 work units + (247 × 10) additional work units for definitive testing. This ratio yields 1.6 active compounds per 1000 work units.

Table 24.4 Illustrative Antitumor Screen (60 mg)

	Truth		
Screening result	Active	Not Active	Total
Active	25	434	459
Not active	15	9,526	9,541
Total	40	9,960	10,000
	$s = 25/40 = .625$	$v = 25/459 = .054$	
	$f = 9526/9960 = .956$	$g = 9526/9541 = .998$	

Table 24.5 Illustrative Antitumor Screen (Two-stage: Δ_1 = 40 mg, Δ_2 = 50 mg)

Screening result	Truth		Total
	Active	Not Active	
Interesting	31	199	230
Not interesting	9	9,761	9,770
Total	40	9,960	10,000

$$s = 31/40 = .775 \qquad v = 31/230 = .135$$
$$f = 9761/9960 = .980 \qquad g = 9761/9770 = .999$$

When the magnitude of the selection criterion is changed, for example, from 70 mg to 60 mg, then 434 false positives are accepted as interesting and 25 actives are expected to be passed. This produces a yield of 1.7 actives per 1000 work units (25/14,590). Table 24.4 presents the results after this change in selection criterion.

For a single-stage test, based on the assumptions presented, it can be shown that 1.7 actives per 1000 work units are about optimum. If the selection criterion for this antitumor screen is increased or decreased from 60 mg, the overall actives per 1000 work units are smaller than 1.7. In the example presented, considerable absolute improvement was made in the sensitivity with only minor changes in the other three indices. It may be a useful exercise for the reader to study Dunnett's paper and decrease the selection criterion to gain an appreciation of how the four measures of an ideal screen will change.

B. Sequential Screening

By utilizing a two-stage sequential procedure with equal sample sizes in both stages and a selection criterion of 40 mg at the first stage and 50 mg for the second stage, the actives per 1000 work units can be increased to 2.15. Of course, additional screening must be done to achieve this result, but the advantage is that this tends to be concentrated on the compounds more likely to be interesting ones. The optimum three-stage procedure will yield 2.3 actives per 1000 work units. The two-stage procedure produces Table 24.5.

As a general rule, it appears that a two-stage procedure, by optimizing decision rules and rescreening compounds before declaring compounds interesting, increases both sensitivity and positive accuracy. However, in the authors' experience the administrative details of data and record handling, the uncertainties regarding the amount of compound necessary, the time delays for results, and the laboratory scheduling difficulties have made it difficult to interest many screening investigators in the merits of sequential screening techniques.

C. Other Approaches to Screening

Davies (1958) is credited with first suggesting statistical methods in the design of screening studies. Later, Davies (1963) viewed the problem in a broad sense and considered the optimum strategy as one that would provide the highest expected net dollar gain per unit of effort subject to restrictions imposed by staff, organization, and compound availability. He presented screening problems as only a part in a much larger context consisting of the allocation of Research and Development resources to all projects and the appropriate balance of various activities of research laboratories.

Dunnett (1961, 1971) discussed a model to take into account the costs and benefits of any particular screening procedure in order to develop an optimum screening method. The model postulates a prior distribution $g(\varphi)$ of the activity levels for the screened compounds, for which it is often possible to make some reasonable assumptions even though, in practice, it is unknown. Then, based on the probability $P(\varphi)$ of acceptance for a compound having activity j (the OC curve) and the screening costs of the procedure, a screening procedure can be developed that optimizes the output of the screen relative to the total cost. Bergman and Gittens (1985) have reviewed some aspects of this approach. Without doubt, this approach is the most general one. Limitations centered on estimates of the quantitation of market size for drug discoveries without knowledge of specific compound advantages and disadvantages. In addition, the time requirement from screening to development of a new product, including final government approval, is often 7 or more years. Thus, the market estimates can change dramatically. However, in defense of Davies' and Dunnett's proposals, such judgments are usually made by research management on an intuitive basis, beginning with the hiring of specific scientific talent.

Other approaches have been to consider optimizing on various criteria, in most cases using a multistage or modified sequential approach. King (1963) offered a screening design that attempted to maximize the proportion of unknowns possessing interesting activity among those compounds passed to definitive testing. He discussed the situation where there were existing restrictions on available resources, such as a limited supply of desired animals, restrictions of scientific manpower and physical facilities, and conflicting demands of other research programs. King concluded in this paper that a two-stage screening procedure was more efficient than a one-stage screen.

In a slightly different vein, King (1964) studied the optimal replication required in sequential drug screening. He found that there was little to be gained in efficiency by using more than three stages. In situations where the screen was sensitive to interesting activity equal to or greater than 2 relative standard errors, he suggested that no replication was needed until the final stage. King's optimization criterion was equivalent to maximizing the ratio $(1 - \beta)/\alpha$, where $1 - \beta$ and α had the risk level interpretation presented in Table 24.2. These criteria suggested that values of $\alpha = .001$ and $\beta = .30$ for a ratio $(1 - \beta)/\alpha$ of 700 appeared optimum. Such a choice would tend to overlook approximately 30% of the active agents submitted but it would compensate by providing a very high concentration of truly active compounds among those accepted for definitive testing. Thus, King proposed a strategy where both sensitivity and positive accuracy were approximately equal to .7, as illustrated in Section II.

Dunnett and Lamm (1962) and Colton (1963) have considered multistage screening procedures. Both papers stipulated fixed probability levels α and β for rejecting interesting compounds (false negatives) and accepting not-interesting compounds (false positives) and minimized the expected number of observations required to classify a compound with a specific activity level. Colton showed that the meaningful values for the two risks were in the range of .01, .05, .10, and .20, thereby suggesting that sensitivity and positive accuracy were the parameters to optimize. Multistage sampling plans were presented which compare the properties of equal and unequal sample size.

D. Group Mixtures Screening

Redman and King (1965) suggested group screening utilizing balanced and partially balanced incomplete block designs as a novel method to increase the rate of compound screening without reducing necessary replication. Their proposal presented a technique which tested pools of several compounds at a time, rather than testing each one singly. A group scheme was used which permitted the identification of active compounds based on the initial results.

For each design, let

m = Number of experimental units (animals or cages in feeding experiments)
r = Number of replicates of each compound or agent
v = Number of agents tested per experiment
g = Group size (number of agents per group or mixture)

For example, if the experimental plan in the tumor screening examples were 10 treatments of subcutaneous injections of 3 animals per cage and 3 control treatments (placebo or positive control) of 3 animals per cage, then our resources required 13 cages and 39 rodents per experiment. Under such circumstances, a group screening plan of $m = 13$, $r = 3$, $v = 13$, and $g = 3$ repeated 3 times might be a valid recommendation. Utilizing this plan with the same resources, 39 compounds could be screened rather than the 10 in the conventional design.

Table 24.6 Design Layout

Cage	Rodent	Compound			Cage	Rodent	Compound		
A	1	1	3	9	H	22	22	24	17
	2	2	4	10		23	23	25	18
	3	3	5	11		24	24	26	19
B	4	4	6	12	I	25	25	14	20
	5	5	7	13		26	26	15	21
	6	6	8	1		27	27	29	35
C	7	7	9	2	J	28	28	30	36
	8	8	10	3		29	29	31	37
	9	9	11	4		30	30	32	38
D	10	10	12	5	K	31	31	33	39
	11	11	13	6		32	32	34	27
	12	12	1	7		33	33	35	28
E	13	13	2	8	L	34	34	36	29
	14	14	16	22		35	35	37	30
	15	15	17	23		36	36	38	31
F	16	16	18	24	M	37	37	39	32
	17	17	19	25		38	38	27	33
	18	18	20	26		39	39	28	34
G	19	19	21	14					
	20	20	22	15					
	21	21	23	16					

For this example, 3 sets of 13 compounds are selected and randomly assigned numbers from 1 to 39. Compounds of similar chemical structure are not selected for each set of 13 compounds. The mixtures are made following the design layout illustrated in Table 24.6.

There are $_{13}C_3 = 286$ possible mean values in this situation. Each of compounds 1 to 13 can be associated with one mean of 3 rodents' tumor weights accounting for 13 of the 286 means. For example, if we let X_i represent the tumor weight of the ith rodent, then the compound 1 response is measured as $\frac{1}{3}(X_1 + X_6 + X_{12})$ and the compound 13 response is evaluated as $\frac{1}{3}(X_5 + X_{11} + X_{13})$.

Three separate randomizations (1 through 13, 14 through 26, and 27 through 39) are required to evaluate the 39 compounds. The randomizations are easily programmed and the ranking of compounds becomes a routine matter for the screening investigator. Moran (1973) reviewed additional properties of the randomization procedure and the normal approximation.

These designs are based on the assumption that the majority of compounds tested act as placebos and, in addition, the grouping of compounds does not produce toxic effects. Experience has shown that when positive controls are coded as compounds, they are readily detected by using these grouping techniques. Another concept that is fostered by this technique is that the experimental control average is a fair baseline to use for making comparisons of statistical significance. In the illustrated example the mean of each set of 13 rodents can be considered as the basis for statistical comparisons.

Group screening has been successfully used as a technique in the Lilly Research Laboratories. One example is presented by DeLong et al. (1970) for an antiviral screening model. Group screening seems best suited for new screening models where there is a large backing of old compounds to be screened rapidly. Additional technical help is usually required for mixing the compounds, for detailed record keeping, and for animal color coding when the animals are individually injected.

Partially balanced designs are readily available in many textbooks. The National Bureau of Standards Applied Mathematics Series 63 has enlarged and revised the set of tables of Clatworthy (1973).

IV. GENERAL CONSIDERATIONS

A. Statistician's Viewpoint

The statistician in the role of consultant should be a good scientist, listener, and evaluator. Almost on a daily basis a statistician is queried by chemists and biologists about certain aspects of a screening procedure, for example, the number of replicates, the assay variation, location effects within cages or incubators, the randomization of treatments to experimental units, or the virulence over time of the experimental infection.

The point at which a new experimental model is being conceived is the best time for biologists and statisticians to work together on techniques of randomization, replication, blocking, dosage schedules, analyses, and a vigorous program of blind positive controls. The statistician can offer a wide range of experimental designs that will increase the sensitivity, specificity, and positive accuracy and yet not reduce the rate of screening significantly. She may also systematize data handling and computerize the routine analyses.

The statistician is often asked by research management to be a team member to review and evaluate specific screens. The evaluation of screens is a perplexing task because the criteria for evaluation are not obvious. When a screen selects interesting compounds that fail to repeat during definitive testing, is it because the screen has low positive accuracy or because no really active compounds have been submitted to the screen?

A statistician interested in the broader aspects of screening may find opportunities for statistical research. First and foremost, he may be concerned that there are not enough compounds screened to insure the probability of finding an interesting chemical series; in addition, he may find this major failure difficult to illustrate. We call this a Type III error, failure to screen enough compounds to find interesting activity. Hence, a statistician committed to screening as a philosophy is usually interested in methods to increase the rate of screening. In addition, the statistician becomes more aware that screening must be balanced by a strong complementary definitive testing program.

B. Screening Investigator's Viewpoint

A screening investigator may be a microbiologist, virologist, immunologist, pharmacologist, physiologist, or one of the many other specialists in a modern pharmaceutical research laboratory. Specialists will be called screening investigators to describe their function rather than their disciplines.

The task may be viewed by management as routine, requiring only organizational ability to handle massive amounts of data and to run an orderly and systematic laboratory. On the other hand, we believe that a screening investigator is a scientist who understands the relationship between experimental models and human illness and who is knowledgeable about the modes of action of various chemical series upon experimental models. In addition, he can attempt to interpret why in vitro screens and in vivo screens do or do not always provide consistent data. The thought processes required to develop a screening model that is related to the real world are difficult to master. Investigators who can develop a reproducible disease-related in vitro or in vivo screening model are expert scientists.

The screening investigator is primarily interested in the best possible model, a screen that is biologically sensitive to few extraneous causes. On the negative side, some investigators may sensitize their screens so that all positive controls will consistently pass the selection criterion. They may do this at the expense of slowing the screening rate to a point where the capacity to screen compounds can be questioned.

To insure positive accuracy and also sensitivity, the screening investigator frequently screens too few compounds, requires more replicates than necessary, and is too demanding in selection criterion for interesting. These latter philosophies are consistent with being a good scientist but fail to be good science if the task is to screen compounds for potential activity.

In establishing a new model, the investigator considers various administration routes: subcutaneous, intramuscular, intraperitoneal, and oral. Other variables studied are the time of administration and dosing amounts. In experimentally induced infections, for example, the investigator must decide to administer compounds before or after infection. The number of days for dosing must also be considered.

Working with animal models, an investigator is a good husbandryman. She is cognizant of unexpected effects and her judgment is often significant toward understanding initial activity of a compound. Decisions made by the screening investigator are based on less data and information than most scientists are normally accustomed to having. The decision of interesting or not-interesting can, upon subsequent investigation, be proven incorrect. This can lead to a further dilemma in that the model becomes suspect of producing inaccurate data. Therefore, the biologist strives for a screen with positive accuracy. Sensitivity may be measured by allowing positive controls to be submitted blindly to the screen.

C. Chemist's Viewpoint

A medicinal chemist synthesizes compounds within several series of compounds in which he has expertise. A chemical series is a group of compounds composed of the same basic structure with each analog having one or more minor modifications that often are such as to alter activity in a screening model.

A chemist often has primary interest in a certain therapeutic drug class and submits his compounds to a specific set of screens. For example, a chemist may have interest in cardiovascular research. He learns the chemistry of drugs that have antiarrhythmic, hypertensive, hypotensive, vasoconstrictor, and vasodilator activity and works closely with the pharmacologist who screens compounds for cardiovascular activity.

The initial quantity of compound that is synthesized is often limited, and a chemist prefers a screen that uses as little compound as possible, usually less than 500 mg. If he has enough compound, a chemist may choose to send his compounds for screening to 10 or more screens and if he is not working with an active chemical series, he may screen more broadly. After finding an active series, he seeks the best compound in the series by definitive testing for structure-activity and dose-response relationships. He also screens for additional beneficial effects, as well as detrimental side effects. The evaluation of compounds within a series is usually done by determining a therapeutic index, a ratio of the beneficial and serious detrimental effects dosages. Often interesting compounds fail the definitive testing stage for reasons other than no activity, for example, poor absorption, too short a half-life, or detrimental side effects. Therefore, chemists would like to have more than one active series.

Time becomes critical for a chemist. He may produce a few hundred compounds within a series to adequately convince himself and the research team that the best compound is selected and that sufficient data are provided for patent protection. Hence, the chemist must ask many questions: What is the optimum amount of compound to make? What is the required capacity of a screen (compounds screened per week, month, or year). What is the optimum number of different screens to which a compound should be sent?

In the search for a new drug, chemists are able to synthesize, isolate, and characterize new compounds on a rational basis. The role of a serendipitous discovery (an unexpected finding of activity in a biological system), however, is often the basis for more rational syntheses. Today, new useful drugs must be found by a combined knowledge of medicinal chemistry and biology and serendipitous discovery. There are many examples where a rational compound synthesized for one animal model has led to a serendipitous discovery in an entirely different biological system.

As mentioned earlier, chemists often synthesize compounds in what is called a chemical series once an active lead is found in that series. Although chemists do not disagree with the statistician's estimates of the probability of activity, P, they certainly do not believe that each compound synthesized has an equal probability of success, such as $P = .004$. Statisticians also do not believe that each compound has equal probability of success. One example of what a chemist might find more plausible in a collection of 10,000 compounds is a 100 chemical series of 100 compounds, with a distribution of P as shown in Table 24.7.

Table 24.7 Probability Distribution (Example)

Compound series	Number of compounds	P	Number of actives
1–50	5,000	0	0
51–70	2,000	.001	2
71–80	1,000	.003	3
81–90	1,000	.010	10
91–95	500	.020	10
96–100	500	.030	15
	10,000		40

D. Research Management's Viewpoint

Research management may allocate up to 15% of the research budget to screening for clinically active compounds. In many companies, this can be millions of dollars per year and, therefore, management is interested in how this money is spent. There are at least two basic screening philosophies that management can support. One can be called blind (indiscriminate) screening with limited sensitivity. The second is selective (discriminate) screening with high sensitivity and positive accuracy.

These philosophies differ on two basic issues. One is what constitutes a screening model. Management ascribing to blind screening feel that screening models should be able to detect inactive and active compounds quickly but with some risk of error. On the other hand, selective screening proponents believe that good screening includes the detection of inactive and active compounds plus the ability to discriminate among active compounds. Discriminate screening may also imply a screening model that provides some idea of mechanism of activity. In essence, management people who support selective screening with no blind screening believe that testing of chemical-biological relationships is the preferred method of compound discovery.

The second issue is: How rare are interesting compounds? The proportion of interesting compounds among compounds synthesized or those that could be made is not known; however, for each drug marketed, usually between 1000 and 5000 chemical entities are screened. Selective screening advocates contend that frequency of activity is not valuable information because good medicinal chemists and biologists can rule out many chemical structures before they are synthesized on grounds of toxicity or suspected adverse side effects. In addition, they believe that structure-activity relationships exist which reduce the need for screening compounds blindly.

In defense of screening, some management people claim that innovative blind screening of compounds affords the best opportunity for finding an unpredictable discovery. For example, blind screening advocates agree that a compound with expected activity as an antiinflammatory agent may have unexpected activity as an analgesic, or a series of compounds expected to have activity as central nervous system agents may have beneficial effects on the gastrointestinal system.

Although the viewpoints of blind and selective screening appear to be in conflict, both are valid. The task for research management is to foster a balance. In an attempt to reach this balance, management recognizes that there are over 40 therapeutic classes of drugs, some with many subclasses. Most classes have appropriate in vivo and in vitro models which screen for biological activity. The question becomes: In how many therapeutic classes should a compound be screened? This particular question has not received much discussion in existing literature. Some work of the authors shows that the amount of compound required by screens often controls the amount of screening, because most research laboratories have a limited capacity to scale up compound quantities beyond a few grams. If the optimum screening effort appears to be 10 to 20 screens of diverse therapeutic classes, then a plan to solve administrative difficulties in a screening program is required. To keep an even screening load from week to week or from month to month, it may be necessary to set priorities for compounds and to create backlogs of lower priority compounds.

Another consideration is to properly balance the number of compounds made and screening capacity. The relationship between the number of synthesized compounds, the number of screens, and the capacity of each screen should be well coordinated. However, as more specific biological models are developed it may be as equally productive to screen old compounds as well as new ones.

Management in making screening decisions also recognizes that each company has developed the following:

1. Expertise in medicinal chemistry in certain therapeutic classes
2. A large number of chemical compounds which are a unique chemical series
3. An ability to synthesize compounds in these chemical series
4. A number of screening systems that have proven reliable as experimental models
5. The ability to modify or add new screens in complimentary therapeutic classes

It is essential that all these factors be considered when making recommendations about a company's screening program.

In summary, in the evaluation of overall effectiveness of a screen, the following criteria apply:

1. The screen should be inexpensive.
2. The screen should produce a result in 10 days or less.
3. The screen should require small amounts of compound (e.g., under 500 mg).
4. The screen should have an annual capacity of hundreds of compounds and should be able to handle all compounds that chemists may wish to submit for screening, including both old and new compounds.
5. The screen should be capable of statistically reproducible results and positive controls submitted at random should meet the test criterion 50–80% of the time.
6. The screen should have a logical biological basis such that the model provides an initial profile of information.
7. The screen should be standardized so that it can function routinely with technical personnel.
8. The screen should have an efficient data-handling system.

9. The screen should receive random blind negative and positive controls.
10. The screen should have an agreed-on decision rule for passing interesting compounds.
11. The screen should have the capability to reject uninteresting compounds.
12. The screen should be made as accurate and precise as possible by routes of administration, excellent environmental conditions, healthy animals, rigid disease control, and temperature control.
13. The screen should be designed to precede logical definitive testing.
14. The screen does not require:
    a. Providing a dose response curve of activity
    b. Providing the capacity to distinguish between levels of activity in an active compound series
    c. Being correct on every compound.

REFERENCES

Bergman, S. W. and Gittens, J. C. (1985). Screening procedures for discovering active compounds. In *Statistical Methods for Research Planning*, Marcel Dekker, Inc., New York, pp. 105-139.

Clatworthy, W. H. (1973). *Tables of Two-Associate-Class Partially Balanced Designs*, National Bureau of Standards Applied Mathematics Series 63. Washington, D. C.

Colton, T. (1963). Optimal drug screening plans. *Biometrika*, **50**:31-45.

Davies, O. L. (1958). The design of screening tests in the pharmaceutical industry. *Bull. Inst. Int. Stat.*, **36**:226-241.

Davies, O. L. (1963). The design of screening tests. *Technometrics*, **5**:481-490.

DeLong, D. C., Baker, L. A., Doran, W. J., and Redman, C. E. (1970). Use of drug and disease pooling in the detection of an antiviral compound. In *Progress in Antimicrobial and Anticancer Chemotherapy, vol. 1*, University of Tokyo Press, Tokyo, pp. 485-488.

Dunnett, C. W. (1961). Statistical theory of drug screening. In *Quantitative Methods in Pharmacology*, (H. DeJonge, Ed.). North Holland Publishing Company, Amsterdam.

Dunnett, C. W. (1971). Optimum procedures for drug screening. 38th Session of the International Statistical Institute, Washington D.C., Contributed Papers: pp. 118-122.

Dunnett, C. W. (1972). The never-ending search for new and better drugs. In *Statistics: A Guide to the Unknown*, (J. M. Tannur, Ed.). Holden-Day Inc., San Francisco pp. 23-33.

Dunnett, C. W., and Lamm, R. A. (1962). Sequential procedures for drug screening. Presented at Amer. Stat. Assoc. meeting, Minneapolis, March 1962.

King, E. P. (1963). A statistical design for drug screening. *Biometrics*, **19**:429-440.

King, E. P. (1964). Optimal replication in sequential drug screening. *Biometrika*, **51**:110.

Lau, T-S. (1991). On dependent repeated screening tests. *Biometrics*, **47**:77-86.

Moran, M. A. (1973). The analysis of incomplete block designs as used in the group screening of drugs. *Biometrics*, **29**:131-142.

Morrison, Alan S., *Screening in Chronic Disease*, Oxford University Press, New York, 1985.

Redman, C. E., and King, E. P. (1965). Group screening utilizing balanced and partially balanced incomplete block designs. *Biometrics*, **21**:865-874.

25

Survey of Statisticians in the Pharmaceutical Industry

C. Ralph Buncher

University of Cincinnati Medical Center,
Cincinnati, Ohio

Bob Wilkinson

Robert Wilkinson Associates,
Blauvelt, New York

A survey of statistical units in the U.S. Pharmaceutical Industry which employed four or more statisticians, at the masters or doctoral level, was carried out by the authors. Each pharmaceutical company in the U.S. was queried and larger companies which have dispersed units that qualified were queried at each of the units. The letter requesting cooperation was sent to the Department Director to whom all of the departmental statisticians report at each company/ location. Clinical and preclinical units were queried separately if they reported to different areas in the organization.

This is the largest and most comprehensive survey of statisticians in the U.S. Pharmaceutical Industry known. A prior survey was done by Stuart Bessler and was presented in 1976 with a total of 38 replies. Another survey emphasizing the midwest was done by Christine Trautwein and was first presented in 1979. Published results are found in the first edition of this book using a total of 23 replies [1].

A total of 54 departmental level replies to this survey representing 708 statisticians and 44 different companies was tabulated; there were only three

547

Table 25.1   Distribution of "White" Statisticians Who Were Born in the U.S.

	MS	PhD	Total
Male	172	148	320
Female	132	36	168
Total	304	184	488

$\chi^2 = 28.90$
Observed − Expected = 27.3

non-respondents (response rate = 95%). Of the total statisticians reported in the survey, there were 316 statisticians at the Ph.D. level (45%) and 392 at the M.S. level (55%). The gender breakdown was 251 female (35%) and 457 male (65%) statisticians but there was also an interaction in that 78% of the Ph.D.s were male compared to 54% of the M.S. persons who were male. Every company employed at least one female and one male with the exception of one very small group which did not report any females. Only one small company reported not having any Ph.D. level statisticians (i.e., the whole group was at the M.S. level).

The organizations were also asked to classify the statisticians by whether they were U.S. born or not and by their racial classification into white, black, Oriental, other Asian, or other. A total of 671 of the statisticians was so classified. There was a total of 515 persons classified as white with 95% of them listed as born in the United States, 18 were black (50% U.S. born), 92 were Oriental (11% U.S. born), 33 were other Asian (15% U.S. born), and 13 were other (15% U.S. born).

Table 25.1 shows the white U.S.-born statisticians by gender and statistics degree. One can see an excess of male Ph.D. and female M.S. statisticians compared to a random distribution. Table 25.2 shows the Oriental, not U.S.-born statisticians by gender and statistics degree. The same excess cells are seen and if anything to a greater extent than for white U.S. born. One can hypothesize that this simply reflects the statisticians available in the pool or alternatively that there is some selectivity into the industry. We note that comparable numbers are not available for the total population of statisticians to test these hypotheses.

The respondents were asked the question of how long have the statisticians worked for the company. The responses showed that 59.5% reported 0–4 years, 21.3% reported 5–9 years, 12.2% reported 10–14 years, 4.5% reported 15–19 years, and 2.5% reported 20 or more years. The question of the number of years in the industry was not asked to keep the questionnaire simple, but in retrospect

Table 25.2 Distribution of "Oriental" Statisticians Who Were Not Born in the U.S.

	MS	PhD	Total
Male	6	42	48
Female	21	13	34
Total	27	55	82

$\chi^2 = 21.87$
Observed − Expected = 9.8

that information would have helped to characterize the statisticians with respect to experience in the field rather than just at the same company. It is well known that statisticians move from company to company in this industry and thus short term employees might have many years of experience in the industry.

There was also an attempt to characterize those who are the heads of statistics units in the industry so respondents were asked to describe themselves. Four of these 54 persons (7%) were female. Seven of the 54 (13%) were Oriental or other Asian while the remainder were white. Eight of 53 were not U.S. born (15%), one was a nonrespondent, while the rest were U.S. born.

The gender results above can be contrasted to the Trautwein survey [1] in which none of the 23 statistical managers was female. That survey did not ask about ethnic background.

Organizational structure was queried to ascertain where in the organization the statisticians were located and the characteristics of the persons to whom the statistical group reported. Therefore the supervisor of the head of statistics was characterized. Twenty-six report to someone with a title of Vice President; fourteen report to a Director and another twelve to an executive director or comparable title.

This supervisor of the head of statistics in turn reports to the president (or a president) in thirteen instances, to a vice president in 34 instances, and to a director in the other seven cases. The supervisor of the head of statistics was trained as a statistician in seventeen instances (31%), was an M.D. in 21 instances (39%), was a nonstatistician Ph.D. in twelve instances (22%), and had other training in the remaining 4 instances (7%). The implication that statisticians are now moving up the corporate ladder to positions above director of statistics, although almost exclusively to that position which supervises both statistics and clinical data processing, is clear.

In addition fewer statisticians report to a medical person than in prior surveys. Trautwein [1] reported that 55% of the statistical groups reported directly to a clinical/medical group.

The statistical manager was asked with which other groups do the statisticians work. The clear winner was "clinical/medical trials" with 79.5% of statisticians working in that category. "Pre-clinical/Toxicology" involved 10.8% of statisticians with another 5.2% in "Nonclinical." The remaining categories in order of frequency with fewer than 2% in each category were "Computers/Information systems," "Quality control," "Statistics," "Other," "Post-marketing surveillance," and "Marketing/Business statistics." The categories on the questionnaire, which may or may not have coincided with the designations within the company, were listed in the order of their suspected frequency and this turned out to be almost the same order as listed here. It is possible that there was good prediction or alternatively that some order of choice effect was present in that a respondent may have been more likely to select the first choice listed rather than one later in the list when both were appropriate.

In addition these data are compared to prior surveys for time trends. The prior 1979 survey of 23 statistical managers [1] reported that all were male and that 78% held the Ph.D./Sc.D. degree and 22% the M.A./M.S. degree. In addition the current survey shows four of the managers (7%) are females and that virtually all of the managers hold a Ph.D. or comparable degree. White males born in the U.S. constitute 77% of the current managers but females, nonwhites, and those born outside the U.S. comprise one-fourth of the managers. Data are not available from this survey to adjust this comparison for time in the industry or years of experience.

Some questions on the computer equipment used were also included. The groups were asked whether they used a mainframe computer, a minicomputer, a personal computer, or a laptop computer always, sometimes, or never.

The results were: mainframe—63% always, 33% sometimes, and 4% never; a minicomputer—18% always, 29% sometimes, and 53% never; a PC—22% always, 75% sometimes, and 4% never; and a laptop—0% always, 35% sometimes, and 65% never (Table 25.3). When asked about the quantity of PCs available per statistician, 47% reported about one PC per statistician, 43% reported less than one, and 10% reported more than one.

REFERENCE

Trautwein, C. A. (1981). A survey of pharmaceutical managers, in *Statistics in the Pharmaceutical Industry, 1st ed.*, (Buncher, C. R. and Tsay, J-Y, Eds.), Marcel Dekker Inc., New York, 1981. Chapter 18.

Index